Plant Tissue Culture Concepts *and* Laboratory Exercises

Second Edition

Plant Tissue Culture Concepts *and* Laboratory Exercises

Second Edition

Edited by
Robert N. Trigiano, Ph.D.
Dennis J. Gray, Ph.D.

CRC Press
Taylor & Francis Group
Boca Raton London New York

CRC Press is an imprint of the
Taylor & Francis Group, an **informa** business

Cover Photograph: Somatic embryo of grape (see Chapter twenty-one). Courtesy of D. J. Gray.

CRC Press
Taylor & Francis Group
6000 Broken Sound Parkway NW, Suite 300
Boca Raton, FL 33487-2742

First issued in hardback 2017

CRC Press is an imprint of Taylor & Francis Group, an Informa business

Library of Congress Cataloging-in-Publication Data

Plant tissue culture concepts and laboratory exercises / edited by
Robert N. Trigiano, Dennis J. Gray — 2nd ed.
p. cm.
Includes bibliographical references and index.
ISBN 0-8493-2029-1 (alk. paper)
1. Plant tissue culture Laboratory manuals. I. Trigiano, R. N.
(Robert Nicholas), 1953- . II. Gray, Dennis, J. (Dennis John),
1953-
QK725.P584 1999
571.5′.382—dc21 99-34987
CIP

Library of Congress Card Number 99-34987

ISBN 13: 978-1-138-40776-3 (hbk)
ISBN 13: 978-0-8493-2029-3 (pbk)

Overview and scheduling of laboratory exercises

The book is arranged such that the "typical" or "common" supplies and media formulations for all of the laboratory exercises are listed in Chapter three, "Getting started with tissue culture — media preparation, sterile technique, and laboratory equipment." However, unique supplies and media are given in each chapter when necessary. In particular, student supplies, such as scalpels and blades, are described in Chapter three under "Equipment and supplies for a tissue culture laboratory"; those items are not listed in individual chapters. We recommend that each student or team of students be provided with a kit containing those instruments and supplies. Reference to specific products or suppliers does not constitute endorsement or criticism of similar ones not mentioned.

Although arranged by topic, several laboratory exercises from different topic areas use the same plant species. Therefore, it might be most efficient to conduct laboratories on a range of topics concurrently in order to efficiently utilize certain plant material. To assist in planning, the plant species are referenced to laboratory chapters in Table 1.

We encourage instructors to take the time to peruse the laboratory exercises well in advance of the anticipated laboratory starting date. Some experiments will require starting stock plants. Other experiments will depend on cultures started or obtained from outside sources weeks or months ahead of time. For example, most of the genetic transformation and in vitro plant pathology exercises require plant pathogenic bacteria, e.g., *Agrobacterium tumefaciens* and/or *Pseudomonas syringae*. Note that regardless of the source of the pathogens (the American Type Culture Collection, a colleague at your institution, etc.), be sure to acquire the proper documentation and/or permits for the cultures. Preliminary planning will save time and disappointment later on and will make for a better experience for students and instructors.

Robert N. Trigiano

Dennis J. Gray

Table 1 Correlation of Plants to Laboratory Exercises for Scheduling Purposes

Plant	Chapter
Ajuga	38
Cabbage	34
Carnation	36
Carrot	32
Chrysanthemum	15, 28, 33
Cineraria	23
Dieffenbachia	10
Fern	41
Lilac	12
Melon	21, 34
Mint	36
Orchardgrass	20, 28
Peanut	22
Petunia	18
Spruce	25
Potato	11, 27, 36
Rose	13
Soybean	39
Syngonium	9
Tobacco	4, 27, 30, 32, 39
Torenia	17
Watermelon	16
Yellow poplar	24

Editors

Robert N. Trigiano, Ph.D., is Professor of Ornamental Plant Biotechnology in the Institute of Agriculture, Department of Ornamental Horticulture and Landscape Design at the University of Tennessee at Knoxville.

Dr. Trigiano received his B.S. degree with an emphasis in Biology and Chemistry from Juniata College, Huntingdon, PA in 1975 and an M.S. in Biology (Mycology) from the Pennsylvania State University in 1977. He was an Associate Research Agronomist, mushroom culture and plant pathology, for Green Giant Co., Le Sueur, MI until 1979 and then a Mushroom Grower for Rol-Land Farms, Ltd., Blenheim, Ontario, Canada during 1979 and 1980. He completed a Ph.D. in Botany and Plant Pathology (co-majors) at North Carolina State University at Raleigh in 1983. After concluding postdoctoral work in the Plant and Soil Science Department at the University of Tennessee, he was an Assistant Professor in the Department of Ornamental Horticulture and Landscape Design at the same university in 1987, promoted to Associate Professor in 1991 and to Professor in 1997.

Dr. Trigiano is a member of the American Society for Horticultural Science and the Mycological Society of America, and the honorary societies of Gamma Sigma Delta, Sigma Xi, and Phi Kappa Phi. He has been an Associate Editor for the journals of the American Society of Horticultural Science and *Plant Cell, Tissue and Organ Culture* and Editor for *Plant Cell Reports.* He has received the T.J. Whatley Distinguished Young Scientist Award (University of Tennessee, Institute of Agriculture, 1991) and the Gamma Sigma Delta Research Award of Merit (University of Tennessee, 1991).

Dr. Trigiano has been the recipient of several research grants from the U.S. Department of Agriculture (USDA), the Horticultural Research Institute, and from private industries. He has published more than 100 research papers, book chapters, and popular press articles. He teaches undergraduate/graduate courses in Plant Tissue Culture, Plant Disease Fungi, DNA Analysis, Protein Gel Electrophoresis, and Plant Microtechnique. Current research interests include somatic embryogenesis and micropropagation of ornamental species, fungal physiology, and population analysis.

Dennis J. Gray, Ph.D., holds the University of Florida Research Foundation Professorship and is a member of the Institute of Food and Agricultural Sciences' Horticulture Department. He directs the plant biotechnology program at the Central Florida Research and Education Center where he also serves as the Assistant Center Director.

Dr. Gray graduated with a B.A. degree in Biology from California State College, Stanislaus, in 1976 and received an M.S. degree in Mycology, with a minor in Botany from Auburn University in 1979. He earned a Ph.D. degree in Botany, with a minor in Plant Pathology from North Carolina State University in 1982. After a postdoctoral fellowship at the University of Tennessee, he joined the faculty of the University of Florida in 1984, reaching the rank of Professor in 1993.

Dr. Gray has been a member of the American Association for the Advancement of Science, the American Institute of Biological Sciences, the American Society for Horticultural Science, the Botanical Society of America, the Council for Agricultural Science and Technology, the Society for In Vitro Biology, the International Association for Plant Tissue Culture and Biotechnology, the International Horticultural Society, and Sigma Xi. He was Associate Editor, then Managing Editor, of the internationally recognized refereed journal *Plant Cell, Tissue and Organ Culture* from 1988 through 1994.

Dr. Gray has received research grants and support from the Binational Research and Development Fund, the Florida Department of Agriculture and Consumer Affairs, the Florida High Technology and Industry Council, the U.S. Department of Agriculture, and private industry. He has been an author or co-author of more than 200 publications and holds several patents. His current interests include the developmental biology of regenerative plant cells and the integration of contemporary and newly emerging technologies for crop improvement.

Contributors

Stephen M. Attree
Department of Biology
University of Saskatchewan
Saskatoon, Saskatchewan, Canada

C. Jacyn Baker
Microbiology and Plant Pathology Laboratory
U.S. Department of Agriculture
Beltsville, Maryland

Charleen Baker
DNA Plant Technology
Oakland, California

Robert M. Beaty
Department of Botany
University of Tennessee
Knoxville, Tennessee

Caula A. Beyl
Alabama A & M University
Department of Plant and Soil Science
Normal, AL

Mark P. Bridgen
Department of Plant Science
University of Connecticut
Storrs, Connecticut

James D. Caponetti
Department of Botany
College of Arts and Sciences
University of Tennessee
Knoxville, Tennessee

Michael E. Compton
School of Agriculture
University of Wisconsin-Platteville
Platteville, Wisconsin

Bob V. Conger
Department of Plant and Soil Science
Institute of Agriculture
University of Tennessee
Knoxville, Tennessee

Andrew J. Daun
Knight Hollow Nursery, Inc.
Middleton, Wisconsin

Wanda Ellis
Agri-Starts II, Inc.
Plymouth, Florida

Larry C. Fowke
Department of Biology
University of Saskatchewan
Saskatoon, Saskatchewan, Canada

Victor P. Gaba
Department of Virology
Institute of Plant Protection
The Volcani Center
Bet Dagan, Israel

Lee Goode
Agri-Starts II, Inc.
Plymouth, Florida

Effin T. Graham
Department of Ornamental Horticulture
and Landscape Design
Institute of Agriculture
University of Tennessee
Knoxville, Tennessee

Dennis J. Gray
Central Florida Research
and Education Center
Institute of Food and Agricultural Sciences
University of Florida
Leesburg, Florida

R. J. Henny
Central Florida Research
and Education Center
Institute of Food and Agricultural Sciences
University of Florida
Apopka, Florida

Leslie G. Hickok
Department of Botany
University of Tennessee
Knoxville, Tennessee

Ernest Hiebert
Plant Pathology Department
University of Florida
Gainesville, Florida

S. Jayasankar
CFREC, IFAS
University of Florida
Leesburg, Florida

Michael E. Kane
Environmental Horticulture Department
University of Florida
Gainesville, Florida

Michael J. Klopmeyer
Ball FloraPlant
West Chicago, Illinois

Chia-Min Lin
Plant Pathology Department
University of Florida
Gainesville, Florida

Melina Lopez-Meyer
Department of Biology
Texas A & M University
College Station, Texas

Ignacio E. Maldonado-Mendoza
Department of Biology
Texas A&M University
College Station, Texas

Kathleen R. Malueg
Department of Ornamental Horticulture
and Landscape Design
Institute of Agriculture
University of Tennessee
Knoxville, Tennessee

Roger A. May
Department of Horticulture
Michigan State University
East Lansing, Michigan

Deborah D. McCown
Knight Hollow Nursery, Inc.
Middleton, Wisconsin

Scott A. Merkle
Daniel B. Warnell School of Forest
Resources
University of Georgia
Athens, Georgia

Norton M. Mock
Microbiology and Plant Pathology
Laboratory
U.S. Department of Agriculture
Beltsville, Maryland

Craig L. Nessler
Department of Biology
Texas A & M University
College Station, Texas

Elizabeth W. Orlandi
Microbiology and Plant Pathology
Laboratory
U.S. Department of Agriculture
Beltsville, Maryland

John E. Preece
Department of Plant and Soil Science
Southern Illinois University
Carbondale, Illinois

Sandra M. Reed
Floral and Nursery Plant Research Unit
U.S. National Arboretum
USDA/ARS
McMinnville, Tennessee

Patricia J. Rennie
Department of Biology
University of Saskatchewan
Saskatoon, Saskatchewan, Canada

Randy B. Rogers
Department of Natural Resources
and Environmental Sciences
University of Illinois
Urbana, Illinois

James A. Saunders
USDA/ARS
Climate Stress Laboratory
Natural Resources Institute
Beltsville, Maryland

Otto J. Schwarz
Department of Botany
University of Tennessee
Knoxville, Tennessee

Mary Catherine Scott
Department of Ornamental Horticulture
and Landscape Design
Institute of Agriculture
University of Tennessee
Knoxville, Tennessee

Mary Ann L. Smith
Department of Natural Resources
and Environmental Sciences
University of Illinois
Urbana, Illinois

Gayle R. L. Suttle
Microplant Nurseries, Inc.
Gervais, Oregon

Leigh E. Towill
USDA/ARS
National Seed Storage Laboratory
Fort Collins, Colorado

Robert N. Trigiano
Department of Ornamental Horticulture
and Landscape Design
Institute of Agriculture
University of Tennessee
Knoxville, Tennessee

Richard E. Veilleux
Department of Horticulture
Virginia Polytechnic Institute
and State University
Blacksburg, Virginia

Thomas R. Warne
Department of Botany
University of Tennessee
Knoxville, Tennessee

Hazel Y. Wetzstein
Department of Horticulture
University of Georgia
Athens, Georgia

Acknowledgments

We would like to thank the contributing authors for their efforts and patience throughout this lengthy endeavor; Joy E. Nevine (JEN) was indispensable in producing the excellent computer drawings; the University of Tennessee's Institute of Agriculture and the University of Florida's Institute for Food and Agricultural Sciences for providing financial support for this project; and Bob Lyons for helping to develop the format of the "procedure boxes." We thank our spouses, Kay Trigiano and Amy Gray, for assistance with proofreading, production, and, of course, their undying support. Finally, we are indebted to Andrew N. Trigiano and Bob D. Gray who provided constant inspiration and a welcomed source of diversions during the completion of the second edition.

Contents

Part I: Introduction

Chapter one — Concept....3
Introduction to plant tissue culture
Dennis J. Gray and Robert N. Trigiano

Part II: History of plant tissue culture

Chapter two — Concept....11
History of plant tissue and cell culture
James D. Caponetti, Dennis J. Gray, and Robert N. Trigiano

Part III: Supporting methodologies

Chapter three — Concept....21
Getting started with tissue culture — media preparation, sterile technique, and laboratory equipment
Caula A. Beyl

Chapter four — Laboratory Exercise....39
Nutrition of callus cultures
James D. Caponetti

Chapter five — Concept/Laboratory Exercise....45
Histological techniques
Robert N. Trigiano, Kathleen R. Malueg, and Effin T. Graham

Chapter six — Concept....55
Photographic methods for plant cell and tissue culture
Dennis J. Gray

Chapter seven — Concept....61
Statistical analysis of plant tissue culture data
Michael E. Compton

Part IV: Propagation techniques

Chapter eight — Concept....75
Propagation from preexisting meristems
Michael E. Kane

Chapter nine — Laboratory Exercise....87
Micropropagation of *Syngonium* by shoot culture
Michael E. Kane

Chapter ten — Laboratory Exercise....97
Micropropagation of *Dieffenbachia*
R.J. Henny, Lee Goode, and Wanda Ellis

Chapter eleven — Laboratory Exercise....103
Micropropagation of potato by node culture and microtuber production
Michael E. Kane

Chapter twelve — Laboratory Exercise....111
Micropropagation of lilacs
Deborah D. McCown and Andrew J. Daun

Chapter thirteen — Laboratory Exercise....119
Micropropagation and in vitro flowering of rose
Michael E. Kane

Chapter fourteen — Concept....125
Organogenesis
Otto J. Schwarz and Robert M. Beaty

Chapter fifteen — Laboratory Exercise....139
Direct shoot organogenesis from leaf explants of chrysanthemum
Robert N. Trigiano and Roger A. May

Chapter sixteen — Laboratory Exercise....149
Shoot organogenesis from cotyledon explants of watermelon
Michael E. Compton and Dennis J. Gray

Chapter seventeen — Laboratory Exercise....159
Direct and indirect shoot organogenesis from leaves of *Torenia fournieri*
Mark P. Bridgen

Chapter eighteen — Laboratory Exercise....167
Shoot organogenesis from petunia leaves
John E. Preece

Chapter nineteen — Concept....175
Nonzygotic embryogenesis
Dennis J. Gray

Chapter twenty — Laboratory Exercise....191
Embryogenic callus and suspension cultures from leaves of orchardgrass
Dennis J. Gray, Robert N. Trigiano, and Bob V. Conger

Chapter twenty-one — Laboratory Exercise....205
Somatic embryogenesis from seeds of melon
Dennis J. Gray

Chapter twenty-two — Laboratory Exercise....213
Somatic embryogenesis from mature peanut seed
Hazel Y. Wetzstein and Charleen Baker

Chapter twenty-three — Laboratory Exercise 223
Direct somatic embryogenesis from leaves and flower receptacles of cineraria
Robert N. Trigiano, Mary Catherine Scott, and Kathleen R. Malueg

Chapter twenty-four — Laboratory Exercise 231
Somatic embryogenesis from immature seeds of yellow poplar
Scott A. Merkle

Chapter twenty-five — Laboratory Exercise 239
Induction of somatic embryogenesis in conifers
Stephen M. Attree, Patricia J. Rennie, and Larry C. Fowke

Part V: Crop improvement techniques

Chapter twenty-six — Concept 249
Use of protoplasts for plant improvement
Michael E. Compton, James A. Saunders, and Richard E. Veilleux

Chapter twenty-seven — Laboratory Exercise 263
Isolation, culture, and fusion of tobacco and potato protoplasts
Richard E. Veilleux and Michael E. Compton

Chapter twenty-eight — Laboratory Exercise 277
Isolation of protoplasts from leaves of chrysanthemum and orchardgrass
Robert N. Trigiano

Chapter twenty-nine — Concept 285
Haploid cultures
Sandra M. Reed

Chapter thirty — Laboratory Exercise 291
Production of haploid tobacco plants using anther culture
Sandra M. Reed

Chapter thirty-one — Concept 297
Transformation
Melina López-Meyer, Ignacio E. Maldonado-Mendoza, and Craig L. Nessler

Chapter thirty-two — Laboratory Exercise 305
Transformation of tobacco and carrot using *Agrobacterium tumefaciens*
and expression of the β-glucuronidase (GUS) reporter gene
Ignacio E. Maldonado-Mendoza, Melina López-Meyer, and Craig L. Nessler

Chapter thirty-three — Laboratory Exercise 321
Transformation of chrysanthemum leaf explants using *Agrobacterium tumefaciens*
Robert N. Trigiano

Chapter thirty-four — Laboratory Exercise 329
Construction and use of a simple gene gun for particle bombardment
Dennis J. Gray, Michael E. Compton, Ernest Hiebert, Chia-Min Lin, and Victor P. Gaba

Chapter thirty-five — Concept 337
Germplasm preservation
Leigh E. Towill

Chapter thirty-six — Laboratory Exercise 345
Vitrification as a method to cryopreserve shoot tips
Leigh E. Towill

Chapter thirty-seven — Concept 355
Secondary product expression in vitro
Mary Ann L. Smith

Chapter thirty-eight — Laboratory Exercise 361
Pigment production in *Ajuga* cell cultures
Mary Ann L. Smith and Randy B. Rogers

Chapter thirty-nine — Concept/Laboratory Exercise 373
In vitro plant pathology
C. Jacyn Baker, Norton M. Mock, and Elizabeth W. Orlandi

Part VI: Special topics

Chapter forty — Concept 387
Variation in tissue culture
S. Jayasankar

Chapter forty-one — Concept 397
Short-circuiting the fern life cycle: apospory in *Ceratopteris richardii*
Leslie G. Hickok and Thomas R. Warne

Chapter forty-two — Concept 407
Commercial laboratory production
Gayle R. L. Suttle

Chapter forty-three — Concept 417
Indexing for plant pathogens
Michael J. Klopmeyer

Chapter forty-four — Laboratory Exercise 427
Culture indexing for bacterial and fungal contaminants
Michael E. Kane

Part VII: Appendix

Appendix I 435
List of some suppliers, addresses, and toll-free telephone numbers

Part VIII: Index

Index 437

Part I

Introduction

chapter one

Introduction to plant tissue culture

Dennis J. Gray and Robert N. Trigiano

The purpose of this chapter is to provide an initial focus to *Plant Tissue Culture Concepts and Laboratory Exercises* and to begin the process of defining terms and ideas unique to this subject area. Keeping in mind that the mission of the entire book is to introduce, define, and provide training, we will use this "introduction" primarily to orient the reader on the book's structure and to highlight information that is discussed in depth in subsequent chapters. We have taken a "minimalist" approach to development of definitions and use of terminology for the simple reason that this serves, in part, the role of an introductory textbook; in our experience, the specific meaning of many terms varies depending on whom is asked and it is often impossible to arrive at a globally satisfactory definition. Rather than confuse the beginning student (and ourselves) with "verbal gymnastics," we chose to simplify terminology. This results in the emphasis being placed on assimilation of key concepts and allows the student to become aware of the richness and natural controversy that involves terminology of such an emerging field after having acquired preliminary background knowledge and confidence. An example of our minimal approach to terminology can be seen in the very title of the book — "... plant tissue culture ...." To be painstakingly exact, the title should have been "... plant cell, tissue, organ, and organ system culture...." However, we achieved the desired effect of announcing the subject of the book without being tedious or confusing. This same approach represents our goal throughout the text.

Meant to serve as a primary text for both introductory and advanced courses, *Plant Tissue Culture Concepts and Laboratory Exercises* furnishes instructors and students alike with a broad consideration of the field — providing historical perspectives, discussing state-of-the-art techniques and methodologies, and looking forward to future advances and applications. It also presents many useful protocols and procedures and, thus, should serve as a valuable reference. The book is intentionally written to be rather informal — it provides the reader with a minimum number of references but does not sacrifice essential information or accuracy. Broad-topic chapters are authored by specialists with considerable experience in the field and are supported by one or more laboratory exercises illustrating central concepts of the topic. Collectively, the laboratory exercises are exceptionally diverse in nature, providing something for everyone — from beginning to advanced students. Importantly, the authors have successfully completed the exercises many times, often with either tissue culture classes or in their own research laboratories. A unique feature of all the laboratory exercises is that the authors have provided in general terms what results should be expected from each of the experiments. At the end of each exercise, there are a series

0-8493-2029-1/00/$0.00+$.50

of questions designed to provoke individual thought and critical examination of the experiment and results. Our intentions are that instructors will not attempt to do all the experiments, but rather select one or two for each concept that serves the needs and interests of their particular class. Thus, exercises for different crop types, such as grasses, ornamentals, trees, and vegetables, are variously provided to allow tailoring of a class to the department or discipline in which it is taught. For an advanced class, different experiments may be assigned among resourceful students. More advanced experiments following the general or beginning class exercises are embedded within some of the laboratory chapters.

The laboratory projects are executed with many different agronomic and horticulturally important plants including chrysanthemum, orchardgrass, syngonium, ajuga, petunia, watermelon, peanut, and lilac, all of which are easy to grow, can be ordered from common sources for nominal fees, or can be purchased at the local market. This range of plants is substantially broader than that used in any previous text on the subject and should stimulate interest in students from many botanical and agricultural disciplines.

The various chapters necessarily assume that the student has a good understanding of botany and botanical terminology. As such, it is recommended that companion plant anatomy, genetics, morphology, and physiological textbooks be available as needed in order to provide access to basic botanical knowledge.

In addition to this introductory section, the textbook is divided into the following five primary parts: History of Plant Tissue Culture, Supporting Methodologies, Propagation Techniques, Crop Improvement Techniques, and Special Topics. Each section combines related facets of plant tissue culture and includes one to several concept chapters, usually with accompanying laboratory exercises.

Part II, "History of Plant Tissue Culture," is an abbreviated account of how the field developed from an early theoretical base to the highly technical discipline that exists today. This treatment is somewhat unique because it documents research progress as modulated by significant world events. In addition to recording the people, places, and dates for pertinent discoveries, we feel it is interesting for students to see the challenges encountered by researchers that result in the often uneven pace of research. We end this chapter on a contemporary "hot" topic by discussing progress being made with genetically engineered plants in light of society's reaction, much of which information is gleaned from news reports. Beyond a past historical account, we hope that this chapter will provide a "snapshot" of the controversial present.

Part III, "Supporting Methodologies," begins the process of teaching tissue culture methodology. Chapter three covers several key topics that must be assimilated to accomplish subsequent laboratory exercises. It describes basic equipment needed and discusses the nature of nutrient culture medium and provides formulas for the most commonly used types. Chapter three also discusses various methods to prepare medium, and provides complete and logical examples of how to make solutions and dilutions and to accomplish sterile culture work. An experienced teacher with laboratory resources and methodologies already in place may choose to pay less attention to this chapter, whereas the instructor of a newly established course will find it indispensable. Chapter four is meant to bring the student into first contact with actual "tissue culture" by demonstrating the essential need for specific nutrients to grow "callus cultures." "Callus" is probably the most common term used to describe tissue cultures, but it only pertains to a certain type of unorganized tissue. As previously mentioned, a common term like "callus" can be defined in several, often conflicting, ways depending on opinion, but for here we consider a callus to be an unorganized tumor-like growth of cells. Part III ends with three chapters designed to emphasize common methods to visualize and document studies (histology and microscopy/photography, respectively) and to quantify responses (statistical analysis) of tissue culture in research.

Part IV, "Propagation Techniques," encompasses the essential foundation of plant tissue culture. In this section the three types of commonly used culture regeneration systems are introduced, then discussed and illustrated in depth.

We begin Part IV by discussing "Propagation from Preexisting Meristems" (Chapter eight), a process that is more commonly termed "micropropagation." Students tend to particularly enjoy some of the exercises in this section because they culminate in "house plants" that can be taken home and reared. Micropropagation is the simplest and most commercially useful tissue culture method. The tissue culture industry uses micropropagation almost exclusively for ornamental plant production. This is reflected by the range of ornamental plants used in the laboratory exercises. However, a unique exception to micropropagation of ornamental plants is the exercise on tissue culture production of potato (Chapter eleven); tissue culture-derived microtubers now are commonly used to establish potato fields. Other points of interest in this section are given in Chapter ten, which demonstrates the exact methodology used by industry to produce Dieffenbachia, and Chapter thirteen, which shows the production of rose flowers in culture.

The second propagation system to be discussed in Part IV is "Organogenesis" (Chapter fourteen). Organogenesis is the development of organs, typically shoots and/or roots, from cells and tissues that would not normally form them. The term "adventitious" has also been used to describe the plant parts formed by the process of organogenesis. Shoot organogenesis is another means of propagating plants. While not used much in commercial production, organogenesis is used extensively in genetic engineering as a means to produce plants from genetically altered cells. In Chapter fourteen, both the theory and developmental sequences of how cells are induced to follow such a developmental pathway are discussed. This chapter is followed by laboratory exercises that show the induction of shoots from leaves of chrysanthemum (Chapter fifteen), watermelon (Chapter sixteen), *Torenia* (Chapter seventeen), and petunia (Chapter eighteen).

The third propagation system discussed in Part IV is "Nonzygotic Embryogenesis" (Chapter nineteen). Nonzygotic embryogenesis is a broadly defined term meant to cover all instances where embryogenesis occurs outside of the normal developmental pathway found in the seed. One type of nonzygotic embryogenesis is termed somatic embryogenesis. This is a unique phenomenon exhibited by vascular plants, in which somatic (nonsexual) cells are induced to behave like zygotes. Such induced cells begin a complex, genetically programmed series of divisions and eventual differentiation to form an embryo that is more or less identical to a zygotic embryo. This type of propagation system is important since the embryos develop from single cells, which can be genetically engineered, and are complete individuals that are capable of germinating directly into plants. Thus, potentially, somatic embryogenesis also represents an efficient propagation system. Chapter nineteen discusses the developmental processes and significance of nonzygotic embryogenesis. Laboratory exercises illustrate nonzygotic embryogenesis for a range of crop types, including a grass (orchardgrass) (Chapter twenty), a vegetable (cantaloupe) (Chapter twenty-one), an agronomic crop (peanut) (Chapter twenty-two), a flowering ornamental plant (cineraria) (Chapter twenty-three), and both angiosperm (yellow poplar) and gymnosperm (white spruce) tree species (Chapters twenty-four and twenty-five, respectively).

In Part V, "Crop Improvement Techniques," the aforementioned propagation techniques are integrated with other methodologies in order to modify and manipulate germplasm. Chapter twenty-six discusses the use of plant protoplasts. Protoplasts are plant cells from which the cell wall has been enzymatically removed, making the cells amenable to cell fusion and other methods of germplasm manipulation. Two laboratory exercises concerning tobacco and potato (Chapter twenty-seven) and chrysanthemum and orchardgrass (Chapter twenty-eight) follow. Chapter twenty-nine details the use of haploid culture in plant improvement. Haploid cultures usually are derived from microspore mother cells and result in cells, tissues, and plants with half the normal somatic cell chromosome number. Such plants are of great use in genetic studies and breeding, since all of the recessive genes are expressed and by doubling the haploid plants back to the diploid ploidy level, dihaploid plants are produced, which are completely homozygous. True homozygous plants are time consuming and often impossible to produce by conventional breeding. Chapter thirty is a laboratory exercise detailing haploid plant production from the microspores contained in tobacco anthers.

Concept Chapter thirty-one discusses genetic transformation (also known as genetic engineering), which is a current hot topic in agriculture. Transformation wherein genes from unrelated organisms can be integrated into plants without sexual reproduction, resulting in "transgenic plants," is the most significant application for plant tissue culture when considering its impact on humankind (see Part II). Two laboratory exercises discuss transformation of tobacco and carrot (Chapter thirty-two) and chrysanthemum (Chapter thirty-three) using *Agrobacterium tumefaciens,* nature's own and the original genetic engineer. This soil-inhabiting bacterium is ubiquitous and strains infect a wide range of host plants including angiosperms (monocots and dicots), gymnosperms, and ferns. *Agrobacterium* causes a tumor-like proliferation of cells, hence the common name of the disease is "crown gall," by transferring some of its plasmid (T_i) DNA to the host cell. Researchers have cleverly learned how to disarm (can no longer cause disease) *Agrobacterium* by deleting the genes that cause tumor growth. They can substitute genes that we want to transfer to our host species. Among these useful genes are those for disease resistance, herbicide tolerance, flower color, and fruit ripening. Chapter thirty-four describes the construction of a device for particle bombardment of plants, an alternate method of transformation. In this method, DNA is coated onto small tungsten or gold particles and literally shot into the plant cells.

Chapter thirty-five describes the use of cryopreservation of germplasm and Chapter thirty-six is a laboratory exercise illustrating its use. Cryopreservation is an efficient means of safeguarding valuable plant germplasm by freezing all metabolic activity. Cryopreserved cells and tissue can be kept for extended periods of time without mutations occurring or any physiological decline. Concept Chapter thirty-seven describes the production of secondary products by plant cells in culture. This subject is somewhat unique in the context of previous topics, because the end product is not a regenerated plant, but rather a chemical. Use of plants as biofactories to produce complex pharmaceuticals is of great interest, particularly due to its potential application in health care. In Chapter thirty-eight, the methods for production of an intensely colored pigment from *Ajuga* cell cultures are described. A final topic Part V is that of in vitro plant pathology. This topic (Chapter thirty-nine) is introduced to demonstrate the use of in vitro systems to mimic whole plants in the development of disease symptoms — overall, a convenient means of studying host-pathogen interactions.

Part VI, "Special Topics," is a bit of a "catch-all" section where we placed topics that we considered important enough to warrant inclusion in the book, but which did not quite fit in the other sections. Chapter forty discusses reasons that genetic and/or phenotypic variation occurs as a result of culture and presents up-to-date supporting research findings. Tissue culture of ferns (Chapter forty-one) illustrates the complete fern life cycle in a culture vessel. Beyond a prelude to propagation, it demonstrates alternation of generations, a basic biological principle. Chapter forty-two is a look into the mechanics of commercial plant production, which is the primary method of producing many house plants and is an important industry worldwide. This chapter validates the commercial application of many exercises presented in Part IV, "Propagation from Preexisting Meristems." Chapter forty-three describes the importance of clean cultures and problems associated with maintaining in vitro cultures and Chapter forty-four is a companion laboratory exercise showing how plant pathogens are detected in cultured plants.

In addition to the information included in this book, explosive expansion of internet access in the last few years has provided new and almost indispensable learning resources for the student. We mention two useful listservers here and encourage students to log onto them for background information and additional support. One is the Plant-TC listserv, which was conceived and maintained by Dr. Mark Galatowitsch at the University of Minnesota. The Plant-TC listserv is an international forum for discussing scientific, educational, hobby, and other aspects related to plant tissue culture. To subscribe, send an e-mail message to "listserv@tc.umn.edu". Leave the subject line blank and enter the following as a message: "sub plant-tc YOUR NAME". Subscription is then automatic. Dr. Galatowitsch also maintains a list of plant tissue culture-related links at "http://www.agro.agri.unm.edu/planttc/optc.htm". Another internet service is Agnet, which is maintained by Dr. Douglas Powell at the University of Guelph. Agnet provides current news that focuses on agriculture and, particularly, the development and impact of transgenic crops worldwide. To subscribe to Agnet, send mail to: "listserv@listserv.uoguelph.ca". Leave the subject line blank and in the body of the message enter the following: "subscribe agnet-L YOUR NAME".

Based on the many telephone calls, letters, and e-mails that we received, the first edition of *Plant Tissue Culture Concepts and Laboratory Exercises* successfully facilitated training of students in current principles and methodologies of our rapidly evolving field. We asked for constructive criticisms from users of the book and employed them to make a number of improvements in this second edition. The most notable improvement is the addition of listed steps as procedure boxes in the laboratory exercises, which will make procedures easier to follow. Several chapters have been added, including this introduction, which was sorely lacking in the first edition. We believe that these improvements will ensure that the second edition will be even better than the first and, hopefully, will enjoy at least the same level of success! As always, we welcome comments from colleagues and students as they put the textbook to use.

Part II

History of plant tissue culture

chapter two

History of plant tissue and cell culture

James D. Caponetti, Dennis J. Gray, and Robert N. Trigiano

The field of plant tissue culture is based on the premise that plants can be separated into their component parts (organs, tissues, or cells), which can be manipulated in vitro and then grown back to complete plants. This idea of handling higher plants with the ease and convenience of microorganisms has conjured up many wonderful possibilities for their study and use; these possibilities, in turn, have been the long-standing stimuli driving research and development in the field. Plant tissue culture along with molecular genetics is a core technology for genetic engineering. After many years of overly optimistic promise, genetically engineered plants have finally reached the marketplace. At the end of this chapter, we discuss how the greatly anticipated societal benefits of this technology have been modulated by a good deal of concern regarding its safety.

The first successful plant tissue and cell culture was accomplished by Gottlieb Haberlandt near the turn of the 20th century when he reported the culture of leaf mesophyll tissue and hair cells (see Steward, 1968; Krikorian and Berquam, 1969). This was a remarkable accomplishment considering that little was known about plant physiology at the time. In retrospect, however, Haberlandt must have drawn on a body of previous knowledge in plant biology. We must assume that he was familiar with the writings of early philosophers such as Aristotle, Theophrastus, Pliny the Elder, Dioscorides, Avicenna, Magnus, Angelicus, and Goethe and that his studies surely must have included the anatomical observations of Hooke, Malpighi, Grew, Nageli, and Hanstein. His own research must have led him to the investigations by early plant physiologists such as van Helmont, Mariotte, Hales, Priestly, Ingenhousz, and Senebier and he must have had access to the morphological and physiological investigations of 19th century botanical researchers such as Schleiden, Schimper, Pringsheim, Unger, Hedwig, Hofmeister, Vochting, Sachs, Goebel, Bower, and Farlow. The information available from these sources coupled with improved light microscopes must have given Haberlandt the insight to culture plant cells and to predict that they could not only grow, but divide and develop into embryos and thence to whole plants, a scenario referred to as totipotency by Steward (1968).

Unfortunately, the cells that Haberlandt cultured did not divide and, thus, his ideas of plant development and totipotency did not come to fruition. The probable reason for the failure was that the plant growth regulators (PGRs) needed for cell division, proliferation, and embryo induction were not present in the culture medium. Indeed, PGRs had not yet been discovered. Apparently, Haberlandt became discouraged and pursued other

0-8493-2029-1/00/$0.00+$.50

physiological investigations. However, his ideas of embryo induction in culture did not go unnoticed in the scientific world. For example, Hannig (1904) cultured nearly mature embryos excised from seeds of several species of crucifers. Moreover, Haberlandt's lack of success was not in vain because one of his students, Kotte (1922), reported the growth of isolated root tips on a medium consisting primarily of inorganic salts. At the same time, and quite independently, Robbins (1922) reported a similar success with root tips and stem tips, and White (1934) reported that not only could cultured tomato root tips grow, but they could be repeatedly subcultured to fresh medium of inorganic salts supplemented with yeast extract, unbeknown to him a good source of B vitamins.

Innovative plant tissue culture techniques progressed rapidly during the 1930s due to the discovery that B vitamins and natural auxin were necessary for the growth of isolated tissues containing meristems. The growth-promoting effects of thiamine on isolated tomato root tips was reported by White (1937). A series of ingenious experiments with oat seedlings by Fritz Went in the 1920s plus other plant physiologists, including Kenneth Thimann, led to the discovery of the first PGR, indoleacetic acid (IAA). IAA is a naturally occurring member of a class of PGRs termed "auxins." The events leading to the discovery of IAA are well-documented in a report published in 1937. Duhamet (1939) reported the stimulation of growth of excised roots by IAA.

The avenue was now open for rapid progress in the successful culture of plant tissues during the 1930s. The culture of meristems other than root and shoot tips was explored. With improved culture media, La Rue (1936) achieved better success at culturing plant embryos compared to the efforts of Hannig 32 years earlier. Gautheret (1934) reported the successful culture of the cambium of several species of trees to produce callus on medium containing B vitamins and IAA. Further research with other meristematic tissues led Nobecourt (1937) and Gautheret (1939) to obtain callus from carrot root cambium and White (1939) to obtain tobacco tumor (crown gall) tissue. The rapid progress of plant tissue culture, however, was soon abruptly curtailed by the start of World War II in 1939.

Since one theater of World War II was concentrated in Europe, further progress in plant tissue culture by the European originators was almost impossible in the midst of the disruption and destruction and the shortages of laboratory supplies and equipment. Scientists in other locations also suffered shortages of laboratory supplies, equipment, and personnel. During the war years (1939 to 1945), nevertheless, some progress was made in plant tissue culture. Johannes Van Overbeek and associates reported they were able to obtain seedlings from heart-shaped embryos by enriching culture media with coconut milk besides the usual salts, vitamins, and other nutrients. Also by this time, Panchanan Maheshwari, along with associates and students in India, were very active in angiosperm embryology research that began before the war and progressed into the 1940s as described in his book (1950). Armin Braun reported on tumor induction related to crown gall disease. Experiments on tobacco tissues by Folke Skoog demonstrated organ formation in cultured tissues and organs. In Western Europe, some progress occurred in that Guy Camus was the first to report grafting experiments in tissue cultures and Georges Morel reported on developing techniques to culture parasitized plant tissues.

With war's end in the spring of 1945, a resumption of prewar tissue culture activity could not occur immediately. But some progress in plant tissue culture technology occurred. Ernest Ball reported on greatly improving the potential for culturing shoot tips, beginning with those of nasturtium and lupine. Albert Hildebrandt and co-workers made improvements on the medium for the growth of tobacco and sunflower tissues and Morel was well into applying tissue culture techniques to the study of parasites associated with plant tissues. Herbert Street and colleagues began a series of extensive studies on the nutrition of excised tomato root tips.

After 1950, rapid progress was made in plant tissue culture techniques. Also, many advancements were accomplished in the knowledge of plant development, especially in the area of the effects of PGRs. While intensive studies on the in vivo and in vitro effects of auxins continued, other classes of PGRs were now recognized. One such class, the cytokinins, emerged from the investigations of Skoog and associates on the nutritional requirements of tobacco callus, which extended to include study of the induction of bud formation on tobacco stem segments by adenine sulfate (Skoog and Tsui, 1951). These initial investigations led to the discovery of the cytokinin, kinetin, a cell division promoter, by Skoog and Carlos Miller. This research subsequently led to the discovery of other cytokinins during the latter half of the 1950s and into the 1960s. Intensive research then began with adenine, kinetin, and several other newly discovered cytokinins on their in vitro shoot-promoting effects for the rapid propagation of plants, especially the economically important horticultural and agronomic cultivars.

Meanwhile, according to Stowe and Yamaki (1957), the "western world" became familiar with a third class of PGR, the gibberellins, through the efforts of a group at Imperial Chemical Industries in Britain and a group at the USDA. These groups "discovered" research on gibberellins that had been conducted by many Japanese plant physiologists beginning with Kurosawa in what was then Formosa. After the "western world" became aware of the gibberellins in the early 1950s, research on the in vivo effects of gibberellins accelerated at a rapid pace worldwide. In vitro research with gibberellins was slow to begin in the late 1950s and increased substantially in the 1960s and 1970s.

A similar situation occurred with the two other classes of PGRs, each of which contains a single representative, ethylene and abscisic acid. Crocker et al. (1935) were the first to propose that ethylene is involved in fruit ripening. In vivo investigations in the 1960s confirmed the role of ethylene in fruit ripening, as first proposed in the 1930s, and also in early seedling growth, leaf abscission and epinasty, senescence, and other growth-promoting and inhibitory effects. In vitro studies with ethylene did not begin until the 1970s. The in vivo effects of abscisic acid on plant development were not recognized until the 1960s. Studies into the 1970s were directed toward leaf and fruit abscission and bud dormancy. Other plant growth-promoting and inhibiting studies soon followed. In vitro investigations with abscisic acid began in the 1970s with experiments involving zygotic and somatic embryogenesis.

By the mid 1950s, plant tissue culture methods had progressed to the point of making a major impact on research in plant developmental biology. Those investigators who had studied plant development over the previous 50 years using the conventional techniques of morphology, anatomy, physiology, biochemistry, cytology, and genetics progressively incorporated the "tool" of plant tissue culture into their research. Major breakthroughs of knowledge in plant development came sooner than would have been possible without the techniques of sterile culture. The progress also was hastened by innovations in laboratory equipment and supplies, and by improved methods of worldwide transportation and communication between developmental botanists around the world. In particular, the development of HEPA filters, which screen fungal spores and bacteria from air, made laminar flow transfer hoods possible. Laminar flow hoods became commonly available in the late 1960s to early 1970s and, finally, made sterile culture a routine task. Some of the European and American originators of the basic techniques of plant tissue culture from the 1930s and 1940s traveled around the world after the 1950s to many research laboratories as visiting botanists or to attend meetings, and became the "teachers" of the basic methods to numerous colleagues and students.

Concomitantly, another aspect of plant tissue culture greatly aided the increased research activity in plant development. The methods of obtaining plant callus tissue from several sources were well developed by the mid-1950s. The study of the causes of mammalian cancer (especially human cancers) became very popular during this time period. It was logical, then, that plant callus development was equated to mammalian tumor development. Thus, research on plant callus and cell suspension cultures intensified in many research laboratories because several agencies that were interested in mammalian cancer research awarded generous grants to plant developmental botanists to study "plant cancers."Among these agencies were the American Cancer Society, the National Cancer Institute, the National Institutes of Health, and the U.S. Department of Health, Education and Welfare. A few notable pioneer investigators, among many, in the study of plant callus and cell suspensions in the early 1950s to the mid-1960s included Morel and Wetmore (1951), Henderson and Bonner (1952), Steward et al. (1952; 1954; 1958), Muir et al. (1954), Braun (1954), Nickell (1955), Partanen et al. (1955), Das et al. (1956), Nitsch and Nitsch (1956), Torrey (1957), Reinert (1958), Klein (1958), Bergmann (1960), and Murashige and Skoog (1962).

In the 1950s, the early prediction that somatic plant cells could undergo embryogenesis finally was validated by Steward et al. (1958) and Reinert (1958), who showed that somatic cells of carrot would differentiate into embryos when cultured within a proper nutrient — PGR regime. The ability to regenerate plants from single somatic cells through such a "normal" developmental process was envisioned to have great applications in propagation and genetic engineering (see below). Nonzygotic embryogenesis now has been demonstrated in most species of higher plants that have been tested (see Chapter nineteen).

In the early 1960s, Murashige completed a study while working in Skoog's laboratory leading to a commercial application for tissue culture. A culture medium developed by Murashige was originally devised for the rapid growth and bioassays with tobacco callus (Murashige and Skoog, 1962). However, research by several nurseries in California and elsewhere, through the advice of Murashige and knowledge of the work of Morel on orchid propagation, showed that practical shoot tip propagation of several ornamental plants could be accomplished on Murashige and Skoog medium (1962). Today, the ornamental plant industry depends heavily on tissue and organ culture micropropagation to supply high-quality, low-cost stocks of many species (see Chapters eight, ten, and forty-two).

Reports concerning the recovery of plants from haploid cells began to appear in the 1960s (see Chapter twenty-nine). The first successes were obtained with *Datura* (Guha and Maheshwari, 1966) and tobacco (Bourgin and Nitsch, 1967). This discovery received significant attention from plant breeders, since plants recovered from doubled haploid cells are homozygous and express all recessive genes, making them ideal pure breeding lines. Haploid-based breeding programs now are in place for several major agronomic crops.

Plant protoplasts were isolated and began to be cultured in the 1960s (Cocking, 1960). The removal of the plant cell wall allowed many novel experiments on membrane transport to be undertaken and, with totipotent protoplasts, allowed the somatic hybridization of sexually incompatible species to be accomplished. Somatic hybridization now has been used successfully in a number of cultivar development programs (Chapter twenty-six). In addition, the uptake of DNA into the plant cell was facilitated by removal of the wall, leading to the first reports of plant genetic transformation (Chapter thirty-one and below).

Perhaps the greatest stimulus and change in the direction of plant cell and tissue culture research occurred after the discovery of restriction endonuclease enzymes in the early 1970s. These enzymes cleave DNA molecules at predictable sites and allow specific genes to be removed, modified, and inserted into other DNA strands. The watershed of technological development that occurred after this breakthrough, and still is occurring today, has provided the basic tools needed for genetic engineering (Chapter thirty-one). The term "biotechnology" was coined to denote this new research field. The availability of totipotent plant cells that, conceivably, could be altered by insertion of specific genes caused a revolution in plant research, because the obvious implications to agriculture of such genetically altered plants were so great. The potential commercial value of such genetically modified (or GM) plants attracted an unprecedented amount of industrial and investor interest such that, in the late 1970s and 1980s, a number of new plant biotechnology-based companies sprang up around the world, but primarily in the U.S. Some of these companies exist today, but many either failed or were absorbed by larger companies.

Since the publication of the first edition of *Plant Tissue Culture Concepts and Laboratory Exercises* in 1996, a renaissance in agriculture has occurred due to the commercialization of GM crops. Establishment of GM crops has been very rapid; to document growth of this new sector, the following statistics were gleaned from news reports from the past year. In 1995, there were no commercial plantings of GM crops. However, out of 80 million acres of corn planted yearly in the U.S., 400,000 acres of GM corn were planted in 1996, increasing to 3 million acres in 1997 and 17 million acres in 1998. Similarly, of 71 million acres of soybean planted in 1997, 20 million acres were genetically modified and some estimates are that nearly 100% of the soybean crop will be composed of GM varieties within the next few years. Approximately 45% of the U.S. 1998 cotton crop was genetically modifed. Corn and soybean account for about 75% of all GM crops planted, followed by cotton (12%), canola (11%), and potato (<1%). While almost 75% of all GM crops are grown in the U.S., production is increasing in Argentina, Canada, Mexico, Spain, France and South Africa. Although China grows GM crops, data as to the extent are not available.

In 1998, over 99% of the GM crop acreage was devoted to just two transgenic traits, herbicide and insect resistance (Bt). Less than 1% of GM acreage was devoted to "quality" traits, such as controlled ripening and enhanced oil profiles. Globally, herbicide-resistant soybean dominates GM crop acreage (52%), followed by Bt corn (24%). The success of these crops is evident in that it is estimated that $200 million worth of pesticide was eliminated by implementation of insect-resistant GM crops alone. Herbicide usage has dropped because applications can be delayed until weeds are commingled with the crop plant and are most susceptible.

When considering these GM crops, along with other additives in use today, such as enzymes from microorganisms in processed products like soft drinks, cakes, cheese, bread, fish, and meats, up to 90% of common foodstuffs already may contain GM components. This rapid integration of GM crops into the food supply has caused concern in certain parts of the world. The concern is based on the perception that there has been inadequate testing of GM crops and centers on several issues, including the following: (1) unwanted spread of transgenes into wild species by pollination, resulting in, for example, herbicide-resistant weeds; (2) development of resistant insects, for example, due to overuse of Bt crops; (3) potential for increased use of herbicides in herbicide-resistant GM crop production; (4) human health concerns, for example, transfer of antibiotic resistance genes from digested GM food to bacteria in the gut, and possible unrecognized toxicity of transgenic proteins. While controversial, most experts consider such concerns to be overreaction and often politically motivated, since GM crops already have been subjected to rigorous testing. Currently, several countries are considering multiyear moratoriums on the planting of GM crops. However, it is expected that the overall benign impact of GM crops on the environment, including reduction of chemical inputs, along with consumer recognition of resultant increased quality of foodstuffs will be combined with better education to solve these concerns.

The use of plant tissue culture technology to enable the development of GM crops illustrates its key, but often overlooked, importance in our lives. Plant tissue culture also is an increasingly important tool for advanced studies of plant development, especially with the integration of molecular biological techniques. However, as will become evident throughout this book, much remains to be learned in terms of the basic methods and procedures needed for efficient manipulation of plants in vitro.

Literature cited

Ball, E. 1946. Development in sterile culture of stem tips and subadjacent regions of *Tropaeolum majus* and of *Lupinus albus* L. *Amer. J. Bot.* 33:301–318.

Bergmann, L. 1960. Growth and division of single cells of higher plants in vitro. *J. Gen. Physiol.* 43:841–851.

Bhojwani, S. S., V. Dhawan, and E. C. Cocking. 1986. *Plant Tissue Culture: A Classified Bibliography.* p. 789. Elsevier Science Publishers, Inc., New York.

Bourgin, J. P. and J. P. Nitsch. 1967. Obtention de *Nicotiana* haploides a partir d'etamines cultivees in vitro. *Ann. Physiol. Veg.* 9:377–382.

Braun, A.C. 1954. Studies on the origin of the crown gall tumor cell. In: Abnormal and Pathological Plant Growth, Brookhaven National Lab. Symp. No. 6:115–127.

Camus, G. 1943. Sur le greffage des bourgeons d'endive sur des fragments de tissues cultives in vitro. *C. R. Acad. Sci.* 137:184–185.

Cocking, E. C. 1960. A method for the isolation of plant protoplasts and vacuoles. *Nature* 187:962–963.

Crocker, W., A. E. Hitchcock, and P. W. Zimmerman. 1935. Similarities in the effects of ethylene and the plant auxins. *Contributions Boyce Thompson Inst.* 7:231–248.

Das, N.K., K. Patau, and F. Skoog. 1956. Initiation of mitosis and cell division by kinetin and indoleacetic acid in excised tobacco pith tissue. *Physiol. Plant.* 9:640–651.

Dormer, K.J. and H.E. Street. 1949. The carbohydrate nutrition of tomato roots. *Ann. Bot. N.S.* 13:199–217.

Duhamet, L. 1939. Action de l'heteroauxine sur la croissance de racines isolees de *Luipinus albus*. *C. R. Acad. Sci.* 208:1838–1840.

Durosawa, E. 1926. Experimental studies on the secretions of *Fusarium heterosporum* on rice plants. *Trans. Nat. Hist. Soc. Formosa* 16:123-127.

Gautheret, R. J. 1934. Culture de tissu cambial. *C. R. Acad. Sci.* 198:2195–2196.

Gautheret, R. J. 1939. Sur la possibilite de realiser la culture indefinie des tissus de tubercules de carotte. *C. R. Acad. Sci.* 208:118–120.

Guha, S. and S. C. Maheshwari. 1966. Cell division and differentiation in the pollen grains of *Datura* in vitro. *Nature* 212:97–98.

Hannig, E. 1904. Zur physiologie pflanzlicher embryonen. I. Uber die kultur von cruciferen-embryonen ausserhalb des embryosacks. *Bot. Ztg.* 62:45–80.

Henderson, J.H.M. and J. Bonner. 1952. Auxin metabolism in normal and crown gall tissue of sunflower. *Amer. J. Bot.* 39:444–451.

Hildebrandt, A.C., A.J. Riker, and B.M. Duggar. 1946. The influence of the composition of the medium on growth in vitro of excised tobacco and sunflower tissue culture. *Amer. J. Bot.* 33:591–597.

Hogland, D.R. and D.I. Arnon. 1938. The Water Culture Method for Growing Plants Without Soil. Calif. Agric. Exp. Stat. Circ. 347, Berkeley, CA.

Klein, R.M. 1958. Activation of metabolic systems during tumor cell formation. *Proc. Natl. Acad. Sci. (U.S.A.)* 44:349–354.

Kotte, W. 1922. Kulturversuche mit isolierten wurzelspitzen. *Beitr. Allg. Bot.* 2:413–434.

Krikorian, A.D. and D.L. Berquam. 1969. Plant cell and tissue culture — the role of Haberlandt. *Bot. Rev.* 35:59-88.

La Rue, C. D. 1936. The growth of plant embryos in culture. *Bull. Torrey Bot. Club* 63:365- 382.

Maheshwari, P. 1950. *An Introduction to the Embryology of Angiosperms*. p. 453. McGraw-Hill, New York.

Morel, G. 1944. Le developpement du mildou sur des tissues de vigne cultives in vitro. *C. R. Acad. Sci.* 218:50-52.

Morel, G. 1948. Recherches sur la culture associee de parasites obligatoires dt de tissus vegetaux. *Ann. Epiphyt. N. S.* 14:123-234.

Morel, G. 1965. Clonal propagation of orchids by meristem culture. *Cymbidium Soc. News* 20:3-11.

Morel, G. and R.H. Wetmore. 1951. Fern Callus Tissue cultures. *Amer. J. Bot.* 39:141–143.

Muir, W.H., A.C. Hildebrandt, and A.J. Riker. 1954. Plant tissue cultures produced from single isolated cells. *Science* 119:877–878.

Murashige, T. and F. Skoog. 1962. A revised medium for rapid growth and bioassays with tobacco tissue cultures. *Physiol. Plant.* 15:473–497.

Nickell, L.G. 1955. Nutrition of pathological tissues caused by plant viruses. *Ann. Biol.* 31:107–121.

Nitsch, J.P. and C. Nitsch. 1956. Auxin dependent growth of excised *Helianthus tuberosus* tissues. *Amer. J. Bot.* 43:839–851.

Nobecourt, P. 1937. Cultures en serie de tissus vegetaux sur milieu artificiel. *C. R. Seanc. Soc. Biol.* 205:521–523.

Partanen, C.R., I.M. Sussex, and T.A. Steeves. 1955. Nuclear behavior in relation to abnormal growth in fern prothalli. *Amer. J. Bot.* 42:245–256.

Reinert, J. 1958. Morphogenese und ihre kontrolle an gewebekulturen aus carotten. *Naturwiss.* 45:344–345.

Robbins, W. J. 1922. Cultivation of excised root tips and stem tips under sterile conditions. *Bot. Gaz.* 73:376–390.

Skoog, F. 1944. Growth and organ formation in tobacco tissue cultures. *Amer. J. Bot.* 31:19–24.

Skoog, F. and C.O. Miller. 1957. Chemical regulation of growth and organ formation in plant tissues cultured in vitro. *Symp. Soc. Exp. Biol.* 11:118–131.

Skoog, F. and C. Tsui. 1951. Growth substances and the formation of buds in plant tissues. In: *Plant Growth Substances.* Skoog, F., Ed. pp. 263–285. University of Wisconsin Press, Madison.

Steward, F.C. 1968. *Growth and Organization in Plants.* p. 564. Addison-Wesley Publishing, Reading, MA.

Steward, F.C. and S.M. Caplin. 1954. The growth of carrot tissue explants and its relation to the growth factors present in coconut milk. I(A). The development of the quantitative method and the factors affecting the growth of carrot tissue explants. *Ann. Biol.* 30:385–394.

Steward, F.C., S.M. Caplin, and F.K. Millar. 1952. Investigations on growth and metabolism of plant cells. I. New techniques for the investigation of metabolism, nutrition and growth of undifferentiated cells. *Ann. Bot. N. S.* 16:57–77.

Steward, F.C., M.O. Mapes, and M.J. Smith. 1958. Growth and organized development of cultured cells. I. Growth and division of freely suspended cells. *Amer. J. Bot.* 45:693–703.

Stowe, B. B. and T. Yamaki. 1957. The history and physiological action of the gibberellins. *Annu. Rev. Plant Physiol.* 8:181–216.

Street, H.E. and J.S. Lowe. 1950. The carbohydrate nutrition of tomato roots. II. The mechanism of sucrose absorption by excised roots. *Ann. Bot. N.S.* 14:307–329.

Torrey, J.G. 1957. Cell division in isolated single plant cells in vitro. *Proc. Natl. Acad. Sci. (U.S.A.)* 43:887–891.

Van Overbeek, J., M.E. Conklin, and A.F. Blakeslee. 1942. Cultivation in vitro of small *Datura* embryos. *Amer. J. Bot.* 29:472–477.

Went, F.W. and K.V. Thimann. 1937. *Phytohormones.* p. 294. Macmillan, New York.

Wetmore, R.H. 1954. The use of in vitro cultures in the investigation of growth and differentiation in vascular plants. In: *Abnormal and Pathological Plant Growth,* Brookhaven National Lab. Symp. No. 6:22–40.

White, P. R. 1934. Potentially unlimited growth or excised tomato root tips in liquid medium. *Plant Physiol.* 9:585–600.

White, P. R. 1937. Vitamin B_1 in the nutrition of excised tomato roots. *Plant Physiol.* 12:803–811.

White, P. R. 1939. Potentially unlimited growth of excised plant callus in an artificial medium. *Amer. J. Bot.* 26:59–64.

Part III

Supporting methodologies

chapter three

Getting started with tissue culture — media preparation, sterile technique, and laboratory equipment

Caula A. Beyl

A plant tissue culture laboratory has several functional areas, whether it is designed for teaching or research and no matter what its size or how elaborate it is. It has some elements similar to a well-run kitchen and other elements that more closely resemble an operating room. There are areas devoted to preliminary handling of plant tissue destined for culture, media preparation, sterilization of media and tools, a sterile transfer hood or 'clean room' for aseptic manipulations, a culture growth room, and an area devoted to washing and cleaning glassware and tools (see Chapter forty-two). The following chapter will serve as an introduction to what goes into setting up a tissue culture laboratory — what supplies and equipment are necessary and some basics concerning making stock solutions, calculating molar concentrations, making tissue culture media, preparing a transfer (sterile) hood, and culturing various cells, tissues, and organs.

Equipment and supplies for a tissue culture teaching laboratory

Ideally, there should be enough bench area to allow for both preparation of media and storage space for chemicals and glassware. In addition to the usual glassware and instrumentation found in laboratories, a tissue culture laboratory needs an assortment of glassware, which may include graduated measuring cylinders, wide-necked Erlenmeyer flasks, medium bottles, test tubes with caps, petri dishes, volumetric flasks, beakers, and a range of pipettes. In general, glassware should be able to withstand repeated autoclaving. Baby food jars are an inexpensive alternative tissue culture container well suited for teaching. Ample quantities can be obtained by preceding the recycling truck on its pickup day (provided you are not embarrassed by the practice). Some tissue culture laboratories find presterilized disposable culture containers and plastic petri dishes to be convenient, but the cost may be prohibitive for others on a tight budget. There are also reusable plastic containers available but their longevity and resistance to wear, heat, and chemicals varies considerably.

0-8493-2029-1/00/$0.00+$.50

It is also good to stock metal or wooden racks to support culture tubes both for cooling and later during their time in the culture room, metal trays (such as cafeteria trays) and carts for transport of cultures, stoppers and various closures, nonabsorbent cotton, cheese cloth, foam plugs, metal or plastic caps, aluminum foil, Parafilm™, and plastic wrap.

To teach tissue culture effectively, some equipment is necessary such as a pH meter, balances (one analytical to four decimal places and one analytical to two decimal places), bunsen burners, alcohol lamps or electric sterilizing devices, several hot plates with magnetic stirrers, a microwave oven for rapid melting of large volumes of agar medium, a compound microscope and hemocytometer for cell counting, a low-speed centrifuge, stereomicroscopes (ideally with fiber optic light sources), large (10- or 25-L) plastic carboys to store high-quality (purity) water, a fume hood, an autoclave (or, at the very least, a pressure cooker), and a refrigerator to store media, stock solutions, plant growth regulators (PGRs), etc. A dishwasher is useful, but a large sink with drying racks, pipette and acid baths, and a forced air oven for drying glassware will also work. Also, deionized distilled water for the final rinsing of glassware is needed. Aseptic manipulations and transfers are done in multistation laminar flow hoods (one for each pair of students).

Equipment used in the sterile transfer hood usually includes a spray bottle containing 70% ethanol, spatulas (useful for transferring callus clumps), forceps (short, long, and fine-tipped), scalpel handles (#3), disposable scalpel blades (#10 and #11), a rack for holding sterile tools, a pipette bulb or pump, bunsen burner, alcohol lamp or other sterilizing device, and a sterile surface for cutting explants (see below). If necessary for the experimental design, uniform-sized leaf explants can be obtained using a sterile cork borer (see Chapter four).

There are a number of options for providing a sterile surface for cutting explants. A previously autoclaved stack of paper towels wrapped in aluminum foil is effective and as each layer becomes messy, it can be peeled off and the next layer beneath it used (Figure 3.1). Others prefer reusable surfaces such as ceramic tiles (local tile retailers are quite generous and will donate samples), metal commercial ashtrays, or glass petri dishes (100 × 15 mm). Sterile plastic petri dishes also can be used but the cost may outweigh the advantages. A container is needed to hold the alcohol used for flaming instruments, if flame sterilization is used. An ideal container for this purpose is a slide staining Coplin jar with a small wad of cheesecloth at the bottom to prevent breakage of the glass when tools are dropped in. It has the advantage that it is heavier glass and since the base is flared, it is not prone to tipping over. Other containers can also serve the same purpose such as test tubes in a rack or placed in a flask or beaker to prevent them from spilling. Plastic containers, which can catch fire and melt, never should be used to hold alcohol.

Water

High-quality water is a required ingredient of plant tissue culture media. Ordinary tap water contains cation, anions, particulates of various kinds, microorganisms, and gases that make it unsuitable for use in tissue culture media. Various methods are used to treat water including filtration through activated carbon to remove organics and chlorine, deionization or demineralization by passing water through exchange resins to remove dissolved ionized impurities, and distillation that eliminates most ionic and particulate impurities, volatile and nonvolatile chemicals, organic matter, and microorganisms. The process of reverse osmosis, which removes 99% of the dissolved ionized impurities, uses a semipermeable membrane through which a portion of the water is forced under pressure and the remainder containing the concentrated impurities is rejected. The most universally reliable method of water purification for tissue culture use is a deionization treatment

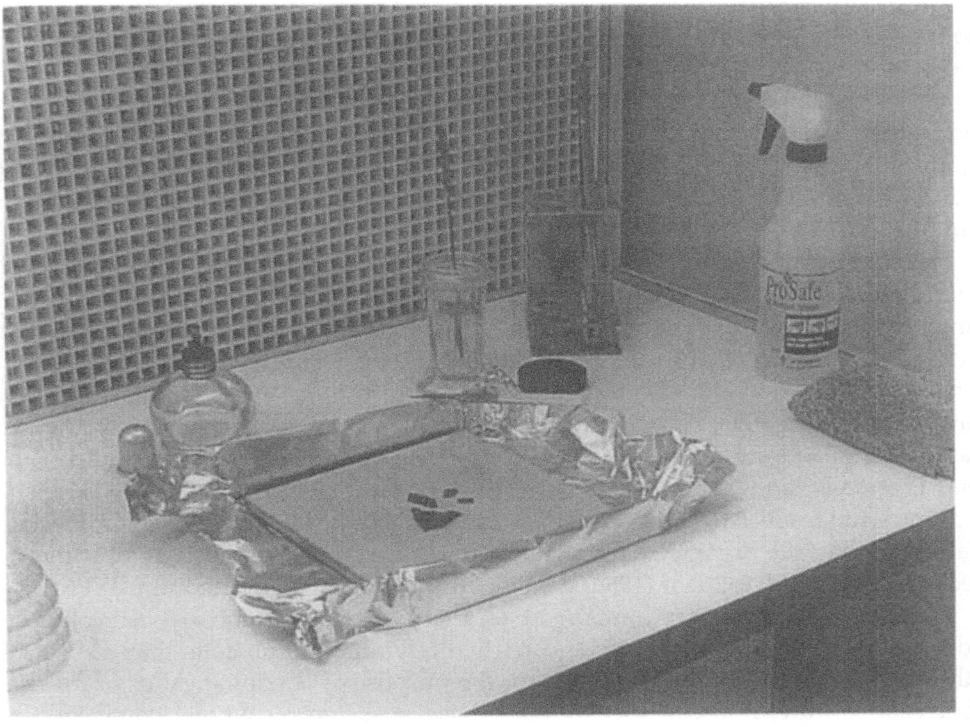

Figure 3.1 A typical layout of materials in the hood showing placement of the sterile tile work surface, an alcohol lamp, spray bottle containing 80% ethanol, a cloth for wiping down the hood, and two different kinds of tool holders — a glass staining (Coplin) jar and a metal rack for holding test tubes.

followed by one or two glass distillations, although simple deionization alone is sometimes successfully used. In some cases, newer reverse osmosis purifying equipment (Milli-RO™, Millipore™, RO pure™, Barnstead™, Bion™, Pierce™), combined with cartridge ion exchange, adsorption, and membrane filtering equipment, has replaced the traditional glass distillation of water.

The culture room

After the explants are plated on the tissue culture medium under the sterile transfer hood, they are moved to the culture room. It can be as simple as a room with shelves equipped with lights or as complex as a room with intricate climate control. Most culture rooms tend to be rather simple, consisting of cool white fluorescent lights mounted to shine on each shelf. Adjustable shelves are an asset and allow for differently sized tissue culture containers and for moving the light closer to the containers to achieve higher light intensities. Putting the lights on timers allows for photoperiod manipulation. Some cultures grow equally as well in dark or light. Temperatures of 26 to 28°C are usually optimum. Heat buildup can be a problem if the room is small, so adequate air conditioning is required. Good air flow also helps to reduce condensation occurring inside petri dishes or other vessels. Some laboratories purchase incubators designed for plant tissue culture. If a liquid medium is used, the culture room should be equipped with a rotary or reciprocal shaker to provide sufficient oxygenation. The optimal temperature, light, and shaker conditions vary depending on the plant species being cultured.

Characteristics of some of the more common tissue culture media

The type of tissue culture medium selected depends on the species to be cultured. Some species are sensitive to high salts or have different requirements for PGRs. The age of the plant also has an effect. For example, juvenile tissues generally regenerate roots more readily than adult tissues. The type of organ cultured is important; for example, roots require thiamine. Each desired cultural effect has its own unique requirements such as auxin (see below) for induction of adventitious roots and altering the cytokinin-to-auxin ratio for initiation and development of adventitious shoots.

Development of culture medium formulations was a result of systematic trial and experimentation. Table 3.1 gives a comparison of the composition of several of the most commonly used plant tissue culture media with respect to their components in milligrams/liter and molar units. Murashige and Skoog (1962) medium (MS) is the most suitable and most commonly used basic tissue culture medium for plant regeneration from tissues and callus. It was developed for tobacco based primarily on the mineral analysis of tobacco tissue. This is a "high salt" medium due to its content of K and N salts. Linsmaier and Skoog medium (1965) is basically Murashige and Skoog (1962) medium with respect to its inorganic portion, but only inositol and thiamine HCl are retained among the organic components. To counteract salt sensitivity of some woody species, Lloyd and McCown (1980) developed the woody plant medium (WPM).

Gamborg's B5 medium (Gamborg et al., 1968) was devised for soybean callus culture and has lesser amounts of nitrate and particularly ammonium salts than MS medium. Although B5 was originally developed for the purpose of obtaining callus or for use with suspension culture, it also works well as a basal medium for whole plant regeneration. Schenk and Hildebrandt (1972) developed SH medium for the callus culture of monocots and dicots. White's medium (1963), which was designed for the tissue culture of tomato roots, has a lower concentration of salts than MS medium. Nitsch's medium (Nitsch and Nitsch, 1969) was developed for anther culture and contains a salt concentration intermediate to that of MS and White's media.

Many companies (see Appendix I) sell packaged prepared mixtures of the better-known media recipes. These are easy to make because they merely involve dissolving the packaged mix in a specified volume of water. These can be purchased as the salts, the vitamins, or the entire mix with or without PGRs, agar, and sucrose. These are convenient, less prone to individual error, and make keeping stock solutions unnecessary. However, they are more expensive than making media from scratch.

Components of the tissue culture medium

Growth and development of explants in vitro are products of its genetics, surrounding environment, and components of the tissue culture medium, the last of which is easiest to manipulate to our own ends. Tissue culture medium consists of 95% water, macro- and micronutrients, PGRs, vitamins, sugars (because plants in vitro are often not photosynthetically competent), and sometimes various other simple-to-complex organic materials. All in all, about 20 different components are usually needed.

Inorganic mineral elements

Just as a plant growing in vivo requires many different elements from either soil or fertilizers, the plant tissue growing in vitro requires a combination of macro- and micronutrients. The choice of macro- and microsalts and their concentrations is species dependent. MS medium is very popular because most plants react to it favorably, however, it

Table 3.1 Composition of Five Commonly Used Tissue Culture Media in Milligrams per Liter and Molar Concentrations

Compounds	Murashige and Skoog	Gamborg B-5	WPM	Nitsch and Nitsch	Schenk and Hildebrandt	White
			Macronutrients in mg/L (mM)			
NH_4NO_3	1650 (20.6)	—	400 (5.0)	—	—	—
$NH_4H_2PO_4$	—	—	—	—	300 (2.6)	—
NH_4SO_4	—	134 (1.0)	—	—	—	—
$CaCl_2 \cdot 2H_2O$	332.2 (2.3)	150 (1.0)	96 (0.7)	166 (1.1)	151 (1.0)	—
$Ca(NO_3)_2 \cdot 4H_2O$	—	—	556 (2.4)	—	—	288 (1.2)
$MgSO_4 \cdot 7H_2O$	370 (1.5)	250 (1.0)	370 (1.5)	185 (0.75)	400 (1.6)	737 (3.0)
KCl	—	—	—	—	—	65 (0.9)
KNO_3	1900 (18.8)	2500 (24.8)	—	950 (9.4)	2500 (24.8)	80 (0.8)
K_2SO_4	—	—	990	—	—	—
KH_2PO_4	170 (1.3)	—	170 (1.3)	68 (0.5)	—	—
NaH_2PO_4	—	130.5 (0.9)	—	—	—	16.5 (0.12)
Na_2SO_4	—	—	—	—	—	200 (1.4)
			Micronutrients in mg/L (mM)			
H_3BO_3	6.2 (100)	3.0 (49)	6.2 (100)	10 (162)	5 (80)	1.5 (25)
$CoCl_2 \cdot 6H_2O$	0.025 (0.1)	0.025 (0.1)	—	—	0.1 (0.4)	—
$CuSO_4 \cdot 5H_2O$	0.025 (0.1)	0.025 (0.1)	0.25 (1)	0.025 (0.1)	0.2 (0.08)	0.01 (0.04)
Na_2EDTA	37.3 (100)	37.3 (100)	37.3 (100)	37.3 (100)	20.1 (54)	—
$Fe_2(SO_4)_3$	—	—	—	—	—	2.5 (6.2)
$FeSO_4 \cdot 7H_2O$	27.8 (100)	27.8 (100)	27.8 (100)	27.8 (100)	15 (54)	—
$MnSO_4 \cdot H_2O$	16.9 (100)	10.0 (59)	22.3 (132)	18.9 (112)	10.0 (59)	5.04 (30)
KI	0.83 (5)	0.75 (5)	—	—	0.1 (0.6)	0.75 (5)
$NaMoO_3$	—	—	—	—	—	0.001 (0.001)
$Na_2MoO_4 \cdot 2H_2O$	0.25 (1)	0.25 (1)	0.25 (1)	0.25 (1)	0.1 (0.4)	—
$ZnSO_4 \cdot 7H_2O$	8.6 (30)	2.0 (7.0)	8.6 (30)	10 (35)	1 (3)	2.67 (9)
			Organics in mg/L (mM)			
Myo-inositol	100 (550)	100 (550)	100 (550)	100 (550)	1000 (5500)	—
Glycine	2.0 (26.6)	—	2.0 (26.6)	2.0 (26.6)	—	3.0 (40)
Nicotinic acid	0.5 (4.1)	1.0 (8.2)	0.5 (4.1)	5 (40.6)	5.0 (41)	0.5 (4.1)
Pyridoxine HCl	0.5 (2.4)	0.1 (0.45)	0.5 (2.4)	0.5 (2.4)	0.5 (2.4)	0.1 (0.45)
Thiamin HCl	0.1 (0.3)	10.0 (30)	1.0 (3.0)	0.5 (1.5)	5.0 (14.8)	0.1 (0.3)
Biotin	—	—	—	0.2 (0.05)	—	—

may not necessarily result in the optimum growth and development for every species since the salt content is so high.

The macronutrients are required in millimolar (mM) quantities in most plant media. Nitrogen (N) is usually supplied in the form of ammonium (NH_4^+) and nitrate (NO_3^-) ions, although sometimes more complex organic sources, such as urea and amino acids like glutamine or casein hydrolysate, which is a complex mixture of amino acids and ammonium, are also used. Although most plants prefer NO_3^- to NH_4^+, the right balance of the two ions for optimum in vitro growth and development for the selected species may differ.

In addition to nitrogen, potassium, magnesium, calcium, phosphorus, and sulfur are provided in the medium as different components referred to as the macrosalts. $MgSO_4$

provides both magnesium and sulfur; $NH_4H_2PO_4$, KH_2PO_4, or NaH_2PO_4 provides phosphorus; $CaCl_2 \cdot 2H_2O$ or $Ca(NO_3)_2 \cdot 4H_2O$ provides calcium; and KCl, KNO_3, or KH_2PO_4 provides potassium. Chloride is provided by KCl and/or $CaCl_2 \cdot 2H_2O$.

Microsalts typically include boron (H_3BO_3), cobalt ($CoCl_2 \cdot 6H_2O$), iron (complex of $FeSO_4 \cdot 7H_2O$ and Na_2EDTA or rarely as $Fe_2(SO_4)_3$), manganese ($MnSO_4 \cdot H_2O$), molybdenum ($NaMoO_3$), copper ($CuSO_4 \cdot 5H_2O$), and zinc ($ZnSO_4 \cdot 7H_2O$). Microsalts are needed in much lower (micromolar, μM) concentrations than the macronutrients. Some media may contain very small amounts of iodide (KI), but sufficient quantities of many of the trace elements inadvertently may be provided because reagent-grade chemicals contain inorganic contaminants.

Organic compounds

Sugar is a very important part of any nutrient medium and its addition is essential for in vitro growth and development of the culture. Most plant cultures are unable to photosynthesize effectively for a variety of reasons including insufficiently organized cellular and tissue development, lack of chlorophyll, limited gas exchange and CO_2 in the tissue culture vessels, and less than optimum environmental conditions, such as low light. A concentration of 20 to 60 g/L sucrose (a disaccharide made up of glucose and fructose) is the most often used carbon or energy source, since this sugar is also synthesized and transported naturally by the plant. Other mono- or disaccharides and sugar alcohols such as glucose, fructose, sorbitol, and maltose may be used. The sugar concentration chosen is dependent on the type and age of the explant in culture. For example, very young embryos require a relatively high sugar concentration (>3%). For mulberry buds in vitro, fructose was found to be better than sucrose, glucose, maltose, raffinose, or lactose (Coffin et al., 1976). For apple, sorbitol and sucrose supported callus initiation and growth equally as well but sorbitol was better for peach after the fourth subculture (Oka and Ohyama, 1982).

Sugar (sucrose) that is bought from the supermarket is usually adequate, but be careful to get pure cane sugar as corn sugar is primarily fructose. Raw cane sugar is purified and according to the manufacturer's analysis consists of 99.94% sucrose, 0.02% water, and 0.04% other material (inorganic elements and also raffinose, fructose, and glucose). Nutrient salts contribute approximately 20 to 50% to the osmotic potential of the medium and sucrose is responsible for the remainder. The contribution of sucrose to the osmotic potential increases as it is hydrolyzed into glucose and fructose during autoclaving. This may be an important consideration when performing osmotic-sensitive procedures such as protoplast isolation and culture.

Vitamins are organic substances that are parts of enzymes or cofactors for essential metabolic functions. Of the vitamins, only thiamine (vitamin B_1 at 0.1 to 5.0 mg/L) is essential in culture as it is involved in carbohydrate metabolism and the biosynthesis of some amino acids. It is usually added to tissue culture media as thiamine hydrochloride. Nicotinic acid, also known as niacin, vitamin B_3, or vitamin PP, forms part of a respiratory coenzyme and is used at concentrations between 0.1 and 5 mg/L. MS medium contains thiamine HCl as well as two other vitamins, nicotinic acid and pyridoxine (vitamin B_6) in the HCl form. Pyridoxine is an important coenzyme in many metabolic reactions and is used in media at concentrations of 0.1 to 1.0 mg/L. Biotin (vitamin H) is commonly added to tissue culture media at 0.01 to 1.0 mg/L. Other vitamins that are sometimes used are folic acid (vitamin M, 0.1 to 0.5 mg/L), riboflavin (vitamin B_2, 0.1 to 10 mg/L), ascorbic acid (vitamin C, 1 to 100 mg/L), pantothenic acid (vitamin B_5, 0.5 to 2.5 mg/L), tocopherol (vitamin E, 1 to 50 mg/L), and para-aminobenzoic acid (0.5 to 1.0 mg/L).

Inositol is sometimes characterized as one of the B complex vitamin group but it is really a sugar alcohol involved in the synthesis of phospholipids, cell wall pectins, and membrane systems in cell cytoplasm. It is added to tissue culture media at a concentration of about 0.1 to 1.0 g/L and has been demonstrated to be necessary for some monocots, dicots, and gymnosperms.

In addition, other amino acids are sometimes used in tissue culture media. These include L-glutamine, asparagine, serine, and proline, which are used as sources of reduced organic nitrogen, especially for inducing and maintaining somatic embryogenesis (see Chapter nineteen). Glycine, the simplest amino acid, is a common additive since it is essential in purine synthesis and is a part of the porphyrin ring structure of chlorophyll.

Complex organics are a group of undefined supplements such as casein hydrolysate, coconut milk (the liquid endosperm of the coconut), orange juice, tomato juice, grape juice, pineapple juice, sap from birch, banana puree, etc. These compounds are often used when no other combination of known defined components produces the desired growth or development. However, the composition of these supplements is basically unknown and may also vary from lot to lot causing variable responses. For example, the composition of coconut milk (used at a dilution of 50 to 150 ml/L), a natural source of the PGR, zeatin (see below), not only differs between young and old coconuts, but also between coconuts of the same age.

Some complex organic compounds are used as organic sources of nitrogen such as casein hydrolysate, a mixture of about 20 different amino acids and ammonium (0.1 to 1.0 g/L), peptone (0.25 to 3.0 g/L), tryptone (0.25 to 2.0 g/L), and malt extract (0.5 to 1.0 g/L). These mixtures are very complex and contain vitamins as well as amino acids. Yeast extract (0.25 to 2.0 g/L) is used because of the high concentration and quality of B vitamins.

Polyamines, particularly putrescine and spermidine, are sometimes beneficial for somatic embryogenesis. Polyamines are also cofactors for adventitious root formation. Putrescine is capable of synchronizing the embryogenic process of carrot.

Activated charcoal is useful for absorption of the brown or black pigments and oxidized phenolics. It is incorporated into the medium at concentrations of 0.2 to 3.0% (w/v). It is also useful for absorbing other organic compounds including PGRs such as auxins and cytokinins, vitamins, iron, and zinc chelates (Nissen and Sutter, 1990). Carry-over effects of plant growth regulators are minimized by adding activated charcoal when transferring explants to media without PGRs. Another feature of using activated charcoal is that it changes the light environment by darkening the medium so it can help with root formation and growth. It may also promote somatic embryogenesis and may enhance growth and organogenesis of woody species.

Leached pigments and oxidized polyphenolic compounds and tannins can greatly inhibit growth and development. These are formed by some explants as a result of wounding. If charcoal does not reduce the inhibitory effects of polyphenols, addition of polyvinylpyrrolidone (PVP, 250 to 1000 mg/L), or antioxidants such as citric acid, ascorbic acid, or thiourea can be tested.

Plant growth regulators

PGRs exert dramatic effects at low concentrations (0.001 to 10 μM). They regulate the initiation and development of shoots and roots on explants and embryos on semisolid or in liquid medium cultures. They also stimulate cell division and expansion. Sometimes a tissue or an explant is autotrophic and can produce its own supply of PGRs. Usually PGRs must be supplied in the medium for growth and development of the culture.

The most important classes of the PGRs used in tissue culture are the auxins and cytokinins. The relative effects of auxin and cytokinin ratio on morphogenesis of cultured tissues were demonstrated by Skoog and Miller (1957) and still serve as the basis for plant tissue culture manipulations today. Some of the PGRs used are hormones (naturally synthesized by higher plants) and others are synthetic compounds. PGRs exert dramatic effects depending on the concentration used, the target tissue, and their inherent activity even though they are used in very low concentrations in the media (from 0.1 to 100 µM). The concentrations of PGRs are typically reported in milligrams/liter or in micromolar units of concentration. Comparisons of PGRs based on their molar concentrations are more useful because the molar concentration is a reflection of the actual number of molecules of the PGR per unit volume (see Table 3.2).

Auxins play a role in many developmental processes, including cell elongation and swelling of tissue, apical dominance, adventitious root formation, and somatic embryogenesis. Generally, when the concentration of auxin is low, root initiation is favored; and when the concentration is high, callus formation occurs. The most common synthetic auxins used are 1-naphthaleneacetic acid (NAA), 2,4-dichlorophenoxyacetic acid (2,4-D), and 4-amino-3,5,6-trichloro-2-pyridinecarboxylic acid (picloram). Naturally occurring indoleacetic acid (IAA) and indolebutyric acid (IBA) are also used. IBA was once considered synthetic, but has also been found to occur naturally in many plants including olive and tobacco (Epstein et al., 1989). Both IBA and IAA are photosensitive so that stock solutions must be stored in the dark. IAA is also easily broken down by enzymes (peroxidases and IAA oxidase). IAA is the weakest auxin and is typically used at concentrations between 0.01 and 10 mg/L. The relatively more active auxins such as IBA, NAA, 2,4-D, and picloram are used at concentrations ranging from 0.001 to 10 mg/L. 2,4-D and picloram are examples of auxins used primarily to induce and regulate somatic embryogenesis.

Cytokinins promote cell division and stimulate initiation and growth of shoots in vitro. The cytokinins most commonly used are zeatin, dihydrozeatin, kinetin, benzyladenine, thidiazuron, and 2-iP. In higher concentrations (1 to 10 mg/L) they induce adventitious shoot formation, but inhibit root formation. They promote axillary shoot formation by opposing apical dominance regulated by auxins. Benzyladenine (BA) has significantly stronger cytokinin activity than the naturally occurring zeatin. However, a concentration of 0.05 to 0.1 µM thidiazuron, a diphenyl-substituted urea, is more active than 4 to 10 µM BA, but thidiazuron may inhibit root formation, causing difficulties in plant regeneration. Adenine (used at concentrations of 2 to 120 mg/L) is occasionally added to tissue culture media and acts as a weak cytokinin by promoting shoot formation.

Gibberellins are less commonly used in plant tissue culture. Of the many gibberellins thus far described, GA_3 is the most often used, but it is very heat sensitive (after autoclaving 90% of the biological activity is lost). Typically it is filter sterilized and added to autoclaved medium after it has cooled. Gibberellins help to stimulate elongation of internodes and have proved to be necessary for meristem growth for some species.

Abscisic acid is not normally considered an important PGR for tissue culture except for somatic embryogenesis and in the culture of some woody plants. For example, it promotes maturation and germination of somatic embryos of caraway (Ammirato, 1974) and spruce (Roberts et al., 1990).

Organ and callus cultures are able to produce the gaseous PGR, ethylene. Since culture vessels are almost entirely closed, ethylene can sometimes accumulate. Many plastic containers also contribute to ethylene content in the vessel. There are contrasting reports in the literature concerning the role played by ethylene in vitro. It appears to influence embryogenesis and organ formation in some gymnosperms. Sometimes in vitro growth can be promoted by ethylene. At other times, addition of ethylene inhibitors results in better initiation or growth. For example, the ethylene inhibitors, particularly silver nitrate, are used to enhance embryogenic culture initiation in corn. High levels of 2,4-D can induce ethylene formation.

Table 3.2 Plant Growth Regulators, Their Molecular Weights, Conversions of mg/L Concentrations into µM Equivalents, and Conversion µM Concentrations into mg/L Equivalents

			mg/L equivalents for these µM concentrations				µM equivalents for these mg/L concentrations			
Plant growth regulator	Abbreviation	M.W.	0.1	1.0	10.0	100.0	0.1	0.5	1.0	10.0
Abscisic acid	ABA	264.3	0.0264	0.264	2.64	26.4	0.38	1.89	3.78	37.8
Benzyladenine	BA	225.2	0.0225	0.225	2.25	22.5	0.44	2.22	4.44	44.4
Dihydrozeatin	2hZ	220.3	0.0220	0.220	2.20	22.0	0.45	2.27	4.53	45.3
Gibberellic acid	GA3	346.4	0.0346	0.346	3.46	34.6	0.29	1.44	2.89	28.9
Indoleacetic acid	IAA	175.2	0.0175	0.175	1.75	17.5	0.57	2.85	5.71	57.1
Indolebutyric acid	IBA	203.2	0.0203	0.203	2.03	20.3	0.49	2.46	4.90	49.0
Potassium salt of IBA	K-IBA	241.3	0.0241	0.241	2.41	24.1	0.41	2.07	4.14	41.4
Kinetin	Kin	215.2	0.0215	0.215	2.15	21.5	0.46	2.32	4.65	46.7
Naphthaleneacetic acid	NAA	186.2	0.0186	0.186	1.86	18.6	0.54	2.69	5.37	53.7
Picloram	Pic	241.5	0.0242	0.242	2.42	24.2	0.41	2.07	4.14	41.4
Thidiazuron	TDZ	220.3	0.0220	0.220	2.20	22.0	0.45	2.27	4.54	45.4
Zeatin	Zea	219.2	0.0219	0.219	2.19	21.9	0.46	2.28	4.56	45.6
2-Isopentenyl adenine	2-ip	203.3	0.0203	0.203	2.03	20.3	0.49	2.46	4.92	49.2
2,4-Dichlorophenoxyacetic acid	2,4-D	221.04	0.0221	0.221	2.21	22.1	0.45	2.26	4.52	45.2

Agar and alternative culture support systems

Agar is used to solidify tissue culture media into a gel. It enables the explant to be placed in precise contact with the medium (for example, on the surface or embedded), but remain aerated. Agar is a high molecular weight polysaccharide that can bind water and is derived from seaweed. It is added to the medium in concentrations ranging from 0.5 to 1.0% (w/v). High concentrations of agar result in a harder medium. If a lower concentration of agar is used (0.4%) or if the pH is low, the medium will be too soft and will not gel properly. The consistency of the agar can also influence the growth. If it is too hard, plant growth is reduced. If it is too soft, hyperhydric plants may be the result (Singha, 1982). To gel properly, a medium with 0.6% agar must have a pH above 4.8. Sometimes activated charcoal in the medium will interfere with gelling. Typical tissue culture agar melts easily at ~65°C and solidifies at ~45°C.

Agar also contains organic and inorganic contaminants, the amount of which varies between brands. Organic acids, phenolics, and long-chain fatty acids are common contaminants. A manufacturer's analysis shows that Difco Bacto agar also contains (amounts in ppm): 0.0 to 0.5 cadmium, 0.0 to 0.1 chromium, 0.5 to 1.5 copper, 1.5 to 5.0 iron, 0.0 to 0.5 lead, 210.0 to 430.0 magnesium, 0.1 to 0.5 manganese, and 5.0 to 10.0 zinc. Generally, relatively pure, plant tissue culture-tested types of agar should be used. Poor quality agar can interfere or inhibit the growth of cultures.

Agarose is often used when culturing protoplasts or single cells. Agarose is a purified extract of agar that leaves behind agaropectin and sulfate groups. Since its gel strength is greater, less is used to create a suitable support or suspending medium.

Gellan gums like Gelrite™ and Phytagel™ are alternative gelling agents. They are made from a polysaccharide produced by a bacterium. Rather than being translucent (like agar), they are clear so it is much easier to detect contamination, but they cannot be reliquified by heating and gelled again, and the concentration of divalent cations like calcium and magnesium must be within a restricted range or gelling will not occur.

Mechanical supports such as filter paper bridges or polyethylene rafts do not rely on a gelling agent. They can be used with liquid media, which then circulates better, but keeps the explant at the medium surface so that it remains oxygenated. The types of support systems that have been used are as varied as the imagination and include rock wool, cheesecloth, pieces of foam, and glass beads.

Steps in the preparation of tissue culture medium

The first step in making tissue culture medium is to assemble needed glassware — for example, a 1-L beaker, 1-L volumetric flask, stirring bar, balance, pipettes — and the various stock solutions (Procedure 3.1 and Table 3.3). The plethora of units used to measure concentration may be confusing when first encountered, so a description of the most common units and what they mean is given below. Once familiar with these, you can confidently proceed to the section on making stock solutions.

Procedure 3.1
Making stock solutions for vitamins and plant growth regulators.

Step	Instructions and comments
1	To make a stock solution for nicotinic acid, look at how much is required for one liter of medium. In this example, we are going to assume we are making MS medium, so we will need 0.5 mg.

Procedure 3.1 continued Making stock solutions for vitamins and plant growth regulators.	
Step	Instructions and comments
2	It would be convenient to be able to add the 0.5 mg of nicotinic acid by adding a volume of the nicotinic acid stock that corresponded to 1 ml. If 0.5 mg of nicotinic acid must be in 1 ml of the stock solution, then (multiplying by 100) use 50 mg for the 100 ml of stock solution. One milliliter may be dispensed into 1.5 ml eppendorf tubes and frozen (–20°C).
3	Now prepare a PGR stock — IBA for a rooting medium. If 1.0 mg/L of IBA is required for rooting medium, then our solution must contain 1 mg in each milliliter of the IBA stock. We must weigh 100 mg of IBA for the 100 ml of IBA stock solution. IBA is poorly soluble in water so first dissolve it in a small amount (~1 ml) of a solvent such as 95% ethanol, 100% propanol, or 1 N KOH. Swirl it to dissolve and then slowly add the remainder of the water to a final volume of 100 ml. Label the stock solution and add 1 ml for every liter of medium to be made. Store in a brown bottle in the refrigerator (4°C). Note: Many growth regulators need special handling to dissolve. Indoleacetic acid, indolebutyric acid, naphthaleneacetic acid, 2,4-D, benzyladenine, 2-iP, and zeatin can be dissolved in either 95% ethanol, 100% propanol, or 1 N KOH. Kinetin and ABA are best dissolved in 1 N KOH but thidiazuron will not dissolve in either alcohol or base, so a small amount of dimethylsulfoxide must be used. By using only 1 ml of each stock, only very small amounts of the solvent are added to the medium and will minimize toxic effects.
4	Now let us make a stock solution of IBA, but this time we want to have a 5-µM solution of IBA in the medium. Look up the molecular weight of IBA in Table 3.2 and find that it is 203.2. Using the same rationale in the two examples above, we wish to have 5 µmol delivered to the medium by using 1 ml of IBA stock solution. So to make 100 ml of stock solution, you must have 500 µmol or 0.5 µmol (mM) in the 100 ml of stock solution. If 1 mol is 203.2 g, 1 mmol is 203.2 mg, so... 0.5 mmol IBA × 203.2 mg/1 mmol = 101.6 mg of IBA needed for 100 ml of IBA stock solution. To prepare the solution, weigh out the IBA, dissolve it in 1 N KOH as described above, and label the stock solution and store in the refrigerator (4°C) in a brown bottle. Add 1 ml of the IBA stock solution to deliver 5 µmol for every liter of the medium.

Units of concentration clarified (Procedure 3.2)

Concentrations of any substance can be given in several ways. The following list gives some of the methods of indicating the concentration commonly found in literature on tissue culture:

- Percentage based on volume percent (v/v): Used for coconut milk, tomato juice, and orange juice. For example, if 100 ml of 5% (v/v) coconut milk was desired, 5 ml of coconut milk would be diluted to 100 mL with water.
- Percentage based on weight percent (w/v): This is often used to express concentrations of agar or sugar. For example, to make 1% (w/v) agar solution, dissolve 10 g of agar in 1 L of nutrient medium.
- Molar solution: A mole is the same number of grams as the molecular weight (Avogadro's number of molecules), so a 1-M solution represents 1 mol of the substance in 1 L of solution and 0.01 M represents 0.01 times the molecular weight in 1 L. A millimolar (mM) solution is 0.01 mol/L and a micromolar (µM) solution is 0.000001 mol/L. Substances like PGR are active at micromolar concentrations. Molar concentration is used to accurately compare relative reactivity among different compounds. For example, a 1-µM concentration of IAA would contain the

Table 3.3 Macro- and Micronutrient 100× Stock Solutions for Murashige and Skoog (MS) Medium (1962)

Stock (100×)	Component	Amount
Nitrate	NH_4NO_3	165.0 g
	KNO_3	190.0 g
Sulfate	$MgSO_4 \cdot 7H_2O$	37.0 g
	$MnSO_4 \cdot H_2O$	1.7 g
	$ZnSO_4 \cdot 7H_2O$	0.86 g
	$CuSO_4 \cdot 5H_2O$	2.50 mg[a]
Halide	$CaCl_2 \cdot 2H_2O$	44.0 g
	KI	83.0 mg
	$CoCl_2 \cdot 6H_2O$	5.0 mg[b]
PBMo	KH_2PO_4	17.0 g
	H_3BO_3	620.0 g
	Na_2MoO_4	25.0 mg
NaFeEDTA[c]	$FeSO_4 \cdot 7H_2O$	2.78 g
	NaEDTA	3.74 g

Note: The number of grams or milligrams indicated in the amount column should be added to 1000 ml of deionized distilled water to make 1 L of the appropriate stock solution. For each liter of the media made, 10 ml of each stock solution will be used.[d]

[a] Because this amount is too small to weigh conveniently, dissolve 25 mg of $CuSO_4 \cdot 5H_2O$ in 100 ml of deionized distilled water, then add 10 ml of this solution to the sulfate stock.

[b] Because this amount is too small to weigh conveniently, dissolve 25 mg of $CoCl_2 \cdot 6H_2O$ in 100 ml of deionized distilled water, then add 10 ml of this solution to the halide stock.

[c] Mix the $FeSO_4 \cdot 7H_2O$ and NaEDTA together and heat gently until the solution becomes orange. Store in an amber bottle or protect from light.

[d] To make 100x stock solutions for any of the media listed in Table 3.1, multiply the amount of chemical listed in the table by 100 and dissolve in 1 L of deionized, distilled water.

same number of molecules as a 1-μM concentration of kinetin, although the same could not be said for units based on weight.

- Milligrams per liter (mg/L): Although not an accurate means of comparing substances molecularly, this is simpler to calculate and use, since it is a direct weight. Such direct measurement is commonly used for macronutrients and sometimes with PGRs; 1 mg/L means placing 1 mg of the desired substance in a final volume of 1 L of solution. One milligram is 10^{-3} g.
- Microgram per liter (μg/L): This is used with micronutrients and also sometimes with growth regulators. It means placing 1 μg of substance in 1 L of solution; 1 μg = 0.001 or 10^{-3} mg = 0.000001 or 10^{-6} g.
- Parts per million (ppm): Sometimes media components are expressed in ppm which means 1 part per million or 1 mg/L.

Instructions for making media can be found in Procedure 3.3. These instructions describe MS medium preparation, but will work just as effectively for any other media

Procedure 3.2
Converting molar solutions to milligrams/liter and milligrams/liter to molar solutions using conversion factors.

Step	Instructions and comments
1	To determine how many mg/liter are needed for 1.0 M concentration, first, look up the molecular weight of the PGR. In this example, we will use kinetin. The molecular weight is 215.2 so a 1 M solution will consist of 215.2 g/L of solution. By using conversion factors and crossing out terms, you cannot go wrong! $$1\text{ M solution} = \frac{\cancel{1\text{ mol}}}{\text{liter solution}} \times \frac{215.2\text{ g}}{\cancel{1\text{ mol}}} = 215.2\text{ g/L}$$
2	To see what a 1.0 mM solution of kinetin would consist of, multiply the grams necessary for a 1 M solution by 10^{-3}, which would give you 215.2×10^{-3} g/L or 215.2 mg/L. $$1\text{ M solution} = \frac{\cancel{1\text{ mmol}}}{\text{liter solution}} \times \frac{215.2\text{ mg}}{\cancel{1\text{ mmol}}} = 215.2\text{ mg/L}$$
3	To see what a 1.0 μM solution of kinetin would consist of, multiply the grams necessary for a 1 M solution by 10^{-6} which would give you 215.2×10^{-6} g/L or 215.2 μg/L. $$1\ \mu\text{M solution} = \frac{\cancel{1\ \mu\text{mol}}}{\text{liter solution}} \times \frac{215.2\ \mu\text{g}}{\cancel{1\ \mu\text{mol}}} = 215.2\ \mu\text{g/L or } 0.215\text{ mg/L}$$
4	How to determine the molar concentration of a solution that is in milligram/liter… Let us assume you are given a 10 mg/L solution of indolebutyric acid (IBA) and you wish to know its molarity. First look up the molecular weight of the PGR. In this example, using IBA, the molecular weight of IBA is 203.2. Again, using conversion factors and crossing out terms: $$\frac{10\ \cancel{\text{mg}}\text{ IBA}}{\text{liter of solution}} \times \frac{\cancel{1\text{ g}}}{1000\ \cancel{\text{mg}}} \times \frac{1\text{ mol}}{203.2\ \cancel{\text{g}}} = \frac{0.0000492\text{ mol}}{\text{liter of solution}} = 49.2\ \mu\text{M IBA}$$

Note: Remember 1 mol = 10^3 mmol = 10^6 μmol.

that you choose. Merely follow the same steps and substitute the macro- and micronutrient stocks that you have made for the desired medium. Omit the agar to produce a liquid medium for use in suspension culture. PGRs can be customized for the medium of your choice whether it is intended for initiation of callus, shoots, roots, or some other purpose.

Procedure 3.3
Step-by-step media making.

Step	Instructions and comments
1	If you are making 1 L of media, put about half the volume of deionized, distilled water into a beaker and add a stirring bar. Place the beaker on a stirrer so that when you add different components, they will be thoroughly mixed.
2	Add 10 ml of the following stock solutions (details on how to make them is provided in Table 3.3): nitrate, sulfate, halide, PBMo, and NaFeEDTA.
3	Add the appropriate amount of the vitamin stocks. How to make the vitamin and PGR stocks is found in Procedure 3.1. At this time, also add the appropriate amounts of inositol (usually 100 mg) and sucrose (usually 20 to 30 g/L).
4	Add the appropriate volume of the PGR stocks that you plan to use (details on how to make them are also found in Procedure 3.1).

Procedure 3.3 continued
Step-by-step media making.

Step	Instructions and comments
5	Adjust the pH using 0.1 to 1.0 N NaOH or HCl depending upon how high or low it is. For most media, the pH should be 5.4 to 5.8, and adjust the final volume (1 L) with deionized, distilled water.
6	If you are going to use liquid medium, you can distribute the media into the tissue culture vessels you plan to use and autoclave. If you plan to use solid medium, weigh out and add the agar or other gelling agent to the liquid. If you are going to put it into tissue culture vessels (tubes or boxes), melt the agar in the medium by putting it on a heat plate while stirring or use a microwave to melt the agar. You may have to experiment with settings, since microwaves differ in output. Once the medium is distributed into the tissue culture vessels, these are capped and placed in the autoclave for sterilization. Autoclave for 15 min at 121°C. If a large quantity is being autoclaved, longer times may be necessary. If you plan to distribute the medium into tissue culture vessels after autoclaving (such as into sterile, disposable petri dishes), cap the vessel you are autoclaving the medium in with a cotton or foam plug and cover the plug with aluminum foil. When the medium is cool enough to handle (about 60°C), it can be moved to the sterile transfer hood, uncapped, and poured into dishes.

Adjusting the medium pH is an essential step. Plant cells in culture prefer a slightly acidic pH, generally between 5.3 and 5.8. When pH values are lower than 4.5 or higher than 7.0, growth and development in vitro generally are greatly inhibited. This is probably due to several factors including PGRs, such as IAA and gibberellic acid, becoming less stable, phosphate and ion salts precipitating, vitamin B_1 and pantothenic acid becoming less stable, reduced uptake of ammonium ions, and changes in the consistency of the agar (the agar becomes liquified at lower pH). Adjusting the pH is the last step before adding and dissolving the agar and then distributing it into the culture vessels and autoclaving. If the pH is not what it should be, it can be adjusted using KOH to raise pH or HCl to lower it (0.1 to 1.0 N), depending on if the pH is too low or too high. While NaOH can be used, it can lead to an undesirable increase in sodium ions. The pH of a culture medium generally drops by 0.3 to 0.5 units after it is autoclaved and then changes throughout the period of culture both due to oxidation and the differential uptake and secretion of substances by growing tissue.

Stock solutions of the mineral salts

Mineral salts can be prepared as stock solutions 10 to 1000 × the concentration specified in the medium. Mineral salts are often grouped into two stock solutions, one for macroelements and one for microelements, but unless these are kept relatively dilute (10×), precipitation can occur. In order to produce more concentrated solutions, the preferred method is to group the compounds by the ions they contain such as nitrate, sufate, halide, P, B, Mo (phosphorus, boron, molybdenum), and iron and make them up as 100× stocks. Table 3.2 lists the stock solutions for MS medium made at 100× final concentration, which means that 10 ml of each stock is used to make 1 L of medium. Some of the stock solutions require extra steps to get the components into solution (for example, NaFeEDTA stock), or may require making a serial dilution to obtain the amount of a trace component for the stock (sulfate and halide stocks).

Sometimes the amount of a particular component needed for a tissue culture medium is extremely small, so that it is difficult to weigh out the amount even for the 100× stock solution. Because such small quantities of a substance cannot be weighed accurately, a serial dilution technique is used. The following example illustrates how a serial dilution can be used to obtain the correct amount of a component (in this case $CuSO_4{\cdot}5H_2O$) of the medium for its appropriate stock solution.

The stock solution calls for 2.5 mg of $CuSO_4{\cdot}5H_2O$ as a part of the sulfate stock of MS medium. Make an initial stock solution by placing 25 mg of $CuSO_4{\cdot}5H_2O$ in 100 ml of deionized/distilled water. After mixing thoroughly, use 10 ml of this solution (which will contain the desired 2.5 mg) and place it into the sulfate stock. This procedure required only one serial dilution but any component can be subjected to one or more dilutions to obtain the desired amount.

Once a stock solution is made, it should be labeled as in the example below to avoid error and inadvertently keeping a stock solution too long.

Nitrate stock (100×)	← Type of stock
Murashige and Skoog, 1962	← Media type
Use 10 ml/L medium	← Instructions
1/08/99	← Date made
Jane Doe	← Who made the stock

Making stock solutions of the PGRs and vitamins

Vitamins and PGR stock solutions can be made up in concentrations of 100 to 1000× of that required in the medium. Determine the desired amount for one liter of medium, the volume of stock solution needed to deliver that dosage of vitamin or PGR, and the volume of stock you wish to make. Procedure 3.1 gives examples of how to make vitamin and PGR stock solutions. Many of the PGRs require special handling to get them into solution. You will also find this information in Procedure 3.1.

Sterilizing equipment and media

Tissue culture media, in addition to providing an ideal medium for growth of plant cells, also are an ideal substratum for growth of bacteria and fungi. So it is necessary to sterilize the media, culture vessels, tools, and instruments, and surface disinfest the explants as well. The most commonly used means of sterilizing equipment and media is by autoclaving at 121°C with a pressure of 15 psi for 15 min or longer for large volumes. Glassware and instruments are usually wrapped in heavy-duty aluminum foil or put in autoclave bags. Media (even that containing agar) are in liquid form in the autoclave, requiring a slow exhaust cycle to prevent the media from boiling over when pressure is reduced. Larger volumes of media require longer autoclave times. Media should be sterilized in tissue culture vessels with some kind of closure such as caps or plugs made of nonabsorbant cotton covered by aluminum foil. This way they do not become contaminated when they are removed from the autoclave for cooling. Use of racks that tilt the tissue culture tubes during cooling can give a slanted surface to the agar medium. These can be purchased or made from scratch using a little ingenuity.

Some components of the medium may be heat labile or altered by the heat so that they become inactive. These are usually added to the medium after it has been autoclaved, but before the medium has solidified. It is filtered through a bacteria-proof membrane (0.22 µm) filter and added to the sterilized medium after it has cooled enough not to harm the heat-labile compound (less than 60°C) and then thoroughly mixed before distributing it into the culture vessels. A rule of thumb is to add the filtered material at a point when the culture flask is just cool enough to be handled without burning one's hands. Some filters are available presterilized and fit on the end of a syringe for volumes ranging from 1 to 200 ml. More elaborate disposable assemblies are also available.

Preparing the sterile transfer hood

Successfully transferring explants to the sterile tissue culture medium is done in a laminar flow transfer hood. A transfer hood is equipped with positive-pressure ventilation and a bacteria-proof high-efficiency particulate air (HEPA) filter. The laminar flow hoods come in two basic types. Generally, air is forced into the cabinet through a dust filter and a HEPA filter, and then it is directed either downward (vertical flow unit) or outward (horizontal flow unit) over the working surface at a uniform rate. The constant flow of bacterial and fungal spore-free filtered air prevents nonfiltered air and particles from settling on the working area which must be kept clean and disinfected. The simplest transfer cabinet is an enclosed plastic box or shield with an UV light and no airflow. A glove box can also be used but both of these low-cost, low-technology options are not convenient for large numbers of transfers.

In the transfer hood, you should have ready some standard tools for use such as a scalpel (with a sharp blade), long-handled forceps, and sometimes a spatula. Occasionally fine-pointed forceps, scissors, razor blades, or cork borers are needed for preparing explants.

Many people prefer doing sterile manipulations on the surface of a presterilized disposable petri dish. Other alternatives that have worked well are stacks of standard laboratory-grade paper towels. These can be wrapped in aluminum foil and autoclaved. When the top sheet of the stack is used, it can be peeled off and discarded leaving the clean one beneath it exposed to act as the next working surface. Another alternative is to use ceramic tiles. These can also be wrapped in aluminum foil and autoclaved but tiles with very slick or very rough surfaces should be avoided.

In Procedure 3.4 is a suggested protocol to follow for preparing the sterile transfer hood. It also contains tips for keeping your work surface clean, eliminating contamination, and avoiding burns when flaming your instruments!

Procedure 3.4
Getting under the hood.

Step	Instructions and comments
1	Turn on the transfer hood so that positive air pressure is maintained. This ensures that all of the air passing over the work surface is sterile. You should feel air flowing against your face at the opening. Make sure there are no drafts such as open windows or air conditioning vents that may interfere with the air flow coming out of the hood.
2	Use a spray bottle filled with 70% (v/v) ethanol and spray down the interior of the hood. Do not spray the HEPA filter! This is more effective than absolute alcohol for sterilizing surfaces, perhaps because 70% (v/v) ethanol denatures DNA. You can also use a piece of cheesecloth saturated with 70% (v/v) ethanol to help to distribute the ethanol more uniformly. Allow it to dry. To maintain the cleanliness of the interior, anything that is now placed inside the hood must be sprayed with 70% (v/v). This includes the alcohol lamp, the slide staining jar filled with 80 to 95% (v/v) ethanol for flaming the instruments, a rack for the tools, racks of the tissue culture vessels containing medium, and all of the presterilized wrapped bundles containing your working surface (tiles, paper towels, ashtrays, etc.) and your tools (forceps, spatula, scalpel, etc.).

Procedure 3.4 continued Getting under the hood.	
Step	Instructions and comments
3	When you are ready to begin, wash your hands thoroughly with soap and water and then just before placing then in the hood, spray them down with 70% ethanol.
4	You may now open your work surface by peeling back the heavy-duty aluminum foil exposing the surface of the tile (or other alternative). Never block the air flow across the surface coming from the filter unit. Also, do not pass your hands across the surface of the tile. Talking while you are in the unit also compromises sterility. If you must talk, turn your head to one side. If you have long hair, fasten it so that it does not dangle onto your work surface.
5	Keep any open sterile containers as far back in the hood as you can. When you open containers that have been sterilized, keep your fingers away from the opening. If you open a glass container or vessel, as general rule, pass the opening through the flame. This creates a warm updraft from the vessel helping to prevent contamination from entering it.
6	Flaming instruments can be hazardous if you forget that ethanol is flammable! When you are flame sterilizing an instrument, for example, forceps, dip it into the jar containing the 80% ethanol and when you lift it out, keep the tip of the instrument at an angle downward (away from your fingers) so that any excess alcohol does not run onto your hand (Figure 3.1). Then pass the tip through the flame of the alcohol burner and hold the instrument parallel to the work surface. When the flame has consumed the alcohol, let the tool cool on the rack until you are ready to use it. Never, never place a hot tool back into the jar containing 80% ethanol! I know of an experienced scientist who momentarily forgot this simple rule and the resulting fire singed his hand and the hair off his forearm!

Storage of culture media

Once culture media have been made, distributed into tissue culture vessels, and sterilized by autoclaving, some formulations can be stored for up to 1 month provided they are kept sealed to prevent excessive evaporation of water from the media. They should also be placed in a dark, cool place to minimize degradation of light-labile components such as IAA. Storage at 4°C prolongs the time that media can be stored, but condensate may form in the container and encourage contamination. By making media 5 to 7 days in advance, you allow time to check for any unwanted microbial contamination before explants are transferred onto the medium. However, media for certain sensitive species or operations, or that contain particularly unstable ingredients, must be used fresh and cannot be stored. For example, media with IAA, which breaks down quickly, should be used as soon as possible.

Surface disinfesting plant tissues

Just as the media, instruments, and tools must be sterilized, so must the plant tissue be disinfested before it is placed on culture medium. Many different materials have been used to surface disinfest explants, but the most commonly used are 1% (v/v) sodium hypochlorite (commercial bleach contains 5% sodium hypochlorite), 70% alcohol, or 10% hydrogen peroxide. Others include using a 7% saturated solution of calcium hypochlorite, 1% solution of bromine water, 0.2% mercuric chloride solution, and 1% silver nitrate solution. If these more rigorous techniques are used, precautions should be taken to minimize health and safety risks, especially with the heavy metal-containing solutions.

The type of disinfectant used, the concentration, and the amount of exposure time (1 to 30 min) vary depending on the sensitivity of the tissue and how difficult it is to disinfest. Woody or field-grown plants are sometimes very difficult to disinfest and may benefit from being placed in a beaker with cheesecloth over the top and placed under running water overnight. In some cases, employing a two-step protocol (70% ethanol followed by bleach) or adding a wetting agent such as Tween 20 or detergent helps to increase the effectiveness. In any case, the final step before trimming the explant and placing onto sterile medium is to rinse it several times in sterile, distilled water to eliminate the residue of the disinfesting agent. Laboratory exercises presented throughout the book will contain detailed directions for various procedures to disinfest tissue.

Final word

Tissue culture is much like good cooking. There are simple recipes and then there is 'haute cuisine.' Cooking is also a very rewarding activity particularly when the end result is delicious. By following the procedures outlined in this chapter and in the succeeding chapters of the book, you should find that with care and attention to detail, you will be a 'chef extraordinaire' and your tissue culture ventures will be successful!

Literature cited

Ammirato, P.V. 1974. The effects of abscisic acid on the development of somatic embryos from cells of caraway (*Carum carvi* L.). *Bot. Gaz.* 135:328–337.

Coffin, R., C. D. Taper, and C. Chong. 1976. Sorbitol and sucrose as carbon source for callus culture of some species of the Rosaceae. *Can. J. Bot.* 54:547–551.

Epstein, E., K.-H. Chen, and J. D. Cohen. 1989. Identification of indole-3-butyric acid as an endogenous constituent of maize kernals and leaves. *Plant Growth Regul.* 8:215–223.

Gamborg, O. L., R. A. Miller, and K. Ojima. 1968. Nutrient requirements of suspension cultures of soybean root cells. *Exp. Cell Res.* 50:151–158.

Kochba, J., P. Spiegel-Roy, H. Neumann, and S. Saad. 1978. Stimulation of embryogenesis in Citrus ovular callus by ABA, Ethephon, CCC and Alar and its supression by GA_3. *Z. Pflanzenphysiol.* 89:427–432.

Lloyd, G. and B. McCown. 1980. Commercially feasible micropropagation of mountain laurel, *Kalmia latifolia*, by use of shoot tip culture. *Intl. Plant Prop. Soc. Proc.* 30:421–427.

Murashige, T. and F. Skoog. 1962. A revised medium for rapid growth and bio-assays with tobacco tissue cultures. *Physiol. Plant.* 15:473–497.

Nissen, S. J. and E. G. Sutter. 1990. Stability of IAA and IBA in nutrient medium to several tissue culture procedures. *HortScience* 25:800–802.

Nitsch, J. P. and C. Nitsch. 1969. Haploid plants from pollen grains. *Science* 163:85–87.

Oka, S. and K. Ohyama. 1982. Sugar utilization in mulberry (*Morus alba* L.) bud culture. In: Plant Tissue Culture. A. Fujiwara, Ed. Proc. *5th Int. Cong. Plant Tiss. Cell Cult., Jap. Assoc. Plant Tissue Culture*, Tokyo, Japan, pp. 67–68.

Roberts, D. R., B. S. Flinn, D. T. Webb, F. B. Webster, and B. C. Sutton. 1990. Abscisic acid and indole-3-butyric acid regulation of maturation and accumulation of storage proteins in somatic embryos of interior spruce. *Physiol. Plant.* 78:355–360.

Schenk, R. V. and A. C. Hildebrandt. 1972. Medium and techniques for induction and growth of monocotyledonous plant cell cultures. *Can. J. Bot.* 50:199–204.

Singha, S. 1982. Influence of agar concentration on in vitro shoot proliferation of *Malus* sp. 'Almey' and *Pyrus communis* 'Seckel'. *J. Amer. Soc. Hort. Sci.* 107:657–660.

Skoog, F. and C. O. Miller. 1957. Chemical regulation of growth and organ formation in plant tissues cultured in vitro. *Symp. Soc. Exp. Biol.* 11:118–131.

White, P. R. 1963. *The Cultivation of Plant and Animal Cells*. 2nd ed. Ronald Press, New York.

chapter four

Nutrition of callus cultures

James D. Caponetti

Callus is a relatively undifferentiated tissue consisting primarily of parenchymatous cells. Callus tissue can serve as an experimental system to investigate and solve a broad range of basic research problems in plant cytology, physiology, morphology, anatomy, biochemistry, pathology, and genetics. It can also be used to resolve applied research problems in organogenesis and embryogenesis related to the propagation of horticultural and agronomic plants.

Tissues from various organs from many species of plants can be induced to form callus. However, many seemingly unrelated factors may determine the ability of a specific tissue to form callus. Among these chemical factors are mineral nutrition and plant growth regulators (PGRs), environmental factors such as light, temperature, and humidity, and the genetic constitution or the genotype of the plant. A medium that induces good callus growth in one genotype or species may fail to do so in another closely related plant. The same holds true for a particular combination and concentration of PGRs added to the medium. Also, tissues from different plant species and genotypes can respond differently from one another under various conditions of light, temperature, and humidity. Therefore, extensive empirical research is required to determine which combination and concentration of medium nutritional ingredients, PGRs, and environmental factors are best for maximizing callus production for each species or genotype. An example of such studies are Murashige and Skoog (1962) and Linsmaier and Skoog (1965).

The following laboratory experiments illustrate the technique of obtaining callus from mature tissues of several angiosperm stems and roots and compare callus growth on a standard medium formulation. Also, callus growth of one species will be contrasted on several media to which ingredients are added in a serial sequence. Specifically, the laboratory experiences will familiarize students with the following concepts: (1) that one medium may be sufficient to support callus growth from some species, but is not able to support growth from others; and (2) that some ingredients in a medium are more vital for callus tissue growth when provided in combination rather than separately. Each of the above-mentioned experiments will be detailed below, including procedures and anticipated results. In all cases, growth will be measured by fresh and dry weight of developed callus.

0-8493-2029-1/00/$0.00+$.50

General considerations

Source of plant material

The tobacco plants (*Nicotiana tabacum* L.) needed for these experiments can be cultivated in a greenhouse from seed obtained from a farmer's co-op store or from a biological supply company (see Chapters twenty-nine and thirty-nine for details). Alternatively, tobacco plants of the appropriate size may be available from other laboratories/greenhouses on campus. Plants that are about 1.5 m tall would be the appropriate size. The tobacco varieties that produce good results include WI 38, Xanthi, Burley 14, Burley 21, KY 9, KY 17, VA 509, SpG-28, and Hicks. The other plant material needed for these experiments can be purchased from a grocery store produce counter. Depending on what is available at the time of year that these experiments are conducted, good results can be obtained with Irish potato tuber (*Solanum tuberosum* L.), turnip tuber (*Brassica rapa* L.), Jerusalem artichoke (*Helianthus tuberosus* L.) rhizome, sweet potato root (*Ipomoea batatas* L.), carrot root (*Daucus carota* L.), and parsnip root (*Pastinaca sativa* L.).

Exercises

Experiment 1. Growth potential of callus from several species of plants on a single complete medium

An important basic concept in plant tissue culture is that tissue from different species will not grow equally on one specific medium. This experiment was designed to illustrate this concept by comparing the growth of tobacco stem pith on a medium designed for it with the growth of stem and root tissues of several other plants on the same medium. The experiment can best be performed by teams consisting of two students each. Refer to Figure 4.1 for a diagrammatic representation of this exercise. The experiment requires 4 to 8 weeks in order to obtain measurable results.

Materials

The following laboratory items are needed for each team of students:

- Several tobacco stem sections
- One or more of the plant tissues listed above (see Figure 4.1)
- Coconut milk
- Several 150 × 25-mm culture tubes with plastic or metallic caps
- Supply of wire and wooden racks
- Number 3 cork borer and glass rod
- 10% solution of commercial bleach to which several drops of dishwashing liquid have been added
- Drying ovens set for 60°C
- Aluminum foil weighing cups

The culture medium is modified from Murashige and Skoog (1962) and Linsmaier and Skoog (1965) and was designed specifically for the culture of tobacco stem pith tissue. Each liter of medium is composed of Murashige and Skoog (1962) major and minor salts (see Chapter three for composition), 100 mg of myo-inositol, 0.4 mg of thiamine HCl, 2.0 mg (10.7 μM) of naphthaleneacetic acid (NAA), 0.5 mg (2.3 μM) of kinetin, 30 g (88 mM) of sucrose, and 10 g of agar. Before adding the agar, adjust the pH of the medium to 5.7. Heat to melt the agar and distribute about 20 ml of medium to each of 50 150 × 25-mm

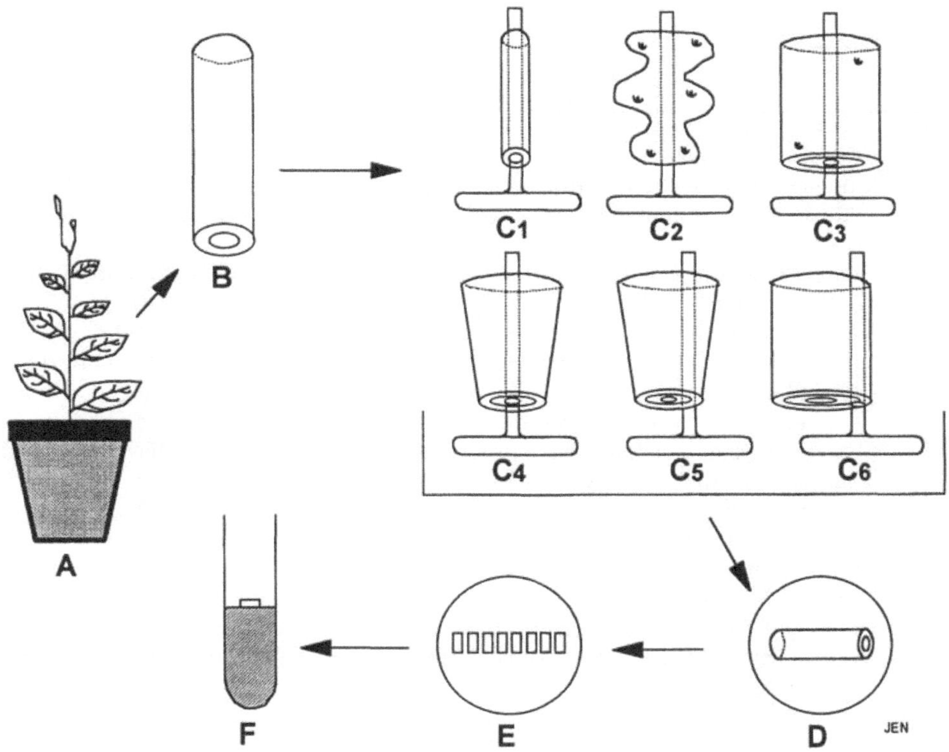

Figure 4.1 Diagrammatic representation for obtaining explants from various plants. (A) Tobacco plant; (B) tobacco internode segment; (C_1 to C_6) all tissue removed using a no. 3 cork borer after surface disinfestation; (C_1) tobacco pith; (C_2) Jerusalem artichoke pith; (C_3) Irish potato pith; (C_4) turnip pith; (C_5) carrot and parsnip cortex and phloem tissues; (C_6) sweet potato cortex and phloem tissues; (D) cylinder of tissue removed from the cork borer and placed into a sterile, plastic petri dish; (E) tissue cut into equally sized disks for culture and determining fresh and dry weights; (F) Tissue placed in culture tubes. (From Trigiano, R. N. and Gray, D. J., Eds., *Plant Tissue Culture Concepts and Laboratory Exercises, First Edition*, CRC Press LLC, Boca Raton, FL, 1996.)

culture tubes. Cover each tube with a plastic or metallic cap, place the tubes in a metal rack, and autoclave for 20 min. For each tissue type used in the experiment, 20 to 30 tubes should be prepared by each team. If facilities are limited, each team could process one tissue.

Follow the protocol in Procedure 4.1 and refer to Figure 4.1 for a diagrammatic representation of tissue explant preparation and culture.

Procedure 4.1
Growth potential of callus from several species of plants on a single complete medium.

Step	Instructions and comments
1	Surface disinfest stems and roots with 10% (v/v) of liquid chlorine bleach solution for 10 min each in large beakers and rinse three times with sterile distilled water.
2	Cut out cylinders of tissue with sterile no. 3 cork borer as illustrated in diagrams C_1, C_2, C_3, C_4, C_5, and C_6 of Figure 4.1.

Procedure 4.1 continued
Growth potential of callus from several species of plants on a single complete medium.

Step	Instructions and comments
3	Express each tissue core from the cork borer into sterile plastic petri dishes with the sterilized metal punch.
4	Cut 5-mm-thick sections with a sterile scalpel and place one section in each of about 30 culture tubes containing modified MS medium composed of MS medium as described.
5	Incubate tubes in a randomized complete block design at 25°C with 25 $\mu mol \cdot m^{-2} \cdot sec^{-1}$ of white light.
6	Weigh 30 explant disks of each tissue type in tared aluminum cups and record the fresh weights. Divide the net weight by 30 to obtain the fresh weight of one disk. Place the cups in an oven set at 60°C for several days to obtain dry weights. Divide the net weight by 30 to obtain the dry weight of one disk.
7	When the tobacco explants have produced callus that almost fills the diameter of the tube (about 4 to 8 weeks), pool uncontaminated calluses of each tissue separately into tared cups. Record the fresh weight and divide by the number of calluses to obtain the fresh weight of one callus. Place the cups in the drying oven for several days to obtain dry weights and then calculate the dry weight of one callus. Don't forget to subtract fresh and dry weights of the explant disks in order to obtain accurate weight gains of calluses.
8	Student teams should share and analyze the data by one of the procedures described in Chapter seven.

Anticipated results

Since the culture medium used was designed for tobacco, the tobacco explant disks will produce the largest and heaviest calluses when the experiment is terminated. The other plant tissues will produce callus, but not to the extent that tobacco disks do.

Questions

- What concept of tissue growth does this experiment demonstrate?
- Why does the experiment work best with tobacco pith tissue?
- Would the experiment work with other explant tissues? Explain.
- Would raising the temperature of the growth chamber to 35°C improve the growth of the tobacco tissues? If so, why? If not, why not?

Experiment 2. Initiation and growth of tobacco callus on serially composed media

Another important basic concept in plant tissue culture is that most tissues require a combination of specific nutrients and PGRs to produce the appropriate growth response of the tissues. This experiment was designed to illustrate this concept by starting with a medium without nutrients, and then preparing eight other media by adding one category of nutrients at a time to each medium, ending with a medium that contains all the essential nutrients. Two additional media test the effects of coconut milk, which supposedly contains complete nutrition and PGRs for growth of callus from explant tissues of tobacco.

All 11 media are different, but are based on the MS medium described previously and should be prepared as directed in Experiment 1. The 11 media all include distilled water and 1% agar and are adjusted to pH 5.7. The media are listed below.

1. Water only
2. MS salts
3. MS salts and 3% sucrose
4. MS salts, 3% sucrose, and 0.4 mg thiamine HCl
5. MS salts, 3% sucrose, and 100 mg myo-inositol
6. MS salts, 3% sucrose, 0.4 mg thiamine HCl, 100 mg myo-inositol
7. MS salts, 3% sucrose, 0.4 mg thiamine HCl, 100 mg myo-inositol, 0.5 mg (2.3 µM) kinetin
8. MS salts, 3% sucrose, 0.4 thiamine HCl, 100 mg myo-inositol, 2.0 mg (10.7 µM) NAA
9. MS salts, 3% sucrose, 0.4 mg thiamine HCl, 100 mg myo-inositol, 0.5 mg kinetin, and 2.0 mg NAA
10. 20% coconut milk (200 ml of coconut milk and 800 ml of water)

However, the amount of nutrients in medium 10 may not be sufficient to produce the full potential of callus growth. Therefore, it is necessary to test the performance of tobacco pith on a medium 11 that consists of all the ingredients in medium 9 and 20% coconut milk. Coconut milk can be obtained from fresh nuts, as a canned concentrate from an ethnic Asian grocery store or Sigma. Use Procedure 4.2 as a guideline for this experiment.

Procedure 4.2
Initiation and growth of tobacco callus on serially composed media.

Step	Instructions and comments
1	Surface disinfest 6-cm stem segments of tobacco with 10% (v/v) of liquid chlorine bleach solution for 10 min in large beakers, and rinse three times with sterile distilled water.
2	Cut out cylinders of tobacco pith tissue with a sterile no. 3 cork borer as illustrated in diagram C_1 of Figure 4.1.
3	Express the cylinders of pith tissue from the cork borer into sterile plastic petri dishes with a sterile metal punch.
4	Cut 5-mm-thick sections with a sterile scalpel and place one section in each of about 30 culture tubes, each containing 1 of 11 serially composed media (see list of 11 media).
5	Incubate tubes in a randomized complete block design at 25°C with 25 $\mu mol \bullet m^{-2} \bullet sec^{-1}$ of white light.
6	When the explants on medium 11 have produced callus that almost fills the diameter of the tube (about 4 to 8 weeks), pool uncontaminated calluses from each medium separately into tared cups. Record the fresh weights and divide by the number of calluses to obtain the fresh weight of one callus. Place the cups in the drying oven for several days to obtain dry weights and then calculate dry weight of one callus. Each team already has fresh and dry weights of tobacco explant tissues from Experiment 1 (Procedure 4.1, step 7). Don't forget to subtract these fresh and dry weights in order to obtain accurate weight gains of calluses from the separate 11 media.
7	Student teams should share and analyze the data by one of the procedures described in Chapter seven.

Anticipated results

Growth of callus from the disks on media 1 through 8 will be relatively poor. Growth from disks on medium 9 will be good because all essential nutrients are present. Callus growth on medium 11 will be best because it contains all the nutrients of medium 9 plus the nutrients from coconut milk. Growth of callus on medium 10 will occur, but will be less than that on medium 9 or 11.

Questions

- What is the function of each ingredient of the medium in the growth of callus?
- Why are fresh and dry weights recorded for calluses of each plant tissue?
- Why is it important that all explant disks be of uniform size?

Literature cited

Linsmaier, E. M. and F. Skoog. 1965. Organic growth factor requirements of tobacco tissue cultures. *Physiol. Plant.* 18:100–127.

Murashige, T. and F. Skoog. 1962. A revised medium for rapid growth and bioassays with tobacco tissue cultures. *Physiol. Plant.* 15:473–497.

chapter five

Histological techniques

Robert N. Trigiano, Kathleen R. Malueg, and Effin T. Graham

Plant histology can be defined simply as the study of the microscopic structures or characteristics of cells and their assembly and/or arrangement into tissues and organs. Several histological techniques are commonly used for examining plant tissues, each providing somewhat similar gross information, but differing in resolution of details and the medium in which the samples are prepared. These techniques include those for brightfield and fluorescent microscopy, in which specimens can be prepared for cutting into thick sections (15 to 40 μm) either without a stabilizing medium (fresh sections), in cryofluids (frozen sections), or embedded in paraffin-like materials, or in various formulations of plastic. Other techniques that employ electron microscopy either do not require specialized embedding medium for specimen preparation (scanning electron microscopy) or have samples embedded in plastics (transmission electron microscopy) that can be cut into ultrathin sections (65 to 100 nm). Explanation of the details involved in all of these techniques is beyond the scope of this chapter.

One might reasonably ask why include a chapter on histology in a plant tissue culture book? In many situations, histological techniques provide essential information that may not be evident upon only visual inspection. Much of our understanding of in vitro developmental processes that are presented throughout this book largely result from detailed histological research. For example, somatic embryos (Chapter nineteen) can be produced on the surface of a leaf explant, but may be so morphologically aberrant that they are unrecognizable. Using histological techniques and scrutinizing anatomical features, the characteristics of somatic embryos can be more readily seen. Another example for using histological techniques is to investigate the origin of specific structures, i.e., adventitious shoots and roots (Chapter fourteen), embryos, etc., that develop in culture. Histological development can be studied over time by periodically sampling tissues and/or the result can be examined in the mature structure. One must always be mindful that growth of tissues is dynamic and changing from moment to moment, whereas histological sections, for example, are static and fixed in time, and only present a very narrow glimpse of the developmental process. Nevertheless, the origin of tissues and organs may be convincingly inferred from serial observation of many samples and sections.

The field of plant histology and microtechnique is quite broad and cannot be conveyed in one simple chapter. Therefore, the intent of this chapter is to briefly summarize the essentials and provide a "primer" of sorts for preparing specimens from plant tissue cultures for histological examination using elementary paraffin and scanning electron microscopy techniques. The techniques presented herein are considerably easier and avoid

0-8493-2029-1/00/$0.00+$.50

many of the toxic materials found in more traditional protocols. For a broader view of general histological methodologies, students should consult one or more of the following references: Berlyn and Miksche, 1976; Sass, 1968; Jensen, 1962; Johansen, 1940.

General considerations

Equipment and supplies

The equipment and materials needed to complete a paraffin histological study are listed below:

- Indelible marker (Scientific Products [SP], Atlanta, GA)
- 500- or 1000-ml plastic beaker
- Aluminum pan
- Rotary microtome with disposable blade holder (SP)
- Warming oven (58 to 60°C)
- Two slide warming trays (40 and 50°C)
- Water bath (50°C)
- Hot/stir plate
- 1% agar or agarose in a 150-ml beaker
- Disposable microtome blades (SP)
- Paraplast® embedding medium (SP)
- Disposable plastic base molds and embedding rings (SP)
- 5-ml microbeakers (SP)
- Alcohol lamp and metal spatula
- Screw-capped Coplin jars
- Histochoice® fixative (Amresco, Inc., Solon, OH) plus 20% ethanol
- Disposable snap cap plastic 10-ml specimen vials
- Wooden applicator sticks
- Clean glass slides and coverglasses
- Camel or sable hair artist brush
- Test tube racks
- Eukitt or similar coverglass resin (EMS, Fort Washington, PA)
- Microclear® (Micron Environmental Industries, Fairfax, VA)
- 30, 50, 75, 95 (all aqueous) and 100% isopropanol
- 0.2% aqueous solution of Alcian blue 8GX (Sigma) 200 mg/100 ml of water
- Quick-mixed hematoxylin (Graham, 1991) — Stock A: 750 ml distilled water, 210 ml propylene glycol, 20 ml glacial acetic acid, 17.6 g aluminum sulfate, and 0.2 g sodium iodate; Stock B: 100 ml propylene glycol and 10 g certified hematoxylin (Sigma). Stain is prepared by adding 2 ml of Stock B to 98 ml of Stock A.

General protocol for paraffin studies

The procedure described below will work well with most tissues, including those from stock plants as well as those cultured in vitro. Moreover, it has several important advantages over more traditional paraffin protocols. First, it circumvents the use of toxic aldehydes (formaldehyde) and heavy metals (chromium and mercury) as fixatives. Second, it avoids the use of specimen/slide adhesive agents, such as Haupt's, that require the use of formaldehyde and may produce staining artefacts. Third, it does not use toxic substances, such as xylene, for deparaffinizing sections. Fourth, sections are stained directly through the paraffin, greatly reducing the number of operational steps in the process. The advantages listed above combine to make preparing tissue for histology and slides a safe and less intimidating process for both students and instructors alike.

Fixation

The first step in the protocol is to identify typical specimens (i.e., embryos on leaf sections), remove them from the culture dishes, and place in a small petri dish containing water. Most explants are too large to be adequately fixed and usually the subject (embryos) occupies a very small area. The specimen then should be carefully trimmed using a razor blade or a scalpel to include the subject and a small area of surrounding tissue. Five or fewer specimens are placed in about 5 ml of Histochoice®, a nonaldehyde, nontoxic fixative amended with 20% ethanol, contained in each specimen vial. If possible, the open vials should be aspirated (–1 atm or less) for about 30 min to remove air from the samples and promote infiltration of the fixative. For very small or very delicate specimens, we suggest that fixation without aspiration be tried first. Replace the original fluid with fresh fixative. The vials are closed and the tissue is allowed to remain in the fixative at room temperature for a minimum of 24 hours; samples may be stored at room temperature for weeks or months without degradation of structures. If long-term storage is needed, we suggest using screw-capped rather than snap-capped vials.

Dehydration and infiltration

After fixation, the samples must be dehydrated and placed in a solvent compatible with paraffin. Traditionally, this has been accomplished with graded series of ethanol and transitioning to tertiary butyl alcohol. This process can be considerably shortened by using isopropanol to both dehydrate the tissue and dissolve the paraffin. Histochoice® fixative can be removed from the vials with a long, thin pipette and poured down the drain with plenty of water. Specimens are dehydrated using 30, 50, 75, 95, and three changes of 100% isopropanol, each for about 30 min. After the last change of pure isopropanol, fill the vials about one quarter full with fresh 100 isopropanol, add a few pellets of Paraplast® embedding medium, loosely recap the vials, and incubate at 60°C. Periodically over the next several hours, swirl the contents and add a few more pellets of Paraplast® to each of vials. At the end of the day, remove the caps from the vials to allow the isopropanol to evaporate. At this time, fill a 500- or 1000-ml plastic beaker with Paraplast® pellets, place several pellets into a number of base molds contained in an aluminum pan, and incubate in the 60°C oven. The next morning all of the isopropanol should be dissipated, the specimens completely infiltrated with Paraplast®, and the pellets in the beaker and base molds melted.

Casting specimens into blocks

Using a sharp razor blade, whittle a flattened paddle-shaped end on several wooden applicator sticks and store in a vial with the flattened end immersed in molten paraffin in the oven. Transfer a specimen into the molten paraffin contained in a base mold and position it so that the sectioning plane of the specimen is parallel to the bottom of the base mold. For example, if a transverse section of a stem is required, the stem should be placed perpendicular to the base mold; the long axis of the stem should be "sticking up" straight up in the molten paraffin. If the stem is positioned flatly in the base mold, longitudinal sections will result. With more complex specimens, like single somatic embryos or shoots growing from basal tissue, this orientation step becomes more critical. It is important to visualize the exact desired section plane at this step. Once the sample is oriented correctly, quickly transfer the base mold to a cool surface — an inverted aluminum pan containing ice works well. The paraffin surrounding in the bottom of the base mold will quickly become cloudy (congealed) and specimen will be immobilized. Rapidly attach an embedding ring to the base mold and fill the apparatus with molten paraffin from the plastic beaker. This is a very vulnerable point in the process. If the immobilizing paraffin is allowed to congeal too far, the added molten paraffin will not bond with it and the block will split apart. At this time, an identifying label consisting of white paper and text written in pencil (do not use ink) may be inserted partially into the molten paraffin. Place the casted blocks in the refrigerator or they may be floated in a large beaker containing ice water. The base mold may be removed after the block is hardened.

Microtomy and mounting sections on slides

A word of caution before beginning to section. Regardless of whether a stainless steel knife or disposable razor blade is used to cut sections, they are both extremely sharp and cut flesh very easily. Exercise care while working with the microtome and if the microtome is left unattended, place a note on the knife holder warning in bold print that a knife or blade has been installed. The rotary action should be kept locked at all times except when sectioning.

One of the advantages of using embedding molds is that the cutting face of the block is square. If the block were sectioned as is, the resulting "ribbon of very large sections" would be straight. However, seldom does the specimen occupy the entire block face. Therefore, it is often desirable to trim away some of the excess paraffin from around the sample. The horizontal edges of the block must be parallel with each other or, another way of visualizing it is, the sides of the block must be of equal length. The block edges must also be kept parallel to the knife edge. If the ribbon of sections curves to the right or left, the side of the block opposite the direction of curvation is too long. Remove the block from the chuck and retrim so that the sides are parallel and of equal length. Do not trim the block while mounted in the chuck over the knife!

Mount a new disposable blade in the blade holder — only a small portion of the blade should be clear of the carriage. Typically, new blades are coated with oil and may be cleaned with a small wad of tissue moistened with a little Microclear®; carefully wipe tissue perpendicular to the blade surface and allow to completely dry. Set the microtome to cut 10-μm sections. Now mount the specimen block in the chuck and orient it so the bottom of the cutting face is parallel to the edge of the blade. Position the specimen at the level of the knife and carefully and gently slide the knife holder toward the block until they are very close. With the utmost care, turn the wheel clockwise causing the specimen to advance 10 μm per revolution. The first sections may not include the entire block face — continue advancing the block until a complete section is obtained. Lock the advancing wheel into position, carefully remove the partial sections and clean the knife as before. After the Microclear® has evaporated, continue to operate the microtome and after five or more sections are produced, place a camel hair brush underneath the ribbon for support while it is being held away from the knife holder. A 30- to 50-section ribbon can be produced and easily manipulated. Lay the ribbon out on a clean piece of dark paper and, using a razor blade, divide it into segments (sections) that do not exceed about 75% the length of a glass microscope slide.

Modern Superfrost® slides afford the following well-established advantages: (1) they can be used directly from the manufacturer without cleaning; (2) the glass surface bonds with tissue sections without an adhesive coating; and (3) the special labeling pen resists all common histological solvents (Graham and Trentham, 1998). If these are not available in your laboratory, prewash glass microscope slides in a mild dishwashing detergent solution, rinse well with tap water, and rinse with distilled water. Allow the slides to dry in test tube racks with an aluminum foil base. Washed and dried slides can be stored indefinitely in slide boxes. Label the frosted portion of the slide with appropriate information using an indelible marker or lead pencil and cover specimen surface of the slide with distilled water. Moisten the tip of a brush, touch it to the surface of the first section to the left, and lift the entire ribbon off of the paper. Now touch the first section to the right onto the water and lay the rest of the ribbon down so that air is not trapped underneath the sections. Place the slide on the 50°C warming table to expand the sections. After about 10 min, set the slide on a sponge and slightly tilt it to drain off most of the water. Affix the sections to the glass by placing the slide on the 40°C warming table for 12 to 16 hours or overnight. The sections are now ready to be stained.

Staining sections

The following staining procedure (Procedure 5.1) is adapted from Graham and Joshi (1995).

Procedure 5.1
Staining through paraffin with alcian blue and hematoxylin.

Step	Instructions and comments
1	Preheat the alcian blue solution in a closed Coplin jar and place the slides in the stain for 1 h at 50°C.
2	Gently rinse the slides in a stream of cold tap water and carefully blot dry the remaining water droplets.
3	Transfer the slides to Coplin jars containing hematoxylin stain for 15 min at room temperature, rinse and dry as described before, and place on the 40°C slide table for a minimum of 30 min.
4	The paraffin in the sections is removed with three, 5-min soaks in Micro-Clear® followed by three, 5-min changes of 100% isopropanol.
5	After the slides have air-dried, pool some resin (Eukitt) in the center of the slide and gently lower a coverglass so that the resin spreads evenly over the surface without air bubbles. The resin may be cured overnight on a 40°C warming table.

An alternative to this staining process is offered by Graham and Trentham (1998). Procedure 5.2 is very efficient and provides a triple stain in a single solution.

Procedure 5.2
Triple stain using alcian blue, safranin O, and Bismarck brown.

Step	Instructions and comments
1	Prepare an 0.1 M acetic acid-sodium acetate buffer, pH 5.0 and three 1% stock solutions of alcian blue 8 GX, safranin O, and Bismarck brown, using 0.1 g of dye powder dissolved in 100 ml of absolute ethanol.
2	To 100 ml of acetate buffer, add 5 ml of alcian blue, 2 ml of safranin O, and 1 ml of Bismarck brown stock solutions.
3	Extract paraffin from sections by moving slides through four baths of Micro-Clear®, then through five baths of 100% isopropanol to remove the paraffin solvent and finally place the slides on a rack to air-dry.
4	Immerse slides in the triple stain. Coloring of tissues will be evident within 15 min, however, optimum staining differs widely among various specimens. Progress of staining is monitored easily by rinsing a slide briefly with distilled water, blotting it dry with soft tissue paper, and observing under a compound microscope. The slide may be returned to the stain solution until satisfactory staining is achieved. We suggest reexamining the slide at 15-min intervals.

Procedure 5.2 continued Triple stain using alcian blue, safranin O, and Bismarck brown.	
Step	Instructions and comments
5	When staining is sufficient, dry slides for 1 hour on a 40°C slide warmer and affix a coverslip as described in Procedure 5.1, step 5. The results of this triple stain are the following: polysaccharide cells are blue; lignified walls, nuclei, and chloroplasts are red; and cuticle (if present) is brown or yellow-brown.

Do not become discouraged if your results are less than perfect the first time — don't give up — try again! Like any other technique, the operator becomes more skilled after some experience.

Immobilization of specimens for paraffin sectioning

This sample preparation technique involves immobilization and precise orientation of specimens between two layers or a "sandwich" of agar. It offers several advantages over more conventional methods of preparation, especially when working with small and/or delicate tissues. Small samples, e.g., embryos or callus, are often lost or damaged during the dehydration sequences when fluids are removed and replaced. By immobilizing the specimens in agar, it is nearly impossible to lose samples and the agar minimizes damage from handling. However, the real advantage to the technique is that the sample can be oriented in the agar before fixation, eliminating the need to position it while casting the block.

This procedure is adapted primarily from Hock (1974). Prepare 1% water agar (1 g agar per 100 ml of water in a tall 150-ml beaker) on a hot/stir plate and after the agar is melted, store in a 50°C water bath. Dip a clean glass microscope slide into the molten agar and then place it in a 100 × 15-mm petri dish containing moistened filter paper. The remainder of the immobilization procedure is diagrammatically represented in Figure 5.1. Select a specimen and place it on top of the cool and hardened agar. With a pasteur pipette, cover the sample with a small amount of molten agar. The position of the sample is now secure. Draw an arrow with pencil on a piece of white paper about the width of the sample. Using a stereomicroscope, if necessary, and fine forceps, place the arrow directly opposite or behind the desired cutting face of the tissue and tack it into position using molten agar. Cover the entire preparation with a thin coating of molten agar. Once the agar has hardened, trim any excess agar away using a razor blade. The specimen can be removed from the slide by "slipping" the razor blade under the bottom of the agar

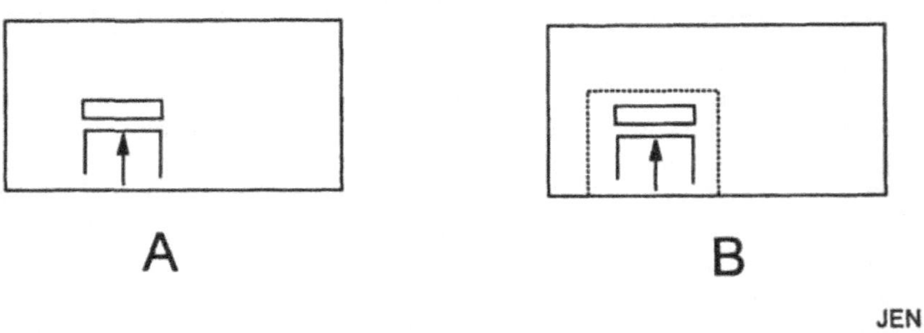

Figure 5.1 Immobilization. (A) A specimen (stem) is placed on an agar-coated slide, then covered with molten agar. A white slip of paper with an arrow drawn in pencil is also affixed with agar. (B) After the agar has hardened, the preparation may be trimmed and removed (dotted lines) from the glass slide. (From Trigiano, R. N. and Gray, D. J., Eds., *Plant Tissue Culture Concepts and Laboratory Exercises, First Edition*, CRC Press LLC, Boca Raton, FL, 1996.)

sandwich and placing it in fixative. Follow the previously described protocols for fixation of tissue and infiltration with paraffin.

Instead of using an embedding mold, use a 5-ml plastic microbeaker to cast the specimen into a paraffin block. Draw a straight line on the bottom of the beaker with an indelible pen. Fill the container with molten paraffin and transfer the sample to the beaker so that the sample is parallel, yet slightly behind, and the arrow is positioned orthogonally (at a right angle) to the line on the bottom (Figure 5.2). Place the microbeaker on a cold surface and label with a paper slip as described before.

JEN

Figure 5.2 Top view of positioned specimen cast in a plastic microbeaker. (From Trigiano, R. N. and Gray, D. J., Eds., *Plant Tissue Culture Concepts and Laboratory Exercises, First Edition*, CRC Press LLC, Boca Raton, FL, 1996.)

Dissection and mounting of specimens

The block of paraffin in which the specimen was cast cannot be mounted on the microtome. Therefore, it is necessary to dissect the specimen and a small amount of the surrounding block and mount it on a block formed by an embedding mold. Usually there are some old blocks around from previous projects or newly cast ones may be used. Before the microbeaker is removed, take a sharp razor blade and cut through the beaker and score the paraffin in the block underlying the line on the microbeaker. This will ensure that the cutting face is identified even if the specimen and arrow are not clearly visible in the block. Dissect the specimen from the block. Create a flat field on the face of the embedding mold block. Melt a thin layer of the embedding block using the flat side of a spatula heated with an alcohol burner. Quickly attach the side opposite of the cutting face of the specimen block to the embedding mold block. Reinforce the attachment by melting small chips of paraffin with a hot spatula tip along all four sides of the specimen block (Figure 5.3). The block may be sectioned and slides prepared as previously described.

Preparation of specimens for scanning electron microscopy (Procedure 5.3)

The sections prepared for paraffin histology present a two-dimensional (flat) interior view of a small area of a much larger three-dimensional object. In contrast, scanning electron microscopy (SEM) allows for three-dimensional topical or internal views of an entire specimen. Samples for SEM do not need to be embedded in paraffin nor do they need to be stained. Moreover, the resolving power of a scanning electron microscope is much greater than a compound light microscope. However, in many ways, preparation of samples for SEM is technically easier than for paraffin histology. Note: methodologies for critical point drying and sputter coating are not presented and should be performed by a competent SEM technician or other person familiar with these techniques.

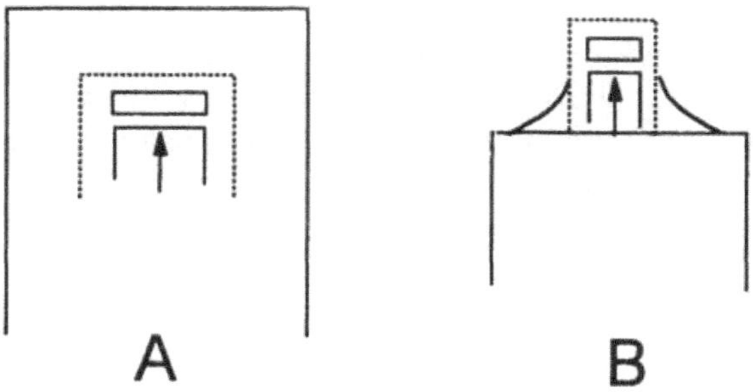

Figure 5.3 Mounting the dissected specimen (A) to a flat block cast using an embedding mold (B). Note the paraffin reenforcement along the sides of the specimen. (From Trigiano, R. N. and Gray, D. J., Eds., *Plant Tissue Culture Concepts and Laboratory Exercises, First Edition*, CRC Press LLC, Boca Raton, FL, 1996.)

The following materials will be needed to prepare samples for SEM:

- 3% glutaraldehyde in 0.05 M potassium phosphate buffer, pH 6.8 — store in the refrigerator
- 1% osmium tetroxide in the same buffer (optional) — store in a shatterproof bottle in the freezer
- Aluminum mounting stubs

All of the above may be purchased from Electron Microscopy Sciences, Fort Washington, PA.

- Double-sided sticky tape
- 0.05 M potassium phosphate buffer, pH 6.8 — prepare by dissolving 0.87 g K_2HPO_4 and 0.68 g KH_2PO_4 in 200 ml of water
- Graded series of ethanol or acetone (10, 30, 50, 75, 95, and 100%)

Note: glutaraldehyde and osmium tetroxide are extremely toxic and should only be handled wearing gloves and in a certified fume hood. For additional information, consult material safety data sheets.

Procedure 5.3
Preparation of specimens for scanning electron microscopy.

Step	Instructions and comments
1	Fix and aspirate small pieces of tissue in cold (4°C) buffered solution of glutaraldehyde for at least 24 h.
2	Dispose of the fixing solution in a waste bottle and rinse specimens at least three times with cold (4°C) phosphate buffer.
3	With great care and working in a fume hood, dispense enough of the cold, buffered osmium tetroxide (osmium) solution to cover the samples and recap the vials. After 2 h, dispose of the osmium in a properly labeled hazardous waste bottle and rinse the tissues with three changes of cold buffer. In many instances, the tissue will be blackened after postfixing with osmium — this is normal.

Procedure 5.3 continued
Preparation of specimens for scanning electron microscopy.

Step	Instructions and comments
4	Dehydrate the tissue for 30 min in each member of the graded ethanol or acetone series except 100%. The final steps of dehydration are three changes of room temperature absolute ethanol or acetone over 30 min. The samples may be stored in absolute ethanol for a few days until they can be critical point dried by the SEM technician. Note: we suggest that if the specimens cannot be dried within a few days, that they be stored in 75% ethanol or acetone.
5	Have an EM technician critically point dry the specimens. Place a piece of double-sided clear tape onto an aluminum stub and mount the dried sample on the tape. The samples are ready to be gold-palladium coated and viewed with the aid of a technician.

Literature cited

Berlyn, G. P. and J. P. Miksche. 1976. *Botanical Microtechnique and Cytochemistry.* Iowa State University Press, Ames.

Graham, E. T. 1991. A quick-mixed aluminum hematoxylin stain. *Biotechnic Histochem.* 66:279–281.

Graham, E. T. and P. A. Joshi. 1995. Novel fixation of plant tissue, staining through paraffin with alcian blue and hematoxylin, and improved slide assembly. *Biotechnic Histochem.* 70:263–266.

Graham, E.T. and W. R. Trentham. 1998. Staining paraffin extracted, alcohol rinsed and air dried plant tissue with an aqueous mixture of three dyes. *Biotechnic Histochem.* 73:178–185.

Hock, H. S. 1974. Preparation of fungal hyphae grown on agar coated microscope slides for electron microscopy. *Stain Technol.* 49:318–320.

Jensen, W. A. 1962. *Botanical Histochemistry.* W. H. Freeman Co., San Francisco.

Johansen, D. A. 1940. *Plant Microtechnique.* McGraw-Hill, New York.

Sass, J. E. 1968. *Botanical Microtechnique,* 3rd ed. Iowa State University Press, Ames.

chapter six

*Photographic methods for plant cell and tissue culture**

Dennis J. Gray

Plant cell and tissue culture inherently is based on direct visual observation, unlike other disciplines, such as molecular genetics, virology, and physiology, where results from experiments often are abstract. With in vitro culture, the confirmation that the basic experimental response has occurred is based primarily on visual proof. For example, whether or not cultured tissue undergoes embryogenesis or organogenesis is confirmed by visualizing the presence of somatic embryos or organs. Thus, the fundamental documentation of plant cell and tissue culture research often is a photograph.

As an essential component of cell and tissue culture research, accurate photography is needed in order to satisfy the doctrine of scientific research, which demands that experimental methods and results be documented and repeatable so that they can be confirmed and built upon by others. Unfortunately, many research reports are plagued by poor or inadequate photographic technique. Due to the pivotal nature of photography in plant cell and tissue culture research, poor photographic documentation severely detracts from or, in many cases, jeopardizes the credibility of a research report. This is unfortunate, since high-quality photography is not difficult given proper equipment and supplies.

In-depth instruction in photographic technique is beyond the scope of this book and will not be attempted. Instead, we will assume that a general knowledge of photography and microscopy preexists or is readily attainable. In this chapter, the use of equipment, supplies, and techniques that are required for basic photographic documentation will be described. Although advanced techniques that utilize photography, such as confocal microscopy and video imaging, are becoming more common, they remain highly specialized and will not be included in this chapter. Similarly, scanning and transmission electron photomicroscopy routinely are used in plant cell and tissue culture research, but generally are taught as separate courses and will not be discussed here.

Types of photography used in plant cell and tissue culture research

Several different photographic techniques are required in order to document various aspects of plant cell and tissue culture. Because objects of interest vary greatly in size and type (e.g., from a cell to a culture vessel), but often are very small, equipment to accomplish both macro- and microphotography is required.

* Florida Agricultural Experiment Station Journal Series No. R-07026.

0-8493-2029-1/00/$0.00+$.50

Macrophotography

When cultures contained in whole petri dishes or groups of petri dishes must be photographed, a simple 35 mm SLR camera equipped with a close-up lens and mounted on a copy stand is preferable. In this instance the photographic field might range from 2 cm (a large callus culture) to 10 cm (a petri dish) to 40 cm (a group of petri dishes).

Microphotography using stereomicroscopes

For characterization of callus cultures, small shoots, nonzygotic embryos, and other objects that range in size from 0.5 mm to 1.5 cm, stereomicrography is most appropriate. A stereomicroscope or "dissecting scope" allows the surface morphology of a specimen to be examined. Because specimens often possess much surface relief, the ability to photograph a wide depth of field is essential in stereomicrography. Depth of field can be adjusted on those stereomicroscopes that incorporate an iris diaphragm between the objective lens and the eyepieces. Depth of field increases as the iris diaphragm is closed (see the example below). Stereomicroscopes often use a separate light source. The ideal type of light source for stereomicrography consists of a fiber optic illuminator with a bifurcated fiber bundle. This arrangement allows for great flexibility in lighting, allowing an optimum view of the specimen to be obtained (Figure 6.1).

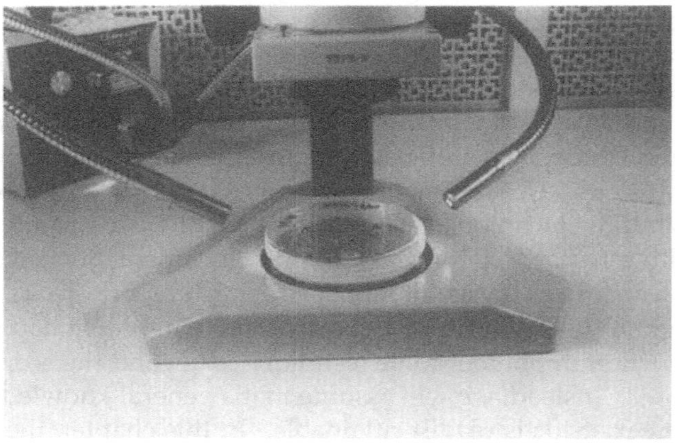

Figure 6.1 Typical setup of specimen in petri dish on stereomicroscope. Note fiber optic illuminator bundles positioned on each side of the specimen. (From Trigiano, R. N. and Gray, D. J., Eds., *Plant Tissue Culture Concepts and Laboratory Exercises, First Edition*, CRC Press LLC, Boca Raton, FL, 1996.)

Microphotography using compound microscopes

A compound microscope is used when smaller objects such as early-stage embryos or cells, or the internal structure of cells and tissues must be examined. A compound microscope can resolve objects from the 10-µm to 1.5-mm size range, depending on optical quality. However, a compound microscope differs from a stereomicroscope in that it utilizes light that is transmitted through the specimen. Therefore, the specimen must be thin and lucent enough to allow light to pass. Typically, histological processing of tissue is employed to produce lucent sections (see Chapter five), or else cells and tissues that can be pressed thin enough for light to pass are utilized. Compound microscopes are configured as either "upright," in which the specimen is below the objective lens, or "inverted," in which the specimen is placed above the objective lens.

Upright microscopes are most common, typically have the highest resolving power, and usually are used to view specimens mounted on microscope slides with cover slips. Inverted microscopes are used to examine cells and tissues in liquid, such as suspension cultures or protoplasts, since they can view specimens through the bottom of the culture vessel. Inverted microscopes are indispensable in protoplast research, where living cultures must be monitored *in situ*.

Focusing and cropping the image

Methods of adjusting the camera to obtain sharply focused photographs differ from camera to camera and between macrophotography and types of microphotography, with stereomicrography generally being the most difficult to accurately focus.

Focusing the subject on a copy stand usually is not difficult and amounts to following the directions for a given SLR camera. One note of caution, however, is to make sure that the camera angle is the same as that of the copy stand bed in order to maintain a consistent focal plane across the field of view. This parameter frequently is out of alignment but can be measured and adjusted by using a small bubble level (obtained from a hardware store) to compare and adjust the angle of camera vs. that of the copy stand bed.

Focusing the subject through microscopes is more difficult, since most photomicrography systems must be adjusted to synchronize the optical focal plane (i.e., the image that the eye sees) with the film focal plane. Since everyone's eyes focus differently, users must learn to adjust the camera to their own preference. Focal systems vary between photomicroscopes (individual owner's manuals should be consulted), but often consists of a photoreticle (an inscribed pattern of black lines visible through one eyepiece) that must be brought into sharp focus along with the specimen by a combination of adjusting both the microscope focus and the focusing ring on the individual eyepiece. Once this is accomplished, the optical and film focal planes are synchronized so that focusing the specimen through the eyepiece results in the camera also being in focus. It is recommended that those who normally wear corrective lenses use them during photomicrography in order to obtain a consistent sharp focus.

Generally, the specimen should be magnified to fill most of the viewing area. A common mistake is to produce a photograph of a large, empty field with only a small specimen in the center. This is not only a waste of film, but does little to illustrate the specimen.

Film types

Film is the recording medium of choice for photography. Typically, both color positive (slide) and black and white negative (B & W) film types in 35-mm format are used. Color slide film is used to provide life-like, detailed images; 35-mm slides are conveniently stored and are the desired medium for oral presentation of results. B & W film provides a negative that can be used to produce B & W prints, which is the most commonly used method of presenting results in a publication. Use of the correct film type often determines whether or not an acceptable photograph can be obtained.

Color slide film

Color film is highly sensitive to color balance of the light used for illumination. Since most light used to illuminate copy stands, stereomicroscopes, and compound microscopes is "tungsten balanced," a compatible color film type must be used. Tungsten-balanced film usually has a "T" or the word "tungsten" printed after the film type. Excellent film for color photography is Kodak Ektachrome ASA 64 T (EPY 135), for use with copy stands or microscopy where abundant light is present, and Kodak Ektachrome ASA 160 T (EPT 135), for use in low-light situations, such as high magnification microscopy or fluorescence microscopy. In all instances, the light source, be it on a copy stand, fiber optic illuminator, or built into the microscope, must be adjusted to the maximum or near-maximum level in order to achieve the proper color balance to be compatible with tungsten film.

Black and white film

While B & W film is not subject to color balance problems, quality of prints produced from color negatives varies tremendously with the film type. For very low-light conditions, any fast (high ASA number) film can be used. However, for high-quality images, Kodak Technical Pan (TP 135) is recommended. This film may be used at several different film speeds (ASA levels) and with several different developers. However, the highest-quality negatives result from utilizing a film speed of ASA 25 and development in Kodak Technidol developer. The resulting negatives may be sent out to have prints produced or, preferably, be printed by the investigator, which allows the ideal enlargement and exposure to be obtained and generally results in the best composition.

Composition

While proper equipment and materials are essential in order to accomplish acceptable photography, the composition of the subject in the viewing area often makes the difference between a poor photograph and an excellent photograph. As mentioned above, the specimen should be carefully cropped so as to fill most of the viewing area. Also, the orientation and lighting of the specimen are critical. The following example illustrates importance of depth of field adjustment and specimen orientation to obtain an optimized micrograph with a stereomicroscope.

Figure 6.1 shows the typical setup of a stereomicroscope and fiber optic illuminator with a bifurcated fiber bundle. A petri dish containing cotyledon bases of *Cucumis melo* L. on agar, which were cultured to produce somatic embryos (see Chapter twenty-one), is on the microscope stage. The petri dish lid is removed briefly during each exposure. The microscope stage has a black surface, which generally provides better contrast. The fiber bundles are positioned to provide optimum lighting of the specimen, as determined by viewing the specimen through the microscope while maneuvering the bundles. The illuminator is turned to the maximum power setting if color film is to be used. The specimen within the dish is adjusted on the agar so as to show the desired view (in this case, embryos) and the magnification is adjusted to fill the viewing area to the greatest extent possible.

Depth of field

With stereomicrography, depth of field, which can be adjusted on those microscopes that possess an iris diaphragm between the objective and eyepiece lenses, is of critical importance. Figure 6.2A shows a typical photomicrograph obtained with the iris diaphragm set to full open. The focus was set on the apical dome and foreground of the cotyledon (large arrow). Note that the background area of the cotyledon (smaller arrow) is out of focus as is the globular stage embryo in the foreground (smaller arrow). This demonstrates the limited focal plane obtained under these conditions due to reduced depth of field. An improved view of the specimen is shown in Figure 6.2B, which was obtained by adjusting the iris diaphragm on the stereomicroscope so that it was nearly closed, thus, increasing the depth of field — no other adjustments were made. Note now that both the fore- and backgrounds are in focus, resulting in a superior micrograph.

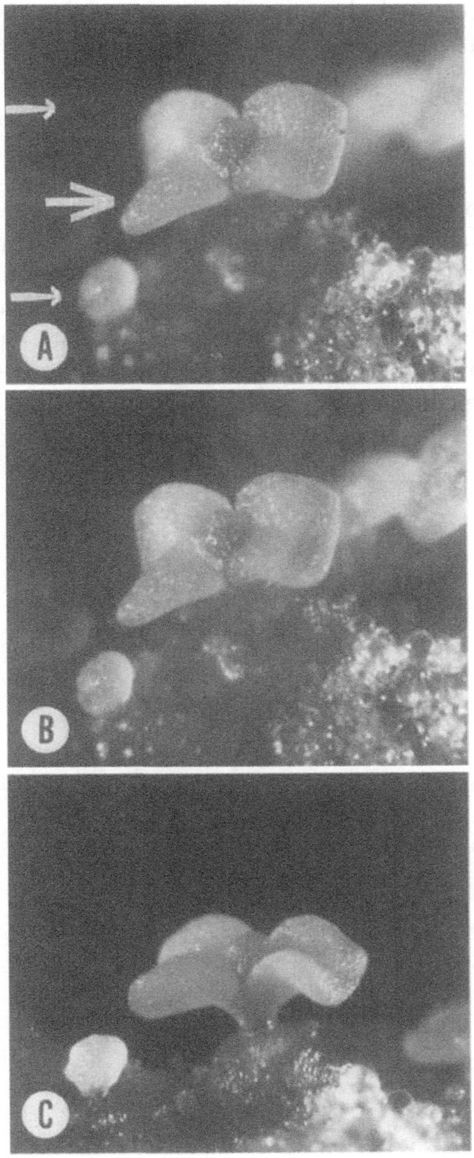

Figure 6.2 Optimization of depth of field and orientation of *Cucumis melo* somatic embryos for photography with a stereomicroscope. (A) Specimen photographed with poor depth of field. Note area that is in focus (large arrow) vs. out-of-focus areas in the fore- and background (small arrows); (B) specimen photographed with increased depth of field obtained by closing the stereomicroscope iris diaphragm. Note that fore- and background now are in focus. (C) Same embryos as in A and B demonstrating improved photograph obtained by careful orientation of the specimen. (From Trigiano, R. N. and Gray, D. J., Eds., *Plant Tissue Culture Concepts and Laboratory Exercises, First Edition*, CRC Press LLC, Boca Raton, FL, 1996.)

Orientation

Specimen orientation, particularly when accomplishing stereomicroscopy, makes a dramatic difference in the clarity of documentation. Figure 6.2C shows the same embryos as in Figures 6.2A and B, except that the specimen was positioned to show the embryos from the side. In this orientation, much more of the important morphological attributes of the larger embryo can be seen, including the attachment of the embryo to cotyledon tissue. Note also that the morphology of the subtending globular-stage embryo can be better determined.

Conclusion

Photography is an important tool in tissue and cell culture research. Due to the ranges in size and detail of tissue and cell culture specimens, several different types of photography usually are needed to provide adequate documentation. While a photograph is relatively simple to obtain, proper photographic and microscopic equipment and supplies are required to accomplish high-quality photography. With equipment in place, photographic technique then becomes of paramount importance. Details of composition often determine the difference between acceptable and unacceptable photography.

chapter seven

Statistical analysis of plant tissue culture data

Michael E. Compton

Statistics is an important tool for biological research because it provides an objective, nonbiased way to evaluate experimental treatments. In statistics, we use the mathematical probability that an event can occur to determine whether or not treatment effects are real. Statistics measure variation both within and between experimental treatments, allowing scientists to select the best treatment. Therefore, statistics should be included as an important part of experimental design and implementation in order to interpret results. This chapter is a shortened version of a review article published previously on this subject (Compton, 1994) and discusses methodologies associated with scientific research, experimental designs, and methods of comparing treatment means, providing examples of how to present data in tables and graphs.

Methodologies associated with scientific research

In science we use a logical and objective approach to problem solving. We examine what is already known by reading literature and formulate a hypothesis that can be tested experimentally. The hypothesis is tested objectively and either accepted or rejected according to the results obtained. The "null" hypothesis is typically used in statistics and states that all experimental treatments elicit a similar response. Rejection of the null hypothesis denotes that the imposed treatments vary in the response that they cause.

Before designing an experiment, the scientist should clearly and concisely define the problem to be solved in order to formulate questions that will lead to a solution. In response, a set of experimental objectives are formulated to help solve the problem. Next, a series of treatments are established that allows the scientist to study the problem. The treatments should be based on prior knowledge of the subject matter and evoke a wide range of response. Experimental materials used in plant tissue culture studies, such as culture vessels, culture medium, growth environment, etc., are selected in accordance with experimental objectives when establishing treatments. Knowledge of previous experiments, successful or not, is essential when selecting treatments and experimental materials. After establishing the treatments and experimental materials, the experimental design along with the observational units, number of replicates per treatment, and randomization scheme are chosen.

0-8493-2029-1/00/$0.00+$.50

Explants typically are selected as the observational unit in plant tissue culture experiments. Selection of explant type and source may influence the outcome. In most tissue culture experiments the experimental unit, or replicate, is the culture vessel. This is because a culture vessel houses explants, which are the unit of observation. When there is one explant per culture vessel, both the vessel and explant are considered a replicate. When there are multiple explants per culture vessel the vessel is considered the unit of replication and the explants within the vessel a subsample. The number of replicates selected per treatment is important because the degree of replication affects precision of the statistical test(s). Generally, precision increases as the number of replicates is increased. However, a plateau is reached where adding replicates no longer improves precision.

When designing an experiment, each replicate is assigned to a treatment. When assigning replicates to treatments, it is important to make assignments in a manner such that all have an equal chance of receiving a given treatment. This is called randomization. Randomization schemes vary according to the experimental design employed. After selecting the treatments and experimental design, one should determine the data to be collected. The data should properly evaluate and explain treatment effects according to experimental objectives.

An analysis of variance (ANOVA) summary table with all the degrees of freedom (DF) and associated sources of variation (treatment, experimental error, etc.), along with graphs and/or tables that show the theoretical results, should be constructed to evaluate the effectiveness of the proposed experiment in answering the desired questions. Consulting a knowledgeable colleague or statistician at this time should determine if the experiment was well conceived.

Data must be statistically analyzed and interpreted as planned before the experiment, according to the hypothesis, experimental conditions, and previous facts. Results that differ from those obtained previously may be due to mistakes in preparation of experimental materials, data entry and analysis, fluctuation of environmental conditions during the experiment, etc. There is always some degree of uncertainty regarding the conclusions from an experiment. For this reason it is important to repeat the experiment to confirm your results.

Experimental designs

The following section will discuss the application of completely randomized, randomized complete block, and split-plot designs in plant tissue culture research.

The completely randomized design

A majority of plant cell and tissue culture studies employ a completely randomized design (CRD), because cell cultures are generally grown in environmental chambers that accurately control light, temperature, and humidity. In the CRD, treatments are assigned to experimental units at random. The numbers of treatments and replicates per treatment that can be tested are not limited by the CRD. It is the easiest design to employ and analyze, even when there are missing data or when treatments are unevenly spaced (Lentner and Bishop, 1986; Little and Hills, 1978). The CRD also is the most precise design because it maximizes the degrees of freedom (DF) used for estimating experimental error. This reduces the F-value required to detect statistical differences among treatments. This is important in small experiments or when outside variation is great.

An experiment designed to identify tomato (*Lycopersicon esculentum* Mill.) cultivars that displayed high rates of shoot regeneration from pedicel explants illustrates use of the CRD in plant tissue culture studies. Explants were cultured in test tubes that contained

15 ml of shoot regeneration medium (Compton and Veilleux, 1991). There were 24 tubes (replicates) per treatment, each with one explant (observational unit), resulting in a total of 264 experimental units. Treatments were assigned to the experimental units completely at random. Experimental data included the number of explants with shoots and the number of shoots per explant.

ANOVA was used to analyze the data in order to determine if the genotypes differed in their ability to produce shoots from pedicel explants. This procedure generates the DF, sums-of-squares (SS) and mean square (MS) values for treatment and experimental error used to perform the statistical test. Treatment SS measures the degree of variation associated with the treatments, whereas experimental error SS measures variation associated with experimental units. In the tomato study, the F-value for treatment (cultivar) was calculated by dividing its MS (7.937) by the mean square error [MSE] 1.35 [Table 7.1]. Significance of the observed F-statistic was determined using 10 and 253 DF. A significant F-statistic indicates that there are differences in shoot regeneration rates among the cultivars but does not determine which genotypes differ. To determine differences among treatment means, a mean separation procedure must be performed. Mean separation procedures should be chosen prior to ANOVA to avoid personal bias in the test. This is because each mean separation test uses a unique formula when calculating differences among means. The types of mean separation tests will be discussed in greater depth later in this chapter.

Table 7.1 Analysis of Variance Summary Table for the Number of Shoots per Explant from Tomato Shoot Regeneration Study

Source	DF	Mean square	F-value
Cultivar	10	7.937	5.87[a]
Error	253	1.352	
Corrected total	263		

Note: Data analyzed using procedures for a CRD.

[a] Significant F-value at the 0.001 level.

From Compton, M.E. 1994. *Plant Cell Tissue Organ Cult.* 37: 217–242. With permission.

The CRD is most efficient when there is little variation among experimental units. Efficiency and precision are reduced in situations where there is a high degree of variation not associated with treatment effects. This is because nontreatment variation is assigned to experimental error, resulting in a large MSE and fewer rejections of the null hypothesis. In situations where there is a high degree of outside error and the source can be identified, designs that employ blocking should be used (Lentner and Bishop, 1978; Little and Hills, 1978).

Single factor CRD with subsampling

Sometimes it is more efficient to culture several explants in a single culture vessel. In this situation the response of each explant is measured, resulting in multiple measurements for each vessel (i.e., subsampling [Lentner and Bishop, 1986]). This type of design is more efficient than the normal CRD with one observational unit per experimental unit when there is a wide range in response among explants. Statistical precision is improved by making several measurements per culture vessel, thus reducing variation among replicates of the same treatment. Subsampling may be used to study variability among explants, which may be useful in future experiments.

An experiment that examined the ability of watermelon (*Citrullus lanatus* [Thunb.] Matsum. & Nakai) cotyledon explants to produce adventitious shoots illustrates the use of subsampling. The objective of this experiment was to identify genotypes with high level adventitious shoot organogenesis. Cotyledons from seedlings of 11 cultivars were cultured in petri dishes that contained 25 ml of shoot regeneration medium (Compton and Gray, 1993). Cotyledons from each cultivar were arranged into 6 dishes, each with 5 explants, resulting in a total of 66 dishes and 330 explants. Explants from each cultivar were assigned to petri dishes at random, making sure that only explants from one cultivar were cultured together. Individual petri dishes were distributed at random in the growth room. The number of explants with adventitious shoots and the number of shoots per explant were recorded.

The ANOVA for CRD experiments with subsampling differs from a CRD with one observational unit per experimental unit in that SS are generated for treatment, experimental error, and subsampling error (Table 7.2). A special error term is used for calculating the treatment F-value and measures variation among experimental units. Subsampling SS represents variation caused by culturing multiple explants within the same experimental unit and is not used in computing treatment differences. This improves statistical precision by removing variation among explants from the experimental error. In this example, significance among treatments was detected by dividing the treatment MS by the new MSE, resulting in a significant F-value of 6.83 and indicating that the cultivars displayed varying levels of shoot regeneration. However, a statistical test must be conducted on the treatment means. When conducting mean separation tests in subsampling experiments, the special MSE must be used to calculate values for the statistical test. Failure to specify the correct error term may result in the grouping of means that should be separated and/or separation of similar means.

Table 7.2 Analysis of Variance Summary Table for the Number of Adventitious Shoots per Explant from Watermelon Shoot Organogenesis Study

Source	DF	Mean square	F-value
Genotype	10	38.312	6.83[a]
Experimental error[b]	55	5.610	
Subsampling error	319	4.644	
Corrected total	329		

Note: Data were analyzed using procedures for a single factor completely randomized design with subsampling.

[a] Significant F-value at the 0.001 level.

[b] Obtained from the nested term replicate within genotype [rep(Genotype)].

Data obtained from Compton, M.E. and D.J. Gray. 1993. *J. Amer. Soc. Hort. Sci.* 118:151–157. With permission.

Subsampling experiments promote efficient use of space and materials, and measures variability among observational units. They are most efficient when there is much variation in response among explants and little outside variation. The design is not efficient when there is little variation among explants or when there is a high degree of variability from some other identifiable source. This is because it reduces the DF for experimental error, which means that a higher F-value is required to detect significant differences among treatments. If there is little variation among explants, a CRD without subsampling should be used. If variation from an identifiable outside source exists, a RCBD with subsampling should be used.

Randomized complete block design (RCBD)

The CRD is only efficient when experimental units are homogeneous (Lentner and Bishop, 1986). This is because all unrecognized variation, regardless of source, is lumped into the experimental error term. Heterogeneity among experimental units results in a high MSE and reduces statistical precision. In experiments with high heterogeneity, it is best to group experimental units into homogeneous units or blocks. Grouping experimental units into uniform blocks provides a better estimate of treatment effects and improves statistical precision.

In RCBDs, treatments are grouped into blocks that contain at least one replicate from each treatment. Experimental units are randomized within blocks, each employing a separate randomization scheme. This minimizes variability within a block while maximizing variability among blocks. For this reason, experimental units within each block should be as uniform as possible. The number of treatments should be limited because variability within each block increases as the number of treatments increase.

The tomato shoot regeneration study described above can be adapted to illustrate a RCBD by arranging treatments into 24 blocks, each containing one test tube with a single pedicel explant from each cultivar. The ANOVA for a RCBD differs from that of a CRD in that block SS is calculated in addition to treatment and experimental error SS. Block SS identifies variation controlled by blocking. Treatment SS represents variability related to the imposed treatment. Experimental error SS measures variability among experimental units. Blocking improves statistical precision by removing variability controlled by blocking from the experimental error. In this example, the F-value for cultivar (5.99) was obtained by dividing the cultivar MS (7.937) by the MSE (1.326) and significance determined using 10 and 230 DF (Table 7.3). A means separation procedure must be performed to determine which treatment means differ.

Table 7.3 Analysis of Variance Summary Table for the Number of Shoots per Explant from Tomato Shoot Regeneration Study

Source	DF	Mean square	F-value
Block	23	1.639	
Cultivar	10	7.937	5.99[a]
Error	230	1.326	
Corrected total	263		

Note: Data were analyzed using procedures for a RCBD.

[a] Significant F-value at the 0.001 level.

From Compton, M.E. 1994. *Plant Cell Tissue Organ Cult.* 37:217–242. With permission.

A RCBD is useful when some variation is caused by something other than the imposed treatment(s). If little difference among blocks occurs, a RCBD will not improve statistical precision and a CRD should be used. Kempthorne (1973) provides a formula that determines if blocking has improved precision. In contrast, if there is much variation among explants, a RCBD with subsampling should be used (see review by Compton, 1994).

Split-plot design

This design is typically used when two treatment factors are superimposed on each other. One treatment is assigned to "main plots" that contain all levels of the second factor

(Lentner and Bishop, 1986). Main plot treatments may be randomized in any of the experimental designs described above. Treatments assigned to subplots are randomized within each main plot. Each subplot is randomized differently. The advantage of employing a split-plot design is that it identifies experimental error associated with each treatment factor. Treatments with large experimental error rates are assigned to main plots and those with reduced experimental error to subplots. Dividing experimental error into main plots and subplots improves statistical precision. The split-plot design is most efficient when treatment factors have different experimental error rates. Statistical precision is lost when there is no difference in experimental error between main plots and subplots. For a more detailed discussion on the use of split-plot designs in plant tissue culture see the review by Compton (1994).

Factorial experiments

In complete factorial experiments, treatments are arranged such that all possible combinations of two or more factors are examined simultaneously. Factorial experiments are useful because they allow the scientist to examine treatment interdependencies and are more powerful than multiple single-factor experiments. Any of the aforementioned treatment randomization schemes can be used for factorial treatment designs. A factorial treatment design was employed in an experiment examining the effects of benzyladenine (BA) and indole-3-butyric acid (IBA) on tobacco (*Nicotiana tabacum* L.) callus growth (Compton, 1986). Treatments were three levels (0, 1, and 10 μM) of BA and IBA arranged in a 3 × 3 complete factorial. Pieces of tobacco callus (1 cm^3) were transferred to test tubes that contained 15 ml of test medium. There were six replicate test tubes per treatment that were randomized in a CRD. Callus height, diameter, and dry weight were recorded 1 month after culture initiation.

In the analysis of factorial experiments, SS are generated for each treatment, treatment interaction, and for experimental error (Table 7.4). F-values for BA, IBA, and their interaction were obtained by dividing the MS for each by the MSE. Significance for the main effects (BA and IBA) was determined using 2 and 98 DF, whereas significance for the BA by IBA interaction was tested at 4 and 98 DF. ANOVA indicated that callus growth in the form of dry weight was influenced by BA and IBA simultaneously. This was determined by the significant BA by IBA interaction, indicating that BA affected callus dry

Table 7.4 Analysis of Variance Summary Table for Tobacco Callus Growth Study

		Callus dry weight (mg)	
Source	DF	Mean square	F-value
BA	2	9.367	5.65[a]
IBA	2	12.209	7.36[a]
BA × IBA	4	4.361	2.63[a]
Error	98	1.761	
Corrected total	106		

Note: Data were analyzed using procedures for a CRD with treatments arranged in a 3 × 3 complete factorial.

[a] Significant F-value at the 0.05 (*), 0.01 (**), or 0.001 (***) level, respectively.

Data obtained from Compton, M.E. 1986. M.S. thesis. Southern Illinois University, Carbondale, IL.

weight differently at each level of IBA and IBA affected callus dry weight differently at each BA concentration. Interaction means for this type of data are presented in a line graph with treatment differences estimated using SE (Figure 7.1). When treatment interactions are significant, the influence of each treatment separately should not be discussed (Lentner and Bishop, 1986; Zar, 1984).

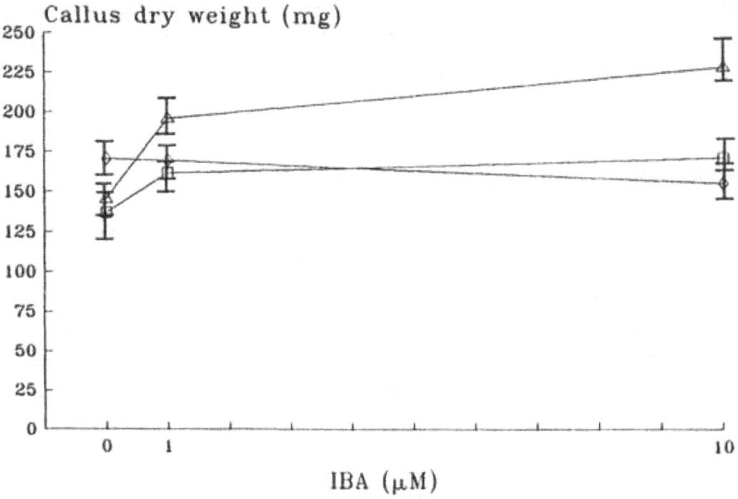

Figure 7.1 Graph of tobacco callus dry weight as affected by various combinations of BA and IBA. Graph demonstrates the presentation of interaction means when treatments were arranged in a 3 × 3 complete factorial. Vertical lines represent standard errors. □ = 0 BA; Δ = 1 μM BA; ◊ = 10 μM BA. (From Compton, M.E. 1994. *Plant Cell Tissue Organ Cult.* 37:217–242. With permission.)

Factorial experiments can become prohibitive with relatively few factors. For example, the above 3 × 3 factorial experiment had 9 treatments. If a third factor with three levels was added the number of treatment combinations would increase to 27. Add to that 10 replicates and there would be 270 observational units. Factorial experiments require a more complicated analysis model and more care in performing mean separation analyses, especially when there are missing data. Interpretation of higher-order interactions may be difficult in any factorial experiment. However, the information to be gained often compensates for the extra work.

Methods for comparing treatment means

Once a significant F-test with a small P-value is obtained, the scientist must elucidate specific differences among treatments. This is not a problem when there are only two treatments because treatment means are simply presented in a table with the associated F-test significance level. When there are more than two treatments, a post-ANOVA analysis is necessary. The easiest way to compare treatment means is to rank them in ascending or descending order and pick the best treatment. The problem with this method is that natural variation that occurs within a treatment is not considered. There are many mean separation procedures that account for within treatment variation. These are standard error of the mean (SE), multiple comparison and multiple range tests, and regression analysis.

Standard error of the mean

SE is obtained by dividing the sample standard deviation by the square root of the number observations for that treatment (Zar, 1984). Hence, the size of SE depends on the magnitude of the data. When using SE for mean separation purposes, treatment means are ranked along with their respective SE and the difference between paired values calculated. Treatments are declared different if the collective values for the paired treatments do not overlap. Using SE provides an indication of variability within treatments and allows the reader to make his/her own comparisons. One disadvantage of using SE for mean separation purposes is that it is considered conservative and may only detect differences between means ranked far apart.

Multiple comparison and multiple range tests

Multiple comparison or multiple range tests should only be used when treatments are unrelated, e.g., different growth regulators, genotypes, etc. (Lentner and Bishop, 1986). The terms "multiple comparison test" and "multiple range test" are used interchangeably. However, the two differ in their method of making comparisons. Multiple comparison tests use the same critical value to compare adjacent and nonadjacent means. Examples of multiple comparison tests are Bonferoni, Fisher's least significant difference (LSD), Scheffe's, Tukey's Honestly Significant Difference Test (Tukey's HSD), and Waller-Duncan K-ratio T test (Waller-Duncan). Multiple range tests employ different critical values to compare adjacent and nonadjacent means, which protects against committing a type I error (instance where treatments are declared different that truly are the same). Examples of multiple range tests include Duncan's New Multiple Range Test (DNMRT), Ryan-Einot-Gabriel-Welsh Multiple F-test (REGWF), Ryan-Einot-Gabriel-Welsh Multiple Range Test (REGWQ), and Student-Newman-Kuels (SNK).

Multiple range or multiple comparison tests are best used only after obtaining a significant ANOVA (Lentner and Bishop, 1986). Most mean separation procedures may be used in situations when a significant ANOVA has not been obtained if specific comparisons were planned prior to data analysis. However, there is a higher probability of creating a type I error, especially when comparing nonadjacent means. DNMRT, REGWQ, SNK, Tukey's HSD, and Waller-Duncan are the most commonly recommended mean separation procedures for unrelated treatments (Lentner and Bishop, 1986; SAS Institute, 1988).

Simple linear regression

Regression, or trend, analysis is used in experiments with quantitative treatments where the primary objective is to develop a model that quantifies the relationship between response variables and treatment levels (Mize and Chun, 1988; Kleinbaum and Kupper, 1978). The model is used to identify which trend (e.g., linear, quadratic, or cubic) best describes the data and may be used for predicting optimum treatment levels. It is assumed that there is a continuous response of the observational units to the imposed treatments when using quantitative levels. Trend analysis is most appropriate when there are three or more evenly spaced, equally replicated treatments. A forward step procedure is most often used to identify the best model. The simplest model (linear) is tested first and more complicated models (quadratic and cubic) tested after rejection of lower-order models. A backward step procedure may be used but is less common. Lack-of-fit (LOF), T, and r-square (r^2) values are used to determine the best model. A significant T-value and nonsignificant LOF signify that the model accurately describes the relationship between treatment and response variables. A significant LOF value indicates that the model does not fit the data and other models should be tested. The r^2 value estimates variation described by the model. A high r^2 suggests that much of the variability is described by the model. If a low r^2 is obtained it may be advantageous to check other models. The most appropriate model generally has a significant T-value, nonsignificant LOF, and high r^2. It is important that extrapolation be limited to within treatment boundaries as the model equation may not accurately describe treatment effects beyond the tested parameters.

Regression analysis was used to determine the effect of sucrose on the growth of grape (*Vitis vinifera* L. 'Thompson Seedless') embryogenic cultures (Compton and Gray, 1995). Embryogenic cells and somatic embryos (heart and globular stage) were incubated on medium containing 60, 90, 120, 150, or 180 g/l sucrose for 3 months. Embryogenic cultures were subcultured monthly to fresh medium of the same composition. At the end of the experiment embryogenic cultures were dried in an oven at 70°C for 72 h and dry weight was determined.

In this example, ANOVA determined that the sucrose concentration in the medium influenced the growth of grape embryogenic cultures. Regression and LOF analyses indicated that the data best fit the cubic model as indicated by a significant T-value and non-significant LOF value ($-1774.06 + 57.92X - 0.4837X^2 + 0.001X^3$ [Table 7.5]). Optimum growth was obtained when embryogenic cultures were incubated on medium with 90 to 120 g/l sucrose (Figure 7.2). The high (0.8815) r^2 obtained for this model indicates that much of the variation in the data was explained by the model. The terms "high" and "low" r^2 are relative to the type of data analyzed. High r^2 values for biological data may range between 0.50 and 0.90, whereas a low r^2 for nonbiological data may be 0.90 (Kleinbaum and Kupper, 1978).

Table 7.5 Analysis of Variance Summary Table for Trend and Lack of Fit Analyses on Dry Weight of Grape Embryogenic Cultures Grown on Medium with Varying Sucrose Concentrations

MODEL 3 (cubic)		Analysis of variance			
Variable	DF	Parameter estimates	Standard error	T for HO: Parameter = 0	Significance
Intercept	1	**−1774.057**	231.967	−7.648	0.0001
Sucrose	1	**57.928**	6.675	8.677	0.0001
Sucrose^2	1	**−0.483**	0.059	−8.166	0.0001
Sucrose^3	1	**0.001**	0.0001	7.508	0.0001
	Root MSE	37.225	**R–square**	**0.8815**	

Lack of fit test

Source	DF	Sum of squares	Mean square	F–value	Significance
Sucrose	0	—	—	—	—
Sucrose^2	0	—	—	—	—
Sucrose^3	0	—	—	—	—
LOF	1	3860.17	3860.17	3.07	0.0956

Data from Compton, M.E. and D.J. Gray. 1996. *Vitis* 35:1–6. With permission.

Categorical data

Categorical data are often based on counts of individuals that can be placed into groups. This occurs when data consist of yes/no values (e.g., yes the explant responded or no the explant failed to respond) or when explant response is placed into a category (e.g., explants

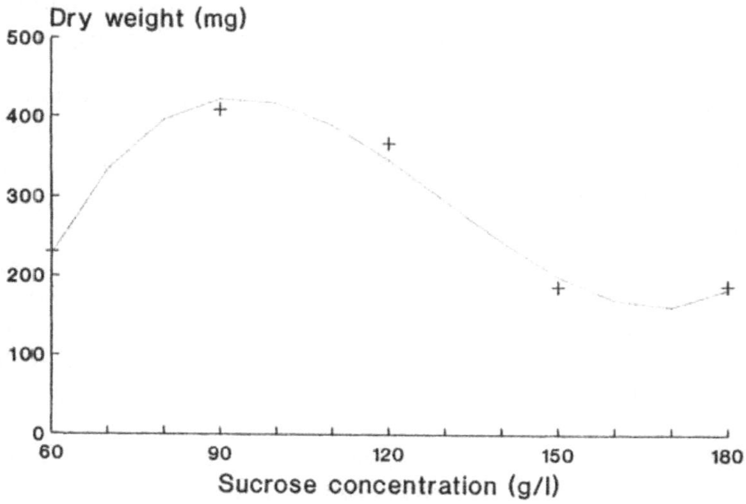

Figure 7.2 Effect of sucrose concentration on dry weight of grape somatic embryogenic cultures. Dotted line represents the predicted regression equation. Treatment means are represented by +. (From Trigiano, R. N. and Gray, D. J., Eds., *Plant Tissue Culture Concepts and Laboratory Exercises, First Edition*, CRC Press LLC, Boca Raton, FL, 1996.)

produced roots, shoots, somatic embryos, or did not respond). Categorical data are not continuous and not normally distributed and, therefore, cannot be analyzed with the procedures mentioned above. Chi-square or Maximum Likelihood are used to analyze categorical data (Lentner and Bishop, 1986; Zar, 1984).

An example of categorical data is an experiment where the investigators were interested in identifying the best dehydration treatment for maize (*Zea mays* L. B73 × A188-19 hybrid) somatic embryos (Compton et al., 1992). In this experiment, somatic embryos were reared on a medium without ABA (medium 1), transferred to medium with 0.1 µM ABA (medium 2) for 2 weeks, or transferred to medium 2 for 2 weeks before transfer to medium with 60 g/l sucrose and no ABA (medium 3) prior to controlled relative humidity dehydration (CRHD) at 70 or 90% relative humidity for 2 weeks. Embryo survival was determined 2 weeks after transfer to germination medium by the ability of somatic embryos to produce chlorophyll (green), roots, coleoptiles, or leaves. Control embryos obtained from each of the above media were transferred directly to germination medium without CRHD. Surviving embryos were assigned a 1, whereas those that failed to survive were assigned 0 (no response).

According to ANOVA, culture medium and dehydration treatment simultaneously influenced the ability of maize somatic embryos to survive dehydration as determined by the significant treatment by medium interaction contrast. The ability of maize somatic embryos to survive dehydration depended on the pretreatment medium and RH level used during CRHD. SE was used to compare treatment means (Table 7.6). Categorical data can be transformed using arc sine prior to ANOVA and converted back to the original scale for demonstration in tables or graphs (Compton, 1994).

Conclusions

This chapter has outlined the steps of experimentation and many of the procedures used in the analysis of plant cell culture data. It is important to remember that any well-conceived

Table 7.6 Percent of Maize Somatic Embryos that Produced Chlorophyll (Greening), Roots, Coleoptiles, and Leaves after Growth on Three Different Embryo Maturation Media Followed by CRHD for 2 Weeks

RH	Medium	No. of embryos	Percent embryos with: Greening			Roots			Coleoptiles		
Control[a]	1[b]	43	58.1	±	7.5	25.6	±	6.7	25.6	±	6.7
	2	147	81.0	±	3.2	32.7	±	3.9	31.3	±	3.8
	3	100	72.0	±	4.5	56.0	±	5.0	31.0	±	4.6
90%	1	43	0.0	±	0.0	0.0	±	0.0	0.0	±	0.0
	2	71	0.0	±	0.0	0.0	±	0.0	0.0	±	0.0
	3	119	0.0	±	0.0	0.0	±	0.0	0.0	±	0.0
70%	1	38	0.0	±	0.0	0.0	±	0.0	0.0	±	0.0
	2	72	0.0	±	0.0	0.0	±	0.0	0.0	±	0.0
	3	82	12.2	±	3.6	14.6	±	3.9	9.8	±	3.3

Note: ± values represent the standard error of the mean.

[a] Control embryos were transferred directly to germination medium without CRHD.

[b] Medium 1 = N6 plus 4.5 μM 2,4-D, 100 mg/l casamino acids, 25 mM proline, and 2 g/l Phytagel; Medium 2 = 2 weeks on MS plus 0.45 μM 2,4-D, 0.1 μM ABA, and 8 g/l agarose; Medium 3 = 2 weeks on medium 2 and 2 weeks on N6 plus 60 g/l sucrose and 2 g/l Phytagel.

Data from Compton et al., 1992. *In Vitro Cell. Dev. Biol. Plant.* 28 p.: 197–201, Copyright © 1992 by the Tissue Culture Association. With permission.

research project should be planned carefully on paper before the experiment is conducted. This helps to ensure that the experiment will answer the desired questions. Planning is probably the most important step in scientific research. Experimental design and analysis provide an objective, nonbias way to accurately evaluate treatment differences. Used properly, statistics is a valuable tool to the scientific researcher and it allows the investigator to declare experimental results with confidence.

Literature cited

Compton, M.E. 1986. Stage I Exudation, Plant Growth Regulators, and Phenolic Compounds Alter the Growth and Development of Tobacco Callus. M.S. thesis, Department of Plant and Soil Science, Southern Illinois University, Carbondale.

Compton, M.E. 1994. Statistical methods suitable for the analysis of plant tissue culture data. *Plant Cell Tissue Organ Cult.* 37:217–242.

Compton, M.E. and D.J. Gray. 1993. Shoot organogenesis and plant regeneration from cotyledons of diploid, triploid and tetraploid watermelon. *J. Amer. Soc. Hort. Sci.* 118:151–157.

Compton, M.E. and D.J. Gray. 1996. Effects of sucrose and methylglyoxal Bis-(guanylhydrazone) on controlling grape somatic embryogenesis. *Vitis* 35:1–6.

Compton, M.E. and R.E. Veilleux. 1991. Shoot, root and flower morphogenesis on tomato influorescence explants. *Plant Cell Tissue Organ Cult.* 24:223–231.

Compton, M.E., C.M. Benton, D.J. Gray, and D.D. Songstad. 1992. Plant recovery from maize somatic embryos subjected to controlled relative humidity dehydration. *In Vitro Cell. Dev. Biol.* 28P:197–201.

Kempthorne, O. 1973. *The Design and Analysis of Experiments.* Robert E. Krieger Publishing Co., Malabar, FL.

Kleinbaum, D.G. and L.L. Kupper. 1978. *Applied Regression Analysis and Other Multivariable Methods.* PWS Publishers, Boston.

Lentner, M. and T. Bishop. 1986. *Experimental Design and Analysis.* Valley Book Company, Blacksburg, VA.

Little, T.M. and F.J. Hills. 1978. *Agricultural Experimentation: Design and Analysis*. John Wiley & Sons, New York.

Mize, C.W. and Y.W. Chun. 1988. Analyzing treatment means in plant tissue culture research. *Plant Cell Tissue Organ Cult*. 13:201–217.

SAS Institute, Inc. SAS/STAT user's guide. 1988. Release 6.03. SAS Institute, Inc., Cary, NC.

Zar, J.H. 1984. *Biostatistical Analysis. Second ed*. Prentice-Hall, Englewood Cliffs, NJ.

Part IV

Propagation techniques

chapter eight

Propagation from preexisting meristems

Michael E. Kane

Micropropagation is defined as the true-to-type propagation of selected genotypes using in vitro culture techniques. Four basic methods are used to propagate plants in vitro. Depending on the species and cultural conditions, in vitro propagation can be achieved by: (1) enhanced axillary shoot proliferation (shoot culture); (2) node culture; (3) de novo formation of adventitious shoots through shoot organogenesis (Chapter fourteen); or (4) nonzygotic embryogenesis (Chapter nineteen). Currently, the most frequently used micropropagation method for commercial production utilizes enhanced axillary shoot proliferation from cultured meristems. This method provides genetic stability and is easily attainable for many plant species. Consequently, the shoot culture method has played an important role in development of a worldwide industry that produces more than 250 million plants yearly. Besides propagation, shoot meristems are cultured in vitro for two other purposes: (1) production of pathogen eradicated plants; and (2) preservation of pathogen eradicated germplasm (see Chapter thirty-five). Concepts related to propagation by shoot and node culture will be discussed in this chapter.

Shoot apical meristems

It is important to briefly review the general structure of shoot meristems. Shoot growth in mature plants is restricted to specialized regions which exhibit little differentiation and in which the cells retain the embryonic capacity for unlimited division. These regions, called apical meristems, are located in the apices of the main and lateral buds of the plant. Cells derived from these apical meristems subsequently undergo differentiation to form the mature tissues of the plant body. Due to their highly organized structure, apical meristems tend to be genetically stable.

There exist significant differences in the shape and size of shoot apices between different taxonomic plant groups (Fahn, 1974). A typical dicotyledon shoot apical meristem consists of a layered dome of actively dividing cells located at the extreme tip of a shoot and measures about 0.1 to 0.2 mm in diameter and 0.2 to 0.3 mm in length. The apical meristem has no connection to the vascular system of the stem. Below the apical meristem, localized areas of cell division and elongation represent sites of newly developing leaf primordia (Figure 8.1). Lateral buds, each containing an apical meristem, develop within the axils of the subtending leaves. In the intact plant, outgrowth of the

0-8493-2029-1/00/$0.00+$.50

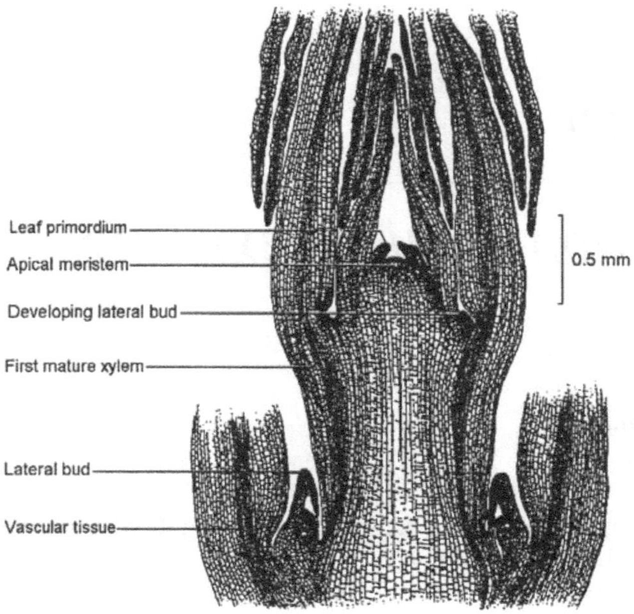

Figure 8.1 Diagrammatic representation of a dicotyledonous shoot tip. The shoot tip comprises the apical meristem, subtending leaf primordia and lateral buds. (From Trigiano, R. N. and Gray, D. J., Eds., *Plant Tissue Culture Concepts and Laboratory Exercises, First Edition*, CRC Press LLC, Boca Raton, FL, 1996.)

lateral buds is usually inhibited by apical dominance of the terminal shoot tip. Organized shoot growth from the apical meristem of plants is potentially unlimited and is said to be indeterminate. However, shoot apical meristems may become committed to the formation of determinate organs such as flowers.

In vitro culture of shoot meristems

The recognized potential for unlimited shoot growth prompted early, but largely unsuccessful, attempts to aseptically culture isolated shoot meristems in the 1920s. By the middle 1940s, sustained growth and maintenance of organization of cultured shoot meristems through repeated subculture were achieved for several species. Ball (1946), however, provided the first detailed procedure for the isolation and production of plants from cultured shoot meristem tips and the successful transfer of rooted plantlets into soil. Ball is often called the Father of Micropropagation because his shoot tip culture procedure is the one most commonly used by commercial micropropagation laboratories today. Although these studies demonstrated the feasibility of regenerating shoots from cultured shoot tips, the procedures typically yielded unbranched shoots.

Several important findings facilitated application of in vitro culture techniques for large-scale clonal propagation from meristems. The discovery that virus-eradicated plants could be generated from cultured meristems led to the widespread application of the procedure for routine fungal and bacterial pathogen eradication as well (Morel and Martin, 1952; Styer and Chin, 1984). Demonstration of rapid production of orchids from cultured shoot tips supported the possibility of rapid clonal propagation in other crops (Morel, 1960; 1965). It should be noted that in vitro propagation in many orchids does not occur via axillary shoot proliferation, but rather the cultured meristems become disorganized and form spheroid protocorm-like bodies that are actually nonzygotic embryos.

The final discovery was the elucidation of the role of cytokinins in the inhibition of apical dominance (Wickson and Thimann, 1958). This finding was eventually applied to enhance axillary shoot production in vitro. Application of this method was expedited by development of improved culture media which supported the propagation of a wide diversity of plant species (Murashige and Skoog, 1962; Lloyd and McCown, 1981).

Meristem and meristem tip culture

Although not directly used for propagation, meristem and meristem tip culture will be briefly described since these procedures are used to generate pathogen-eradicated shoots that subsequently serve as propagules for in vitro propagation. Culture of the apical meristematic dome alone (Figure 8.2) from either terminal or lateral buds, for the purpose of pathogen elimination, is termed meristem culture. In reality, true meristem culture is rarely used because isolated apical meristems of many species exhibit both low survival rates and increased chance of genetic variability following callus formation and indirect shoot organogenesis.

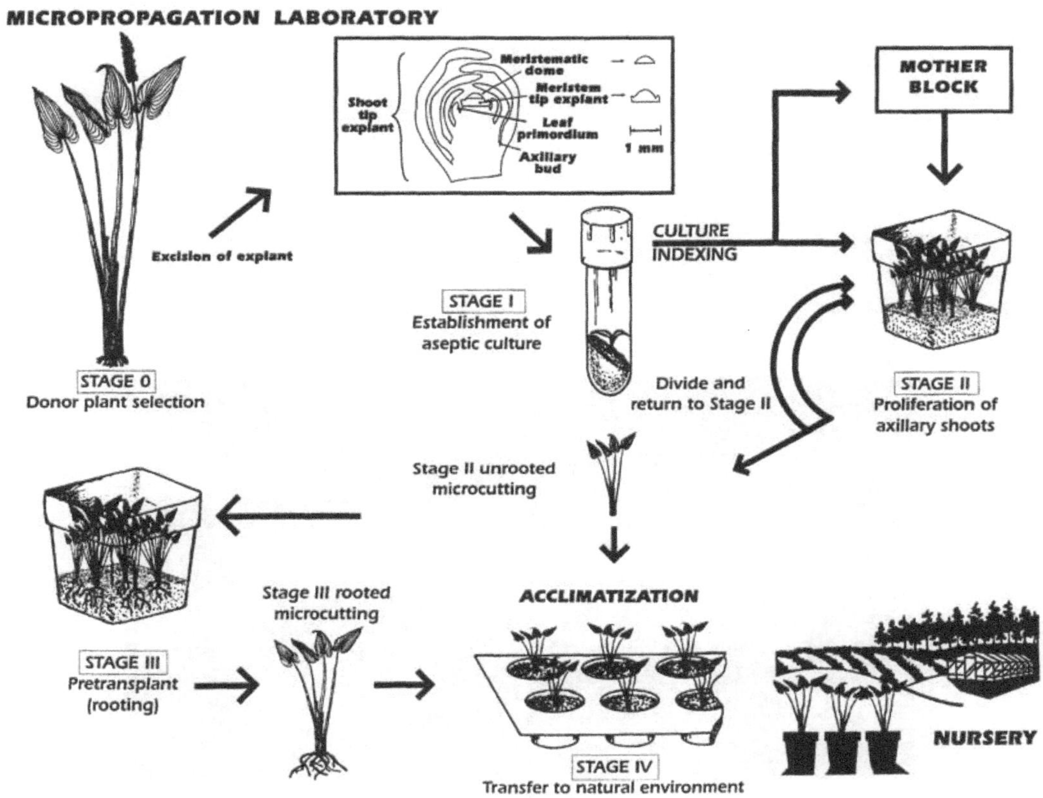

Figure 8.2 Micropropagation stages for production by shoot culture. (From Trigiano, R. N. and Gray, D. J., Eds., *Plant Tissue Culture Concepts and Laboratory Exercises, First Edition*, CRC Press LLC, Boca Raton, FL, 1996.)

Pathogen elimination can often be accomplished by culture of relatively larger (0.2 to 0.5 mm long) meristem tip explants excised from plants that have undergone thermo- or chemotherapy. The meristem tip is comprised of the apical meristem plus one or two subtending leaf primordia (Figure 8.2). This procedure is therefore termed *meristem tip culture*. Caution should be taken when interpreting much of the early published literature of successful "meristem" culture since, in many instances, meristem tip or even larger shoot tip explants were actually used. The term "meristemming" commonly used in the orchid literature is equally ambiguous.

Shoot and node culture

Although not the most efficient procedure, propagation from axillary shoots has proved to be a reliable method for the micropropagation of a large number of species (Kurtz et al., 1991). Depending on the species, two methods, shoot and node culture, are used. Both methods rely on the stimulation of axillary shoot growth from lateral buds following disruption of apical dominance of the shoot apex. Shoot culture (shoot tip culture) refers to the in vitro propagation by repeated enhanced formation of axillary shoots from shoot tips or lateral buds cultured on medium supplemented with growth regulators, usually a cytokinin (George, 1993). The axillary shoots produced are either subdivided into shoot tips and nodal segments that serve as secondary explants for further proliferation (see Figure 8.3a) or are treated as microcuttings for rooting.

Figure 8.3 (a) Stage II shoot multiplication is achieved by repeated formation of axillary shoot clusters from explants containing lateral buds (*Aronia arbutifolia* L.). Depending on the species, individual microcuttings or shoot clusters may be rooted and acclimatized ex vitro. Scale bar = 1 cm. (b) For maximum survival Stage III rooting may be required prior to acclimatization to ex vitro conditions. (c) Rooted and acclimatized Stage IV plantlets. (From Trigiano, R. N. and Gray, D. J., Eds., *Plant Tissue Culture Concepts and Laboratory Exercises, First Edition*, CRC Press LLC, Boca Raton, FL, 1996.)

When either verified pathogen-free stock plants are used or when pathogen elimination is not a concern, relatively larger (1 to 20 mm long) shoot tip or lateral bud primary explants (Figure 8.2) can be used for culture establishment and subsequent shoot culture. Advantages of using larger shoot tips include greater survival, more rapid growth responses, and the presence of more axillary buds. However, these larger explants are more difficult to completely surface sterilize and can potentially harbor undetected latent systemic microbial infection. Compared to other micropropagation methods, shoot cultures: (1) provide reliable rates and consistency of multiplication following culture stabilization; (2) are less susceptible to genetic variation; (3) may provide for clonal propagation of periclinal chimeras.

Node culture, a simplified form of shoot culture, is another method for production from preexisting meristems. Numerous plants such as potato (*Solanum tuberosum* L.) do not respond well to the cytokinin stimulation of axillary shoot proliferation observed in the micropropagation of many crops. Axillary shoot growth is promoted by the culture of either intact shoots (from meristem tip cultures) positioned horizontally on the medium (in vitro layering) or single or multiple node segments. Typically, single elongated unbranched shoots, comprised of multiple nodes, are rapidly produced (see Figure 11.2). These shoots (microcuttings) are either rooted or acclimatized to ex vitro conditions or repeatedly subdivided into nodal cuttings to initiate additional cultures. In some species, modified shoot storage organs such as miniaturized tubers or corms may develop from axillary shoots under inductive culture conditions. Although node culture is the simplest method, it is associated with the least genetic variation.

Micropropagation stages

Murashige (1974) originally described three basic stages (I to III) for successful micropropagation. Recognition of the contamination problems often associated with inoculation of primary explants prompted Debergh and Maene (1981) to include a Stage 0. This stage described specific cultural practices, which maintained the hygiene of stock plants and decreased the contamination frequency during initial establishment of primary explants. As a result of our increased information base, currently it is now agreed that there are five stages (Stage 0 to IV) critical to successful micropropagation. These stages not only describe the procedural steps in the micropropagation process, but also represent points at which the cultural environment is altered (Miller and Murashige, 1976). This system has been adopted by most commercial and research laboratories as it simplifies production scheduling, accounting, and cost analysis (Kurtz et al., 1991). Requirement for completion of each stage will depend on the plant material and specific method used. Diagrammatic representation of the micropropagation stages for propagation by shoot culture is provided in Figure 8.2.

Stage 0: Donor plant selection and preparation

Explant quality and subsequent responsiveness in vitro are significantly influenced by the phytosanitary and physiological conditions of the donor plant (Debergh and Maene, 1981; Read, 1988). Prior to culture establishment, careful attention is given to the selection and maintenance of the stock plants used as the source of explants. Stock plants are maintained in clean, controlled conditions that allow active growth but reduce the probability of disease. Maintenance of specific pathogen-tested stock plants under conditions of relatively lower humidity, and use of drip irrigation and antibiotic sprays have proved effective in reducing the contamination potential of candidate explants. Such practices also allow excision of relatively larger and more responsive explants often without increased risks of contamination.

Numerous practices are also employed to increase explant responsiveness by modifying the physiological status of the stock plant. These practices include the following: (1) trimming to stimulate lateral shoot growth; (2) pretreatment sprays containing cytokinins or gibberellic acid; and (3) use of forcing solutions containing 2% sucrose and 200 mg/L 8-hydroxyquinoline citrate for induction of bud break and delivery of growth regulators to target explant tissues (Read, 1988). Currently, information on the effects of other factors such as stock plant nutrition, light, and temperature treatments on the subsequent in vitro performance of meristem explants is lacking.

Stage I: Establishment of aseptic cultures

Initiation and aseptic establishment of pathogen eradicated and responsive terminal or lateral shoot meristems explants are the goal of this stage. The primary explants obtained from the stock plants may consist of surface-disinfested shoot apical meristems or meristem tips for pathogen elimination or shoot tips from terminal or lateral buds (Figure 8.2).

The presence of microbial contaminants can adversely affect shoot survival, growth, and multiplication rate. Bacteria and fungal contaminants often persist within cultured tissues that visually appear contaminant free. Consequently, it is essential that Stage I cultures be indexed (screened) for the presence of internal microbial contaminants prior to serving as sources of shoot tip and nodal explants for Stage II multiplication.

The following factors may affect successful Stage I establishment of meristem explants: (1) explantation time; (2) position of the explant on the stem; (3) explant size; and (4) polyphenol oxidation. Time of explantation can significantly affect explant response in vitro. In deciduous woody perennials, shoot tip explants collected at various times during the spring growth flush may vary in their ability for shoot proliferation. Shoot tips collected during or at the end of the period of most rapid shoot elongation exhibited weak proliferation potential. Explants collected before or after this period are capable of strong shoot proliferation in vitro (Brand, 1993). Conversely, the best results are obtained with herbaceous perennials that form storage organs, such as tubers or corms, when explants are excised at the end of dormancy and after sprouting.

Explants also exhibit different capacities for establishment in vitro depending on their location on the donor plant. For example, survival and growth of terminal bud explants are typically greater than lateral bud explants. Often similar lateral meristem explants from the top and bottom of a single shoot may respond differently in vitro. In woody plants exhibiting phasic development, juvenile explants typically are often more responsive than those obtained from the often nonresponsive mature tissues of the same plant. Sources of juvenile explants include the following: (1) root suckers; (2) basal parts of mature trees; (3) stump sprouts; and (4) lateral shoots produced on heavily pruned plants.

The excision of primary explants often promotes the release of polyphenols and stimulates polyphenol oxidase activity within the damaged tissues. Oxidized polyphenol products often blacken the explant tissue and medium. Accumulation of these products can eventually kill the explants. Procedures used to decrease tissue browning include (1) use of liquid medium with frequent transfer; (2) adding antioxidants such as ascorbic acid or polyvinylpyrrolidone (PVP); and (3) culture in reduced light or darkness.

There clearly is no one universal culture medium for establishment of all species. However, modifications to the Murashige and Skoog (Murashige and Skoog, 1962) basal medium formulation are most frequently used (see Chapter three). Cytokinins or auxins are most frequently added to Stage I media to enhance explant survival and shoot development (Hu and Wang, 1983). The types and levels of growth regulators used in Stage I media are dependent on the species, genotype, and explant size.

Knowledge of the specific sites of hormone biosynthesis in intact plants provides insight into the relationship between explant size and dependence on exogenous growth regulators in the medium. Endogenous cytokinins and auxins are synthesized primarily in root tips and leaf primordia, respectively. Consequently, smaller explants, especially cultured apical meristem domes, exhibit greater dependence on medium supplementation with exogenous cytokinin and auxin for maximum shoot survival and development (Shabde and Murashige, 1977). Larger shoot tip explants usually do not require the addition of auxin in Stage I medium for establishment. Rapid adventitious rooting of shoot tip explants often provides a primary endogenous cytokinin source. Most Stage I media are agar-solidified and supplemented with at least a cytokinin (Wang and Charles, 1991). The most frequently used cytokinins are N^6-benzyladenine (BA), kinetin (Kin), and N^6-(2-isopentenyl)-adenine (2-iP). Due to its low cost and high effectiveness, the cytokinin BA is most widely used (Gaspar et al., 1996). Substituted urea compounds, such as thidiazuron, exhibit strong cytokinin-like activity and have been used to facilitate the shoot culture of recalcitrant woody species (Huetteman and Preece, 1993).

Many types of auxins are used. The naturally occurring auxin indole-3-acetic acid (IAA) is the least active, whereas the stronger and more stable compounds α-naphthalene acetic acid (NAA), a synthetic auxin, and indole-3-butyric acid (IBA), a naturally occurring auxin, are more often used. Stage I medium PGR levels and combinations that promote explant establishment and shoot growth but limit formation of callus and adventitious shoot formation are selected.

A commonly held misconception is that primary explants exhibit immediate and predictable growth responses following inoculation. For many species, particularly herbaceous and woody perennials, consistency in growth rate and shoot multiplication is achieved only after multiple subculture on Stage I medium. Physiological stabilization may require from 3 to 24 months and four to six subcultures. Failure to allow culture stabilization, before transfer to a Stage II medium containing higher cytokinin levels, may result in diminished shoot multiplication rates or production of undesirable basal callus and adventitious shoots. With some species, the time required for stabilization can be reduced by initial culture in liquid medium.

In many commercial laboratories, stabilized cultures, verified as being specific pathogen tested and free of cultivable contaminants, are often maintained on media that limit shoot production to maintain genetic stability. These cultures, called mother blocks, serve as sources of shoot tips or nodal segments for initiation of new Stage II cultures (Figure 8.2).

Stage II: Proliferation of axillary shoots

Stage II propagation is characterized by repeated enhanced formation of axillary shoots from shoot tips or lateral buds cultured on a medium supplemented with a relatively higher cytokinin level to disrupt apical dominance of the shoot tip. A subculture interval of 4 weeks with a three- to eightfold increase in shoot number is common for many crops propagated by shoot culture. Given these multiplication rates, conservatively, more than 4.3×10^7 shoots could be produced yearly from a single starting explant.

Stage II cultures are routinely subdivided into smaller clusters, individual shoot tips, or nodal segments that serve as propagules for further proliferation (Figure 8.3a). Additionally, axillary shoot clusters may be harvested as individual unrooted Stage II microcuttings or multiple shoot clusters for ex vitro rooting and acclimatization (Figure 8.2). Clearly, Stage II represents one of the most costly stages in the production process.

Both source and orientation of explants can affect Stage II axillary shoot proliferation. Subcultures inoculated with explants that had been shoot apices in the previous subculture often exhibit higher multiplication rates than lateral bud explants. Inverting shoot explants in the medium can double or triple the number of axillary shoots produced on vertically oriented explants per culture period in some species.

The number of subcultures possible before initiation of new Stage II cultures from the mother block is required depends on the species or cultivar and its inherent ability to maintain acceptable multiplication rates while exhibiting minimal genetic variation and off-types (Kurtz et al., 1991). Some species can be maintained with monthly subculture from 8 to 48 months in Stage II. In contrast, in some ferns (*Nephrolepsis*), as few as three subcultures may only be possible before the frequency of off-types increases to unacceptable levels.

Increased production of off-types is often attributed to production of adventitious shoots particularly through an intermediary callus stage (Jain and De Klerk, 1998). Stage II cultures, originally regenerating from axillary shoots, often begin producing adventitious shoots at the base of axillary shoot clusters after a number of subcultures on the same medium. These so-called mixed cultures can develop without any morphological differences being apparent. Selecting only terminal shoots of axillary origin for subculture, instead of shoot bases, decreases the frequency of off-types including the segregation of periclinal chimeras.

Selection of Stage II cytokinin type and concentration is made on the basis of shoot multiplication rate, shoot length, and frequency of genetic variation. Although shoot proliferation is enhanced at higher cytokinin concentrations, the shoots produced are usually smaller and may exhibit symptoms of hyperhydricity. Depending on the species, exogenous auxins may or may not enhance cytokinin-induced axillary shoot proliferation (Figure 8.4). Addition of auxin in the medium often mitigates the inhibitory effect of cytokinin on shoot elongation, thus increasing the number of usable shoots of sufficient length for rooting (Figure 8.5). This benefit must be weighed against the increase chance of callus formation.

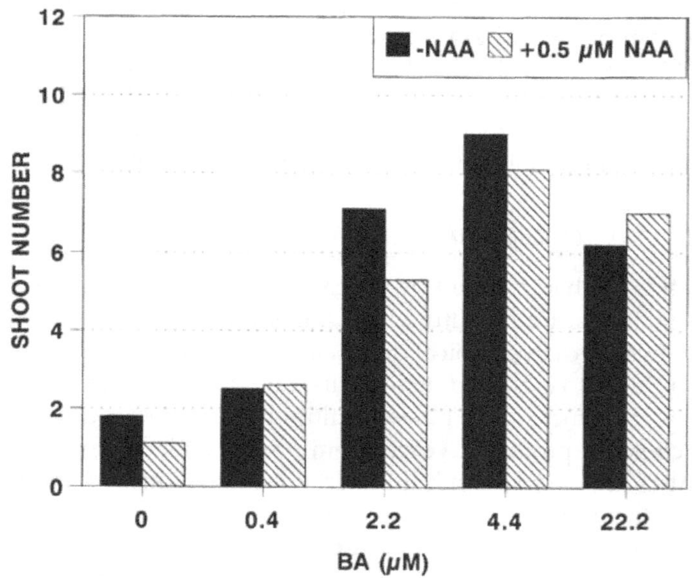

Figure 8.4 Effect of BA concentration on Stage II axillary shoot proliferation produced from two-node explants of *A. arbutifolia* after 28-day culture in presence and absence of 0.5 μM NAA.

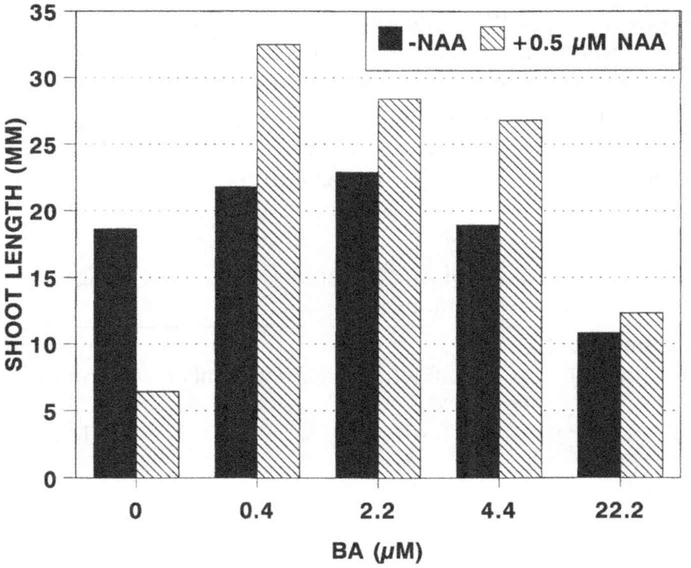

Figure 8.5 Inclusion of auxin may reduce the inhibitory effect of BA on axillary shoot elongation. Data shown for shoots generated from two-node explants of *A. arbutifolia* after 28-day culture in presence and absence of 0.5 μM NAA.

The possibility of adverse carryover effects on the survivability and rooting of plantlets in Stage IV should be evaluated when selecting a Stage II cytokinin. For example, with some tropical foliage plants species, the use of BA in Stage II can significantly reduce Stage IV plantlet survival and rooting to as low as 10% (Griffis et al., 1981). Use of Kin or 2-iP instead of BA yields survival rates in excess of 90%. In some plants the adverse effect of BA on Stage IV survival and rooting has been attributed to production of an inhibitory BA metabolite (Werbrouck et al., 1995). Conceivably, switching the Stage II cytokinin from BA may also eliminate the requirement for Stage III rooting in some species.

Stage III: Pretransplant (rooting)

This step is characterized by preparation of Stage II shoots or shoot clusters for successful transfer to soil. The process may involve: (1) elongation of shoots prior to rooting; (2) rooting of individual shoots or shoot clumps; (3) fulfilling dormancy requirements of storage organs by cold treatment; or (4) prehardening cultures to increase survival. Where possible, commercial laboratories have developed procedures to transfer Stage II microcuttings to soil, thus bypassing Stage III rooting (Figure 8.2).

There are several reasons for eliminating Stage III rooting. Estimated costs for Stage III range from 35 to 75% of the total production costs. This reflects the significant input of labor and supplies required to complete Stage III rooting. Considerable cost saving can be realized if Stage III is eliminated. Furthermore, it is often observed that in vitro-formed root systems are largely nonfunctional and die following transplanting. This results in a delay in transplant growth prior to production of new adventitious roots.

For various reasons, however, it may not always be feasible to transplant Stage II microcuttings directly to soil. Given the aforementioned limitations of Stage III rooting, Debergh and Maene (1981) proposed using Stage III solely to elongate Stage II shoot clusters prior to separation and rooting ex vitro. Elongated shoots may be further pretreated in an aqueous auxin solution prior to transplanting. Usually, Stage III rooting of herbaceous plants can be achieved on medium in the absence of auxins. However, with many woody species, the addition of an auxin (IBA or NAA) in Stage III medium is required to enhance adventitious rooting (Figure 8.3b). Optimum auxin concentration is determined based on percent rooting, root number, and length (Table 8.1). It is critical that the roots not be allowed to elongate to prevent root damage during transplanting. Care must be taken when selecting an auxin. For example, use of NAA for Stage III rooting has been shown to decrease survival rates or suppress posttransplant growth (Connor and Thomas, 1981).

Table 8.1 Addition of Auxin Promotes Stage III Rooting of Shoot Cuttings

Treatment µM IBA (mg/L)	% Rooting	Root number	Root length (mm)
0 (0)	37	2.4	23.2
0.27 (0.05)	43	3.5	18.1
0.54 (0.1)	55	4.1	14.2
2.7 (0.5)	71	5.7	6.5
5.4 (1.0)	84	7.1	4.3

Note: Effects of IBA on Stage III rooting of 10-mm shoots of *Aronia arbutifolia* after 4-week culture are shown.

From Trigiano, R. N. and Gray, D. J., Eds., *Plant Tissue Culture Concepts and Laboratory Exercises, First Edition*, CRC Press LLC, Boca Raton, FL, 1996.

Stage IV: Transfer to natural environment

The ultimate success of shoot or node culture depends on the ability to transfer and re-establish vigorously growing plants from in vitro to greenhouse conditions (Figure 8.3c). This involves acclimatizing or hardening-off plantlets to conditions of significantly lower relative humidity and higher light intensity. Even when acclimatization procedures are carefully followed, poor survival rates are frequently encountered. Micropropagated plants are difficult to transplant for two primary reasons: (1) a heterotrophic mode of nutrition and (2) poor control of water loss.

Plants cultured in vitro in the presence of sucrose and under conditions of limited light and gas exchange exhibit no or extremely reduced capacities for photosynthesis. Reduced photosynthetic activity is associated with low RubPcase activity. During acclimatization, there is a need for plants to rapidly transition from the heterotrophic to photoautotrophic state for survival (Preece and Sutter, 1991). Unfortunately, this transition is not immediate. For example, in cauliflower, no net increase in CO_2 uptake is achieved until 14 days after transplantation. This occurs only following development of new leaves since the leaves produced in vitro in the presence of sucrose never develop photosynthetic competency. Interestingly, before senescencing, these older leaves function as "lifeboats" by supplying stored carbohydrate to the developing and photosynthetically competent new leaves. This is not the rule with all micropropagated plants since the leaves of some species become photosynthetic and persist after acclimatization.

A composite of anatomical and physiological features, characteristic of plants produced in vitro under 100% relative humidity, contributes to the limited capacity of micropropagated plants to regulate water loss immediately following transplanting. These features include reductions in leaf epicuticular wax, poorly differentiated mesophyll, abnormal stomate function, and poor vascular connection between shoots and roots.

To overcome these limitations, plantlets are transplanted into a well-drained "sterile" growing medium and maintained initially at high relative humidity and reduced light (40 to 160 μmol·m^{-2}·s^{-1}) at 20 to 27°C. Relative humidity may be maintained with humidity tents, single tray propagation domes, intermittent misting, or fog systems. However, use of intermittent mist often results in slow plantlet growth following waterlogging of the medium and excessive leaching of nutrients. Transplants are acclimatized by gradually lowering the relative humidity over a 1- to 4-week period. Plants are gradually moved to higher light intensities to promote vigorous growth.

Conclusion

Propagation from preexisting meristems through shoot and node culture is the most reliable and widely used procedure. However, the need for multiple subcultures on different media makes shoot and node culture extremely labor intensive. Total labor costs, typically ranging from 50 to 70% of production costs, limit expansion of the micropropagation industry. Current application of the technology is restricted to high value horticultural crops such as ornamental plants. Expansion of the industry to include production of vegetable, plantation, and forest crops depends on development of more efficient micropropagation systems. Cost-reduction strategies including elimination of production steps and development of reliable automated micropropagation systems will facilitate this expansion (Aitken-Christie et al., 1995).

Literature cited

Aitken-Christie, J., T. Kozai, and S. Takayama. 1995. Automation in tissue culture — general introduction and overview. In: *Automation and Environmental Control in Plant Tissue Culture.* J. Aitken-Christie, T. Kozai, and M.A.L. Smith, Eds., pp.1–18. Kluwer Academic Publishers, Dordrecht.

Ball, E.A. 1946. Development in sterile culture of shoot tips and subjacent regions of *Tropaeolum majus* L. and of *Lupinus albus* L. *Amer. J. Bot.* 33:301–318.

Brand, M. 1993. Initiating cultures of *Halesia* and *Malus*: influence of flushing stage and benzyladenine. *Plant Cell Tissue Organ Cult.* 33:129–132.

Conner, A.J. and M.D. Thomas. 1981. Re-establishing plantlets from tissue culture: a review. *Comb. Proc. Intl. Plant Prop. Soc.* 31:342–357.

Debergh, P.C. and L.J. Maene. 1981. A scheme for commercial propagation of ornamental plants by tissue culture. *Sci. Hort.* 14:335–345.

Debergh, P.C. and P.E. Read. 1991. Micropropagation. In: *Micropropagation Technology and Application.* P.C. Debergh and R.H. Zimmerman, Eds., pp. 1-13. Kluwer Academic Publishers, Boston.

Fahn, A. 1974. *Plant Anatomy.* Pergamon Press, New York.

Gaspar, T., C. Kevers, C. Penel, H. Greppin, D.M. Reid, and T. Thorpe. 1996. Plant hormones and plant growth regulators in plant tissue culture. *In Vitro Cell. Dev. Biol. Plant* 32:272–289.

George, E.F. 1993. *Plant Propagation by Tissue Culture. Part 1. The Technology.* Exegetics, London.

Griffis, J.L., Jr., G. Hennen, and R.P. Oglesby. 1981. Establishing tissue-cultured plant in soil. *Comb. Proc. Intl. Plant Prop. Soc.* 33:618–622.

Hu, C.Y. and P.J. Wang. 1983. Meristem, shoot tip and bud cultures. In: *Handbook of Plant Cell Culture. Vol. 1. Techniques for Propagation and Breeding.* D.F. Evans, W.R. Sharp, P.V. Ammirato, and Y. Yamada, Eds., pp. 177–227. Macmillan Publishing, New York.

Huetteman, C.A. and J.E. Preece. 1993. Thidiazuron: a potent cytokinin for woody plant tissue culture. *Plant Cell Tissue Organ Cult.* 33:105–119.

Jain, S.M. and G.J. De Klerk. 1998. Somaclonal variation in breeding and propagation of ornamental crops. *Plant Tissue Culture Biotech.* 4:63–75.

Kurtz, S., R.D. Hartmann, and I.Y.E. Chu. 1991. Current methods of commercial micropropagation. In: *Scale-up and Automation in Plant Propagation.* Cell Culture and Somatic Cell Genetics of Plants, Vol. 8. I.K. Vasil, Ed., pp. 7–34. Academic Press, San Diego.

Lloyd, G. and B. McCown. 1981. Commercially-feasible micropropagation of Moutain laural, *Kalmia latifolia*, by use of shoot-tip culture. *Intl. Plant Prop. Soc. Proc.* 30:421–427.

Miller, L.R. and T. Murashige. 1976. Tissue culture propagation of tropical foliage plants. *In Vitro* 12:797–813.

Morel, G. 1960. Producing virus-free *Cymbidium*. *Amer. Orchid Soc. Bull.* 29:495–497.

Morel, G. 1965. Clonal propagation of orchids by meristem culture. *Cymbidium Soc. News* 20:3–11.

Morel, G. and Martin, C. 1952. Guerison de Dahlias atteints d'une maladie a virus. *C.R. Acad. Sci. Ser. D.* 235:1324–1325.

Murashige, T. and F. Skoog. 1962. A revised medium for rapid growth and bioassays with tobacco tissue cultures. *Physiol. Plant.* 15:473–497.

Murashige, T. 1974. Plant propagation through tissue culture. *Ann. Rev. Plant Physiol.* 25:135–166.

Preece, J.E. and E.G. Sutter. 1991. Acclimatization of micropropagated plants to greenhouse and field. In: *Micropropagation Technology and Application*. P.C. Debergh and R.H. Zimmerman, Eds., pp. 71–93. Kluwer Academic Publishers, Boston.

Read, P.E. 1988. Stock plants influence micropropagation success. *Acta Hort.* 226:41–52.

Shabde, M. and T. Murashige. 1977. Hormonal requirements of excised *Dianthus caryophyllus* L. shoot apical meristem in vitro. *Amer. J. Bot.* 64:443–448.

Styer, D.J. and C.K. Chin. 1984. Meristem and shoot-tip culture for propagation, pathogen elimination, and germplasm preservation. *Hort. Rev.* 5:221–277.

Wang, P.J. and A. Charles. 1991. Micropropagation through meristem culture. In: *Biotechnology in Agriculture and Forestry, Vol. 17, High-Tech and Micropropagation*. I. Y.P.S. Bajaj, Ed., pp. 32–52. Springer-Verlag, Berlin.

Werbrouck, S.P.O., B. van der Jeugt, W. Dewitte, E. Prinsen, H.A. Van Onckelen, and P.C. Debergh. 1995. The metabolism of benzyladenine in *Spathiphyllum floribundum* 'Schott Petite' in relation to acclimatisation problems. *Plant Cell Rep.* 14:662–665.

Wickson, M. and K.V. Thimann. 1958. The antagonism of auxin and kinetin in apical dominance. *Physiol. Plant.* 11:62–74.

chapter nine

Micropropagation of Syngonium *by shoot culture*

Michael E. Kane

One of the most successful and reliable commercial applications of micropropagation has been the production of tropical foliage plants (house plants) using micropropagation (Henny et al., 1981). In fact, most of the foliage plants purchased in garden shops are produced using micropropagation. Shoot culture is characterized by the establishment of Stage I cultures using isolated surface-disinfested meristem tips, shoot tips, or lateral buds as primary explants (Chapter eight). Shoots that develop from these explants usually are first screened (indexed — see Chapter forty-three) for microbial contamination. Aseptic cultures, once physiologically adapted to culture, are then subdivided into nodal or shoot tip secondary explants and transferred to a medium supplemented with a cytokinin to promote axillary shoot proliferation. The axillary shoots produced are either subcultured for repeated proliferation (Stage II) or rooted as microcuttings in vitro (Stage III) or ex vitro (Stage IV) to produce plantlets (George, 1993); see Chapter forty-one.

Syngonium podophyllum Schott., commonly known as nephthytis or arrowhead vine, is one of the most commercially important tropical foliage plants propagated in vitro. Plants produce arrowhead-shaped variegated leaves in the juvenile state and climbing vines with three-parted leaves when mature. Each year more than 6 million plants of the various *Syngonium* cultivars are produced in the U.S. Like other members of the Araceae, *Syngonium* is susceptible to infection by numerous systemic pathogens that negatively affect plant vigor and quality. However, use of meristem tip culture followed by shoot culture has enabled rapid production of highly branched specific pathogen-eradicated plants. It is conservatively estimated that 5000 *Syngonium* can be produced yearly from a single primary shoot tip explant (Miller and Murashige, 1976).

The primary objective of these laboratory exercises is to illustrate the sequential steps required for the micropropagation of *Syngonium* by shoot culture. The laboratory exercises will familiarize students with the following specific procedures used to: (1) establish Stage I cultures using isolated shoot tip explants; (2) index Stage I cultures for cultivable bacterial and fungal contaminants; (3) rapidly propagate plants by axillary shoot proliferation (Stage II); and (4) successfully root and acclimatize microcuttings ex vitro (Stage IV). These exercises also offer the opportunity to exemplify several important concepts: (1) relationship between initial shoot tip explant size, response, and culture contamination; (2) requirement for physiological adaption of Stage I cultures prior to Stage II axillary shoot proliferation; and (3) direct rooting and acclimatization of microcuttings. The two

0-8493-2029-1/00/$0.00+$.50

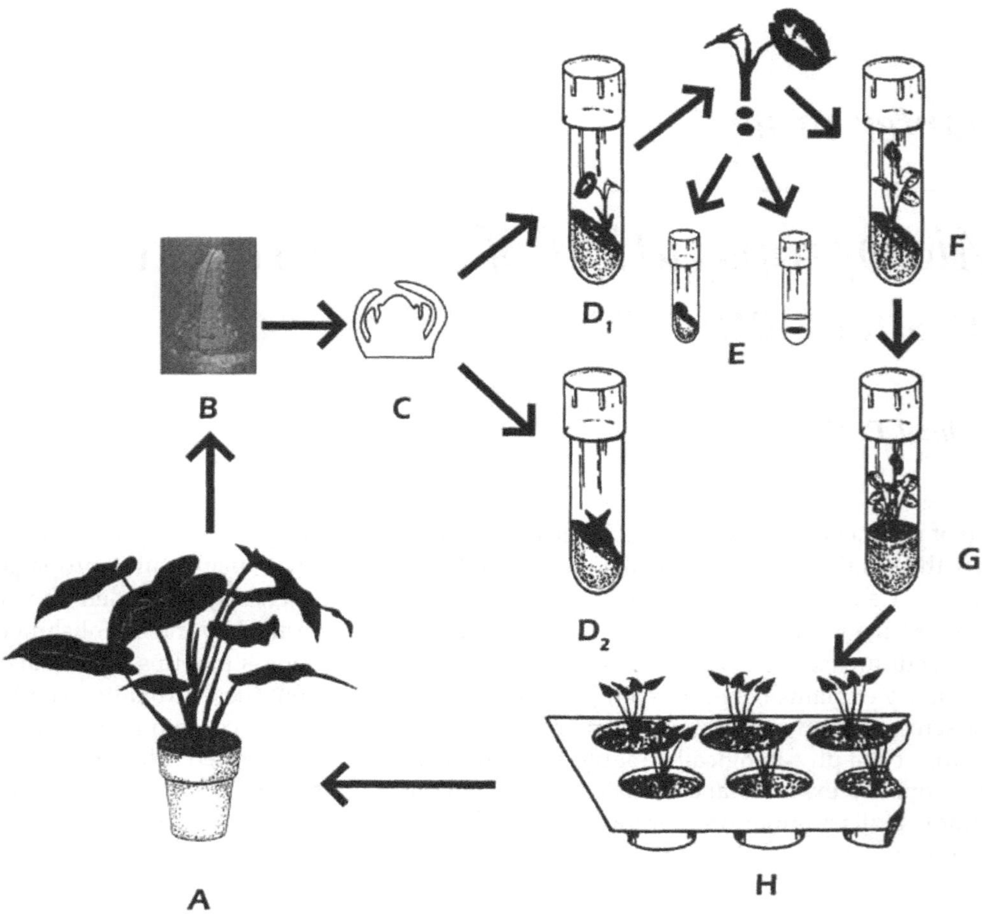

Figure 9.1 Diagrammatic representation of the *Syngonium* shoot culture laboratory exercise protocol. (A) Nodal and terminal vine tip explants are obtained from potted plants and then surface disinfested. (B) Lateral and terminal bud scales are aseptically removed by making shallow horizontal and vertical excisions (dotted lines). (C) Successive inner leaf layers are removed to isolate the meristem and shoot tip explants. (D) Primary explants are inoculated onto SEM (D_1) and SMM (D_2). Growth responses are compared after 6 weeks of culture. (E) Thin basal stem cross sections, excised from 6-week-old shoots generated on SEM, are cultured in both liquid and agar-solidified AC Broth for 7 days to screen for the presence of cultivable contaminants. (F) Shoots are transferred to SEM. (G) Cultures screening positive for contaminants are discarded. Sterile cultures are transferred to SMM and cultured for 4 weeks. (H) Microcuttings are rooted and acclimatized ex vitro (Stage IV). (From Trigiano, R. N. and Gray, D. J., Eds., *Plant Tissue Culture Concepts and Laboratory Exercises, First Edition*, CRC Press LLC, Boca Raton, FL, 1996.)

laboratory exercises are outlined below including maintenance of stock plants, media preparation, procedures, and anticipated results. Refer to Figure 9.1 for a diagrammatic representation of the procedures used to complete both exercises.

General Considerations

Maintenance of Syngonium *stock plants*

Shoots tip explants excised from lateral buds and terminal shoot tips from *Syngonium* vines (Figure 9.2a) are placed in culture. It is recommended that vining stems be used

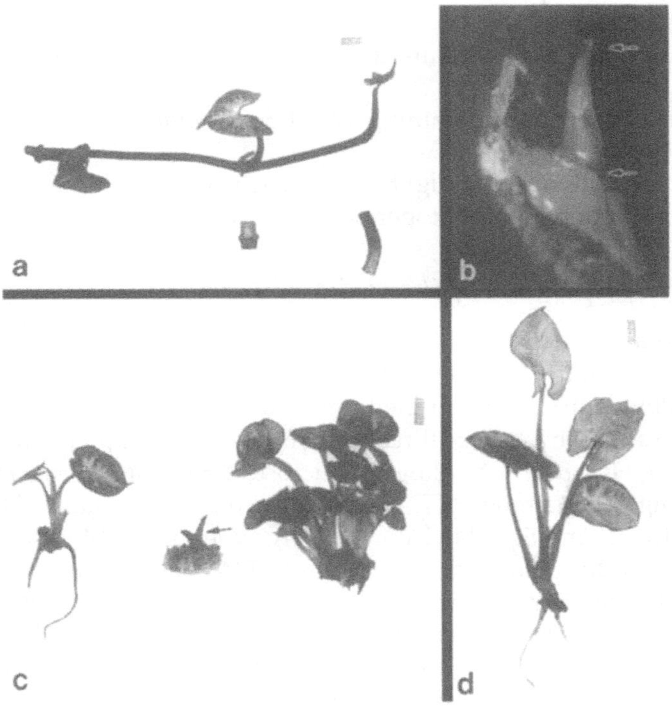

Figure 9.2 Shoot tip culture of *Syngonium*. (a) *Syngonium* vine used as the source of excised meristem tip and shoot tip explants from surface-disinfested nodal segments (lower center) and terminal tips (lower right). Scale bar = 1 cm. (b) Exposed shoot meristem tip within a lateral bud prior to excision and inoculation. Distance between arrows is 3 mm. (c) Left: Stage I shoot culture established from a primary shoot tip explant cultured on SEM for 6 weeks. Center: Inhibition of shoot growth and promotion of basal shoot organogenic callus following 6-week culture of primary shoot tip (arrow) explanted directly on SMM. Right: Typical shoot multiplication response after 4 weeks on SMM from a shoot explant initially established on SEM for 6 weeks. Scale bar = 1 cm. (d) Rooted acclimatized plantlet 3 weeks following transfer to ex vitro conditions. (From Trigiano, R. N. and Gray, D. J., Eds., *Plant Tissue Culture Concepts and Laboratory Exercises, First Edition*, CRC Press LLC, Boca Raton, FL, 1996.)

because the lateral bud scales are usually larger and easier to remove without damaging the underlying meristem tips. However, if vines are not available, terminal shoot tips and lateral buds on nonvining plants can be used. Potted plants can be inexpensively purchased from most retail garden shops. Larger plants will produce vines more rapidly. Most *S. podophyllum* cultivars, such as 'White Butterfly', produce climbing vines, and *S. wendlandii* Schott, another nephthytis species, is an especially fast vine producer. Plants can be grown in most soilless potting mix contained in 20-cm-diameter hanging baskets. Plants should be maintained in a shaded (70 to 80% light reduction) greenhouse or similar environment where they will receive about 150 $\mu mol \cdot m^{-2} \cdot sec^{-1}$ light at 22 to 30°C. Plants should be fertilized twice weekly (150 mg/L N) to promote vining. Depending on growing conditions and plant size, vines will be produced within 6 to 8 weeks. Two plants should produce sufficient numbers of explants for each student or team of students.

Materials

The following are needed for each student or student team:

- *Syngonium* vines to provide 15 nodal explants (including terminal vine tip)
- 20% commercial bleach
- Ten culture tubes each containing 12 ml *Syngonium* Stage I establishment medium (SEM)
- Ten culture tubes each containing 12 ml *Syngonium* Stage II multiplication medium (SMM)
- Incubator or a fluorescent light on a bench top providing approximately 30 $\mu mol \cdot m^{-2} \cdot sec^{-1}$ for a 16-h photoperiod and 25°C

Composition and preparation of the media

Two culture media, SEM and SMM, are required. These media are identical to those described by Miller and Murashige (1976) with the exception that both media are solidified with Phytagel. The basal medium of both consists of Murashige and Skoog (1962) inorganic salts (see Chapter three for composition) supplemented with 1.25 mM (170 mg/L) $K_2PO_4 \cdot H_20$, 87.6 mM (30 g/L) sucrose, 0.55 mM (100 mg/L) myo-inositol, and 1.2 µM (0.4 mg/L) thiamine-HCL, solidified with 2 g/L Phytagel. SEM also contains 14.8 µM (3 mg/L) 2iP and 5.7 µM (1 mg/L) IAA, whereas SMM is supplemented with 98.4 µM (20 mg/L) 2iP alone. Both SEM and SMM can be purchased prepackaged without Phytagel from Sigma Chemical, St. Louis, MO (M8400 and M8525, respectively). Media are adjusted to pH 5.7 before addition of Phytagel and dispensed as 12-ml volumes into 150 × 25-mm culture tubes and autoclaved. The autoclaved media should be cooled and solidified as 45° slants. For Exercise 1, five culture tubes of each medium are prepared for each student or team of students. Additional SEM and SMM will be needed immediately before Exercise 2 is performed.

The AC Broth indexing medium used in Exercise 2 is manufactured by Difco Laboratories (Detroit, MI) and consists (g/L) of 20 Proteose Peptone No. 3, 3 Bacto Beef Extract 3, Bacto Malt Extract, 3 Bacto Yeast Extract, 5 Bacto Dextrose, and 0.2 ascorbic acid. The medium is distributed through Fisher Scientific Supply (Norcross, GA). The solid indexing medium is prepared by the addition of 10 g/L agar. Both media are prepared without pH adjustment and dispensed as 10-ml volumes into 20-ml glass scintillation vials. Culture tubes can be substituted. However, using relatively deeper 150 × 25-mm culture tubes will make it more difficult for the students to perform the stab-and-streak indexing procedure. At least six vials of each medium for each student should be prepared and dispensed immediately before Experiment 2 is performed. After autoclaving, the agar-supplemented medium is cooled and solidified as 45° slants.

Exercises

Experiment 1. Stage I establishment of shoot tip meristems

The purpose of this exercise is to familiarize students with the procedures used to successfully establish Stage I cultures from excised *Syngonium* meristem tip and shoot tip explants. The goal of Stage I is to establish axenic, responsive shoot cultures. Cultures are indexed for specific pathogens or microbial contamination and allowed to become physiologically adapted to in vitro culture conditions before being clonally multiplied by repeated proliferation of axillary shoots in Stage II (see Figure 8.2). Before proceeding, the structure of a generalized shoot apex should be briefly reviewed (see Figure 8.1).

The size of the primary meristem explant will depend on the purpose of culture establishment. For example, when virus elimination is a goal, excision and culture of the apical meristematic dome alone (meristem culture) may be attempted. In reality, cultured apical meristems of many species exhibit low survival rates in vitro and their use may increase the probability of genetic variability due to the tendency to form organogenetic callus (Styer and Chin, 1984); see Chapter fourteen. Consequently, relatively larger (0.2 to 1.0 mm) and more responsive meristem tip explants (meristem tip culture) are usually used. When pathogen eradication is not a concern, relatively larger (1 to 2 mm long) terminal or lateral bud shoot tip explants are used to establish cultures.

Follow the experimental protocol provided in Procedure 9.1.

Procedure 9.1
Establishment of Stage I cultures from *Syngonium* meristem and shoot tip explants.

Step	Instructions and comments
1	*Syngonium* vines consist of a terminal shoot apex and numerous lateral buds each enclosed by a sheathing petiole (Figure 9.2a). Remove the leaves with a scalpel to expose the terminal and lateral buds. Any adventitious roots should also be excised.
2	The defoliated vines are cut to yield 15 nodal segments, each about 1 cm long and consisting of a single bud. It can be noted that the smoothly cutinized epidermis of *Syngonium* should facilitate effective surface disinfestation of the explants. Nodes exhibiting bud break should not be used. Rinse nodal segments in tap water for 15 min and then surface disinfest in 20% commercial bleach with constant agitation for 10 min followed by three 5-min rinses in sterile deionized water.
3	In the transfer hood, under 15 to 30 × magnification provided by a dissecting microscope, a 1 to 5-mm-long meristem tip or shoot tip explant is excised from each surface-disinfested bud with a sterile scalpel. To accomplish this, remove the outer bud scales by making a very shallow, but continuous, incision around the base of the bud and then a continuous median longitudinal incision down the front and rear of the bud scale (see Figure 9.1B). Remove bud scale sections with forceps. The inner leaves are successively removed to expose the meristem tip (Figure 9.2b). Excisions should be performed in a sterile petri dish containing a small volume of sterile water to prevent tissue desiccation. The potential for cross contamination is decreased by using a clean petri dish for each nodal segment. Students should attempt to isolate meristem tip and shoot tip explants of various sizes ranging from 1 to 5 mm in length. Final explant length measurements are made using a thin plastic ruler positioned under the petri dish.
4	Use a stainless steel, tapered microspatula to transfer a single meristem tip or shoot tip explant into five culture tubes, each containing either SEM or SMM. The cut base of each explant should be positioned in the center of the slant in firm contact with the medium, but not completely embedded. Each tube should be labeled with the size of the explant inoculated and maintained in an incubator or under a fluorescent light on a bench top at 25°C and provided with approximately 30 $\mu mol \cdot m^{-2} \cdot sec^{-1}$ for 16 h.
5	Make weekly observations and note differences in the growth responses of the various size explants cultured on SEM and SMM. After 6 weeks, determine the following: (1) percentage of cultures with visible contamination; (2) percentage of responsive explants on each medium; (3) the mean number of axillary shoots produced per responsive explant on each medium; and (4) presence of shoot organogenic callus at the base of each explant. Data should be collected without removing the plants from the culture vessels. Individual student or student team data should be compiled to provide more representative treatment responses.

Anticipated results

Syngonium primary explant size should significantly affect culture survival and responsiveness. Meristem tip explants about 2 mm in length or smaller usually exhibit phenolic browning and high mortality when cultured on either SEM or SMM. When cultured on SEM, larger explants enlarge rapidly and produce leaves and adventitious roots during the 6-week culture period (Figure 9.2c). However, cultures inoculated with larger explants may exhibit a greater frequency of visible contamination. It is important to note that the absence of visible contamination is not an assurance that a culture is aseptic or axenic.

It is often asked why primary explants are not directly inoculated onto Stage II multiplication medium to promote axillary shoot proliferation and save time. The first and most obvious reason for this is the danger of mass propagation of contaminated cultures. The second reason involves a requirement for physiological adjustment of Stage I explants to in vitro culture conditions. Often this physiological adaptation must be completed before enhanced clonal multiplication at higher cytokinin levels is possible. For some plants, repeated subculture on Stage I medium is required before physiological stabilization is achieved.

The requirement for Stage I stabilization in *Sygonium* is exemplified by the poor response of the meristem tips explanted directly on SMM. Unlike explants cultured on SEM, shoot development from primary explants on SMM should be inhibited, whereas formation of shoot organogenetic callus should be promoted at the base of each explant (Figure 9.2c). This response is commercially undesirable since indirect shoot organogenesis may lead to genetic variability in the plants produced and will definitely result in separation of plant chimeras. It is interesting to note that both axillary shoot and adventitious shoot development may occur simultaneously in the same culture. These are often referred to as *mixed cultures*. Prevention of mixed cultures is usually controlled by the type, level, and combinations of plant growth regulators (PGRs) used.

Questions

- If your donor plant was virus infected, which explant tip source (lateral bud or vine tip) do you predict might yield the greatest number of virus-eradicated Stage I cultures?
- For commercial micropropagation (see Chapter forty-one), why is it unwise to attempt to save time by eliminating Stage I and immediately culturing surface-disinfested primary explants (shoot tips) on SMM?
- In their study on *Syngonium* micropropagation, Miller and Murashige (1976) reported that both primary explant establishment and subsequent Stage II shoot multiplication were greatest when explants were cultured in liquid rather than on solid media. Give possible explanations for these observations.

Experiment 2. Procedures for Stage I culture indexing, Stage II shoot multiplication, and Stage IV acclimatization

Rapid in vitro (contaminant-free) production of pathogen-eradicated plants is a fundamental goal of the micropropagation process. The surfaces of donor plants are normally colonized and/or inhabited with diverse bacteria and fungi, most of which are nonpathogenic, and surface disinfestion procedures are used to specifically eliminate this microflora. The presence of microbial contaminants in culture can adversely affect growth, multiplication rate, and survival of shoot cultures. Bacteria and fungal contaminants often persist within cultured shoots that visually appear contaminant free (Knauss, 1976). Consequently, it is essential that Stage I shoot cultures be indexed (screened) for the presence of internal infections before being multiplied in Stage II (see indexing Chapter forty-two).

The first objective of this exercise is to familiarize students with a general indexing procedure used to screen for cultivable bacteria and fungal contaminants in the Stage I *Syngonium* shoot cultures generated in Exercise 1. This indexing exercise can be completed during the same laboratory period that the final data collection for Exercise 1 is made. The second objective is to examine the response of secondary shoot explants excised from 6-week-old established and indexed Stage I shoot cultures following culture on SMM for 4 weeks. Last, procedures for successful rooting and Stage IV acclimatization of Stage II microcuttings will be examined.

Materials

The following supplies are needed for each student or student team:

- Six scintillation vials or 150 × 25-mm culture tubes containing liquid AC Broth
- Six scintillation vials or 150 × 25-mm culture tubes containing agar-solidified AC Broth (slants)
- Five culture tubes each containing 12 ml freshly prepared SEM
- Five culture tubes each containing 12 ml freshly prepared SMM
- Plug trays with soilless growing medium and propagation dome or mist bench

Follow the experimental protocol listed in Procedure 9.2 to complete this exercise.

Procedure 9.2
Stage I culture indexing, Stage II shoot multiplication, and Stage IV acclimatization.

Step	Instructions and comments
1	Each student or team should first consecutively number each of their five Stage I *Syngonium* cultures established on SEM in Exercise 1. One scintillation vial each of the liquid and solidified indexing medium should be labeled "Control." The other corresponding pairs of indexing media should be labeled "Syngonium #1 – 5."
2	Starting with culture #1, remove the shoot (see Figure 9.2c) from the culture vessel and place in a sterile petri dish. Trim roots from the shoot and make several thin cross sections of the stem base using a sterile scalpel. Inoculate the liquid AC Broth in the scintillation vials with one or two tissue sections.
3	Using forceps, inoculate the solidified AC Broth slant by gently stabbing the tissue partially into the medium at the base of the slant. Lift the tissue from the medium, drag it upward along the medium surface, and then position it at the top of the slant. Trim off leaves and transfer the shoot onto freshly prepared SEM and label the tube "Syngonium #1."
4	Repeat the indexing procedure with the other Stage I *Syngonium* cultures. Stage I subcultures are maintained under the same culture conditions as described in Procedure 9.1.
5	In the event that all shoot cultures index negative for contamination, several visibly contaminated cultures should also be indexed. This can be accomplished without removing the contaminated tissue from the tube, by dipping the end of the forceps into the contaminated medium and then inoculating the indexing media as described above. Cultures containing sporulating fungal contaminants should not be used. Vials or tubes containing inoculated indexing media are cultured at 25°C. After 7 days, examine the indexing media for signs of microbial growth. Stage I shoot cultures that index positive for contamination should be discarded. Transfer the axenic Stage I cultures onto SMM and maintain cultures under the same culture conditions as described above.

Procedure 9.2 continued
Stage I culture indexing, Stage II shoot multiplication, and Stage IV acclimatization.

Step	Instructions and comments
6	After 4 weeks, Stage II shoot clusters are divided into single microcuttings and rooted directly in soilless growing medium contained in plug trays or any other small container placed under intermittent mist (5 sec every 10 min) for 3 weeks under lightly shaded conditions. Alternatively, if a mist system is not available, plastic propagation domes can be placed over the plug trays to maintain humidity. After 10 days, the domes are gradually removed over a period of 7 days to acclimatize the plantlets to lower humidity.

Anticipated results

Rapid clouding of the inoculated liquid AC Broth is a positive indication of the presence of cultivable contaminants in the tissue sample. The presence of contaminants on solid AC Broth is confirmed by development of colonies on the surface of the medium and/or development of a halo in the medium where the tissue sample was stabbed. The uninoculated controls should not display microbial growth. Although AC Broth will promote the growth of many microorganisms, it is important to note that growth of all bacteria and fungi will not be supported on this medium. Verified contaminated Stage I cultures are discarded.

After 4 weeks, each Stage II *Syngonium* culture should consist of three to five axillary shoots with no or minimal development of basal organogenic callus (Figure 9.2c). This is in contrast to the inhibition of shoot growth and promotion of organogenic callus observed on the primary explants inoculated directly onto SMM in Exercise 1. *Syngonium* microcuttings should readily root and become acclimatized to ex vitro conditions without the need for Stage III rooting. Rooted and vigorously growing acclimatized plantlets will be produced by week three (Figure 9.2d).

Questions

- Why is it important to both stab into and streak the Stage I tissue sample across the surface of the agar-solidified indexing medium?
- How is it possible to have a contaminated Stage I shoot culture that indexes negative for contamination?
- Provide a possible explanation for the differences in growth response of primary explants cultured first on SEM and then SMM vs. those inoculated directly onto SMM.
- What are the two primary environmental factors that must be regulated during acclimatization?

Literature cited

George, E.F. 1993. *Plant Propagation by Tissue Culture*. Part 1. The Technology. Exegetics, Ltd., Edington.

Henny, R.J., J.F. Knauss, and A. Donnan, 1981. Foliage plant tissue culture. In: *Foliage Plant Production*, J.N. Joiner, Ed., pp. 137–178, Prentice-Hall, Englewood Cliffs, NJ.

Knauss, J. F. 1976. A tissue culture method for producing *Dieffenbachia picta* cv. Perfection free of fungi and bacteria. *Proc. Fla. State Hort. Soc.* 89:293–296.

Miller, L.R. and T. Murashige. 1976. Tissue culture propagation of tropical foliage plants. *In Vitro* 12:797–813.

Murashige, T. and F. Skoog. 1962. A revised medium for rapid growth and bioassays with tobacco cultures. *Physiol. Plant.* 15:473–497.

Styer, D.J. and C.K. Chin. 1984. Meristem and shoot-tip culture for propagation, pathogen elimination, and germplasm preservation. *Hort. Rev.* 5:221–277.

chapter ten

Micropropagation of Dieffenbachia

R. J. Henny, Lee Goode, and Wanda Ellis

The genus *Dieffenbachia*, a member of the aroid family (Araceae), is composed of approximately 30 species of upright herbaceous tropical plants indigenous to South and Central America (Bailey et al., 1976). Commonly called dumbcane, *Dieffenbachia* is one of the most important ornamental tropical foliage plant genera. Their beautifully variegated leaves, ease of growth, and tolerance to interior environments are reasons that *Dieffenbachia* routinely are in the top five highest selling ornamental foliage plant genera in Florida. A Florida Department of Agriculture and Consumer Services, Division of Marketing report indicated the *Dieffenbachia* accounted for 7% of the total product mix of nurseries surveyed during 1993 (Sheehan, 1994).

An estimated 45 to 50 *Dieffenbachia* cultivars are produced by Florida nurseries. Most of the early popular cultivars originated from the species *D. maculata* (Lodd.) G. Don as sports or hybrids. Once selected, new cultivars were propagated asexually from stem cuttings or stem node segments. Using this process, 5 to 10 years were required for new cultivars to become commercially important.

An important early use of micropropagation was for the purpose of producing aroids free of dasheen mosaic virus and was first reported to be successful with *Caladium*, *Xanthosoma*, and *Colocasia* (Hartman, 1974). Subsequently, tissue culture was seen as a method whereby *Dieffenbachia* stock could be freed of systemic viral and bacterial pathogens (Knauss, 1976; Taylor and Knauss, 1978). Systemic diseases made *Dieffenbachia* susceptible to deterioration and loss of quality, since plants in old stock beds were harvested continually over years using conventional cutting propagation. *Dieffenbachia* 'Perfection-137B' (Chase et al., 1981) was the first improved cultivar free of systemic pathogens. As tissue culture techniques were refined, a second benefit to become apparent was the speed with which new cultivars could be multiplied. With micropropagation, new cultivars could now become commercially important within 2 to 3 years from development rather than the 5 to 10 years needed using conventional propagation.

Tissue culture propagation also made possible production of salable plants in a greater range of pot sizes than through conventional cutting propagation. The tendency of tissue-cultured *Dieffenbachia*, which are in a juvenile growth stage, to initially branch more freely made this feasible. The smaller size of tissue culture microcuttings compared to standard stem cuttings kept plants in proportion to the pot over a greater size range. Additionally,

0-8493-2029-1/00/$0.00+$.50

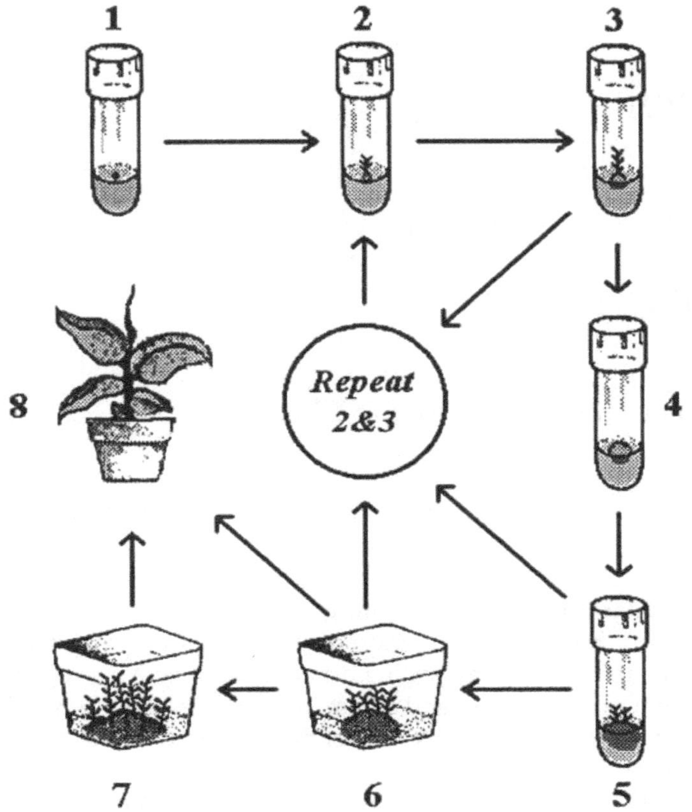

Figure 10.1. Diagrammatic representation of the micropropagation protocol for *Dieffenbachia*. Lateral bud or terminal growing point after surface disinfestation and excision on shoot initiation medium (1). Depending on cultivar differences, after 12 to 18 weeks (two or three transfer cycles) the developing plant is placed on multiplication medium for 6 weeks (2) when the enlarged callus base is separated from the apex (3). The shoot tip is used to repeat steps 2 and 3 or it can be planted out to test for genetic integrity. The callus base is left on multiplication medium to develop adventitious shoots (4). Increasing size of callus base and newly developing adventitious shoots (5) requires transfer to larger culture vessel (i.e., babyfood jar or GA-7 magenta vessel). Several shoots developing from callus base (6) may be harvested and planted out or returned to steps 2 and 3 to renew the cycle. Callus base is maintained in culture to produce additional adventitious buds and cuttings on a 6-week transfer cycle (7). Unrooted cuttings are transferred directly to soil; rooting occurs within 6 to 8 weeks (8). Transfer of *Dieffenbachia* cultures is normally on a 4- or 6-week cycle.

growers were able to eliminate large stock beds and convert them to production areas, which increased the percent of space capable of generating income. Tissue culture propagation of *Dieffenbachia* (Figure 10.1) is a true success story and is becoming ever more important with each passing year.

The following two experiments illustrate the exact methods used by industry to initiate cultures, which is not necessarily recommended for a laboratory exercise because it may be too time consuming, and a culture maintenance and propagation system. The latter experiment uses tissue cultures obtained commercially and comprises a useful laboratory exercise, since it is rapid and reliable.

General considerations

Plant sources

For Experiment 1, any healthy, vigorously growing *Dieffenbachia* plant may be used. For Experiment 2, plants established in culture can be obtained from any of the following companies: (1) Agri-Starts II, Inc., 4341 Round Lake Road, Apopka, FL 32712 (407-889-8487); (2) Twyford Plant Labs Inc., 11850 Twitty Road, Sebring, FL 33870 (941-655-3700); (3) Oglesby Plant Labs Inc., Route 2, Box 9, Altha, FL 32421 (352-762-3806).

Exercises

Experiment 1. Initiation of Dieffenbachia cultures for micropropagation

The following procedure is used commercially to initiate cultures. Axillary buds from potted plants are used as explants and are induced to undergo adventitious bud production. Instructors should be aware that a prolonged time period is needed for individual explants to develop into proliferating cultures; thus, Experiment 1 may not adapt to the time schedule of some courses.

Materials

The following items are needed for each student or student team:

- A healthy, rapidly growing potted plant
- Magenta GA-7™ vessels and 25 × 150-mm culture tubes containing culture medium (Table 10.1) modified as specified in the experiments
- 10% bleach solution and sterile beakers
- Sterile deionized or distilled rinse water
- Scalpel with a supply of #10 blades
- Bunsen burner and quench bottle containing 95% ethanol or another method of sterilizing instruments

Table 10.1 Basic *Dieffenbachia* Micropropagation Medium in Distilled Water and Consisting of the Following Components (Knauss, 1976)

Murashige and Skoog (1962) salts (see Chapter three) plus the following

Component	Concentration (mg/L)
Sucrose	30000.0
Phytagel™	1500.0
$NaH_2PO_4 \cdot H_20$	170.0
Myo-inositol	100.0
Adenine sulfate	80.0
Glycine	2.0
Nicotinic acid	0.5
Pyridoxine-HCl	0.5
Thiamine-HCl	0.4

After Knauss, J. F. 1976. *Proc. Fla. State Hort. Soc.* 89:293–296.

To complete the following experiment, follow the protocols listed in Procedure 10.1.

Procedure 10.1
Initiation of *Dieffenbachia* micropropagation cultures.

Step	Instructions and comments
1	First, cut and remove the lateral buds including a little of the surrounding tissue (one bud is located behind each leaf petiole where it attaches to the main stem) from one plant and place in sterile deionized water containing a wetting agent for 5 min while stirring constantly.
2	Rinse buds three times each for 15 sec in separate containers of sterile deionized water.
3	Immerse buds with constant stirring in a 10% commercial bleach solution for 15 min and rinse again three times in separate containers of sterile deionized water.
4	Using a stereomicroscope, carefully remove one to two layers of tissue from over the bud, being careful not to damage the bud (with larger buds it may be possible to remove three to four tissue layers).
5	Place the bud in 10% commercial bleach for 5 sec, then, lightly holding the bud with sterile forceps, dip it into sterile deionized water for 5 sec. Remove the bud and blot it dry on a sterile paper towel before placing on the culture medium.
6	Place approximately 2/3 of the basal part of the bud below the medium surface.
7	Screen cultures for contamination.

Anticipated results

Contamination may or may not be a problem, depending on the condition of stock plants. The initial culture will grow slowly and take up to 6 months to reach a stage suitable for a multiplication stage. During that time careful culture maintenance and regular transfers are required.

Questions

- Why are axillary buds used as explants?
- What is the purpose of the double surface disinfestation step used on axillary buds?

Experiment 2. Multiplication of Dieffenbachia *micropropagation cultures*

Using small shoots from established tissue cultures is the quickest way to start a laboratory experiment. Significant growth can be obtained within 12 to 16 weeks. Rapidly growing and multiplying *Dieffenbachia* cultures obtained from commercial tissue culture laboratories listed above are used as a convenient explant source. These laboratories produce many cultivars that would be adaptable to such an exercise. The two most important commercial *Dieffenbachia* cultivars, *D. maculata* 'Compacta' and *D. maculata* 'Camille,' would be good test plants.

The experiment demonstrates the effects of increasing cytokinin levels on growth and development of adventitious buds as well as root and shoot formation in *Dieffenbachia*. The experimental procedure is described in Procedure 10.2. Five concentrations of the plant growth regulators N_6-(2-isopentenyl)-adenine (2iP) at 0, 2.5 (12.3 μM), 5.0 (24.6 μM), 10.0 (49.2 μM), and 20.0 mg/L (98.4 μM) are variously added to the medium described in Table 10.1 to comprise an adequate range of treatments. Add indole-3-acetic acid (IAA) at 2.0 mg/L (11.4 μM) to all five 2iP levels. Use at least ten replications (i.e., individual 25 mm × 150-mm test tubes containing a single shoot) per growth regulator level. In each tube start with a single shoot tip approximately 2 cm in length bearing two to three leaves. Follow the protocols outlined in Procedure 10.2 to complete this experiment.

Procedure 10.2
Multiplication of *Dieffenbachia* micropropagation cultures.

Step	Instructions and comments
1	Excise shoots approximately 2 cm long and place on culture medium (Table 10.1) with the base immersed 2 to 3 mm below the surface. Use ten tubes (25 × 150 mm) of each medium with different PGRs.
2	Incubate the test tubes at 25°C under 25 $\mu mol \cdot m^{-2} \cdot sec^{-1}$ of fluorescent light for 16 h per day.
3	After 4 weeks, transfer the developing buds to new tubes of the same medium. At this time note any differences in number of roots, basal diameter of the shoot's base, or presence of any basal shoots. *Note*: The transfer interval may be 5 or 6 weeks instead of 4 weeks, depending on the time available to complete the study.
4	Incubate the test tubes under the same environmental conditions as in step 10.2 for another 4 weeks.
5	Make observations on number of roots, basal diameter of the original shoot, and presence or absence of new basal shoots. Compare heights of the shoots. Before transferring, decapitate the original shoot approximately 10 mm above the base. If a callus base has formed, make the cut just above its top. This will serve to stimulate growth and development of secondary adventitious shoots. Transfer the bases to larger culture vessels such as baby food jars or GA-7 magenta vessels containing 30 ml of medium. Maintain the same levels of PGRs.
6	Incubate the culture vessels for an additional 4 weeks under the same environmental conditions as steps 10.2 and 10.4.
7	Record the final data including fresh weight, total number of shoots and roots, length of longest shoot and root, and diameter of the enlarged callus base, make observations regarding the presence of developing adventitious buds.

Anticipated results

Differences between treatments will be observed within 8 weeks. Shoots on medium with no, or low levels of 2iP will continue to slowly grow and develop roots. As 2iP levels increase, root production will decrease. Simultaneously, enlargement of the shoot base will become noticeable due to proliferation of a chalk-colored callus-like growth. The swollen base will increase in size and begin to develop small islands of green tissue within 4 to 6 weeks, which represents early adventitious bud development, and the callus base will nearly fill the 25-mm-diameter test tubes. The callus base is transferred regularly to foster continuous adventitious bud development and enlarging shoots are harvested for plant production. The original shoot may be transferred to a new tube to begin the cycle again or it can be placed in the new container beside its base.

The transfer intervals for this experiment may be on a 4-, 5-, or 6-week cycle. Significant differences can be observed in 12 weeks. Differences are magnified with longer transfer intervals. Commercially, a 6-week transfer period is often employed with *Dieffenbachia* cultures.

Questions

- Compare the frequency, type, and distribution of contaminants between Experiments 1 and 2. What factors account for the differences?
- How did the plant growth regulators influence the observed responses? Speculate on how even higher levels of either or both PGR might influence growth.

- What biological and commercial factors make micropropagation of ornamental plants like *Dieffenbachia* practical when compared to other types of plants, such as food crops (see Chapter sixteen, for example)?

Literature cited

Bailey Hortorum staff. 1976. *Hortus Third*. Macmillan, New York.

Chase, A.R., F.W. Zettler, and J.F. Knauss. 1981. Perfection-137B, a pathogen-free selection of *Dieffenbachia maculata* derived through tissue culture. *Circular S-280*, Florida Agricultural Experiment Stations, IFAS, University of Florida. 7 pp.

Hartman, R.D. 1974. Dasheen mosaic virus and other phytopathogens eliminated from caladium, taro, and cocoyam by culture of shoot tips. *Phytopathology* 64:237–240.

Knauss, J.F. 1976. A tissue culture method for producing *Dieffenbachia maculata* cv. Perfection free of fungi and bacteria. *Proc. Fla. State Hort. Soc.* 89:293–296.

Murashige, T. and F. Skoog. 1962. A revised medium for rapid growth and bioassays with tobacco tissue cultures. *Physiol. Plant.* 15:473–497.

Sheehan, P. 1994. Foliage facts. *Fla. Dept. Agric. Consumer Services*. 3 pp.

Taylor, M.E. and J.F. Knauss. 1978. Tissue culture multiplication and subsequent handling of known pathogen-free *Dieffenbachia maculata* cv. Perfection. *Proc. Fla. State Hort. Soc.* 91:233–235.

chapter eleven

Micropropagation of potato by node culture and microtuber production

Michael E. Kane

The potato (*Solanum tuberosum* L.), a crop of worldwide importance, is an integral part of the diet of a large proportion of the population of the world. Potato production ranks fourth among major food crops and, unlike other major food crops, potatoes are vegetatively propagated and, unfortunately, susceptible to infection by many viral, fungal, and bacterial pathogens. At least 23 viruses that decrease tuber quality and yield are known to infect the potato. Reliance on vegetative propagation from field-produced seed tubers results in multiplication and spread of infected tubers. Routine production of pathogen-eradicated seed tubers is necessary to maintain adequate yield in the field (Miller and Lipschultz, 1984). This is accomplished using meristem tip culture to produce *specific pathogen-free tested* (SPT) potato tuber seed stocks (Cassells and Long, 1982). Micropropagation techniques are subsequently used to clonally propagate these pathogen-eradicated stock plants (Hussey and Stacey, 1991; Dodds et al., 1992). For discussions concerning culture indexing for plant pathogens and microbial contaminants, see Chapters forty-three and forty-four.

Two very simple, but reliable, micropropagation techniques can be demonstrated using the potato. The first technique, called single- or multiple-node culture, involves production of shoots from cultured single or multiple nodes (in vitro layering) positioned horizontally on the medium. In potato, the initial sources of these nodes are SPT shoots established in vitro from meristem tips (Figure 11.1). Interestingly, the potato cannot be multiplied by using the shoot tip culture method (see Chapter eight) because axillary shoot proliferation is not promoted by the addition of cytokinin to the medium. Typically, single, elongated, unbranched shoots comprising multiple nodes are rapidly produced. These shoots (microcuttings) are either rooted and acclimatized to field conditions or are subdivided into single-node cuttings to initiate additional cultures (Figure 11.1). Although not the most efficient method, node culture is the most reliable micropropagation procedure to produce plants that are true to type.

The second — and perhaps more fascinating — micropropagation technique involves production of miniaturized tubers (microtubers) on potato shoots cultured in vitro (Figure 11.1). Microtuber production is a very useful method to propagate and store valuable potato stocks and may be adaptable for automated commercial propagation and large-scale

0-8493-2029-1/00/$0.00+$.50

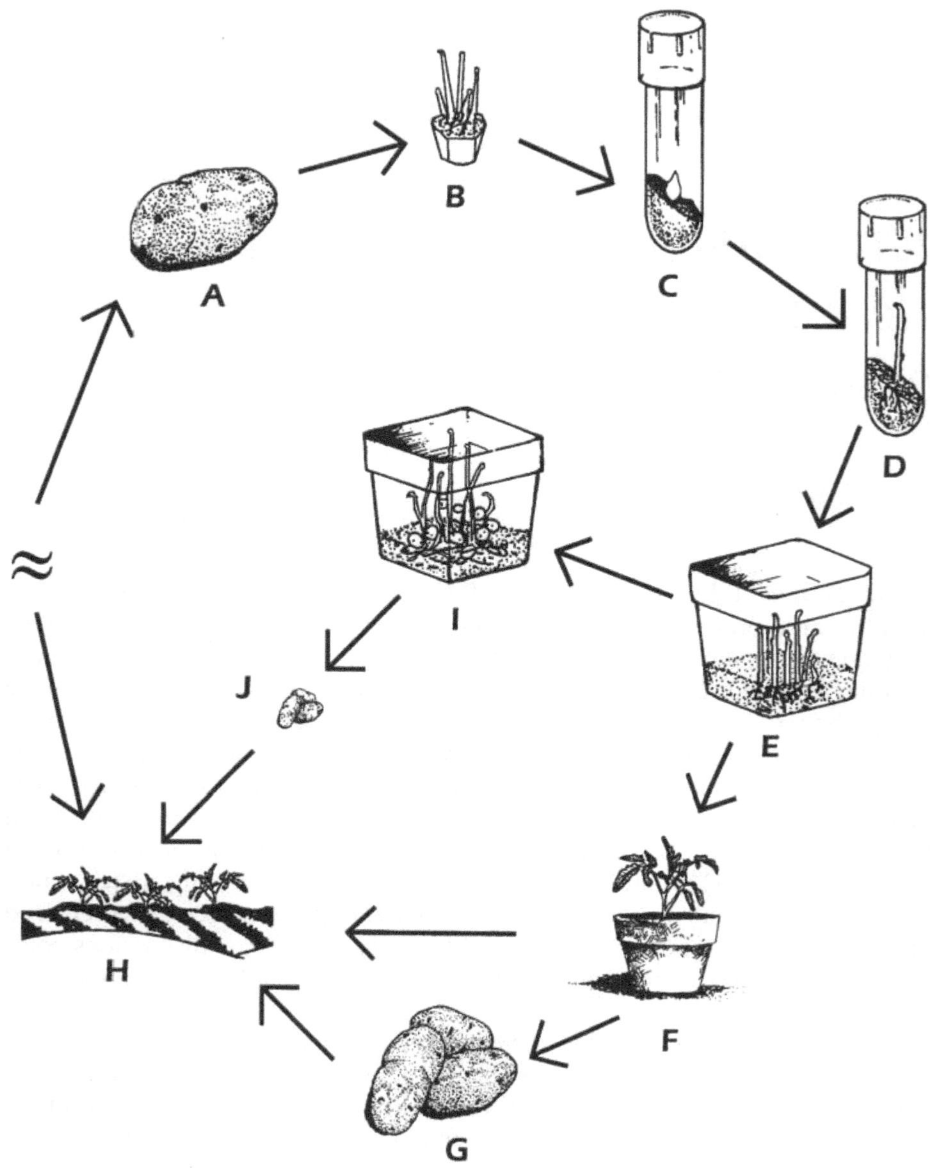

Figure 11.1 Diagrammatic representation of potato micropropagation by node culture and microtuber production. (A) Tuber obtained from a multigeneration pathogen-infected field potato population; (B) meristem tip explants excised from sprouted tuber sections; (C) establishment of meristem tip in culture; (D) culture is SPT and indexed for other contaminants; (E) shoots determined to be specific pathogen free are propagated by node culture to increase numbers; (F) microcuttings are rooted and acclimatized ex vitro; (G) tubers produced ex vitro from STP plants are used as planting stock; (H) field planting of STP seed tuber; (I) shoots may be transferred to a microtuber induction medium and promotive conditions; (J) microtubers produced are stored or used as STP seed tubers for field planting. (From Trigiano, R. N. and Gray, D. J., Eds., *Plant Tissue Culture Concepts and Laboratory Exercises, First Edition*, CRC Press LLC, Boca Raton, FL, 1996.)

mechanized field planting (McCown and Joyce, 1991). Depending on the cultivar, potato microtuber development may be promoted under short-day photoperiods and following addition of cytokinin and high levels of sucrose to the medium (Seabrook et al., 1993; Gopal et al., 1998). Under ideal planting and early growth conditions, field-planted microtubers perform as well as regular seed tubers (McCown and Wattimena, 1987).

The primary objective of this laboratory exercise is to illustrate micropropagation of potato by node culture and microtuber production. This exercise will familiarize students with the following concepts: (1) efficient shoot production is possible from propagules with preexisting shoot meristems in the absence of cytokinin; (2) potato shoot meristems that normally produce elongated shoots in vitro can be induced to form miniature tubers following alteration of environmental and cultural conditions; and (3) microtuberization is a valuable method for potato propagation, storage, and distribution. The laboratory exercise is detailed below, including establishment and maintenance of stock shoot cultures, media preparation, procedures, and anticipated results.

General considerations

Establishment and maintenance of stock potato shoot cultures

Students will use potato nodal cuttings and intact shoots from shoot cultures established from surface-sterilized shoot tips excised from sprouted tuber sections. This eliminates the need for students to perform an often-unsuccessful surface sterilization step and decreases culture response time. Most potato cultivars (genotypes) can be induced to form tubers in vitro; however, the capacity for microtuberization is influenced by genotype. Potato cultivars such as 'Kennebec,' 'Superior,' 'Red Pontiac,' or 'Russet Burbank' have proven to be responsive (Leclerc et al., 1994) and are available in most produce stores. Only use fresh potato tubers that are free of decay.

Wash potato tubers under flowing tap water and then soak for 10 min in 20% (v/v) commercial bleach (1.05% NaOCl). Cut tubers into five to ten pieces, each about 2.0 cm^2, and then soak for 1 h in 0.29 mM (100 mg/L) gibberellic acid (GA_3) to break bud dormancy and promote sprouting. Place treated tuber sections on moist vermiculite in an incubator maintained at approximately 25°C under a 16-h photoperiod provided by cool white fluorescent lamps at about 30 $\mu mol \cdot m^{-2} \cdot s^{-1}$. Excise sprouted shoots when they are approximately 3 to 5 cm in length. Large shoot tip explants 5 to 10 mm long should be used to initiate stock cultures. Rinse excised shoot tips in tap water for 10 min and then surface disinfest in 20% commercial bleach with constant agitation for 10 min followed by three 5-min rinses in sterile distilled water. Trim away bleached tissues and inoculate shoot tip explants into separate 150 × 25-mm glass culture tubes containing 15 ml Murashige and Skoog (1962) basal salts (see Chapter three for composition) supplemented with 87.6 mM (30 g/L) sucrose, 0.56 mM (100 mg/L) myoinositol, and 1.2 µM (0.4 mg/L) thiamine-HCL, solidified with 8 g/L agar. This medium can be purchased prepackaged without the sucrose and agar from Sigma Chemical, St. Louis, MO (M-6899). Cultures are maintained under previously described culture conditions. Shoots should be about 4 to 6 cm long after 30 to 40 days culture. Subculture nodal cuttings from sterile cultures onto the same medium at about 4- to 6-week intervals to generate additional shoot cultures.

Materials

The following items are needed for each team of students:

- Twenty-four 5-week-old stock potato shoot cultures (in culture tubes)
- Long (29-cm) forceps
- Four GA-7 vessels each containing 60 ml node culture medium (NCM)
- Eight GA-7 vessels each containing 60 ml microtuber induction medium (MIM)

- Plug trays with soilless growing medium and propagation dome or mist bench
- Incubator or a fluorescent light on a bench top providing approximately 30 $\mu mol \cdot m^{-2} \cdot s^{-1}$ for a 16-h photoperiod and 20°C
- Incubator set to maintain 20°C and provide approximately 30 $\mu mol \cdot m^{-2} \cdot s^{-1}$ for an 8-h photoperiod (for microtuberization exercise)

Composition and preparation of the exercise media

Two culture media, NCM and MIM, are required for the laboratory exercise. GA-7 magenta vessels (Magenta Corp., Chicago, IL), each containing 60 ml medium, should be used if available (Figure 11.2). However, baby food jars containing 30 ml medium can be substituted. Each student team will require four GA-7 vessels containing NCM and eight GA-7 vessels containing MIM. The NCM is identical in composition to the stock shoot culture medium described above. The MIM contains the same components as the NCM except that the sucrose concentration is increased to 234 mM (80 g/L) and the medium is supplemented with 22.2 μM (5 mg/L) benzyladenine (BA). Both media are adjusted to pH 5.7 before addition of agar and autoclaving.

Figure 11.2 Typical development of potato shoots from two-node cuttings cultured for 7 days in a GA-7 magenta vessel containing NCM. Scale bar = 10 mm. Corresponds to part E of Figure 11.1. (From Trigiano, R. N. and Gray, D. J., Eds., *Plant Tissue Culture Concepts and Laboratory Exercises, First Edition*, CRC Press LLC, Boca Raton, FL, 1996.)

Exercises

Experiment 1. Potato micropropagation by node culture

The objective of this exercise is to demonstrate that node culture, the simplest micropropagation method, can be used to generate large numbers of plants in a short period of time. Each shoot produced in vitro consists of multiple nodes that potentially can be used as a propagule to produce another shoot. However, some potato cultivars may be less responsive or produce smaller shoots in vitro when single-node cuttings are used. During routine subculture of the stock shoot cultures it may be useful to compare the response of single- and two-node cuttings. The typical response of two-node cuttings after only 7 days culture is shown in Figure 11.2.

Follow the protocols listed in Procedure 11.1 to complete this experiment.

Procedure 11.1
Micropropagation of potato by nodal culture.

Step	Instructions and comments
1	Remove intact shoot from stock cultures and place in a sterile petri dish. Add a small volume of sterile water to prevent tissue desiccation and subdivide into single-node cuttings using a sterile scalpel. Trim leaf blades from each nodal cutting, leaving only the petiole bases.
2	Orient four single-node cuttings horizontally onto NCM medium in each of the four GA-7 vessels. Place cultures in an incubator or on a laboratory bench top under fluorescent lighting providing approximately 30 $\mu mol \cdot m^{-2} \cdot s^{-1}$ for a 16-h photoperiod and 20°C. The percent responsive nodal cuttings, average number of shoots, and total nodes produced per nodal cutting are determined after 3 weeks.
3	After data collection, the potato shoots produced may be treated as microcuttings and rooted directly in soilless growing medium contained in plug trays or any other small container and placed under intermittent mist (5 sec every 10 min) for 2 weeks under lightly shaded conditions. Alternatively, if a mist system is not available, plastic humidity domes can be placed over the plug trays to maintain humidity. After 10 to 12 days, the domes are gradually removed over a period of 7 days to acclimatize the plantlets.

Anticipated results

Shoot growth from the nodes should occur rapidly. Typically, an unbranched shoot approximately 5 cm long and consisting of three to five nodes, plus the shoot tip, will be produced from a nodal cutting after 3 weeks. Determine the average number of new nodes (plus the shoot tip) produced per nodal cutting after 3 weeks. Using this value, the students can then estimate the number of shoots generated from the original nodal cutting after 1 year. This is accomplished by calculating Y^x, where Y= node number (including shoot tip) produced per nodal cutting after the 3-week culture cycle and X = 17.3, the number of culture cycles per year. This yearly production rate can be quite large and its calculation provides an opportunity to illustrate the power of this simple micropropagation technique. Potato microcuttings should quickly root and acclimatize to ex vitro (Stage IV) condition.

Questions

- Given a 3-week culture cycle and based on your calculated average total nodes produced per nodal cutting, how many potato shoots can be produced from a single node cutting in a year?
- Why is node culture considered the most reliable method to in vitro propagate plants that are true to type?

Experiment 2. Microtuberization of potato

Follow the protocols listed in Procedure 11.2 to complete this experiment.

Procedure 11.2
Microtuberization of potato.

Step	Instructions and comments
1	Transfer four 4.0-cm-long microcuttings with shoot tips intact into each of the eight GA-7 vessels containing MIM. Position the microcuttings vertically by embedding the cut end of each about 5 mm into the medium.

Procedure 11.2 continued
Microtuberization of potato.

Step	Instructions and comments
2	Place four cultures in an incubator set to provide 30 $\mu mol \cdot m^{-2} \cdot s^{-1}$ fluorescent lighting for an 8-h photoperiod at 20°C and the remaining cultures in an incubator set to maintain a 16-h photoperiod at the same light level and temperature. Usually microtuberization is promoted at temperatures ranging from 15 to 20°C. However, higher culture temperatures can be used if lower temperature control is not possible.
3	Depending on culture response, the number of microtubers produced per shoot and mean microtuber fresh weight can be determined after 6 to 10 weeks.

Anticipated results

Potato tubers normally form underground but are produced in vitro from axillary meristems along the shoot (Figure 11.3). Usually only one microtuber fully develops on each shoot. During the tuberization process, there is apparently strong competition between axillary meristems along the shoot such that the first meristem to tuberize inhibits tuberization of the other meristems (McCown and Joyce, 1991). This has been a factor limiting efficient microtuber production and makes it difficult to design exercises examining factors affecting microtuber production per propagule. However, it is important to point out that each microtuber, being a modified stem, contains numerous shoot meristems, the so-called eyes of the potato. Microtuber meristems exhibit varying degrees of dormancy that can be broken by storing at lower temperature for several weeks. Depending on the cultivar used, microtubers produced under an 8-h photoperiod usually weigh more than those produced under a 16-h photoperiod. One variation to this exercise would be to examine the influence of sucrose concentration (0, 60, 120, 180, and 240 mM) on size and fresh weight of microtubers produced under only an 8-h photoperiod.

Figure 11.3 Typical microtuber formation from cultured shoots 10 weeks after culture initiation. Note the presence of only one primary microtuber on each shoot. Scale bar = 10 mm. Corresponds to part H of Figure 11.1. (From Trigiano, R. N. and Gray, D. J., Eds., *Plant Tissue Culture Concepts and Laboratory Exercises, First Edition*, CRC Press LLC, Boca Raton, FL, 1996.)

Questions

- What are several advantages of producing potato in vitro via formation of microtubes?
- Provide a possible explanation why so few microtubers develop on each shoot.

Literature cited

Cassells, A.C. and R.D. Long. 1982. The elimination of potato viruses X, Y, S and M in meristem and explant cultures of potato in the presence of Virazole. *Potato Res.* 25:165–173.

Dodds, J.H., D. Silva-Rodriguez, and P. Tovar. 1992. Micropropagation of potato *Solanum tuberosum* L.). In: *Biotechnology in Agriculture and Forestry, Vol. 19. High Tech and Micropropagation III,* Y.P.S. Bajaj, Ed., pp. 91–106. Springer-Verlag, New York.

Gopal, J., J.L. Minocha, and H.S. Dhaliwal. 1998. Microtuberization in potato (*Solanum tuberosum* L.). *Plant Cell Rep.* 17:794–798.

Hussey, G. and N. J. Stacey. 1981. In vitro propagation of potato (*Solanum tuberosum* L.). *Ann. Bot.* 48:787–796.

Leclerc, Y., D.J. Donnelly, and J.E.A. Seabrook. 1994. Microtuberization of layered shoots and nodal cuttings of potato: the influence of growth regulators and incubation periods. *Plant Cell Tissue Organ Cult.* 37:113–120.

McCown, B.H. and P.J. Joyce. 1991. Automated propagation of microtubers of potato. In: *Scale-Up and Automation in Plant Propagation,* I.K. Vasil, Ed., pp. 95–109. Academic Press, San Diego.

McCown, B.H. and G.A. Wattimena. 1987. Field performance of micropropagated potato plants. In: *Biotechnology in Agriculture and Forestry. Vol. 3,* Y.P.S. Bajaj, Ed., pp. 80–88. Springer-Verlag, New York.

Miller, S.A. and L. Lipschultz. 1984. Potato. In: *Handbook of Plant Cell Culture. Vol. 3. Crop Species,* P.V. Ammirato, D.A. Evans, W.R. Sharp, and Y. Yamada, Eds., pp. 291–326. Macmillan, New York.

Murashige, T. and F. Skoog. 1962. A revised medium for rapid growth and bioassays with tobacco cultures. *Physiol. Plant.* 15:473–497.

Seabrook, J.E.A., S. Coleman, and D. Levy. 1993. Effect of photoperiod on in vitro tuberization of potato (*Solanum tuberosum* L.). *Plant Cell Tissue Organ Cult.* 34:43–51.

chapter twelve

Micropropagation of lilacs

Deborah D. McCown and Andrew J. Daun

Collectively, the species of the genus *Syringa* are some of the more easily micropropagated woody plants. The French hybrid lilac, *S. vulgaris* L., is among the easiest to establish in culture and has a relatively high multiplication rate. Other species, generally identified as shrub lilacs (e.g., *S. microphylla* and *S. julianae* Schneid.), are slightly more recalcitrant. The tree lilacs, such as *S. reticulata* and *S. pekinensis* Rupr., can be successfully micropropagated, but generally are more difficult to establish and acclimatize to the culture system (McCown and McCown, 1987). However, once established, they perform much like the shrub lilacs, but with slightly lower multiplication rates.

For the horticultural industry, rapid introduction of new superior cultivars is important. Desirable traits may include disease resistance, improved flower color, or more compact growth habit. For commercial purposes, successful isolation (or culture) of vegetative tissue in late winter or early spring can result in significant numbers of propagules the following year. We have routinely produced several thousand rooted microcuttings of a number of clones in a single year; lilacs have excellent axillary shoot proliferation potential in culture.

These laboratory exercises have the following two objectives: (1) isolation and culture of vegetative shoots of a woody plant in the microculture system and (2) exploration of the effect of cytokinin type and concentration on axillary shoot multiplication and development. Cytokinins comprise a specific class of plant growth regulator (PGR).

The specific cultivar of *S. vulgaris* used in these exercises is probably a minor concern as the number in commercial propagation is approaching 100. There is, however, one important caveat — a few lilacs are apparently chimeras, notably 'Sensation' and 'Primrose' (McCulloch, 1989). There are reports that these cultivars may be unstable in culture and yield plants that are not identical to the parental type; thus, avoid known chimeral lilacs. Some suggested cultivars that are readily available through nurseries and are known to be easily micropropagated include 'Charles Joly', 'Ludwig Spaeth', 'Madame Lemoine', and 'President Grevy'.

Exercises

Experiment 1. Isolation and culture of terminal and axillary buds

Materials

The tissue available for isolation for microculture (tissue culture) is an important factor and, along with a number of other variables, can influence success rates. The ideal tissue

0-8493-2029-1/00/$0.00+$.50

is obtained from a small vegetative plant maintained in a "relatively" clean environment such as a greenhouse. If using greenhouse plants, maintain them in soilless potting mixture and fertilize with a liquid preparation of 20-20-20 + micronutrient solution on a regular basis. Supplemental lighting in the greenhouse may be necessary to maintain long days (greater than 16-h daylength). Cutting plants back on a regular basis will encourage production of new shoots. Vegetative shoots from plants growing outside can also be successfully used to initiate cultures, but the number of fungi and bacterial spores from rain/soil splash etc. makes surface disinfestation more difficult. In either case, the tissues should be vegetative and actively growing.

The following items are needed for each student or student team:

- Actively growing vegetative shoots (ten pieces per student or team)
- 15% commercial bleach (0.79% NaOCl) in distilled water (tap water may reduce bleach activity) plus one or two drops of liquid dishwashing detergent
- Two 200-ml baby food jars of sterile water
- Ten 125-ml baby food jars or GA-7 magenta vessels containing culture medium

The amount of bleach solution and number of vessels of rinse water and culture medium are dependent on the number of shoots or nodal sections available for culture. For both the bleach and rinse water, the volume should be sufficient to fully immerse the tissue.

Number of vessels for liquids or culture medium will depend both on the amount of tissue to be disinfested and on the number of different cultivars being cultured. If more than one cultivar is used, then each will require a separate set of vessels, each carefully labeled with the cultivar name. It is nearly impossible to distinguish cultivars once the shoots have been prepared for culture or even after shoots have begun to proliferate. Each student or team of students should have a minimum of ten pieces of tissue (terminal buds and/or nodal sections) to work with. Baby food jars with B-caps (Magenta Corp., Chicago, IL) are conveniently sized vessels; large jars are for bleach and rinse water, and smaller jars for culture medium.

The basal culture medium is composed of woody plant medium (WPM) basal salts (Lloyd and McCown, 1980) — see Chapter three for composition. This medium and all other chemicals listed can be purchased from Sigma Chemical Company, St. Louis, MO. The basic medium is amended with 0.4 g (5 mM) NH_4NO_3, 1.3 g (6 mM) D-gluconic acid (hemicalcium salt), 20 g (59 mM) sucrose, 4 g agar, and 1.4 g/L Phytagel™. Five grams/liter AgarGel™, a mixture of the above two gelling agents, can be substituted. The gluconic acid is added to help prevent hyperhydric growth that can be a problem with lilacs. Approximately 30 ml of medium is used per small baby food jar.

Follow the protocols listed in Procedure 12.1 to complete this experiment.

Anticipated results

New growth from terminal bud explants will be apparent in approximately 2 weeks; axillary buds from nodal segments are somewhat slower, but new shoots should form in about 4 weeks. Once the explant tissue has produced new shoots, the experiment to investigate the effects of different cytokinins and concentration on axillary shoot proliferation can begin.

Experiment 2. Shoot production from axillary buds

Each student or student team will need ten jars of the basal medium described above — two jars of each of the five PGR concentrations listed below:

Procedure 12.1
Tissue preparation, disinfestation, and initial establishment of cultures.

Step	Instructions and comments
1	When disinfesting tissue from greenhouse grown plants, vegetative shoots are generally cut into small segments approximately 2 to 3 cm long. This would include the terminal (apical) bud and several segments including a single node each. Using a razor blade (Figure 12.1), remove the leaf blade, but keep a small section of the petiole that is subtending the bud. Removing the entire petiole and leaving the bud exposed results in much greater injury and consequent bud death in the bleach solution. The trade-off is that the petiole-stem-bud junction may harbor spores that are protected from the disinfesting action of the bleach solution; however, we find this to be the lesser problem of the two.
2	Once the shoots are cut into sections and the leaves trimmed, place them into the bleach solution. This operation and all subsequent steps of the procedure are completed in a laminar flow hood using aseptic techniques. Disinfest the tissue in 15% bleach-soap solution for 15 min with some agitation to break the surface tension and to ensure that all portions of the explant are exposed. The concentration of bleach vs. the time in solution is a dose effect and both variables could be altered if successful disinfestation is not achieved.
3	Remove the tissue pieces from the bleach with sterile forceps and place into sterile water contained in the first rinse jar. Agitate gently for a few minutes, then transfer shoots to the second rinse vessel. Following the second rinse, tissue sections are ready to be transferred to solidified, sterile culture medium contained in the small baby food jars.
4	Ideally each student (or team) will have a minimum of ten pieces of tissue to work with. All tissue segments containing buds should be placed in vessels containing medium without PGRs. A single piece of tissue (a node or terminal bud) per jar should be gently pushed, but not submersed, into the surface of the medium. Cultures should be placed under 24-h artificial light (25 $\mu Mol \cdot m^{-2} s^{-1}$) at a temperature of 26 to 27°C. If a culture room is not available, a single 40-W fluorescent light 25 cm above a shelf or bench will provide adequate light. Lilacs like warm temperatures and growth will be much slower if the temperature is lower. Cultures should be checked daily for contamination. Fungal growth is generally apparent within a few days, but bacterial contamination may take a week to 10 days to be visible. Discard contaminated cultures — hopefully they will be few!

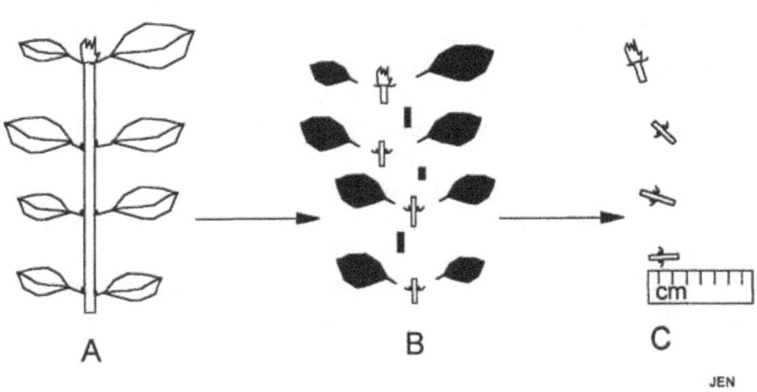

Figure 12.1 Diagram illustrating how to cut shoots (A) into single-node segments 2 to 3 cm long, removing the leaves (B), and keeping a small part of the petiole subtending the bud (C). Shaded sections are discarded. (From Trigiano, R. N. and Gray, D. J., Eds., *Plant Tissue Culture Concepts and Laboratory Exercises, First Edition,* CRC Press LLC, Boca Raton, FL, 1996.)

1. No PGRs
2. 0.22 mg (1 μM) benzyladenine (BA)
3. 1.0 mg (5 μM) BA
4. 0.22 mg (1 μM) zeatin
5. 1 mg/L (5 μM) zeatin

Be sure to have a system for labeling different PGRs and their concentrations. We use a different colored marker for each PGR; a black zero = no PGR, a green one or five equals either 1 or 5 μM BA, and a purple one or five equals either 1 or 5 μM zeatin (another type of cytokinin).

The objective is to obtain good shoot proliferation from axillary buds without promoting adventitious shoots (see Chapter fourteen). The PGR concentrations provided are not definitive, but are only meant as beginning guidelines for these experiments. For commercial production, a more sophisticated experimental matrix consisting of other cytokinins and more concentrations is usually designed. The medium without PGRs is included as a control and shoot proliferation would be expected to be relatively poor.

Follow the protocols outlined in Procedure 12.2 to complete this experiment.

Procedure 12.2
Shoot production from axillary buds.

Step	Instructions and comments
1	As new shoots are produced from the initial isolated buds, they should be transferred to the above media. Shoots should be cut as before so that the terminal bud is in one baby food jar and the nodal sections are in other jars (Figure 12.2). With ten original explants, the shoots in two jars can be subcultured onto each of the five different PGRs containing media. Therefore, you should be able to generate ten new jars — two for each treatment or medium. If some of the initial cultures have been lost to contamination, distribute the remainder among the ten new jars as equally as possible. At this point, only one piece of tissue per jar is needed, but work carefully to avoid contamination!
2	At the warm incubation temperatures suggested above, lilacs can be subcultured about every 3 weeks. Shoots should be maintained on consistent PGR type and concentration (e.g., if cultured on medium containing 1 μM BA, then they should be subcultured to the same medium) and divided so that terminal buds and single node sections are in separate jars. As shoots grow and proliferate, jars will be filled with more tissue and you will begin to accumulate more than two jars per treatment. Six to ten pieces of tissue per jar will be adequate for data collection. For commercial purposes, individual jars will contain around 20 shoots each. At each subculturing, notes should be taken on the number of axillary shoots produced. Apical dominance will keep the terminal shoots unbranched, but you should see differences in the number of axillary shoots produced from the nodal segments among the different treatments. Axillary shoot production from single node sections is the primary method we use for multiplication of lilacs in commercial tissue culture.
3	Construct a table for each cultivar. You will probably want to include the name of the cultivar, PGR type and concentration, date (and perhaps the number of times the shoot has been subcultured), and the number of axillary shoots produced per node segment or terminal bud.

Anticipated results

At Knight Hollow Nursery, we specialize in introducing new and often proprietary cultivars to the market. We have found that 1 μM zeatin provided the best results in terms

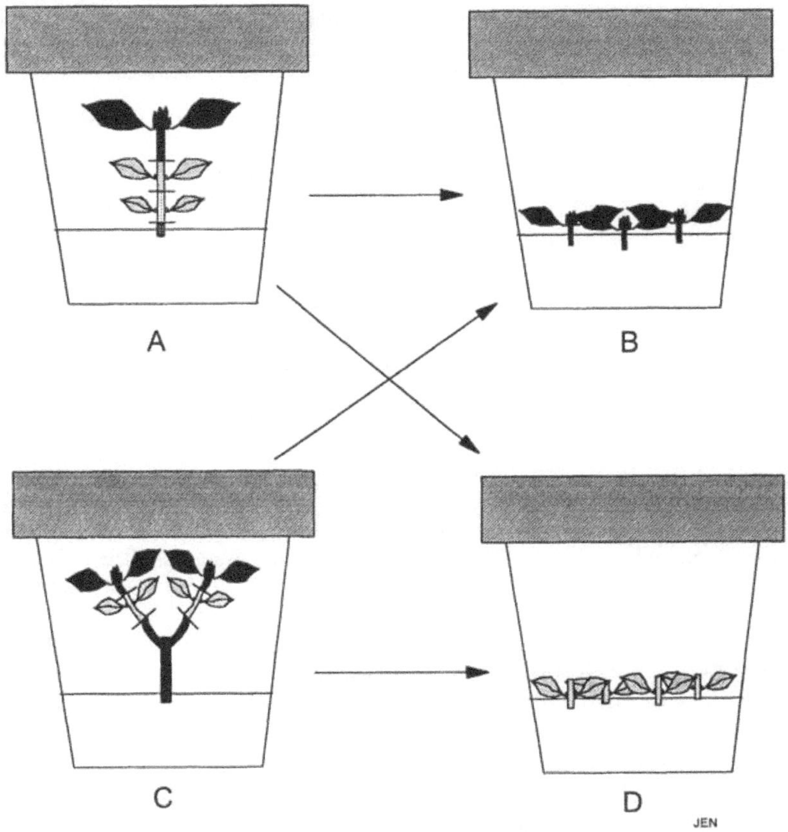

Figure 12.2 Diagram illustrating subculturing the new isolates and separating terminal buds and nodal sections. Jars labeled A and C are the new isolates and jars labeled B and D are fresh culture medium. Terminal buds are placed in one set of jars (B); nodal sections are placed in another set (D). (From Trigiano, R. N. and Gray, D. J., Eds., *Plant Tissue Culture Concepts and Laboratory Exercises, First Edition*, CRC Press LLC, Boca Raton, FL, 1996.)

of axillary bud proliferation in the dozen lilac cultivars we have micropropagated. Many environmental conditions can influence axillary shoot proliferation and you may find that under your laboratory conditions or for your particular cultivar, another medium gives improved multiplication. That is part of the fun of experiments.

Data on three cultivars that have been maintained on 1 μM zeatin are presented in Tables 12.1 and 12.2. They should be used only for comparison with the results of your experiment. Table 12.1 shows the axillary shoot proliferation at 3.5 weeks or at the point we would begin to subculture again. If cultures are allowed to grow an additional 2 weeks, the number of usable (or rootable) microcuttings increases significantly (Table 12.2). Thus, the last subculture round prior to rooting is generally slightly longer than the usual subculture period.

The experiment discussed above can be completed in a normal semester of 16 weeks, but it will be critical to start immediately. If 4 weeks are allowed following initial isolation of tissues for new shoot development and then shoots are subcultured at 3-week intervals, there will be a maximum of five subculture cycles. This experiment works using French hybrid lilacs because they are so easily established in culture, which is unusual for woody plants. We have found that *Celastris* (both American Bittersweet and Chinese Bittersweet) behave in a similar manner and can be substituted for lilacs if they are unavailable.

Table 12.1 Examples of Axillary Shoot Production of *Syringa vulgaris* 'Little Boy Blue' (LBB), 'Marie Frances' (MF), and 'Arch McKean' (AM) Cultivars at the End of a 3.5-Week Growing Cycle

	Initial nodal segments			Usable cuttings generated		
Jar	LBB	MF	AM	LBB	MF	AM
1	22	17	16	34	25	27
2	21	17	13	17	23	10
3	14	13	16	18	35	12
4	17	13	9	19	16	11
5	19	15	10	21	16	20
6	17	15	21	31	28	43
7	22	12	16	27	9	22
8	22	16	12	20	15	19
9	14	14	15	30	10	22
10	20	18	21	8	19	29
Average	19	15	15	20	20	21
Multiplication factor				1.1	1.3	1.4

From Trigiano, R. N. and Gray, D. J., Eds., *Plant Tissue Culture Concepts and Laboratory Exercises, First Edition*, CRC Press LLC, Boca Raton, FL, 1996.

All lilac shoots produced in culture can be successfully rooted ex vitro with no additional treatment. We do not know of any commercial tissue culture laboratory that currently roots lilacs in vitro; culture space is too valuable and better and more functional root systems are produced ex vitro (McCleland and Smith, 1988). Lilac microcuttings generally root at nearly 100% so long as the following conditions are met. First, the cuttings should be actively growing and not hyperhydric. Second, high humidity must be maintained until cuttings are sufficiently rooted, then humidity can be decreased gradually. Third, while the humidity is decreased, the light level should be increased. The rooting and acclimization process generally takes approximately 8 weeks.

Table 12.2 Examples of Axillary Shoot Production of *Syringa vulgaris* 'Little Boy Blue' (LBB), 'Marie Frances' (MF), and 'Arch McKean' (AM) Cultivars at the End of a 5.5-Week Growing Cycle

	Initial nodal segments			Usable cuttings generated		
Jar	LLB	MF	AM	LLB	MF	AM
1	22	17	12	59	35	39
2	17	15	19	45	47	40
3	20	19	15	63	47	52
4	22	13	16	42	48	35
5	21	19	14	46	62	36
6	14	20	18	53	44	58
7	21	14	14	56	41	39
8	22	15	12	42	30	32
9	20	20	12	61	54	30
10	18	17	18	43	28	32
Average	20	17	15	51	44	36
Multiplication factor				2.6	2.6	2.4

From Trigiano, R. N. and Gray, D. J., Eds., *Plant Tissue Culture Concepts and Laboratory Exercises, First Edition*, CRC Press LLC, Boca Raton, FL, 1996.

Questions

- If the 5 μM cytokinin concentration yields the highest axillary shoot proliferation, what follow-up experiment would you design? What problems might be encountered?
- How would you distinguish axillary shoots from adventitious shoots?
- In a commercial operation, terminal buds and nodal sections are often kept in separate vessels. Can you explain the reason for this practice?
- If nodal sections were left as two or three node pieces, would you predict the same or different axillary shoot proliferation? What effect would cytokinin concentration have on this response? How?
- Would orientation of the nodal pieces in the medium (vertical, lateral, upside down) change the response? How?

Literature cited

Lloyd, G. and B. McCown. 1980. Commercially-feasible micropropagation of Mountain Laurel, *Kalmia latifolia*, by use of shoot-tip culture. *Proc. Intl. Plant Prop. Soc.* 30:421–427.

McCleland, M.T. and M.A.L. Smith. 1988. Response of woody plant microcuttings to in vitro and ex vitro rooting methods. *Proc. Intl. Plant Prop. Soc.* 38:593–599.

McCown, D.D. and B.H. McCown. 1987. North American Hardwoods. In: *Cell and Tissue Culture in Forestry*,Vol. III. J.M. Bonga and Don J. Durzan, Eds. Martinus Nijhoff, Boston. pp. 247–260.

McCulloch, S. 1989. Tissue culture of French hybrid lilacs. *Proc. Intl. Plant Prop. Soc.* 39:105–114.

chapter thirteen

Micropropagation and in vitro flowering of rose

Michael E. Kane

Roses (*Rosa* spp.) were among the first ornamental plants to be domesticated. In recent years there has been a significant increase in the demand for rose plants. In the U.S. alone, more than 1.2 billion blooms are sold annually. Roses are propagated for three distinct production sectors: greenhouse cut flower production, pot-grown flowering houseplant sales, and the home gardening market. Many roses cultivars are propagated asexually by budding or grafting of the desired scion cultivar on selected rootstock. Miniature (dwarf) roses are propagated on their own root systems. There have been numerous attempts to develop micropropagation protocols to produce roses that are true-to-type and superior to traditionally propagated rose plants in terms of price, quality, and growth performance (Khui-Khosh and Sink, 1982; Dubois et al., 1988; Campos and Pais,1990; Skirvin et al., 1990).

Although commercial rose micropropagation has not been very successful in the U.S., the technology has clearly facilitated early introduction of new cultivars to the consumer. In vitro techniques have also been used to introduce new genotypes by mutation breeding. One interesting response of in vitro cultured miniature roses is the production of flowers (Campos and Pais, 1990; Kane et al., 1991; 1992). In vitro flowering provides a unique system to also study factors controlling flower production and longevity (Kane et al., 1991; Van Staden and Dickens, 1992).

The purpose of this laboratory exercise is to illustrate the sequential steps required for the micropropagation of miniature roses by shoot culture (see Chapter eight). The laboratory exercise will familiarize students with procedures used to (1) establish Stage I cultures using nodal explants; (2) rapidly propagate plants by axillary shoot proliferation (Stage II); and (3) successfully root and acclimatize microcuttings ex vitro (Stage IV). This exercise also offers the opportunity for students to observe in vitro flowering. The laboratory exercise is outlined below including maintenance of stock plants, media preparation, procedures, and anticipated results.

General considerations

Selection and maintenance of miniature rose cultivars

Many miniature rose cultivars are available and can be purchased as potted plants at most retail garden shops. Cultivars do vary with respect to initial culture establishment and capacity for in vitro flowering. Of 36 miniature rose cultivars screened, 8 (22%) were

0-8493-2029-1/00/$0.00+$.50

unresponsive in vitro (chlorosis, leaf abscission, or severe hyperhydricity) (Kane et al., 1992). Of the remaining 28 responsive cultivars subcultured, 14 (50%) produced flower buds in vitro. Consequently, it is wise to use several miniature rose cultivars. Cultivars such as *Rosa* cv. *Royal Ruby, R.* cv. *Red Minimo, R.* cv. *Rosmarin, R.* cv. *Nancy Hall,* or *R.* cv. *Rosa Rouletti* are easy to establish in vitro and will flower following subculture.

Regardless of the miniature rose cultivar chosen, purchase plants exhibiting new shoot growth if they are to be used immediately. Nodal explants taken from newly formed shoots can be more readily surface disinfested and responsive. If not used immediately, plants should be maintained in full sunlight (greenhouse in winter or outdoors in summer) and fertilized weekly (150 mg/L N). Three weeks before using, plants should be trimmed to promote lateral shoot production. One plant in a 10-cm-diameter pot should yield sufficient numbers of nodal explants for each student or team of students.

Materials

The following are needed for each student or student team:

- Potted miniature rose plants to provide 15 two-node explants per student
- 50% ethanol
- 20% commercial bleach
- Ten culture tubes each containing 12 ml rose establishment/multiplication medium (REM)
- Five GA-7 vessels each containing 60 ml REM
- Incubator or a fluorescent light on a bench top providing approximately 30 $\mu mol \cdot m^{-2} \cdot sec^{-1}$ for a 16-h photoperiod and 25°C

Composition and preparation of the medium

One advantage of this exercise is that the same culture medium can be used for both culture establishment (Stage I) and subsequent axillary shoot multiplication (Stage II) of miniature roses. The basal medium (BM) consists of macro- and micronutrients and vitamins as described by Murashige & Skoog (1962) (see chapter Three for composition) supplemented with 87.6 mM (30 g/L) sucrose, 2.5 µM (0.5 mg/L) BA, 0.6 µM (0.1 mg/L) IAA, 0.24 mM (50 mg/L) citric acid, and 0.28 mM (50 mg/L) ascorbic acid and solidified with 1.5 g/L Phytagel and 4.0 g/L agar. The BM can be purchased prepackaged from Sigma Chemical, St. Louis, MO (M5519). Medium is adjusted to pH 5.5 before addition of gelling agents and dispensed as 12-ml volumes into 150 × 25-mm culture tubes and autoclaved. Ten culture tubes are prepared for each student or team of students.

Exercises

Experiment 1. Stage I establishment and Stage II shoot multiplication from nodal explants

The goals of this two-part exercise are to familiarize students with procedures used to successfully establish Stage I miniature rose shoot cultures from two-node explants (Figure 13.1a) and propagate shoots through production of axillary shoots (Stage II). The goal of Stage I is to establish axenic and responsive shoot cultures. Cultures are indexed for specific pathogens or cultivable microbial contamination and allowed to become physiologically adapted to in vitro culture conditions before being clonally multiplied by repeated proliferation of axillary shoots in Stage II (see Figure 8.2). Before proceeding, the structure of a generalized shoot apex should be briefly reviewed (see Figure 8.1). Follow the protocols outlined in Procedure 13.1 to complete this experiment.

Figure 13.1 Micropropagation and in vitro flowering of the miniature rose. (a) Left: two-node explants (left) are cut from young shoots of potted plants and the leaf blades are removed prior to surface disinfestation. Right: Petiole bases are removed after surface disinfestation to expose lateral buds before inoculation. Scale = 5 mm. (b) Production of axillary shoots and in vitro flowering of the miniature rose following an 8-week subculture. (c) In vitro-produced flowers are smaller and have fewer petals than those produced on plants grown in the greenhouse. Scale bar = 10 mm.

Procedure 13.1
Stage I establishment and Stage II shoot multiplication from nodal explants of rose.

Step	Instructions and comments
1	Cut several newly produced shoots from the donor plant. Remove leaf blades with a scalpel leaving attached only a small basal section (about 3 mm) of each petiole. Cut shoots into two-node explants about 15 mm in length (Figure 13.1a). It can be noted that the smoothly cutinized epidermis, typical of many miniature rose cultivars, should facilitate effective surface disinfestation of the explants. Nodes exhibiting bud break should not be used.
2	Rinse nodal segments in tap water for 15 min and then surface disinfest in 50% ethanol for 1 min followed by a 12-min agitated soak in 20% commercial bleach and three 5-min rinses in sterile deionized water.
3	In the transfer hood, trim off the bleached ends of each two-node explant and the base of each petiole to expose the lateral bud (Figure 13.1a).
4	Transfer a single two-node explant into each of ten culture tubes containing REM. Partially embed the basal end vertically into the medium.
5	Label each tube and maintain in an incubator or under a fluorescent light on a bench top at 25°C and provided with approximately 30 $\mu mol \cdot m^{-2} \cdot sec^{-1}$ for 16 h. To prevent abnormal shoot development and premature leaf abscission, sealing films such as Parafilm should not be used.
6	After 4 weeks, students should determine: (1) the percentage of visibly contaminated cultures; (2) mean number of axillary shoots produced per explants; and (3) percentage of flowering cultures. Data can be collected without removing the plants from the culture vessels.

Procedure 13.1 continued
Stage I establishment and Stage II shoot multiplication from nodal explants of rose.

Step	Instructions and comments
7	In the transfer hood, Stage I shoot clusters are aseptically divided into defoliated two-node secondary explants with five explants inoculated into each of five GA-7 vessels containing 60 mL freshly prepared REM. Cultures are maintained under the conditions described above and observed weekly.
8	After 7 to 8 weeks, students should determine: (1) mean shoot number and length produced per explant and (2) percentage of explants producing flowers. The data collected by each student or student team at 4 and 7 to 8 weeks should be shared to provide more representative responses.

Anticipated results

Many miniature rose cultivars display vigorous growth when established in vitro; the actual response is highly cultivar dependent. Axillary shoots should develop from the cultured primary nodal explants by 2 weeks' culture. By week 4, two to three axillary shoots are produced from each nodal explant. In vitro flowering seldom occurs on shoots produced on primary nodal explants during the first culture period. This suggests that competency for flowering is acquired only following subculture.

Nodal explants subcultured into GA-7 vessels should exhibit vigorous shoot development (Figure 13.1b). Shoots produced in larger vessels, such as GA-7 magenta vessels, will exhibit a greater frequency of flowering. Floral buds are produced by week 6 and the flowers open by week 8 (Figure 13.1b). Approximately 30% of the shoots produce open and very fragrant flowers. Flowers produced in vitro are much smaller and are composed of fewer petals than those produced on plants grown in the greenhouse (Figure 13.1c). Flowers become senescent within 2 weeks of opening. Recurrent flowering does not occur unless shoots are subcultured. Infrequently, vegetative shoots develop from the center of open flowers in vitro. If time constraints exist or demonstration of in vitro rose flowering is the only objective, previously established stock shoot cultures can be used as the source of explants.

Experiment 2. Stage IV — rooting and acclimatization

Depending on the species, from 35 to 75% of the total cost to micropropagate a crop can be reduced by eliminating Stage III rooting (see Figure 8.2). This reflects the significant input of labor and supplies required to complete this stage. The goal of this exercise is to demonstrate direct ex vitro rooting and acclimatization (Stage IV) using the Stage II miniature rose microcuttings produced in Experment 1. Follow the protocols listed in Procedure 13.2 to complete this experiment.

Procedure 13.2
Stage IV rooting and acclimatization.

Step	Instructions and comments
1	Rinse agar off Stage II shoot clusters produced in GA-7 vessels in Experiment 1 and divide into single microcuttings (≥8.0 mm long).
2	Stick microcuttings directly into moistened soilless growing medium contained in plug trays or any other small container placed under intermittent mist (5 sec every 10 min) for 3 weeks under lightly shaded conditions. Alternatively, if a mist system is not available, clear plastic propagation domes can be placed over the plug trays to maintain humidity. After 2 weeks, the domes are gradually removed over a period of 1 week to acclimatize the plantlets to the lower humidity.

Procedure 13.2 continued Stage IV rooting and acclimatization.	
Step	Instructions and comments
3	Evaluate microcutting survival and rooting 3 weeks posttransplant. Plants can be transferred to the greenhouse or laboratory.

Anticipated results

Miniature rose microcuttings should exhibit 100% ex vitro rooting and acclimatization by week three. Acclimatized plants should begin flowering again within 4 weeks posttransplant. Typically acclimatized micropropagated miniature roses are fuller and exhibit growth rates about twice that of plants produced by cutting propagation (Dubois et al., 1988).

Questions

- In terms of disease elimination, what is the disadvantage of using nodal explants?
- Why do shoots produced following subculture exhibit a greater capacity for flowering than those produced on primary explants?
- Which cultural factors could reduce the size of the flowers produced in vitro?
- Premature deterioration and/or wilting of cut rose flowers has been attributed to both the buildup of bacteria in cut flower water and/of products exuded by the stem. How could in vitro flowering of the rose be used to study this problem?

Literature cited

Campos, P.S. and M.S.S. Pais. 1990. Mass propagation of the dwarf rose cultivar 'Rosamini'. *Scientia Hort.* 43:321–330.

Dubois, L.A.M., J. Roggemans, G. Soyeurt, and D.P. De Vries. 1988. Comparison of the growth and development of dwarf rose cultivars propagated in vitro and in vivo by softwood cuttings. *Scientia Hort.* 35:293–299.

Kane, M.E., F. Marousky, and N. Philman. 1991. Producing microorganism-free rose flowers: the tissue culture approach. *Amer. Rose Mag.* 31:16–17.

Kane, M.E., N.L. Philman, and F.J. Marousky. 1992. In vitro flowering of miniature rose cultivars. *HortScience* 27:206.

Khui-Khosh, M. and K.C. Sink. 1982. Micropropagation of new and Old World rose species. *J. Hort. Sci.* 57:315–319.

Murashige, T. and F. Skoog. 1962. A revised medium for rapid growth and bioassays with tobacco cultures. *Physiol. Plant.* 15:473–497.

Skirvin, R.M., M.C. Chu, and H.J. Young. 1990. Rose. In: *Handbook of Plant Cell Culture. Vol. 5.* P.V. Ammirato, D.A. Evans, W.R. Sharp, and Y.P.S. Bajaj, Eds., pp. 716–743. McGraw-Hill, New York.

Van Staden, J. and C.W.S. Dickens. 1992. In vitro induction of flowering and its relevance to micropropagation. In: *Biotechnology in Agriculture and Forestry, Vol. 17. High Tech and Micropropagation 1.* Y.P.S. Bajaj, Ed., pp. 85–115. Springer-Verlag, Berlin.

chapter fourteen

Organogenesis

Otto J. Schwarz and Robert M. Beaty

The ability of plant tissues to form various organs de novo (i.e., organogenesis) has long been an object of interest and practical utility. The ancient Chinese successfully cloned selected genotypes of forest trees through the organogenic process of adventitious rooting of woody cuttings. In a recent historical review of adventitious rooting research, the practice of "cuttage" or rooting of cuttings was said to predate the time of Aristotle (384 to 322 B.C.) and Theophrastus (371 to 287 B.C.) (Haissig and Davis, 1994). The process of organogenesis provides the basis for asexual plant propagation largely from nonmeristematic somatic tissues. One needs only to visit a local garden center in the early spring or fall to witness the many varieties of woody and herbaceous plants offered for sale to the gardening public. Many of these plants are propagated exclusively via asexual means. The cloning of these selected genotypes probably included the de novo organ formation of roots on cuttage; the in vitro multiplication of preexisting shoot meristems followed by de novo root formation on the resultant microshoots; or possibly de novo shoot regeneration on in vitro cultured explant tissues followed by adventive rooting of the microshoots. Whether accomplished in vivo or in vitro using either meristematic or nonmeristematic tissues, the de novo genesis of plant organs is broadly defined here as organogenesis. As stated above, our focus is specifically on "organogenesis in vitro" on nonmeristematic plant tissues. To further understand the problems to be encountered in obtaining an all-encompassing definition of the terms associated with plant organ regeneration in vitro or in vivo, a reading of the reviews by Dore (1965) and Haissig and Davis (1994) is recommended.

The material presented in this chapter represents only a small fraction of the information available in the scientific literature concerning the developmental process of organogenesis. It is hoped that the concepts selected for discussion provide a model framework upon which to build an understanding of the morphogenic events that accompany organogenesis and, in addition, to begin to gain insight into the complex process that directs the morphogenetic process. The production of adventitious apical meristems on conifer cotyledonary tissue is presented in some detail as an example of the organogenic process, followed by a brief discussion of the process of adventitious root initiation.

0-8493-2029-1/00/$0.00+$.50

Organogenesis as a developmental process

In its broadest sense, development is the process that results in a functional, mature organism. According to Fosket (1994) it includes "... all of those events during the life of a plant or animal that produce the body of the organism and give it the capacity to obtain food, to reproduce and to exploit the opportunities and deal with the hazards of its environment." Organogenesis is a developmental process that is in some ways unique to plants. Animal cells follow developmental paths that normally involve irreversible differentiation toward a specific cell or tissue type. In other words, animal cells and tissues remain structurally and functionally committed to an initial developmental end point. Plant cells, however, may retain the ability to dedifferentiate from their current structural and functional state and to begin a new developmental path toward a number of other morphogenetic end points.

Plant tissues in vitro may produce many types of primordia, including those that will eventually differentiate into embryos, flowers, leaves, shoots, and roots. For the purpose of this chapter, nonzygotic embryogenesis will not be discussed; see Chapter nineteen. Although the botanical definition of an "organ" is quite restrictive, in this discussion we regard the development of any of the aforementioned primordia to be encompassed by the defintion of "organogenesis." These primordia originate de novo from a cellular dedifferentiation process followed by initiation of a series of events that results in their formation. The cell or cells thought to be the direct progenitors are somehow stimulated to undergo a number of rapid cell divisions leading to the formation of a meristemoid. Meristemoids are characterized as being an aggregation of meristem-like cells (i.e., smaller, isodiametric, thin-walled, microvacuolated cells with densely staining nuclei and cytoplasm) (Thorpe, 1978; 1982). Early in their development, meristemoids are thought to be morphogenetically plastic and capable of developing into a number of different primordia (i.e., root, shoot, etc.). This unique developmental flexibility has been widely used by plant propagators. Two organogenic events comprise a common approach to in vitro plant propagation. The first event is to regenerate multiple shoot meristems followed by their growth and development into microshoots of a suitable size for the second event, which is the induction of de novo root meristem production. It is believed that under in vitro culture conditions, these organogenic events can be the result of two differing ontogenic pathways.

According to Hicks (1980), two developmental sequences leading to organogenesis emerge from the literature. They differ in the presence or absence of a callus stage in the organogenic sequence of events. A developmental sequence involving an intervening callus stage is termed "indirect" organogenesis.

$$\begin{matrix}\text{PRIMARY}\\ \text{EXPLANT}\end{matrix} \rightarrow \text{CALLUS} \rightarrow \text{MERISTEMOID} \rightarrow \begin{matrix}\text{ORGAN}\\ \text{PRIMORDIUM}\end{matrix}$$

De novo organ formation via indirect organogenesis may increase the possibility of introducing variation in the chromosomal constitution (e.g., ploidy change) of the cells in the callus stage and, hence, the possibility for both physiological and morphogenic variation in the resulting organs. This type of variation has been termed "somaclonal variation" (see Chapter forty). Although some plants are more prone to this problem than others, a general "rule of thumb" for plant propagation is to minimize any stage in the organogenic sequence containing "unorganized" callus growth. For the research scientist concerned with determining the physicochemical mechanism(s) driving the organogenic process, there is a second important reason to avoid or at least minimize this type of developmental event. The presence of a proliferating callus stage has the potential to greatly complicate the analysis of ongoing molecular events that accompany and perhaps drive de novo organ production. Those particular cells or groups of cells that are destined to become organogenic are in relatively small number among the much larger mass of dividing callus cells, reducing the ability to analyze their particular chemistry and physiology.

Direct organogenesis is accomplished without an intervening proliferative callus stage. For de novo organogenesis (i.e., shoot, root, flower) the sequence of events is described as follows:

PRIMARY EXPLANT → MERISTEMOID → ORGAN PRIMORDIUM

The major distinction between these two de novo pathways is, of course, the presence or absence of a discernible callus stage early in the morphogenic sequence of events. The callus tissue, containing cells that have dedifferentiated into a less-determined, morphologically more flexible form, serves as the starting point for de novo organogenesis. In the absence of a callus intermediate stage, cells present within the explant have the capacity to act as the direct precursors of the new primordium. Hicks (1980) presents a substantial listing of in vitro culture systems that provide examples of direct and indirect organogenesis from both meristematic and nonmeristematic explant tissues.

Researchers have utilized the organogenic process as a model system for asking questions about the causal events that direct shoot or root production and, by generalization, other morphogenic processes. In one such model, the process of organogenesis has been broken into several phases illustrated in the following diagram (adapted from Christianson and Warnick, 1985):

COMPETENCE + DETERMINATION +

EXPLANT ————————→ ORGAN

1 2 3

(DEDIFFERENTIATION) (INDUCTION) (DIFFERENTIATION)

Key points of interest to developmental biologists are the initial two phases that precede the differentiation phase. These phases encompass events that begin with dedifferentiation, which results in the attainment of "competence," followed by induction, which culminates in the fully "determined" state. The morphological differentiation and development of the nascent shoot or root then proceed, eventually resulting in a functional organ. The first two phases of the model encompass those events that occur prior to morphogenesis. As shown by Hicks (1980), embryos, seedling parts, apical meristems, primordial organs, cell layers from mature organs, complex fragments from mature organs, intact organs, and protoplasts have been used for primary explants.

Dedifferentiation

The process of dedifferentiation involves reversion to a less committed, more flexible/plastic developmental state that may or may not give rise to callus tissue. In the case of direct organogenesis, competent cells that have not produced callus tissue are thought to be the sole organ progenitors, implying that they undergo the dedifferentiation process and proceed individually or perhaps in small groups to produce a new primordia. Leaf explants of *Convolvulus arvensis* L. produce roots or shoots via an indirect organogenic pathway that involves a dedifferentiation process that produces limited callus growth from which the new organs are generated (Christianson and Warnick, 1983). The result of completion of this first phase is that the primary explant acquires a state of "competence," which is defined by its ability to respond to organogenic stimuli.

Attainment of tissue competence is not always a single-step process involving, for example, the application of plant growth regulators (PGR). When cultured for 4 days on relatively high kinetin and low 2,4-dichlorophenoxyacetic acid (2,4-D) levels and then removed to PGR-free medium, alfalfa (*Medicago sativa* L.) callus produces roots. Shoots are produced when callus is treated to the same culture regime except that the medium contains high 2,4-D and low kinetin levels. However, the PGR pretreatment followed by transfer to basal, PGR-free medium is not sufficient to bring all of the tissue through the second (induction) phase that is required for eventual root or shoot production (Walker et al., 1979). An additional size requirement imposed a lower limit of 105 μm as the minimal diametric size for callus cellular aggregates to be competent for morphogenetic induction. Cellular aggregates allowed to grow to the minimum size became competent for organogenic induction and root or shoot formation.

Induction

The induction phase occurs between the time the tissue becomes "competent" and the time it becomes fully "determined" for primordia production. This phase has been shown to encompass a number of "phenocritical" times postulated to result from the function of an integrated gene pathway that guides the developmental process and precedes morphological differentiation. Landauer (1958) suggested that certain chemical and physical stimuli could interrupt a genetically determined developmental pathway, modifying the morphogenic result, which produces a mutant-like phenotype or "phenocopy." Christianson and Warnick (1984) have identified several chemical agents that intervene during the induction phase of *C. arvensis*. Agents that induce the production of a phenocopy are called phenocopying agents. These phenocopying agents were shown to be only effective during certain experimentally determined times during the induction phase. These "phenocritical times" were postulated to define key steps in the genetic pathway orchestrating the developmental process. The phenocopying agents, or stage-specific inhibitors, were thought in some direct way to block the action of a gene or gene product of a currently active "orchestrating gene." In the *C. arvensis* system, this meant blocking shoot organogenesis and producing the modified phenotype (i.e., phenocopy), in this case, callus.

The end of the induction process is defined as the point when a cell or group of cells becomes fully committed to the production of shoots or roots. Operationally, this end point is attained when explant tissue can be removed from the root- or shoot-inducing medium; can be placed on basal medium without PGR, containing mineral salts, vitamins, and a source of carbon; and can proceed to produce the desired organs. The tissue has now completed the induction process and can be considered to be fully "determined."

There is another term of interest that is used to describe or perhaps measure the extent of developmental commitment toward a particular end point (i.e., bud formation) that a tissue has reached during the induction phase. This term is "canalization," and it is defined by Waddington (1940) as ".... the property of developmental pathways of achieving a standard phenotype in spite of genetic or environmental disturbances." *Convolvulus arvensis* explant tissue that has become fully canalized for shoot production can be placed on media designed to induce roots and still form shoots (Christianson and Warnick, 1983). If the explant is removed from the shoot-inducing medium before it becomes fully canalized, shoot production is severely inhibited, with root production becoming the prominent end point. This developmental flexibility illustrates the degree of morphogenic plasticity that can be exhibited by plant tissue under in vitro culture conditions.

The extent that explant tissues become competent and eventually are determined is, in part, a function of the physical and chemical environment to which they have been exposed. Under in vitro culture conditions, the properties of the medium and associated culture environment play a central role in delivering these organogenic signals. Skoog and Miller (1957) first demonstrated that a key system for the chemical regulation of in vitro organogenesis was the ratio of auxin to cytokinin present in the culture medium. They summarized their findings by concluding that "These tend to show that quantitative interactions between growth factors, especially between IAA and kinetin (auxins and kinins) and between these and other factors, provide a common mechanism for the regulation of all types of growth investigated from cell enlargement to organ formation." A great many other variables, such as day length, light quality and quantity, explant tissue age and size, genotype, and mineral nutrition, to mention a few, have been shown to affect regeneration ability. It is certainly possible that, as Christianson and Warnick (1988) suggest, "… the possibility that at least some cases of nonresponsiveness to culture in vitro are simply due to the failure of one 'happy medium' to elicit competence for induction and to induce organogenesis." In its broadest sense, this idea suggests that a lack of tissue organogenic responsiveness under any particular set of experimental conditions is the result of failure of the explant tissues to achieve the state of competence for induction. This is not a particularly comforting situation for an investigator's state of mind, if one calculates all of the possible combinations and permutations presented by testing even a few of these parameters. Until appropriate biochemical/genetic markers are discovered that clearly indicate the current developmental disposition of the primary explant tissue and until methods are discovered that can override the recalcitrant tissues' current state of genetic programming, our approach will have to remain that of "the educated guess."

Differentiation

Of the three phases proposed by the model, the third phase, or differentiation phase, is the best documented. It is in this phase of the process that morphological differentiation and development of the nascent organ begins. In describing the initial differentiation events of organ initiation, McDaniel (1984) states that "The general picture that emerges is that organ initiation involves an abrupt shift in polarity followed by a smoothing of this shift into a radially symmetrical organization and the concurrent growth along the new axis to form the bulge characteristic of organogenesis." These initial events and those leading to bud formation are shown in pictorial series in Figures 14.1 to 14.3. There has been a great deal of discussion in the literature concerning these initial differentiation events with respect to the tissue types involved and, perhaps more pointedly, to the precise number of cells that take part in meristem initiation. Again the subject remains somewhat obscure in that it seems that there is not an absolute answer. Christianson (1987) states, based on his own experiments and those of Marcotrigiano and Gouin (1984a; 1984b) using plant chimeras, that "… shoots formed in vitro can arise from more than one cell in an explant, but they usually do not." Based on extensive cytological investigation of the cellular events surrounding organogenic shoot initiation on immature zygotic embryos of sunflower (*Helianthus annuus* L.), Bronner et al. (1994) concluded that shoots developed from the "simultaneous divisions of several cells" and, therefore, were multicellular in origin. The epidermis and outer layers of the cortex of the immature embryos were found to contribute these germinal cells. The tissue origin and cell number involved in these very early differentiative events seem

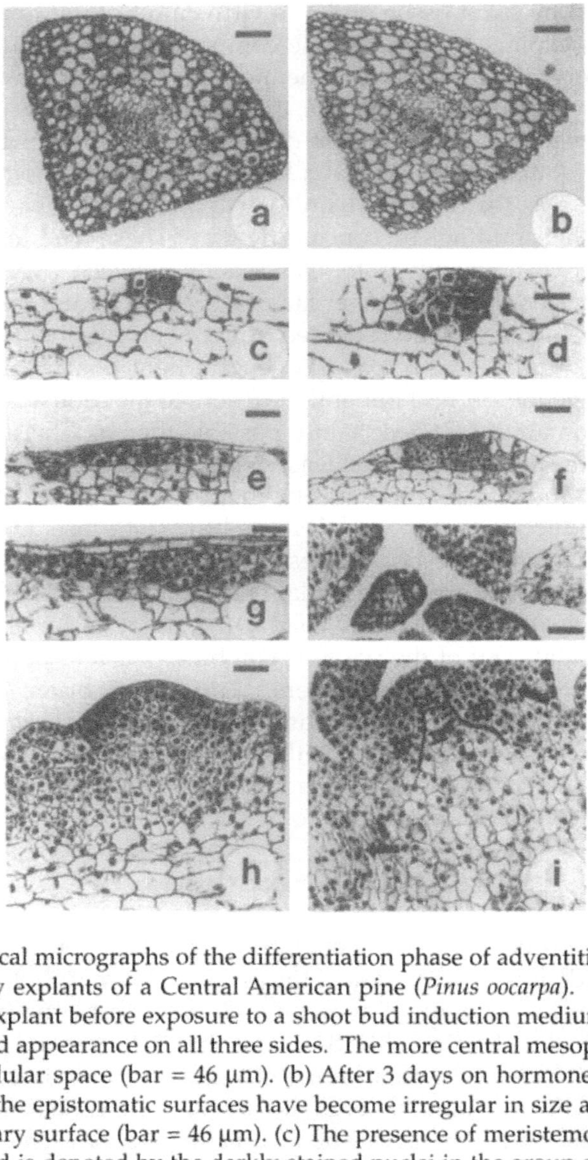

Figure 14.1 Histological micrographs of the differentiation phase of adventitious bud formation on seedling cotyledonary explants of a Central American pine (*Pinus oocarpa*). (a) Transverse section of the cotyledonary explant before exposure to a shoot bud induction medium. The epidermis has a regular, well-ordered appearance on all three sides. The more central mesophyll tissue is compact and has little intercellular space (bar = 46 μm). (b) After 3 days on hormone-containing induction medium, the cells of the epistomatic surfaces have become irregular in size and shape, resulting in an uneven cotyledonary surface (bar = 46 μm). (c) The presence of meristemoids is easily observed after 12 to 18 days and is denoted by the darkly stained nuclei in the group of small cells clustered at the top center of the micrograph (bar = 28 μm). (d) The meristematic region rapidly increases in size. In this case, initial growth extends the meristemoid downward into the mesophyll tissue (bar = 28 μm). (e) Meristematic growth may also occur laterally along the plane of the explant surface (bar = 28 μm). (f) The rapidly developing bud meristem begins to protrude above the surrounding epidermis (bar = 46 μm). (g) The dividing cells of the meristematic regions sometimes do not involve the epidermis. These regions may or may not develop into functional bud meristems. (h) A meristematic dome with several leaf primordia protrudes high above the surrounding epidermis. Note the numerous anticlinal divisions in the epidermis just to the left of the center of the dome and the periclinal divisions just to the right of the center of the dome (bar = 25 μm). (i) Fully developed adventitious buds, complete with enclosing outer primordial leaves. Note the procambial strands in the area of the leaf primordia (arrows). The cytohistological zones associated with pine apical meristems are represented as follows: I, apical zone initials; II, central mother cell zone; III, peripheral tissue zone; and IV, rib meristem (bar = 28 μm). (From Bajaj, Y.P.S., Ed. 1991. *Biotechnology in Agriculture and Forestry and Trees*, Vol. 3, Springer-Verlag, New York p. 312. With permission.)

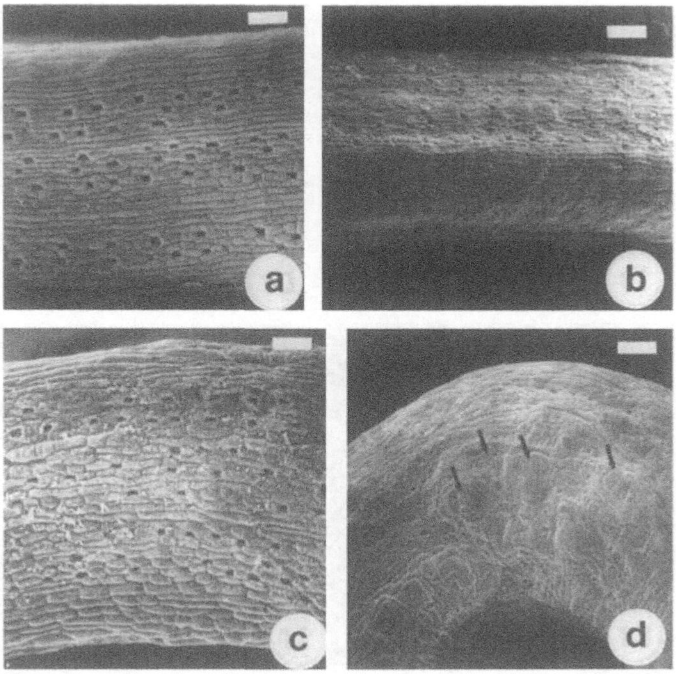

Figure 14.2 Scanning electron micrographs of the differentiation phase of adventitious bud formation on seedling cotyledonary explants of a Central American pine (*Pinus oocarpa*), showing developmental events prior to meristematic dome protrusion. (a) SEM of the explant prior to exposure to shoot bud induction medium. Stomata occur in rows of two and three surrounded by small subsidiary cells. Rows of elongated rectangular epidermal cells are interspersed between the rows of stomata (bar = 83 μm). (b) The explant after 3 days exposure to the induction medium. The abaxial side with almost no stomata and one of the epistomatic sides with its numerous stomata are just beginning to mirror the changes that have already begun in the subsurface tissue by a slight increase in roughness of the epidermal surface (bar = 114 μm). (c) The cells of the epidermis have begun to divide causing an obvious disruption of the regular rows of cells between rows of stomata (bar = 96 μm). (d) The explant surface has now become nodular and uneven. Disruption of the regular rows of cells has increased as the meristematic areas increase in size, producing nodules (bar = 192 μm). (From Trigiano, R. N. and Gray, D. J., Eds., *Plant Tissue Culture Concepts and Laboratory Exercises, First Edition*, CRC Press LLC, Boca Raton, FL, 1996.)

to depend on a number of, as yet, poorly understood variables. Recognizable centers of cell division can occur deep within the explant tissues or, as just described for sunflower, be more superficial in origin.

One can follow, through the careful use of microscopy, the internal histological and surface architectural changes that occur during the differentiation phase of the organogenic process (Figures 14.1 to 14.3). The morphological sequence of events that is depicted tracks the de novo production of shoot buds by the process of direct organogenesis on cotyledons of the Central American pine (*Pinus oocarpa* Schiede). This series is presented as a more or less typical example of the organogenic process. Keep in mind that this is a singular example and that all matter of variations of the developmental theme have been described for in vitro cultured plant species. However, the sequence of events that leads to organogenic shoot bud formation in the cotyledons of this conifer is probably generally applicable to most plant species producing adventitious buds (Thorpe, 1980; Yeung et al., 1981).

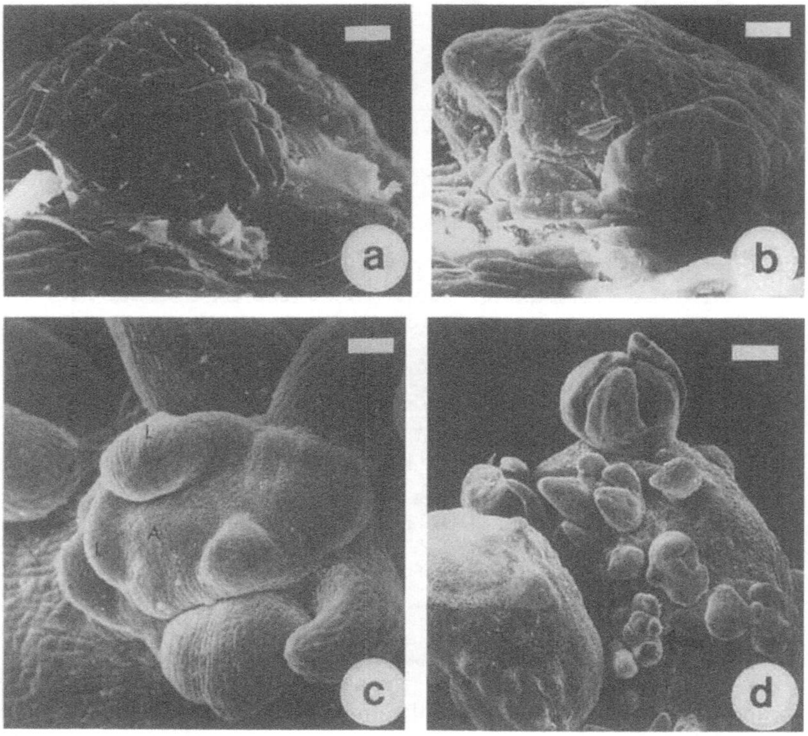

Figure 14.3 Scanning electron micrographs of the differentiation phase of adventitious bud formation on seedling cotyledonary explants of a Central American pine (*Pinus oocarpa*), showing meristematic dome protrusion to a well-developed adventitious bud. (a) Meristematic domes protrude well above the surrounding epidermis of the explant 2 to 3 weeks after beginning exposure to the shoot bud induction medium. Note the continuity of the epidermal surface over the entire dome (bar = 83 μm). (b) A slightly more differentiated dome with newly formed leaf primordia. "L" indicates leaf primordia. "A" indicates the shoot apex (bar = 71 μm). (c) Shoot bud with numerous leaf primordia and elongating leaves about 5 weeks after culture initiation. "L" indicates leaf primordia. "A" indicates the shoot apex (bar = 156 μm). (d) A fully developed shoot bud (B), enclosed by outer primordial leaves. Elongation along the buds central axis (B) has begun. Arrow indicates another bud earlier in development (bar = 714 μm). (From Trigiano, R. N. and Gray, D. J., Eds., *Plant Tissue Culture Concepts and Laboratory Exercises, First Edition*, CRC Press LLC, Boca Raton, FL, 1996.)

The final "differentiation" phase of organogenesis provides the first opportunity to observe the structural genesis of the new organ made possible by the developmental program that was put in place during the preceding "induction" phase. The sequence of morphogenic events observed in de novo bud formation on cotyledons of *P. oocarpa* included changes in surface morphology, the appearance of meristemoids, vertical and/or horizontal expansion of the meristematic region, protrusion of the meristematic region above the surrounding epidermis, the appearance of an organized meristem with leaf primordia, and full development of an adventitious bud. The explant was placed on induction medium containing the PGR benzyladenine (BA) for 10 to 14 days. It was then transferred to basal medium without PGR for the remainder of the experiment. In this particular experiment, the organogenic process was observed for approximately 40 days.

Histological sections of day 0, newly cultured cotyledons illustrate the anatomy of untreated control tissue. In transverse section (Figure 14.1a), the cotyledons are triangular in outline, and the cells of the epidermis are more or less regular in conformation on all three sides of the cotyledon. The round mesophyll cells surrounding the central vascular tissue are compact and exhibit very little intercellular space. Scanning electron microscopy (SEM) micrographs of day 0 control cotyledonary tissue (Figure 14.2a) echo the regularity of the surface and near-surface tissues shown in the histological sections. Even ranks of long, smooth-surfaced rectangular cells lie between irregular double and triple rows of stomata. The subsidiary cells surrounding the stomata are noticeably smaller in surface area than the long ranks of cells between the double and triple rows of epidermal cells. After 3 days on induction medium, the epidermal cells of the epistomatic cotyledonary surfaces have become irregular (Figure 14.1b). Small groups of meristemoids, mainly clustered at or just under the epidermis, are evident by 15 to 18 days (Figure 14.1a, c, d, e, and g). These meristematic zones enlarge both perpendicularly and parallel to the cotyledon surface, taking on a nodular appearance. During this time, as a result of mitotic activity, the epidermal cells lose much of their regularity and become uneven in length, forming small cell clusters (Figure 14.2c and d). These localized areas of meristematic activity begin to form meristematic domes (Figure 14.1f and h) that eventually protrude from the surface of the explant (Figure 14.3a). In this organogenic scheme, the ontogeny of the development of the dome structures involves incorporation of cells of the epidermis. In some conifers, the meristematic domes have been observed to push through the surface epidermal layer, causing the epidermis to rupture. The dome soon achieves a higher degree of organization and begins to produce leaf primordia (Figure 14.3b). Both periclinal and anticlinal divisions occur to the right and left of the central "apical" portion of the newly developed meristem, allowing for the successive production of leaf primordia (Figure 14.1h). The apex continues to grow, giving rise to numerous leaf primordia and culminating in the formation of an adventitious bud (Figures 14.1i and 14.3c).

The cytohistological zones typically associated with conifer seedling apical domes (Sacher, 1954) and apical meristem domes in developing conifer embryos (Fosket and Miksche, 1966) are clearly mirrored in this fully developed shoot bud (see legend, Figure 14.1i) produced in vitro by de novo organogenesis. The simultaneous development of numerous meristemoids on a single explant (Figure 14.3d) is not uncommon in in vitro organogenic culture systems. The degree of differentiation of these nascent bud meristems is highly variable. Buds in early (arrow) to fully mature stages (B) can be present on a single explant (Figure 14.3d).

The series of morphogenic events chronicled for *P. oocarpa* seems to fit the model presented for organogenesis via a "direct" developmental pathway. In practice, well-developed buds are removed from the cotyledons and are placed on a medium formulated to foster their continued growth and development. Elongated shoots, larger than 5 mm for *P. oocarpa* or 1 cm for other *Pinus* species, are transferred to either in vitro or ex vitro conditions designed to promote adventitious rooting, hence completing plantlet regeneration. The initiation of adventitious roots on in vitro-produced microcuttings is often much more than a trivial process. Because of its practical importance, this process is discussed in the following section of this chapter.

Root-specific organogenesis

As mentioned in the opening paragraph of this chapter, the production of root meristems from somatic tissues has a long history, which was driven strongly by the need to produce large quantities of high quality plant materials. Horticulturists select specific plant genotypes for their outstanding floral or vegetative beauty, and foresters mark "plus trees" for their outstanding growth characteristics. Both of these types of selections have the common goal of capturing individual genotypes in order to produce large numbers of economically valuable clones. The ability to produce clones by taking conventional cuttings or by in vitro-based propagation systems, such as the one presented in this chapter, is often controlled by the efficiency with which one is able to induce adventitious rooting of these materials. Certain plant species, as well as many developmentally mature plant materials, are extremely recalcitrant with respect to the production of adventitious roots. Some of the challenges faced by those who wish to asexually propagate woody plants, whether by the ancient art of "cuttage" or by rooting microcuttings produced via in vitro propagation systems, are discussed in detail in several recent review articles (Blakesley et al., 1991; Blakesley and Chaldecott, 1993; Haissig et al., 1992; Wilson, 1994; Wilson and Van Staden, 1990). For those readers interested in this topic, these reviews can provide a ready window into the extensive literature covering this subject area.

The process of de novo root initiation that is adventitious in origin has been generally described as consisting of the following four stages: (1) the formation of a meristematic locus by the dedifferentiation of a stem cell or cells; (2) multiplication of the stem cells into a spherical cluster; (3) further cell multiplication with initiation of planar cell divisions to form a recognizable bilateral root meristem; and (4) cell elongation of those cells located in the basal part of the developing meristem, resulting in the eventual emergence of the newly formed root (Blakesley et al., 1991). The selection of a microcutting of the appropriate developmental stage, coupled with its sequential exposure to an in vitro environment designed to initiate the sequence of events described above, may result in the production of a functional root system of adventitious origin. An example of the successful induction of adventitious roots on hypocotylary explants of pine cultured under in vitro conditions is shown in Figure 14.4. The explants were induced to root as the result of being co-cultured under in vitro conditions with a bacterium (root-stimulating bacterium; RSB) found to stimulate rooting in a variety of woody and herbaceous plants (Schwarz and Burns, 1997). The control explants, cultured without the rooting bacterium, remained rootless 7 weeks after culture initiation.

Figure 14.4 Adventitious roots produced on conifer (*P. strobus*) explants after 7 weeks of in vitro coculture with an adventitious root-inducing bacterium (RSB) compared to nontreated controls.

The site of root meristem initiation has been shown to vary widely; however, according to Esau (1960), it usually occurs "... in the vicinity of differentiating vascular tissues in the organ which gives rise to them." This site of adventitious root initiation places the newly formed root close to both the xylem and phloem of the parental axis, which, again according to Esau, "... facilitates the establishment of vascular connection between the two organs."

Root primordia may also occur in relatively undifferentiated cells, such as those found in callus tissue sometimes produced at the basal/cut end of a microcutting. Usually these roots do not develop the vascular connections needed to sustain their above-ground microshoot partner. Production of adventitious roots in callus tissue almost always results in a regenerated microplant with little or no prospect for survival when transferred to ex vitro conditions.

If the four stages that chronicle the organogenesis of functional adventitious roots are considered in light of the developmental model proposed by Christianson and Warnick (1985), both the dedifferentiation and induction phases of the model can be fit into the first and perhaps second of the four stages. The extent to which the newly induced root meristem division centers (stage 2) are canalized has not often been shown experimentally. However, in a careful study of adventitious root production in *P. radiata*, Smith and Thorpe (1975a; 1975b) showed that the presence of the root-stimulating agent (indolebutyric acid, IBA) was required until meristemoid appearance (i.e., similar to stage 2, as described above). After this point, IBA was no longer required to complete adventitious root production. It is apparent that the *Convolvulus* leaf disk system, described earlier in the chapter, is not completely analogous to adventive root initiation in *P. radiata*. The adventitious process in the *P. radiata* is thought to represent an example of direct organogenesis, its root meristem arising without an intervening callus stage. Remember that the *Convolvulus* example of Christianson and Warnick (1983) involved an indirect organogenic pathway in which the dedifferentiation process produced a limited callus growth from which the new organs were generated. Another interesting difference is that the *Convolvulus* tissues became fully determined prior to the initiation of any observable changes in the morphology of the dedifferentiated cells of the callus. This is not the case in *P. radiata*, as the first event of root initiation was the formation of a discernable meristematic locus defined by the expansion of a single cell and the swelling of its nucleus. One or more of the cells surrounding the meristematic locus (peripheral cell) repartitions its cytoplasm and initiates unequal cell division, cutting off the smallest of the two daughter cells toward the meristematic locus. These anatomical/morphological events take place prior to the point of full canalization, which may be associated with the appearance of fully developed meristemoids. These meristemoids or spheres of meristematic tissues are the result of the continued division of the peripheral cells. Continued development of the meristemoid (stages 3 and 4) gives rise to fully developed adventitious roots. As with the genesis of shoot apical meristems, much work remains to uncover the mechanisms that control the sequence of morphological events required to produce a functional adventitious root. A recent approach to solving the riddle of whole-plant root development involves the use of the step-by-step process of genetic analysis of single gene developmental mutants of *Arabidopsis* (Schiefelbein et al., 1997). These studies have lead to the hypothesis that the patterns observed for root development are "... largely guided by positional cues that are established during embryogenesis ..." Simply stated, the cells within the developing root are reacting to cues resulting from the cells' position in relation to the other cells of the newly forming root. The molecular nature of these signals and their regulation is unknown at this time. Molecular-based genetic analysis, applied to the genesis of adventitious roots, may be "most fruitful" with respect to increasing our understanding of the mechanisms controlling this process.

Summary

The vast majority of published information concerning in vitro plant organogenesis has been based on the intuitive application of technical procedures established through the venerable technique called "trial and error." However, by necessity of a descriptive nature, many careful investigations have had as their underlying purpose the elucidation of the "... basic causes of cellular differentiation and organized development" (Earle and Torrey, 1965). For the developmental biologist, the investigative goal is, therefore, no less than the establishment of a "mechanism" for this "inducible" organogenic process. A great many of the almost infinite number of combinations and permutations of experimental variables have been tried, and results have been largely positive. That is, a great many plant species have responded with the de novo production of shoot and/or root meristems. There remains a large number of instances in which experimental trial and error have been to no avail. The key needed to unlock the developmental potential of these recalcitrant explant tissues must await the uncovering of the link between genetics and the production of three-dimensional form, i.e., the phenomenon of morphogenesis.

Literature cited

Bajaj,. Y.P.S., Ed. 1991. *Biotechnology in Agriculture and Forestry and Trees*, Vol. 3, Springer-Verlag, New York, P. 312.

Blakesley, D. and M.A. Chaldecott. 1993. The role of auxin in root initiation. II. Sensitivity, and evidence from studies on transgenic plant tissues. *Plant Growth Regul.* 13:77–84.

Blakesley, D., G.D. Weston, and J.F. Hall. 1991. The role of endogenous auxin in root initiation. I. Evidence from studies on auxin application, and analysis of endogenous levels. *Plant Growth Regul.* 10:341–353.

Bronner, R., G. Jeannin, and G. Hahne. 1994. Early cellular events during organogenesis and somatic embryogenesis induced on immature zygotic embryos of sunflower (*Helianthus annuus*). *Can. J. Bot.* 72:239–248.

Christianson, M.L. 1987. Causal events in morphogenesis. In: *Plant Tissue and Cell Culture,* Alan R. Liss, New York, pp. 45–55.

Christianson, M.L. and D.A. Warnick. 1983. Competence and determination in the process of in vitro shoot organogenesis. *Dev. Biol.* 95:288–293.

Christianson, M.L. and D.A. Warnick. 1984. Phenocritical times in the process of in vitro shoot organogenesis. *Dev. Biol.* 101:382–390.

Christianson, M.L. and D.A. Warnick. 1985. Temporal requirement for phytohormone balance in the control of organogenesis in vitro. *Dev. Biol.* 112:494.

Christianson, M.L. and D.A. Warnick. 1988. Organogenesis in vitro as a developmental process. *Hort. Sci.* 23:515.

Dore, J. 1965. Physiology of regeneration in carmophytes. In: *Encyclopedia of Plant Physiology,* Vol. XV/2, W. Ruhland, Ed., Springer-Verlag, Berlin, pp. 1–91.

Earle, E.D. and J.G. Torrey. 1965. Morphogenesis in cell colonies grown from *Convolvulus* cell suspensions plated on synthetic media. *Amer. J. Bot.* 52:891–899.

Esau, K. 1960. *Plant Anatomy,* John Wiley & Sons, New York, p. 735.

Fosket, D.E. 1994. *Plant Growth and Development — A Molecular Approach,* Academic Press, New York, p. 580.

Fosket, D.E. and J.P. Miksche. 1966. A histochemical study of the seedling shoot meristem of *Pinus lambertiana. Amer. J. Bot.* 53:694–702.

Haissig, B.E. and T.D. Davis. 1994. A historical evaluation of adventitious rooting research to 1993. In: *Biology of Adventitious Root Formation,* T.D. David and B.E. Haissig, Eds., Plenum Press, New York, pp. 275–332.

Haissig, B.E., T.D. Davis, and D.E. Riemenschneider. 1992. Researching the controls of adventitious rooting. *Physiol. Plant.* 84:310–317.

Hicks, G.S. 1980. Patterns of organ development in tissue culture and the problem of organ determination. In: *The Botanical Review*, Vol. 46, A. Cronquist, Ed., New York Botanical Garden, New York, pp. 1–23.

Landauer, W. 1958. On phenocopies, their developmental physiology and genetic meaning. *Amer. Nat.* 92:201-213.

Marcotrigiano, M. and F.R. Gouin. 1984a. Experimentally synthesized plant chimeras. I. In vitro recovery of *Nicotiana tabacum* L. chimeras from mixed callus cultures. *Ann. Bot.* 54:503–511.

Marcotrigiano, M. and F.R. Gouin. 1984b. Experimentally synthesized plant chimeras. II. A comparison of in vitro and in vivo techniques for the production of interspecific *Nicotiana* chimeras. *Ann. Bot.* 54:513–521.

McDaniel, C.N. 1984. Competence, determination, and induction in plant development. In: *Pattern Formation, a Primer in Developmental Biology*, G.M. Malacinski and S.V. Bryant, Eds., Macmillan, New York, pp. 393–411.

Sacher, J.A. 1954. Structure and seasonal activity of the shoot apicies of *Pinus lambertiana* and *Pinus ponderosa*. *Amer. J. Bot.* 41:749–759.

Schiefelbein, J.W., J.D. Masucci, and H. Wang. 1997. Building a root: the control of patterning and morphogenesis during root development. *Plant Cell* 9:1089–1098.

Schwarz, O.J. and J.A. Burns. 1997. Bacterial Stimulation of Adventitious Rooting in Conifers. U.S. Patent #5,629,468.

Skoog, F. and C.O. Miller. 1957. Chemical regulation of growth and organ formation in plant tissues cultured in vitro. *Symp. Soc. Exp. Biol.* 11:118–140.

Smith, D.R. and T.A. Thorpe. 1975a. Root initiation in cuttings of *Pinus radiata* seedlings. I. Developmental sequence. *J. Exp. Bot.* 26:184–192.

Smith, D.R. and T.A. Thorpe. 1975b. Root initiation in cuttings of *Pinus radiata*. II. Growth regulator interactions. *J. Exp. Bot.* 26:193–202.

Thorpe, T.A. 1978. Physiological and biochemical aspects of organogenesis in vitro. In: *International Congress of Plant Tissue and Cell Culture* (4th, 1978, University of Calgary). International Association for Plant Tissue Culture, Calgary, pp. 49–58.

Thorpe, T.A. 1980. Organogenesis in vitro: structural, physiological, and biochemical aspects. In: *Intl. Rev. Cyt. Suppl.* 11a, I.K. Vasil, Ed., Academic Press, New York, pp. 71–111.

Thorpe, T.A. 1982. Physiological and biochemical aspects of organogenesis in vitro. In: *Proc. 5th Intl. Cong. Plant Tissue and Cell Culture*, Tokyo (Japanese Association for Plant Tissue Culture, distributed by Maruzen Co.), pp. 121–124.

Waddington, C.H. 1940. The genetic control of wing development of *Drosophila*. *J. Genet.* 41:75–139.

Walker, K.A., M.L. Wendeln, and E.G. Jaworski. 1979. Organogenesis in callus tissue of *Medicago sativa*, the temporal separation of induction processes from differentiation processes. *Plant Sci. Lett.* 16:23.

Wilson, P.J. 1994. Review article. The concept of a limiting rooting morphogen in woody stem cuttings. *J. Hort. Sci.* 69:591–600.

Wilson, P.J. and J. Van Staden. 1990. Rhizocaline, rooting co-factors, and the concept of promoters and inhibitors of adventitious rooting — a review. *Ann. Bot.* 66:479–490.

Yeung, E.C., J. Aitken, S. Biondi, and T.A. Thorpe. 1981. Shoot histogenesis in cotyledon explants of *radiata* pine. *Bot. Gaz.* 142:494–501.

chapter fifteen

Direct shoot organogenesis from leaf explants of chrysanthemum

Robert N. Trigiano and Roger A. May

Chrysanthemums or "mums" (*Dendranthema grandiflora* Tzvelev. synonymous with *Chrysanthemum morifolium* Ramat.) are among the most popular floricultural species and are grown as garden and pot crops or as cut flowers. Mums originated in the Far East and are now a major floricultural crop in many countries throughout the world. The majority of the hundreds of different cultivars are propagated commercially by stem cuttings, but many have been successfully micropropagated also by adventitious shoot formation (organogenesis) from a variety of tissue and callus cultures (e.g., Lu et al., 1990). Some of these regeneration protocols can be used in eliminating viruses from cultivars (meristem tip cultures), for rapid increase of the number of plants from unique sports, for recovery of mutations induced by either chemical or irradiation processes, and in genetic transformation of selected genotypes for flower color and disease resistance. Although infrequently encountered in exercises for plant tissue culture classes, chrysanthemums can be used to demonstrate many methodologies and concepts in regeneration protocols that are based on direct shoot organogenesis.

The following laboratory exercises illustrate the technique of direct shoot organogenesis using leaf sections of chrysanthemum. Specifically, the laboratory experiences will familiarize students with the following concepts: (1) differences in the ability of genotypes to regenerate shoots and roots; (2) effects of plant growth regulator (PGR) combinations on shoot and root formation; and (3) pulse treatments for initiation and growth of shoots. These exercises are adapted primarily from a publication by Trigiano and May (1994).

General considerations

Growth of plants

All of the cultivars of chrysanthemum mentioned in the experiments may be obtained from Yoder Brothers, Barberton,OH as unrooted or rooted cuttings. We suggest growing the following cultivars for the exercises: 'Adorn', 'Goldmine', 'Jessica', 'Hekla', 'Iridon', and 'Rave'. From time to time, companies stop production of specific cultivars. If these are not available, experiment with substitute cultivars. However, if laboratory or greenhouse space is limited, grow only the cultivars 'Goldmine' and 'Iridon'. Plant three to five cuttings in each of at least ten 10-cm-diameter plastic pots containing any soilless media 3 months prior to the beginning of the exercises. This should provide adequate materials for classes of about 15 to 20 students. A good estimate of the number of plants needed per class is to

0-8493-2029-1/00/$0.00+$.50

provide a minimum of one plant per student or student team per exercise. The conditions for growing the plants used for explants is very important — if possible, cultivate the plants in the laboratory or growth room illuminated with about 100 μmol·m^{-2}·sec^{-1} fluorescent and incandescent light for 16 h per day at 22 to 25°C. Alternately, plants grown in the greenhouse under shade and with cooling will provide suitable materials. Supplemental lighting in the greenhouse will be necessary to maintain long days and inhibit flowering if the plants are grown in the spring and fall. Pinch longer shoots often to encourage branching and initiation of new leaves. Fertilize regularly (twice weekly with 300 ppm N) to maintain vigorous growth (see May and Trigiano [1991] for details of growing conditions).

Explant preparation and basal medium

For satisfactory and reproducible results, select only young, partially expanded, light green, 2- to 5-cm-long leaves for explant tissue. Surface disinfest whole leaves (do not remove the petiole) using 5% commercial bleach (0.26% NaOCl) for 5 to 10 min (or 10% bleach for material grown in the greenhouse) with constant agitation followed by three rinses with sterile distilled water. Place a single leaf with the top surface up in a sterile 100 × 15-mm plastic petri dish and drain excess water by tilting and gently tapping the dish on the flow hood table. The top surface of the leaf is easily identified by its darker color and abundant trichomes. Excise four or five 1/2- to 1-cm^2 sections from the midrib area of the leaf (Figure 15.1) using a no. 10 scalpel blade. Place the sections with the abaxial surface (underside) of the leaf in contact with the agar medium.

The basal culture medium is composed of Murashige and Skoog (MS) (1962) basal salts (see Chapter three for composition) amended with 30 g (88 mM) sucrose, 1 g (0.55 mM) myo-inositol, 5 mg (2.9 μM) thiamine-HCl, pH 5.7, and 8 g of agar per liter. After sterilization, the medium is poured into 60 × 15-mm plastic petri dishes. Typically, one or two leaf sections are placed in each of the dishes. PGR requirements will be provided for each variation of the experiment. We usually prepare at least four dishes of each treatment for each student or student team.

Exercises

Experiment 1. Differences between genotypes in the ability to regenerate shoots and roots

An important basic concept in the tissue culture of plants is that not all genotypes (in this case, genotypes are identified as cultivars) will respond equally under similar cultural conditions. In fact, there are some cultivars of chrysanthemum that will not produce either shoots or roots in vitro using any treatment. This experiment was developed to illustrate this concept and is probably the easiest of the following exercises to conduct with large classes. We suggest that the exercise be completed by teams of two or three students. This experiment also provides the basic protocol for the other two experiments in this chapter. Refer to Figure 15.1 for a diagrammatic representation of this exercise. Experiment 1 will require approximately 12 weeks for completion.

Materials

The following items are needed for each student or team of students:

- 100 mL of 5% commercial bleach per cultivar
- Four 60 × 15-mm petri dishes containing MS + 0.25 mg/L (1 μM) BA + 2 mg/L (11.5) μM IAA per cultivar

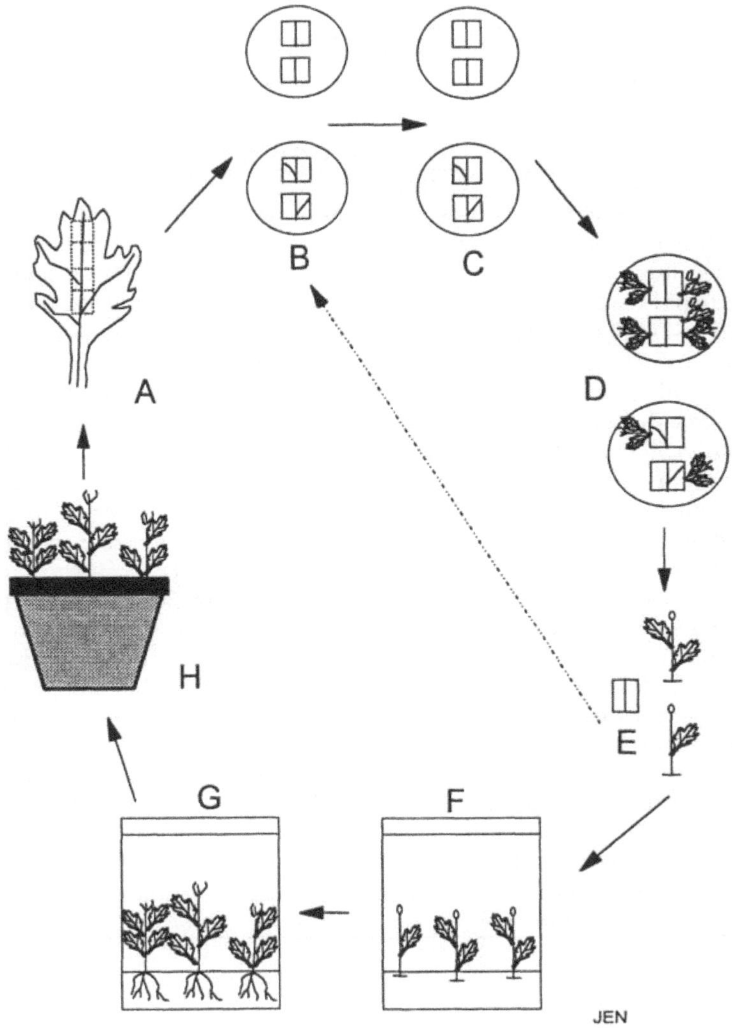

Figure 15.1 Diagrammatic representation of the micropropagation protocol for chrysanthemum. (A) Leaf after surface disinfestation and excision of leaf sections (dotted lines). (B) Two leaf sections incubated on shoot initiation medium. (C) Leaf sections transferred to shoot elongation medium after 2 weeks. (D) Shoots formed on cut edges of the explants 5 weeks after culture initiation. (E) Shoots greater than 2 cm are excised and large lower leafs removed. Leaves produced in vitro may be used as explants are returned to initiation medium to produce more shoots (dotted line). (F) Shoots are transferred to Magenta vessels for rooting. (G) Shoots have elongated and produced roots after 3 weeks. (H) Plantlets are transferred to soilless medium and acclimatized to growing conditions. (From Trigiano, R. N. and Gray, D. J., Eds., *Plant Tissue Culture Concepts and Laboratory Exercises, First Edition*, CRC Press LLC, Boca Raton, FL, 1996.)

- Four 60 × 15-mm petri dishes containing MS medium with 11.5 µM IAA only
- Four 60 × 15-mm petri dishes containing MS medium with 1.0 µM BA only
- Sixteen 60 × 15-mm petri dishes containing MS medium without PGRs per cultivar
- Two to five GA-7 Magenta culture vessels or other similar size vessel (baby food jars) containing 75 ml MS medium without PGRs
- Incubator providing approximately 25 $\mu mol \cdot m^{-2} \cdot sec^{-1}$ and 25°C
- Cell packs with soilless medium, and mist bed or plastic wrap or plastic bag to construct a moist chamber for high humidity

Follow the protocol outlined in Procedure 15.1.

Procedure 15.1
Differences between genotypes in the ability to regenerate shoots and roots.

Step	Instructions and comments
1	Surface disinfest leaves of the different cultivars with 5% clorox (laboratory-grown plants) (10% for greenhouse-grown plants) for 5 to 10 min and rinse three times with sterile distilled water.
2	Dissect leaves as described previously under explant preparation and place two sections in each 60 × 15-mm petri dish containing MS medium supplemented with 11.5 µM IAA and 1.0 µM BA, IAA alone, BA alone, or MS basal medium lacking PGRs. This is the shoot initiation phase. The experiment should be considered a randomized complete block design with four treatments and replications.
3	Incubate dishes at 25°C with 25 $\mu mol \cdot m^{-2} \cdot sec^{-1}$ of fluorescent light for 2 weeks.
4	Transfer all explants to MS basal medium without PGRs and incubate for 3 weeks under the same conditions as stated in step #3. This is the shoot elongation phase.
5	The number of shoots and/or roots within a pertri dish, an experimental unit, should be counted after 3 weeks. Student teams should share and analyze data (standard deviation or ANOVA — see Chapter seven).
6	Excise shoots greater than 2 cm and remove any large leaves that are positioned near the cut surface. Insert the cut end of the shoots into MS basal medium without PGRs contained in GA-7 Magenta vessels. Three rows of five shoots can be placed in each vessel and incubated as in step #3 for the root initiation phase.
7	After 3 weeks, gently wash the agar from the roots with running tap water. Remove large basal leaves and transfer the plantlets to soilless medium in cell packs or individual 1 × 1 cells and place under intermittent mist for 2 to 3 weeks (acclimatization phase). Plastic wrap or a bag may be used to build a "tent" around the tray of cell packs. After a week, make several holes in the "tent" with a sharpened pencil. Add a few holes every 2 to 3 days. The plants should be acclimatized to ambient conditions in 2 to 3 weeks and now can be grown in the greenhouse or laboratory.

Anticipated results

After the first 2 weeks of culture, leaf explants exposed to IAA and BA should have distorted shapes and sparse, light-green crystalline callus may have formed around the cut edges of all cultivars. Darker green meristematic zones or small buds may also be apparent on 'Hekla', 'Iridon', and 'Rave' explants. This ends the initiation phase of the exercise Stages 0 (selection of materials) and I (establishment of cultures) of traditional micropropagation.

Three weeks after transfer to medium without PGRs, shoots and some roots should be formed on leaf sections of 'Hekla', 'Iridon', and 'Rave' that were initially cultured medium containing PGRs. In this case, shoots and roots are formed directly from mesophyll cells without an intervening callus phase or direct organogenesis. Shoots should have formed primarily on the cut margins of the explant. The most shoots (and least number of roots) should be on 'Iridon' explants and decreasing numbers on 'Hekla' and 'Rave'. 'Adorn' and 'Goldmine' explants may have produced a few roots and maybe an occasional shoot. None of the explants of any of the cultivars should have produced appreciable amounts of callus, and those initially incubated on only IAA, BA, or medium lacking PGRs should have formed very few shoots or roots — 'Adorn' and 'Goldmine' may have formed some roots. This is the "shoot elongation" phase of the experiment or Stage II of micropropagation.

Shoots that were excised from leaves should have elongated to almost twice the original length and produced roots (Stage III of micropropagation) in the Magenta vessels containing MS medium without PGRs. Typically, three to five adventitious roots are formed per shoot. These shoots are easily acclimatized to normal growing conditions within 3 weeks using the described procedures (Stage IV of micropropagation).

Questions

- What purpose do the treatments containing individual PGRs (IAA or BA only) and without any PGRs serve in the initiation phase of the experiment? Why are they necessary in the design of the experiment?
- Why is it necessary to transfer the explants from initiation medium (with PGRs) to shoot elongation medium (without PGRs)?
- Why can't the plants be transferred directly from the Magenta vessels to normal growing conditions found in the laboratory or greenhouse?

Experiment 2. The effects of PGRs on formation of shoots and roots

One of the first steps in developing a micropropagation protocol of any plant is to determine the type (or combination) and concentration of PGRs that will provide the desired response — in our case, shoot formation. Leaf tissue of some cultivars of chrysanthemum are very responsive to both the type and concentration of cytokinin in the initiation medium.

This experiment uses the cultivars 'Adorn' and 'Iridon', or, if time and/or materials are limited, the experiment may be completed with 'Iridon' only. Prepare several modifications of MS medium, each containing 11.5 μM IAA combined with the following concentrations and types of cytokinin or cytokinin-like compounds: 0.01, 0.1, 1.0, and 5.0 μM kinetin, 2-iP, BA, and thidiazuron (TDZ). There are a total of 17 treatments including IAA only.

Since there are 17 treatments, it is impossible to include all treatments in every block without using at least four or five different leaves. The inherent variability for the ability to form shoots between different leaves from the same stock plant would probably necessitate increasing the replications to some very large number in order to detect significant differences between treatments. Therefore, from our experience, this experiment works best when designed as a randomized incomplete block design with eight replications. Each incomplete block is composed of four entries or the number of explants obtained from one leaf, thus limiting the variability within an individual block to that of an individual leaf. As matter of fact, explants from within an individual leaf are incredibly uniform in their ability to produce shoots. A more complete treatment of this experimental design with chrysanthemum may be found in Kuklin et al. (1993). A design for this exercise can be found in Table 15.1.

An associated exercise to demonstrate that explants from an individual leaf are equally capable of forming shoots is easily prepared and executed. Culture the four midvein explants from ten leaves on MS medium supplemented with 1.0 μM BA and 11.5 μM IAA for 2 weeks, then transfer to MS medium without PGRs for 3 weeks. Simply compare the number of shoots produced by each explant of an individual leaf — they should be more or less equal.

Materials

The following items are needed for each cultivar in the experiment:

- Eight 60 × 15-mm petri dishes for each treatment
- Eight 60 × 15-mm petri dishes containing MS medium without PGRs

Table 15.1 An Incomplete Block Design with 17 Treatments and 8 Replications

Block	Treatment number				Block	Treatment number			
1	8	17	6	1	**18**	8	9	7	14
2	1	13	11	17	**19**	8	10	12	2
3	8	13	7	17	**20**	2	7	14	6
4	15	8	3	17	**21**	3	14	14	4
5	1	2	17	9	**22**	7	5	12	10
6	11	10	1	5	**23**	6	15	3	9
7	16	17	12	4	**24**	12	7	6	2
8	9	12	15	11	**25**	3	4	11	12
9	11	12	3	9	**26**	14	10	13	15
10	13	1	11	4	**27**	9	5	15	14
11	9	14	2	1	**28**	10	13	16	9
12	15	7	4	13	**29**	8	16	2	14
13	15	1	11	12	**30**	5	6	4	16
14	10	2	14	17	**31**	13	5	2	16
15	16	6	13	3	**32**	7	15	4	6
16	16	11	7	5	**33**	5	16	10	8
17	11	5	17	3	**34**	6	2	8	10

From Trigiano, R. N. and Gray, D. J., Eds., *Plant Tissue Culture Concepts and Laboratory Exercises, First Edition*, CRC Press LLC, Boca Raton, FL, 1996.

Follow the protocols outlined in Procedure 15.2.

Procedure 15.2
The effects of PGRs in the formation of shoots and roots from leaf explants.

Step	Instructions and comments
1	Treatments (PGR type and concentration) should be represented by a number ranging from 1 to 17. Depending on the size of the class, each student or team should be assigned to complete specific blocks, not treatments, as shown in Table 15.1.
2	Surface disinfest leaves and dissect into four midvien sections as previously described under explant preparation and basal medium. One leaf piece should be placed in each of four 60-mm petri dishes (treatments) contained within each incomplete block.
3	Incubate the cultures at 25°C with 25 $\mu mol \cdot m^{-2} \cdot sec^{-1}$ for 2 weeks and then transfer to medium without PGRs for an additional 3 weeks under the same conditions.
4	Count the number of shoots and roots at the end of 5 weeks after the initiation of the experiment. Analyze data for significant differences between the means for each of the treatments.

Anticipated results

After incubation for 2 weeks on shoot initiation medium and an additional 3 weeks on shoot elongation medium, shoots and roots should be formed. Most treatments, except IAA only, will support some shoot formation on explants of 'Iridon' (Figure 15.2a); only roots should form on explants of 'Adorn' regardless of the combination of PGRs employed (Figure 15.2b) Shoots produced on medium containing higher concentrations of TDZ may not elongate after transfer to medium without PGRs. Also, most shoots produced on these treatments will exhibit hyperhydricity. Students should determine which treatment produced (statistically) the most shoots and/or roots.

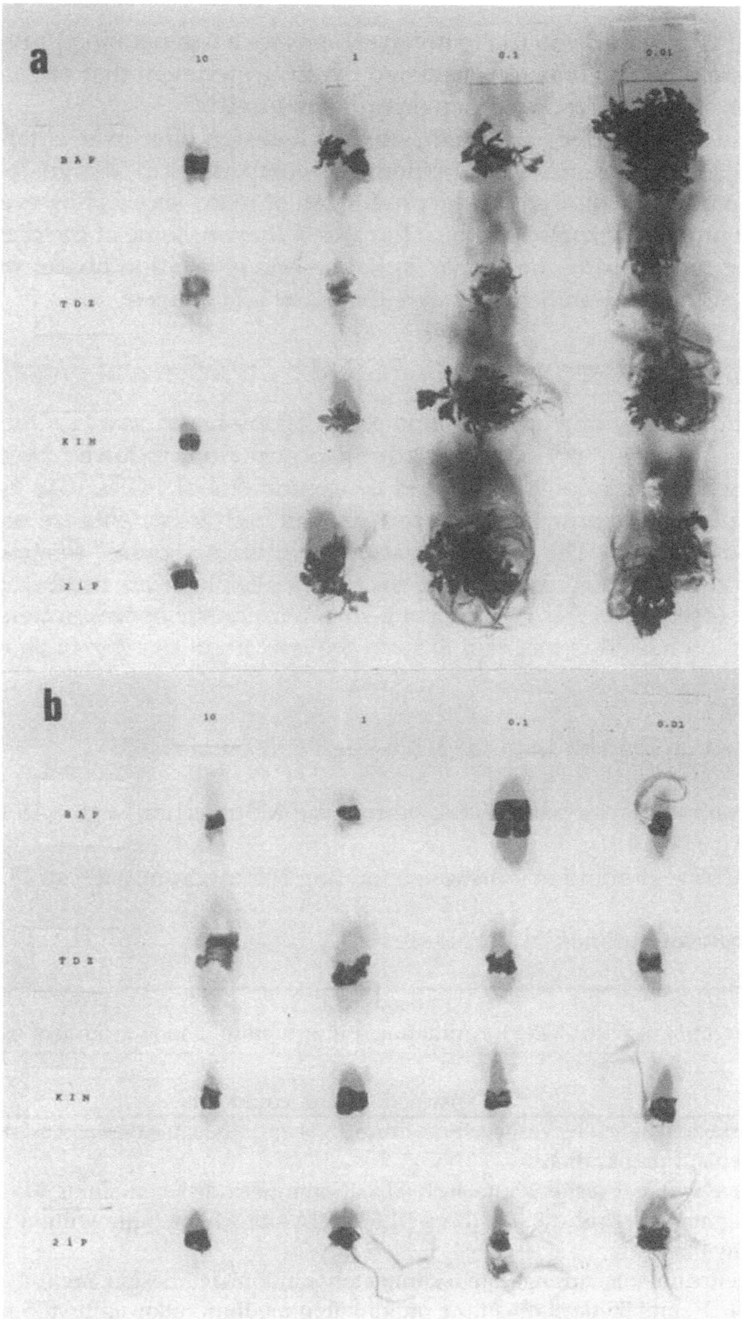

Figure 15.2 Shoots of chrysanthemum 'Iridon' (a) and roots from 'Adorn' (b) formed on the cut edges of explants after an initial incubation of 2 weeks with PGRs followed by transfer to medium without PGRs for 3 weeks. Numbers represent micromolar concentrations. (From Trigiano, R. N. and Gray, D. J., Eds., *Plant Tissue Culture Concepts and Laboratory Exercises, First Edition*, CRC Press LLC, Boca Raton, FL, 1996.)

Questions

- Which PGRs would you use to initiate shoots for a regeneration protocol and why?
- Are there other factors not considered by this experiment that might influence the number of shoots produced per experimental unit?
- What advantages does an incomplete block design offer over other designs such as completely randomized or randomized-complete block designs?
- Some treatments allowed for the production of many shoots; however, a similar or greater number of roots were also formed. What are some of the disadvantages of forming roots during the shoots' initiation and elongation phases of micropropagation of the chrysanthemum? Are there any advantages?

Experiment 3. Pulse treatment with PGRs for initiation and growth of shoots

Most adventitious shoot micropropagation protocols are based on a treatment of explants with PGRs for a specific period of time (initiation period) followed by incubation on medium that either lacks or has reduced concentrations of PGRs. We have used this technique in each of the previous two exercises and such treatments are usually referred to as "pulse" treatments. The practical reasons for utilizing "pulse" treatments are easily demonstrated using 'Iridon' and varying the time the explants are incubated on initiation medium. The objective of this experiment is to determine the optimum time that explants should remain on initiation medium to form the maximum number of shoots.

Materials

For each student or student team the following items are needed:

- Sixteen 60 × 15-mm petri dishes containing MS medium with 11.5 µM IAA + 1.0 µM BA
- Twenty 60 × 15-mm petri dishes containing MS medium without PGRs

Follow the protocol outlined in Procedure 15.3.

Procedure 15.3
Pulse treatment with PGRs for initiation and growth of shoots from leaf explants.

Step	Instructions and comments
1	Surface disinfest leaves (see Procedure 15.1) and dissect leaves as described under explant preparation.
2	Place two leaf sections into each 60 × 15-mm petri dish containing MS medium augmented with 1.0 µM BA + 11.5 µM IAA or MS medium without PGRs (for day 0).
3	The treatments, arranged in a completely randomized design, are 0, 1, 3, 5, 7, 10, 14, 21, and 35 days of culture on initiation medium followed by MS medium without PGRs for the remainder of the 35-day period. The 0-day treatment should be initiated and transferred to MS medium without PGRs after 14 days. All cultures should be incubated at 25°C with 25 $\mu mol \cdot m^{-2} \cdot sec^{-1}$ for the entire 35 days.
4	Count the number of shoots after 35 days and compare the means of the treatments.

Anticipated results

Few, if any, shoots should form on explants from treatments of 0, 1, 3, and 5 days, whereas 10 to 17 days of incubation on medium with PGRs should foster the development of the greatest number of shoots. Fewer shoots are formed on explants from the 21-day treatment, and although some shoots develop on explants from the 35-day treatment, they usually are stunted and/or exhibit hyperhydricity.

Questions

- What is hyperhydricity and why did it occur with shoots produced on the 35-day treatment in this experiment?
- Why were shoots not formed from explants included in the 0-, 1-, 3-, and 5-day treatments?

Literature cited

Kuklin, A. I., R. N. Trigiano, W. L. Sanders, and B. V. Conger. 1993. Incomplete block design in plant tissue culture research. *J. Tissue Cult. Meth.* 15:204–209.

Lu, C.-Y., G. Nugent, and T. Wardley. 1990. Efficient, direct plant regeneration from stem segments of chrysanthemum (*Chrysanthemum morifolium* Ramat. cv. Royal Purple). *Plant Cell Rep.* 8:733–736.

May, R.A. and R. N. Trigiano. 1991. Somatic embryogenesis and plant regeneration from leaves of *Dendranthema grandiflora*. *J. Amer. Soc. Hort. Sci.* 116:366–371.

Murashige, T. and F. Skoog. 1962. A revised medium for rapid growth and bioassays with tobacco tissue cultures. *Physiol. Plant.* 15:473–497.

Trigiano, R. N. and R. A. May. 1994. Laboratory exercises illustrating organogenesis and transformation using chrysanthemum cultivars. *HortTechnology* 4:325–327.

chapter sixteen

Shoot organogenesis from cotyledon explants of watermelon

Michael E. Compton and Dennis J. Gray

Watermelon, *Citrullus lanatus* (Thunb.) Matsum. and Nakai, is a frost tender, vining annual crop with sweet-tasting fruit that is often consumed as a cool desert. The species originated in tropical Africa. In the U.S., watermelon is cultivated mainly in the southern, southwestern, and central states, although some early-maturing cultivars are available for northern climates. The fruit of watermelon varies in size, shape, rind color and pattern, flesh and seed color, and maturity date. Most fruit have a striped rind with red flesh and small black seeds, and weigh 20 to 30 lb at maturity. However, seedless varieties, which have a slightly sweeter taste, also are available. Watermelons are susceptible to a number of bacterial, fungal, and viral diseases, requiring annual field rotation, frequent chemical sprays, and disease-resistant cultivars.

In light of both its importance and susceptibility to diseases, watermelon is an excellent candidate for improvement through genetic engineering. Insertion of bacterial, fungal, and virus resistance genes through recombinant DNA technology would facilitate the development of new disease-resistant genotypes without significantly altering the genetic composition of cultivars with high fruit quality. In addition, the production of polyploids through somaclonal variation in tissue culture could be exploited to produce new parental breeding lines, which are useful in developing new seedless watermelon cultivars.

Cell and tissue culture research has resulted in an efficient plant regeneration system for watermelon (Compton and Gray, 1993; Dong and Jia, 1991; Srivastava et al., 1989). The published protocols use seedling cotyledons as a source of explants, from which adventitious shoots can be obtained by the process of organogenesis. The shoots can be rooted to form complete plants in 6 to 10 weeks from culture initiation. These protocols have been used both to obtain model transgenic plants (Choi et al., 1994) and to obtain tetraploid plants, which can be used to breed improved triploid watermelon cultivars (Compton and Gray, 1994). The purpose of the laboratory exercises in this chapter are to familiarize students with the basic procedures of adventitious shoot regeneration from watermelon cotyledon explants and illustrate differences in shoot regeneration competence from different regions of the cotyledon and among various watermelon cultivars.

0-8493-2029-1/00/$0.00+$.50

Exercises

The basic protocol for preparing explants and transferring cultures is described in Figure 16.1. This is a modification of the procedures described previously by Compton and Gray (1993).

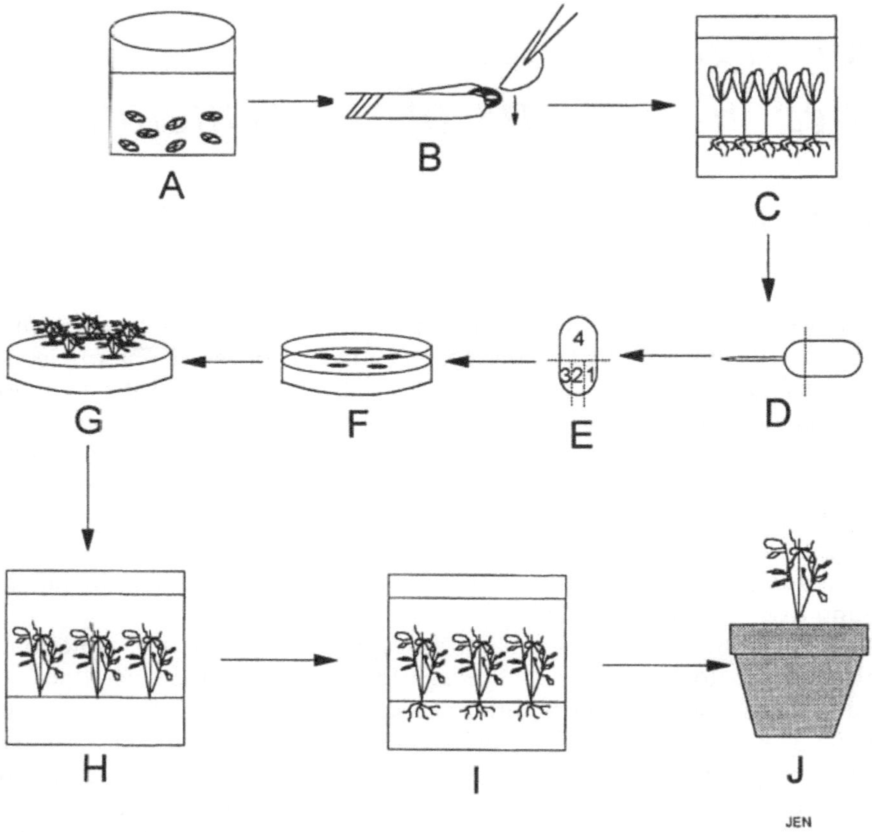

Figure 16.1 Flow chart of the experimental protocol. (A) Imbibe seeds in distilled water for 4-15 h. (B) Remove seed coats using special forceps and scalpel with #10 blade. (C) After surface disinfestation, germinate embryos in vitro for 5 to 6 days. (D) Remove cotyledons from seedlings using a scalpel and a #11 blade. (E) Prepare explants by making three longitudinal cuts (1, 2, and 3) and a final cross-sectional cut (4). (F) Culture explants in 100 × 15-mm petri dishes with regeneration medium. (G) Shoots appear in 3 to 4 weeks. (H) Transfer shoot cultures to GA-7 Magenta vessels containing elongation medium. (I) Excise 1- to 2-cm-long shoots and transfer to rooting medium. (J) Transfer plants to soil and acclimatize. (From Trigiano, R. N. and Gray, D. J., Eds., *Plant Tissue Culture Concepts and Laboratory Exercises, First Edition*, CRC Press LLC, Boca Raton, FL, 1996.)

Experiment 1. Difference in regeneration potential among distinct regions of watermelon cotyledons

Use of the proper explant type is an important factor in successful organ regeneration. Not all explants from the same plant or cotyledon possess an equal regeneration potential. To develop an efficient plant regeneration protocol it is necessary to identify the best region of the plant from which to isolate explants. In watermelon the best explant source is the cotyledon of in vitro-germinated seedlings. However, not all regions of watermelon cotyledons possess an equal regeneration potential. In this experiment, students will be culturing explants from different regions of the watermelon cotyledon to identify the best explant for adventitious shoot regeneration.

Materials

The following will be needed for each student or team:

- Thirty 'Minilee' watermelon seeds
- Toothed forceps (# 6-134 Miltex Instrument Company, Inc., Lake Success, NY) recommended
- Sterile plastic bottles with caps for sterilizing seeds
- 100 ml of 25% bleach solution with one to two drops surfactant
- Six GA-7 magenta vessels containing 50 ml of germination medium
- Twenty-four 100 × 15-mm plastic petri dishes containing 25 ml of regeneration medium
- GA-7 magenta vessels containing 50 ml of shoot elongation medium
- GA-7 magenta vessels containing 50 ml of rooting medium
- Growth chamber providing 25 to 50 $\mu mol \cdot m^{-2} \cdot sec^{-1}$ irradiance and 25°C
- Cell packs with autoclaved soilless medium (equal parts Pro Mix BX and coarse vermiculite), 28 × 53-cm plastic flat without holes and clear plastic dome lid

Follow the protocols in Procedure 16.1 to complete this exercise.

Procedure 16.1
Difference in regeneration potential among distinct regions of watermelon cotyledons.

Step	Instructions and comments
1	Imbibe seeds for 4 to 15 h in sterile distilled water. Use seeds of the cultivar 'Minilee' for this experiment.
2	Remove the seed coat from imbibed seeds using a scalpel with #10 blade. Position the seeds on their side and firmly grasp the pointed end with the special-toothed forceps (see materials list). Cut the seed coat by applying a sharp downward motion at the rounded end. Several cuts may be required to make an opening large enough so that the seed coat can easily be removed. The embryo is removed by gently working it side-to-side using the blunt end of the scalpel. Store excised embryos in sterile distilled water until all have been prepared.
3	Surface disinfest embryos in a 125-ml Erlenmeyer flask containing 100 ml of 25% bleach solution (1.25% NaOCl plus two drops of 'Tween 20' surfactant) for 25 min with constant agitation. Rinse the embryos three to six times with sterile distilled water in the laminar flow hood.
4	Transfer embryos to Magenta GA_7 vessels containing 50 ml of germination medium. Culture six embryos per vessel. Incubate vessels in light (30 $\mu mol \cdot m^{-2} \cdot sec^{-1}$) for 5 to 7 days.
5	Remove germinated seedlings from culture vessels and transfer them to a sterile plastic petri dish, or paper plate, for dissection (seedlings with their cotyledons in close contact are at the best stage). Remove the cotyledons by making a shallow cut about 2 mm above the point of attachment to the hypocotyl. A scalpel with a #11 blade works best. Gently "flick" the cotyledon to detach it from the hypocotyl. Cotyledons have been properly removed when the shoot tip can be seen remaining on the embryo axis after detaching the cotyledons. It is important not to include the shoot apex as it will inhibit adventitious shoot development.

Procedure 16.1 continued Difference in regeneration potential among distinct regions of watermelon cotyledons.	
Step	Instructions and comments
6	Cut the cotyledons into 3 × 5-mm explants by placing them abaxial side (i.e., the "bottom" side) up on the cutting surface as diagrammed in Figure 16.1F or 16.2, depending on the experiment. Culture prepared explants abaxial side down in 100 × 15-mm plastic petri dishes containing 25 ml of regeneration medium (Table 16.1). Gently push the explants into the medium to establish contact with the medium surface. Distribute five explants randomly in each petri dish.
7	Label each plate and seal with parafilm. Arrange prepared plates in a completely randomized design with subsampling in the growth chamber (see Chapter seven).
8	After 4 weeks record the number of explants from each region that have shoots or vegetative buds. Transfer explants with shoots to Magenta vessels containing 50 ml of shoot elongation medium (Table 16.1). Return to the growth chamber.
9	Check cultures at 4 weeks. Remove all shoots that are at least 2 cm long from the primary explant and transfer to Magenta vessels containing rooting medium (Table 16.1). Return to the growth chamber.
10	Check cultures at 10 to 21 days for signs of rooting. Once shoots have produced roots that are at least 1 cm long they may be transferred to soil.
11	For acclimatization, remove plantlets from the culture vessels and gently wash away any agar surrounding the roots. Transplant the plantlets to cell packs containing autoclaved soilless medium (equal parts ProMix BX and coarse vermiculite). Place cell packs with plantlets in a plastic 11 × 22 flat containing about 300 ml of sterile distilled water. Cover flats containing plantlets with a clear plastic lid and seal with wide masking tape. Place the containers with plants in the growth chamber or other well-lit area. Plants should remain covered until signs of new growth are apparent on most of the plants. Begin gradually removing the tape and lid over a period of 3 to 7 days.
12	Analyze data using analysis of variance and compare means using the standard error of the mean or other multiple comparison test as described in Chapter seven.
13	Transfer plants to the greenhouse or garden when conditions are favorable.

Table 16.1 Amendments to MS Medium Used for Regeneration and Rooting of Watermelon Shoots from Cotyledon Explants

	Media (amounts in mM)			
Constituents	G[a]	SR[b]	Ed[c]	R[d]
Sucrose	87.6	87.6	87.6	58.4
BA	—	0.005	—	—
IBA	—	—	—	0.001

Note: Adjust pH of each medium to 5.7 before adding 7 g/L TC agar and autoclaving (see Chapter three for composition of MS medium).

[a] Germination medium.

[b] Shoot regeneration medium.

[c] Shoot elongation medium.

[d] Rooting medium.

From Trigiano, R. N. and Gray, D. J., Eds., *Plant Tissue Culture Concepts and Laboratory Exercises, First Edition*, CRC Press LLC, Boca Raton, FL, 1996.

Anticipated results

Most explant types typically enlarge to about four to five times their original size within the first 3 to 5 days after culture initiation. Buds, leaves, and shoots should be visible 3

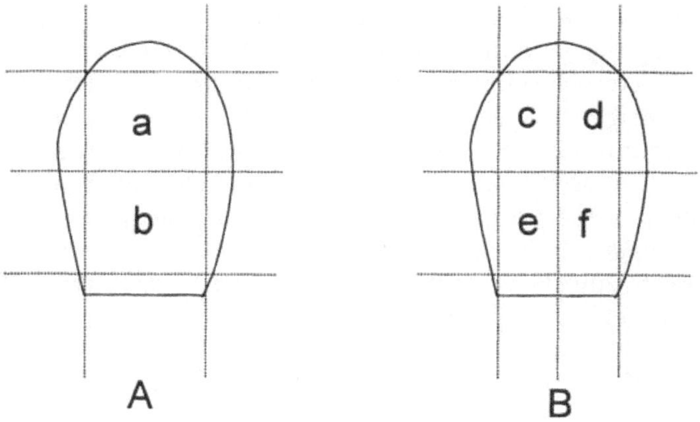

Figure 16.2 Cotyledon explants to be cultured for Experiment 1. (From Trigiano, R. N. and Gray, D. J., Eds., *Plant Tissue Culture Concepts and Laboratory Exercises, First Edition,* CRC Press LLC, Boca Raton, FL, 1996.)

to 4 weeks after culture initiation. Explant types b, e, and f (Figure 16.2) should produce the highest response frequency (50 to 90%). Those from regions a, c, or d may respond but at a much lower frequency (<10%).

Questions

- Which region produced the most explants with shoots and/or buds?
- Why would some regions of the cotyledon respond more readily than others?
- Were there any noticeable abnormalities among the regenerated shoots? If so, were they associated with a particular treatment?

Experiment 2. Effect of plant genotype on the ability of watermelon cotyledon explants to produce shoots

The genotype of the plant often is one of the most important factors in determining explant competence for organ regeneration. Within a species there are often one or more genotypes that readily respond to the imposed plant regeneration protocol, whereas others fail to respond to any devised treatment. It is extremely important to select highly regenerative genotypes for genetic transformation to be successful. Therefore, it is beneficial to test regeneration rates of as many cultivars as feasibly possible. In watermelon, most cultivars are competent for shoot regeneration from cotyledon explants. However, the regeneration potential often varies among cultivars. In this experiment the ability of cotyledon explants from four watermelon cultivars to produce shoots will be examined. The cultivars to be tested include 'Crimson Sweet' (Stokes Seeds, Inc., Buffalo, NY), 'Minilee' (Willhite Seed Co., Poolville, TX), 'Sangria' (Seedway, Inc., Elizabethtown, PA), and 'Yellow Doll' (Otis S. Twilley Seed Co., Inc., Trevose, PA). Other cultivars may be substituted, but those listed should evoke a wide range of responses. Each student or team should work with seeds from a different cultivar.

Materials

The following items are needed for each student or team:

- 15 to 18 watermelon seeds (a different cultivar should be used by each student or team)
- Toothed forceps (# 6-134 Miltex Instrument Company, Inc., Lake Success, NY)

- Sterile plastic bottles with caps for sterilizing seeds (Nalgene® 125-ml wide-mouth polypropylene bottle; lids may be modified for easy rinsing of seeds by cutting a 2-cm-diameter hole and inserting a 1000-µm mesh screen)
- 100 ml of 25% bleach solution with one to two drops surfactant
- Three GA-7 magenta vessels containing 50 ml of germination medium
- Five 100 × 15-mm plastic petri dishes containing 25 ml of regeneration medium
- GA-7 magenta vessels containing 50 ml of shoot elongation medium
- GA-7 magenta vessels containing 50 ml of rooting medium
- Growth chamber providing 25 to 50 $\mu mol \cdot m^{-2} \cdot sec^{-1}$ irradiance and 25°C
- Cell packs with autoclaved soilless medium (equal parts Pro Mix BX and part coarse vermiculite), 28 × 54-cm plastic flat without holes and clear plastic dome lid

The instructions for this experiment are outlined in Procedure 16.2.

Procedure 16.2
Effect of plant genotype on the ability of watermelon cotyledon explants to produce shoots.

Step	Instructions and comments
1	Imbibe, surface disinfest, and germinate seeds of the four cultivars as described in Procedure 16.1, steps 1 to 4.
2	Remove germinated seedlings of each cultivar from the Magenta vessels. Prepare and culture basal cotyledon explants as described in Procedure 16.1 (steps 5 and 6) and Figure 16.1. Each student should culture five petri dishes, each with five explants.
3	Label each plate and seal with parafilm. Arrange prepared plates in a completely randomized design with subsampling in the growth chamber (see Chapter seven). After 9 to 12 days, look at the explants using a stereomicroscope. Some should have developed buds along their cut edges. Record the number of explants with buds.
4	After 4 weeks record the number of explants from each cultivar that have shoots or vegetative buds. Trim away all brown leaves and nonresponding portions of the cotyledon before transferring shoot cultures to Magenta GA_7 vessels containing 50 ml of shoot elongation medium as described in Procedure 16.1. Return cultures to the growth chamber.
5	Shoots may be rooted and acclimatized as described in Procedure 16.1, steps 9 to 11. Shoot cultures can be subcultured to fresh elongation medium at this time to produce more shoots. When shoots have formed roots ≥1 cm in length (10 to 14 days), record the number that rooted and calculate the rooting percentage. Transfer plants to the greenhouse or garden when conditions are favorable.
6	Analyze data using analysis of variance and compare means using the standard error of the mean or other multiple comparison test as described in Chapter seven.

Anticipated results

Most explants will have expanded four to five times their original size after 3 to 5 days on regeneration medium. After 9 to 12 days, some explants should have developed buds along their cut edges (Figure 16.3a). Some of these buds develop into 0.5- to 1-cm-long shoots after 4 weeks on regeneration medium (Figure 16.3b). Explants from the cultivars should respond differently. 'Minilee', 'Yellow Doll', and 'Crimson Sweet' respond better than 'Sangria'. Transferring explants with shoots to elongation medium without BA allows the shoots to elongate (Figure 16.3c) so that they can be easily rooted. Approximately 75 to 95% of shoots will normally root on medium with IBA. Not all plants will survive acclimatization. Typically 60 to 90% of watermelon plants from tissue culture will survive acclimatization. Once the plants have been acclimatized they

Figure 16.3 Organogenesis from cotyledons of watermelon. (a) Shoot buds formed on the cut edge of watermelon cotyledon incubated on shoot regeneration medium for 9 to 12 days. (b) Leaves, buds, and shoots formed on watermelon cotyledon 4 weeks after culture initiation. (c) Elongated shoots on watermelon cotyledon explant 4 weeks after transfer to shoot elongation medium. (d) Acclimatized plant transferred to the greenhouse. (From Trigiano, R. N. and Gray, D. J., Eds., *Plant Tissue Culture Concepts and Laboratory Exercises, First Edition*, CRC Press LLC, Boca Raton, FL, 1996.)

can be transplanted to 10-cm-diameter pots filled with Pro Mix BX and grown in the greenhouse (Figure 16.3d), or moved to the garden when conditions are suitable.

Questions

- Were there any differences in the frequency of explants with shoots, shoots that rooted, or acclimatized plants among the cultivars tested?
- Why should explants from some cultivars respond better than those from others?
- Did you notice any abnormal looking plants? Describe the abnormalities.

Experiment 3. Influence of darkness during seed germination on adventitious shoot organogenesis in watermelon

Dark pretreatments have been shown to improve shoot regeneration from cotyledons, leaves, and petioles of several plant species including bean (*Phaseolus vulgaris* L.), cucumber (*Cucumis sativus* L.), tomato (*Lycopersicon esculentum* Mill.), pear (*Pyrus communis* L.), and potato (*Solanum phureja*). Incubation of plant tissue in darkness is also a common segment of most somatic embryogenic (see Chapters nineteen to twenty-five) and protoplast isolation (see Chapters twenty-six to twenty-eight) protocols. The exact role of darkness is not completely understood, but it is thought that incubating plant tissues in darkness preserves light-sensitive endogenous plant growth regulators and reduces the amount of starch and oxidative polyphenolic compounds. Dark incubation of plant tissues also results in fewer cell wall deposits, thinner cell walls, and less vascular tissue. A reduction in cell wall thickness and the amount of cell wall deposits may promote organ regeneration by facilitating movement of exogenous plant growth regulators to explant regeneration sites. The purpose of this exercise is to observe the effects of dark pretreatment on shoot organogenesis in watermelon.

Materials

The following items are needed for each student or team:

- 18 to 24 watermelon seeds
- Toothed forceps (# 6-134 Miltex Instrument Company, Inc., Lake Success, NY)
- Sterile plastic bottles with caps for sterilizing seeds (Nalgene* 125-ml wide-mouth polypropylene bottle; lids may be modified for easy rinsing of seeds by cutting a 2-cm-diameter hole and inserting a 1000-µm mesh screen)
- 100 ml of 25% bleach solution with one to two drops surfactant
- Four GA-7 magenta vessels containing 50 ml of germination medium
- Six 100 × 15-mm plastic petri dishes containing 25 ml of regeneration medium
- GA-7 magenta vessels containing 50 ml of shoot elongation medium
- GA-7 magenta vessels containing 50 ml of rooting medium
- Two growth chamber environments: one providing 25 to 50 $\mu mol \cdot m^{-2} \cdot sec^{-1}$ irradiance and 25°C and the other providing 25°C and darkness
- Cell packs with autoclaved soilless medium (equal parts Pro Mix BX and part coarse vermiculite), 28 × 54-cm plastic flat without holes and clear plastic dome lid

Follow the protocols in Procedure 16.3 for this experiment.

Procedure 16.3
Influence of darkness during seed germination on adventitious shoot organogenesis in watermelon.

Step	Instructions and comments
1	Imbibe and surface disinfest seeds as described in Procedure 16.1, steps 1 to 3. Transfer embryos to GA-7 magenta vessels containing 50 ml of germination medium. Culture six embryos per vessel. Incubate vessels in light (30 $\mu mol \cdot m^{-2} \cdot sec^{-1}$) or darkness at 25°C for 5 to 7 days.
2	Remove germinated seedlings of each cultivar from the GA-7 magenta vessels. Prepare and culture basal cotyledon explants as described in Procedure 16.1 (steps 5 and 6) and Figure 16.1. Each student should culture six petri dishes (three of each treatment), each with five explants. Label each plate and seal with parafilm. Arrange prepared plates in a completely randomized design with subsampling in the growth chamber (see Chapter seven).
3	After 9 to 12 days, look at the explants using a stereomicroscope. Some should have developed buds along their cut edges. Record the number of explants with buds.
4	After 4 weeks record the number of explants from each treatment that have shoots or vegetative buds. Trim as described in Procedure 16.1. Transfer explants with shoots to GA-7 magenta vessels containing 50 ml of shoot elongation medium as described in Procedure 16.1. Return cultures to the growth chamber.
5	Shoots may be rooted and acclimatized as described in Procedure 16.1, steps 9 to 11.
6	Analyze data using analysis of variance and compare means using the standard error of the mean or other multiple comparison test as described in Chapter seven.

Anticipated results

Seedlings germinated in darkness will be yellow and etiolated, being about twice the height of the green seedlings incubated in light. Explants prepared from seedlings germinated in darkness should become green within hours of exposure to light. These explants should also produce more shoots and buds than cotyledonary explants prepared from seedlings germinated in a light photoperiod. An interesting twist of this exercise would be to germinate seeds from several cultivars in light and darkness and compare their shoot regeneration rates.

Questions

- Was there a difference in shoot organogenesis between explants prepared from seedlings germinated in darkness compared to light photoperiod?
- Speculate on the role of dark pretreatments in stimulating organogenesis.

Literature cited

Choi, P.S., W.Y. Soh, Y.S. Kim, O.J. Yoo, and J.R. Liu. 1994. Genetic transformation and plant regeneration of watermelon using *Agrobacterium tumefaciens*. *Plant Cell Rep.* 13:344–348.

Compton, M.E. and D.J. Gray. 1993. Shoot organogenesis and plant regeneration from cotyledons of diploid, triploid, and tetraploid watermelon. *J. Amer. Soc. Hort. Sci.* 118:151–157.

Compton, M.E. and D.J. Gray. 1994. Regeneration of tetraploid plants from cotyledons of diploid watermelon. *Proc. Fla. State Hort. Soc.* 107:107–109.

Dong, J.Z. and S.R. Jia. 1991. High efficiency plant regeneration from cotyledons of watermelon (*Citrullus vulgaris* Schrad.). *Plant Cell Rep.* 9:559–562.

Srivastava, D.R., V.M. Andrianov, and E.S. Piruzian. 1989. Tissue culture and plant regeneration of watermelon (*Citrullus vulgaris* Schrad. cv. Melitopolski). *Plant Cell Rep.* 8:300–302.

chapter seventeen

Direct and indirect shoot organogenesis from leaves of Torenia fournieri

Mark P. Bridgen

Plant regeneration in tissue culture is a crucial aspect of plant biotechnology as it facilitates the production of genetically engineered plants, the release of disease-free plants from meristem cultures, and the rapid multiplication of difficult-to-propagate species. Plant regeneration through tissue culture can be accomplished by enhanced axillary branching from shoots or lateral buds (micropropagation), somatic embryogenesis, embryo culture, or organogenesis (Dodds and Roberts, 1985). Organogenesis is the initial and subsequent de novo growth of shoots, roots, leaves, or flowers from a tissue and is especially easy to demonstrate with *Torenia fournieri* L. (Bridgen et al., 1994).

Torenia fournieri is commonly called the wishbone flower, bluewings, or torenia. Wishbone flower refers to the shape and position of the stamens that arch together, similar to a chicken wishbone, and fuse at the anther. Torenia has opposite leaves, a rounded growth habit, attains a height of approximately 30 cm, and is well adapted to pot conditions. It is a plant that is easy to obtain from seed and easy to grow under greenhouse, field, or classroom conditions.

Torenia has been used since 1985 to study the process of somaclonal variation (Brand and Bridgen, 1989; Hadi, 1986; Hadi and Bridgen, 1990) and the white-flowering cultivar 'UConn White' was developed through this process (Brand and Bridgen, 1989). Significant work had been completed previously on the organogenesis of torenia from stem segments (Tanimoto and Harada, 1984) and leaf sections (Bajaj, 1972).

In this exercise, techniques will be described to demonstrate direct and indirect shoot organogenesis from leaf sections of torenia; leaf sections will develop shoots and callus very quickly. These exercises can be completed by the end of a 10-week term or 14-week semester. The primary objective of this exercise is to demonstrate the processes of direct and indirect organogenesis with *T. fournieri*. The secondary objective is to give the student practice with aseptic technique and medium preparation. Details of the experiments follow and are adapted primarily from a publication by Bridgen et al. (1994).

0-8493-2029-1/00/$0.00+$.50

General considerations

Growth of plants

Torenia can be propagated by shoot cuttings of established plants or from seed that can be obtained from most commercial seed companies. Germination will occur in 7 to 14 days if seeds are not covered and remain under moist conditions at 21 to 23°C. Plants grow easily in any soilless medium under greenhouse or classroom conditions with adequate moisture, humidity, light (minimum 100 $\mu mol \cdot m^{-2} \cdot sec^{-1}$), and night temperatures of 13 to 15°C. When grown outside, the plants flourish when given shade, cool temperatures, and moist growing media. Plants should be fertilized regularly with a soluble fertilizer to maintain vigorous growth and healthy, green leaf development. Allow the plants to grow for at least 8 weeks after germination before leaf sections are removed. Depending on environmental conditions, flowering should begin 12 weeks after seed germination. Plan on at least one plant per student; a few extra plants for the classroom may be desirable. These plants are especially susceptible to white fly infestation and can be damaged if control measures are not taken. If pesticides are used for control, follow all label recommendations for rates and safety.

Explant preparation and basal medium

Select and remove young leaves from the second and third nodal regions from the top of established torenia plants. Young and vigorously growing plants will respond more quickly in vitro and will give better results than older leaves. Clean leaves gently in tepid running tap water. From this point on, all work should be completed under aseptic conditions. This exercise can be completed with or without the use of a laminar flow hood. In either situation, aseptic procedures should be explained and demonstrated to the students prior to this laboratory. Clean all work surfaces with 70% ethanol and allow to dry when ready to begin the exercise.

Surface disinfest entire leaves by dipping them for 5 to 10 sec in 70% ethanol followed by 10% commercial bleach (0.5% sodium hypochlorite) for 15 min. After the bleach treatment, rinse the leaves three times with sterile deionized water and allow the leaves to remain in the last rinse until they are ready to be cultured. Use only sterile forceps or scalpels to hold plant material after disinfestation. To increase the success rate, limit the number of leaves to three or four in a 200-ml beaker containing the bleach solution. To hasten results and/or to eliminate the previous steps, leaves may be used from previously established aseptic cultures.

Place a single disinfested leaf on a sterile surface such as a sterile 100 × 15-mm petri dish or an autoclaved paper towel. Use a #10 scalpel blade to remove the leaf margins, if they are damaged by the bleach, and discard. Cut leaves into 0.5 × 0.5-cm pieces and place on the appropriate medium with the abaxial surface (underside) in contact with the medium. Each student should culture at least four explants per medium. If contamination is a possibility, have the students wrap culture vessels with PVC film or Parafilm™. Culture vessels should be labeled with the date and individual's initials.

Incubate cultures at 21 to 23°C under a 16-h photoperiod provided by cool-white fluorescent lamps at 50 to 60 $\mu mol \cdot m^{-2} \cdot sec^{-1}$. Observe cultures on at least a weekly basis and compare the explant development on each medium. Look for the initiation of shoots, leaves, and callus.

The three media needed for this experiment have the same basal formula that includes Murashige and Skoog (1962) salts and vitamins (see Chapter three for composition), 88 mM (30 g/L) sucrose, and 7 g/L agar and adjusted to pH 5.7. The callus induction medium and the direct organogenesis medium will be needed at the beginning of this exercise. The indirect organogenesis medium is not needed for 4 to 6 weeks after the initiation of the experiment unless the instructor wants to demonstrate all three types of organogenesis simultaneously. If this latter scenario is desirable, callus will have to be prepared 1 to 2 months prior to this exercise and supplied to the student at the onset of the experiment.

The callus induction medium includes 0.5 mg/L (2.3 μM) 2,4-dichlorophenoxyacetic acid (2,4-D) and the direct organogenesis medium includes 0.25 mg/L (1.1 μM) 6-benzylaminopurine (BA) and 0.05 mg/L (0.25 μM) indole-3-butyric acid (IBA). The indirect organogenesis medium includes 2 mg/L (8.8 μM) BA plus 0.5 mg/L (2.5 μM) IBA. Dispense 20 to 30 ml of media into each 25 × 150-mm borosilicate glass culture tube, cap, and autoclave. If limited space and media are problems, sterile disposable 100 × 15-mm petri dishes can be used for the first 4 weeks instead of culture tubes.

Experiment 1. Direct organogenesis from leaves of Torenia fournieri

Plant regeneration directly from cultured explants can be accomplished by using leaves, bulbs, stems, corms, rhizomes, tubers, or other plant parts. Although direct organogenesis occurs much less frequently than indirect organogenesis, explants of some species can be placed on nutrient media containing cytokinins and low levels of auxins and initiate shoots directly. Multiplication is achieved through subdivision of the shoot-producing clumps.

The objective of this experiment is to demonstrate direct organogenesis from leaf sections of torenia. The experiment will require approximately 4 to 6 weeks until shoots are visible.

Materials

The following items are needed for each student:

- Four culture tubes or 60 × 15-mm petri dishes containing direct organogenesis medium described above
- One healthy, young torenia plant
- 10% commercial bleach

Follow the protocols in Procedure 17.1 to complete this experiment.

Procedure 17.1
Direct organogenesis from leaves of *Torenia fournieri.*

Step	Instructions and comments
1	After the leaves have been surface disinfested and dissected, culture one leaf section in each of the culture vessels with direct organogenesis medium (Figure 17.1). In order to statistically analyze the results at the end of the exercise, place the culture vessels in the growth chamber in a randomized complete block design. There should be four blocks each containing one of the culture vessels from each student.
2	After 4 to 6 weeks of growth, the number of shoots per culture vessel should be counted. Students should give their results to the instructor to collate and summarize the data for the students. Each student will be responsible for analyzing the data with ANOVA etc. (see Chapter seven).
3	If the students would eventually like to obtain a torenia plant ex vitro, individual shoots should be aseptically separated as they are counted. The shoots can be placed on a MS medium without plant growth regulators (PGRs) for further growth and root development (Stage III of micropropagation).

Procedure 17.1 continued
Direct organogenesis from leaves of *Torenia fournieri*.

Step	Instructions and comments
4	After 3 to 5 weeks, the plantlets can be removed from the culture vessel and the agar gently washed from the roots with running tap water. The plantlets may then be transferred to a soilless medium in small pots or plug trays and placed under intermittent mist or inside a plastic bag to acclimatize (Stage IV of micropropagation). The plants should be acclimatized in 2 to 3 weeks and can be grown as "normal" plants.

Anticipated results

Within 3 to 7 days after the initial culture, contamination may be visible, and these cultures should be discarded. Depending on the vigor and age of the original source plant, there may be some bleach damage on the leaves and if it is not extensive, the experiment will work. Newly formed shoots should start to be visible in 2 to 3 weeks. Many shoots will form on the explant except for any of the damaged areas.

Questions

- Would the culture of explants on their abaxial surface affect organogenesis differently than if they were cultured on their adaxial surface?
- What effect does the size of the explant have on organogenesis?
- How does the auxin cytokinin ratio in a medium determine the type of organogenesis?
- What is the difference between adventitious shoots and axillary shoots?
- Why would adventitious shoot proliferation be a poor way to propagate chimeral plants?

Experiment 2. Callus formation and indirect organogenesis of Torenia fournieri

Indirect organogenesis is the production of shoots or roots from callus — an amorphous mass of loosely arranged thin-walled, differentiated parenchyma cells. Types of explants that produce callus include stems, leaves, flower parts, petioles, roots, or the more mitotically active shoot tip or its meristematic region. The cells of the explant divide in vitro and give rise to the organogenic callus (Dodds and Roberts, 1985).

Organogenic callus has meristematically active cells that are the precursors of roots and shoots. The ploidy levels of these cells may vary tremendously due to the fast rate of division. This genetic instability is one of serious problems associated with the use of callus cultures for micropropagation.

The objective of this exercise is to demonstrate indirect organogenesis from torenia callus. It will take approximately 5 to 6 weeks to obtain large clumps of callus from leaf sections and an additional 4 to 6 weeks for shoot production from callus. Students can initiate callus at the same time that Experiment 1 on direct organogenesis is begun. Experiment 2 may take a major part of the term or semester. To decrease the time for this exercise, the instructor can initiate callus for the students 5 to 6 weeks before the exercise begins. Refer to Figure 17.1 for a diagrammatic representation of this exercise.

Materials

In addition to the materials listed for Experiment 1, the following items are needed for each student:

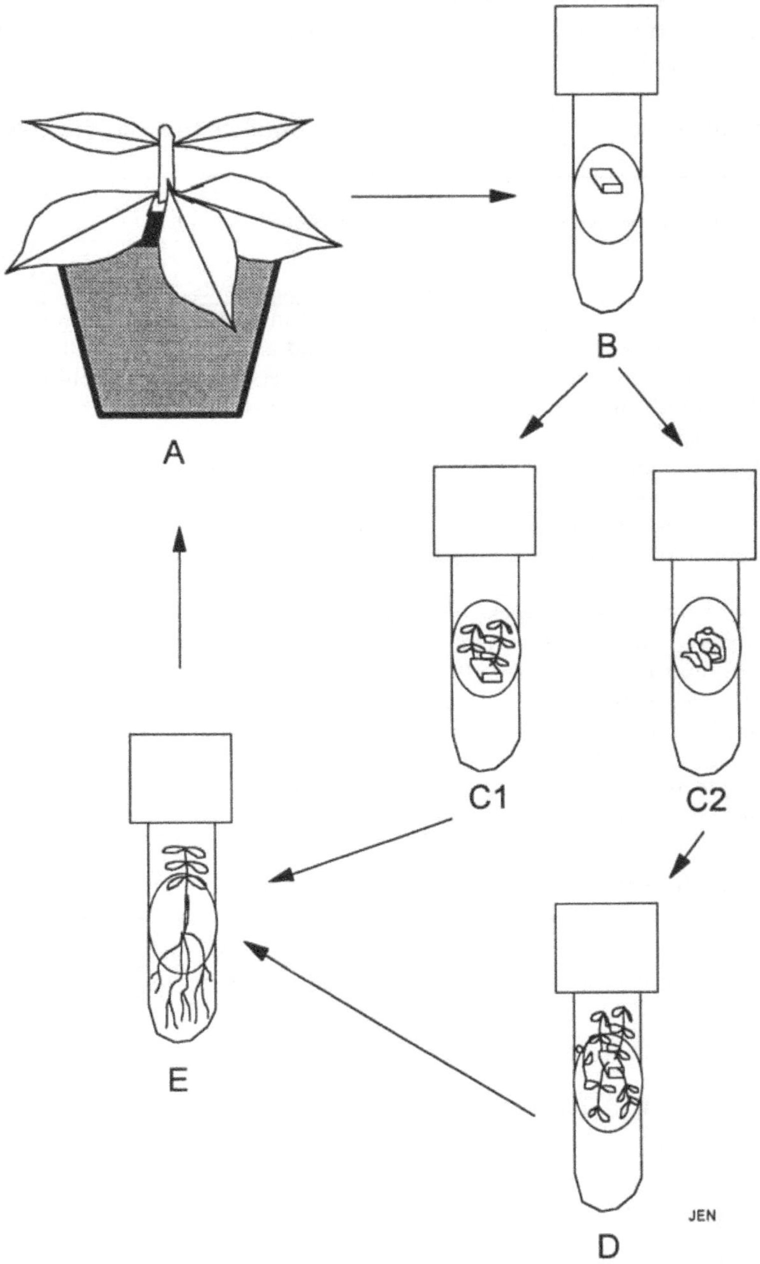

Figure 17.1 Plant regeneration in *Torenia fournieri*. Torenia plants grown in vivo serve as the source of leaf explants (A). Leaf explant sections are established on nutrient medium (B). Explants cultured directly on a cytokinin-containing medium give rise to shoots directly (C1) and callus is derived from explants growing on an auxin medium (C2). Transfer of callus to medium containing cytokinin results in shoot production (D). Shoots derived through either method are rooted on a medium free of plant growth regulators (E). (From Trigiano, R. N. and Gray, D. J., Eds., *Plant Tissue Culture Concepts and Laboratory Exercises, First Edition*, CRC Press LLC, Boca Raton, FL, 1996.)

- Four culture tubes or 60 × 15-mm petri dishes containing callus induction medium or four cultures of torenia callus
- Four culture tubes or 60 × 15-mm petri dishes containing indirect organogenesis medium described above

Follow the protocols set out in Procedure 17.2 to complete this experiment.

Procedure 17.2
Callus formation and indirect organogenesis of *Torenia fournieri.*

Step	Instructions and comments
1	After the leaves have been surface disinfested and dissected, culture one leaf section in each of the culture vessels with callus induction medium. Although callus growth will not be statistically analyzed, the culture vessels should be placed in the growth chamber in a randomized complete block design to distribute variability among the cultures. As in Experiment 1, there should be four blocks each containing one of the culture vessels from each student. For more precise measurements of callus production, fresh weight values can be taken aseptically on a weekly basis and plotted over time at the end of the exercise.
2	Explants should be subcultured onto fresh callus induction medium after 3 to 4 weeks.
3	After a total of 5 to 6 weeks on the induction medium, callus can be aseptically removed and cultured on the indirect organogenesis medium.
4	The number of shoots per culture vessel should be counted after an additional 4 to 6 weeks of growth on the organogenesis medium. Students should give their results to the instructor to be collated and summarized. The summarized data will be returned to the students for analysis with ANOVA etc. (see Chapter seven). If students would like to obtain a torenia plant ex vitro, the procedures that were described in Experiment 1 should be followed.

Anticipated results

Contamination should not be a problem with the indirect organogenesis, since cultures will have been established and growing in vitro. After 4 to 6 weeks on the indirect organogenesis medium, the callus should respond by producing multiple shoots.

Questions

- Where does callus originate on the leaf explants?
- From which locations on the callus do shoots arise?
- Why is indirect organogenesis from callus a poor way to propagate *Torenia* clonally?

Literature cited

Bajaj, Y.P.S. 1972. Effect of some growth regulators on bud formation by excised leaves of *Torenia fournieri*. *Z. Pflanzenphysiol.* 66:284–287.

Brand, A.J. and M.P. Bridgen. 1989. 'UConn White' a white-flowered *Torenia fournieri*. *HortScience* 244:714–715.

Bridgen, M.P., M.Z. Hadi, and M. Spencer-Barreto. 1994. A laboratory exercise to demonstrate direct and indirect shoot organogenesis from leaves of *Torenia fournieri*. *HortTechnology* 4:320–322.

Dodds, J.H. and L.W. Roberts. 1985. *Experiments in Plant Tissue Culture*, 2nd Ed. Cambridge University Press, Cambridge, England.

Hadi, M.Z. 1986. In Vitro Techniques for the Development of Pest Resistant Plants. M.S. thesis. University of Connecticut Storrs, 99 pp.

Hadi, M. and M.P. Bridgen. 1990. Screening of somaclonal variants in vitro to produce insect resistant plants. *HortScience* 25:112.

Murashige, T. and F. Skoog. 1962. A revised medium for rapid growth and bioassays with tobacco tissue cultures. *Physiol. Plant*. 15:473–497.

Tanimoto, S. and H. Harada. 1984. Roles of auxin and cytokinin in organogenesis in *Torenia* stem segments cultured in vitro. *J. Plant Physiol*. 115:11–18.

chapter eighteen

Shoot organogenesis from petunia leaves

John E. Preece

Petunia × *hybrida* Hort. Vilm.-Andr. is the common garden petunia. Genetically, it has a complex background consisting of hybrids of three or four *Petunia* species. It is considered a cultigen because its complex genetics clearly indicate that this hybrid species originated in domestication, not in the wild. Petunia is among the most popular bedding plants and is mainly propagated by seeds; however, some hybrids will not come true from seeds and are propagated by rooting stem cuttings (Liberty Hyde Bailey Hortorium, 1976).

Petunia is a member of the Solanaceae (tobacco) family. There have been many in vitro studies on members of this family because they are not difficult to regenerate in tissue culture. Within this family, petunia is particularly responsive to tissue culture techniques and can be regenerated from anthers, protoplasts, seedling tips, stems, leaves, and other tissues and organs. Petunia is an ideal plant for a tissue culture laboratory exercise because it is responsive in vitro; seeds are easy to obtain, store, and germinate; seedlings are easy to grow; and it can be scheduled year-round. The following exercises use leaves for the initial explant and are designed to familiarize the student with the effects of cytokinin and auxin on the regeneration of shoots and roots and the formation of callus. Additionally, the laboratories will demonstrate both direct and indirect shoot organogenesis. Direct shoot organogenesis occurs when shoots form directly from the original explant without an intervening callus. In contrast, indirect shoot organogenesis happens when callus first forms from the explant, then shoots arise from the callus tissue.

With both direct and indirect shoot organogenesis, cells must return to the meristematic state of growth. Meristems are the least specialized (least differentiated) tissues in the plant. Daughter cells that form in meristems gradually become more specialized as they differentiate into the cells that make up a plant. The balance of plant growth regulators (PGRs) influences the direction that differentiation takes. Specialized or differentiated cells becoming meristematic requires that they become less specialized. This process is the reverse of differentiation and is called dedifferentiation. Therefore, dedifferentiation is necessary before adventitious shoots can form from the initial explant or from callus tissues.

0-8493-2029-1/00/$0.00+$.50

General considerations

Growth of plants

Various petunia cultivars and genotypes will respond in a somewhat dissimilar manner in vitro and will have different optimum requirements for PGRs. However, virtually all *Petunia* × *hybrida* cultivars will respond by regenerating shoots in vitro. In the past, leaf explants from cultivars 'Sugar Daddy', 'Sugar Plum', 'White Pearls', and 'Supermagic Pink' have responded well in vitro. A problem with making recommendations for which cultivar(s) to use in these experiments is that seed companies are continually changing their offerings. Seeds can be purchased from garden centers or directly from seed companies via catalog. Cultivars should be selected from the Grandiflora types and red flowering cultivars should be avoided because several of these have responded poorly in vitro (personal communication with Paul E. Read). For the exercises, two cultivars with a different flower color should be selected. Mixed cultivars should not be selected because they will yield variable results. One of many sources for petunia is W. Atlee Burpee & Co., Warminster, PA 18974 (1-800-888-1447).

About 8 weeks prior to the time of the laboratory exercise, seeds should be sown in any new or disinfested greenhouse soilless medium, but a fine-textured peat-lite medium is preferable. Petunia seeds are tiny and must be sown close to the surface of the medium. A 1-cm-deep furrow should be made, then seeds sown at a density of about two to three per centimeter of row. Be careful not to sow seeds too thickly because it will result in weakened, stretched seedlings that are difficult to separate and transplant. After sowing, the seeds can be covered with a light dusting of medium. Care must be taken during watering because it is easy to wash the seeds out of the furrow. Seeds can be germinated in a growth chamber or in a greenhouse. Because of the varying environmental conditions, extra care must be taken in the greenhouse. The temperature must be maintained at 21 to 24°C.

Fertilization with 100 ppm nitrogen from a water-soluble fertilizer can begin after seedlings emerge. Seedlings, about 1 cm or 1/2 in. tall, can be transplanted into any well-aerated, pest-free soilless medium in 10-cm-diameter pots. The seedlings should be watered immediately after transplanting. There should be one or two plants of each cultivar per student or team of students.

Seedlings should be grown at about 24°C. Leaf explants from seedlings grown at high temperatures >30°C will not respond in vitro. It is best if the seedlings can be subirrigated with an ebb and flood system or by use of a capillary mat. Overhead watering can lead to fungal and bacterial contamination in vitro. Pests, especially white flies, must be controlled or there will be problems with contamination in vitro. Weekly spraying with appropriate pesticides is recommended; use protective clothing and follow label instructions exactly.

Seedlings are sufficiently mature for explants when the leaves touch the sides of the pots, but should still be growing as rosettes. Leaves from young plants that are flowering can be used; however, contamination can be a problem. Avoid using leaves from plants growing outdoors because of contamination and difficulty in disinfestation.

The nutrient medium is Murashige and Skoog (1962) (MS) salts and organics (see Chapter three for composition). The medium should contain 30 g/L (88 mM) sucrose and the pH should be adjusted to 5.8 prior to adding 7 g/L of a high-quality agar (Difco Bacto agar or Sigma agar) per liter. After autoclaving, and before the agar gels, the medium should be poured into 100 × 15-mm-diameter sterile plastic petri dishes. Enough medium should be added to completely cover the bottom of the petri dish (10 to 15 ml). Two leaf explants will be placed right side up (adaxial side up) in each petri dish. At least four petri dishes of each PGR treatment should be used per cultivar per student or team of students.

Exercises

Experiment 1. The effects of cytokinin and auxin on petunia leaf explants

This exercise is designed to familiarize students with the roles that PGRs added to the tissue culture medium play in organogenesis. Although two of the treatments in this exercise consider the effects of cytokinin or auxin used alone, the concept of the effect of cytokinin-to-auxin ratio is illustrated. The cytokinin(s) and auxin(s) that are incorporated into tissue culture media simply add to endogenous PGRs (hormones) that are already present. When a cytokinin or auxin is added singly to a tissue culture medium, the ratio within a plant tissue shifts to a high amount of the PGR that is added and a low amount of the one that is not added (consider that explants will contain both endogenous auxin and cytokinin). It is the changing ratio of cytokinin to auxin that influences whether or not a cell will become determined to undergo a coordinated series of cell divisions followed by differentiation to form shoots, roots, or unorganized callus.

Cultures can be incubated on an open laboratory bench with or without cool white fluorescent lamps or in a lighted growth chamber. Although light is required, the quantity of light and photoperiod are not critical. Results will be evident within 2 weeks and the experiment will be completed in 4 to 8 weeks.

Materials

Items needed for each student or group of students include:

- One or two stock plants of each cultivar
- 10% bleach solution with Tween 20
- Four 100 × 15-mm petri dishes per petunia cultivar containing medium supplemented with 2.5 μM benzyladenine (BA, a cytokinin)
- Four 100 × 15-mm petri dishes per petunia cultivar containing medium supplemented with 7.5 μM indolebutyric acid (IBA, an auxin)
- Four 100 × 15-mm petri dishes per petunia cultivar containing medium supplemented with 2.5 μM BA and 5 μM IBA

Follow the protocols listed in Procedure 18.1 to complete this experiment.

Procedure 18.1
The effects of cytokinin and auxin on petunia leaf explants.

Step	Instructions and comments
1	Remove five or six of the largest, fully expanded leaves from a stock plant of each cultivar and place into an empty petri dish. Fill the dishes with the bleach/Tween 20 solution and cover with the lids; swirl occasionally for 10 min to surface disinfest.
2	Transfer leaves to sterile petri dishes that contain sterile deionized or distilled water. Cover and swirl occasionally for 5 min to remove the bleach solution. Repeat this rinsing step once.

Procedure 18.1 continued The effects of cytokinin and auxin on petunia leaf explants.	
Step	Instructions and comments
3	Dissect and discard the edges and tip of a leaf (Figure 18.1) to produce a strip that is 1 to 1.5 cm wide including the midvein. Cut across the leaf and midvein to yield uniform 0.5-cm explants. Be careful not to injure explants with the tips of the forceps. The midvein of the leaf should be positioned across the width of the explant, near the middle. Place two explants, right side up (adaxial side up), on the surface of medium and wrap each petri dish with parafilm or cling-type plastic food wrap. Label every petri dish and place on the top of a laboratory bench or in a growth chamber. Cultures should be incubated in the light, with cool white fluorescent lamps being best. Light quantity and photoperiod should be uniform for all treatments. This experiment can be arranged as a completely random design with four replications of two samples (explants) per replication.
4	Collect data on root, shoot, and callus formation. Count the number of roots and measure the length of each. Be careful not to confuse root hairs (looks like white fuzz) with fungal contamination. Shoots (stems with leaves) will first appear as bumps (buds) on the edges of the leaf explants near the midveins. They may form directly from the explant tissue or from callus. When leaves first begin to enlarge, distinguishing between stems and leaves can be difficult. Wait until the stems elongate and count the number of shoots that are at least 1 cm long. Callus formation is probably best evaluated using a relative rating scale (i.e., 0 = no callus; 5 = most callus). Look along the veins for the initial formation of callus.

Anticipated results

Most contamination will have appeared after about 1 week of culture. At about this time the first roots will become visible from leaves cultured on medium containing 7.5 μM IBA only. After about 3 weeks, the roots will be well developed and data collection on roots can be terminated. Also after 3 weeks, buds and perhaps small shoots will be visible, especially when the medium contains 2.5 μM BA only. Callus will become visible 7 to 10 days after culture initiation and will continue to grow regardless of the treatment. The most callus and a few adventitious shoots will form when the medium contains both BA and IBA. Some of this callus can be excised and used in the next experiment. Some roots may develop from adventitious shoots because the shoot tips will be producing hormonal auxin.

Questions

- What are the roles of auxin and cytokinin in the formation of plant organs?
- In the context of this experiment, why do plants develop roots and shoots instead of simply growing as large masses of callus?
- How can information gained from this exercise be extrapolated to the process of wound healing in plants?
- Why is it important to have major veins in the leaf explants?

Experiment 2. Regeneration of petunia from callus and monitoring somaclonal variation

This exercise is designed for the student to compare direct and indirect organogenesis and monitor the performance of the resulting plants. Petunia plants that have been regenerated from callus have demonstrated somaclonal variation (Lewis-Smith et al.,

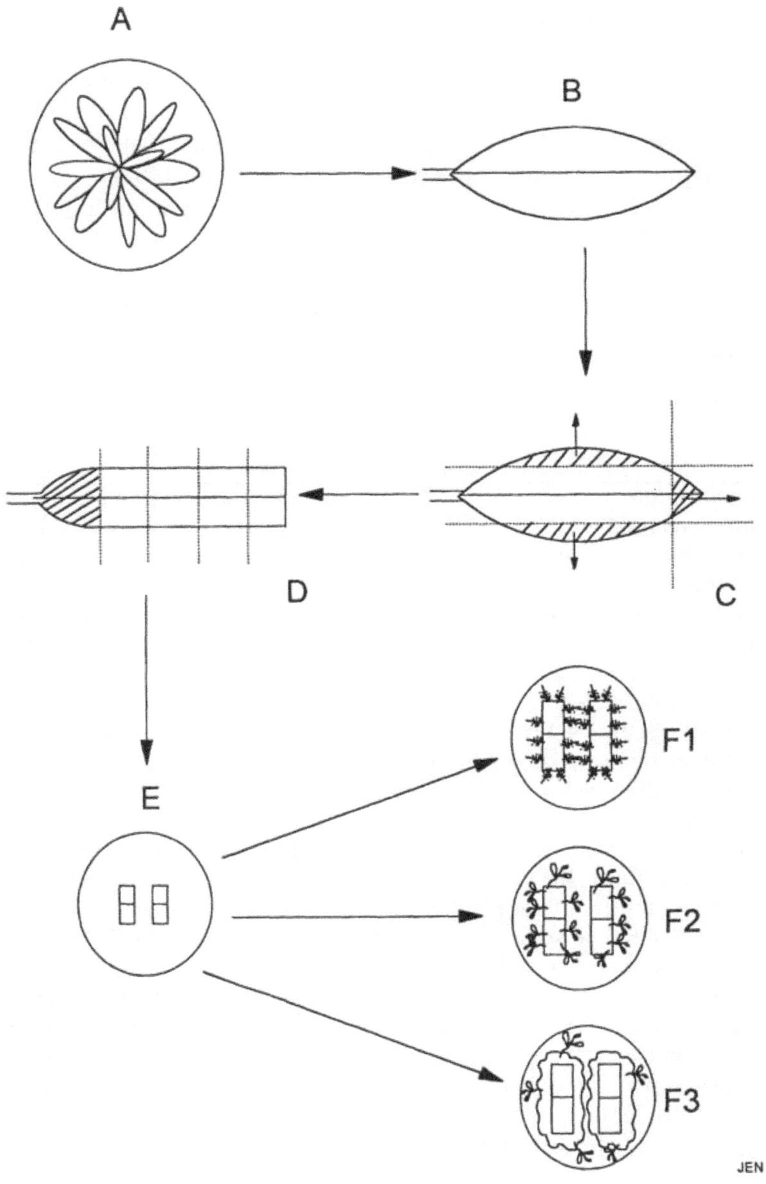

Figure 18.1 Diagrammatic representation of explant preparation and expected results for *Petunia* experiments. (A) Stock plant that is sufficiently mature for collection of leaves for explants; (B) excised leaf; (C) cut away and discard edges and tip of leaf; (D) cut explants so that they are 0.5 cm wide and 1.5 cm long including the midvein going across the width; (E) place two explants of each cultivar into each petri dish. (F1) When the medium contains only auxin, only roots should form; (F2) when the medium contains only BA, only shoots should form; (F3) when the medium contains both BA and IBA, the most callus should form. Some shoots may form from the callus. (From Trigiano, R. N. and Gray, D. J., Eds., *Plant Tissue Culture Concepts and Laboratory Exercises, First Edition*, CRC Press LLC, Boca Raton, FL, 1996.)

1990). This exercise will take 8 to 16 weeks to complete depending on how long the plants are monitored in the greenhouse.

Materials

Items needed for each student or group of students include:

- All materials from Experiment 1 will be needed except for the treatment media
- Four 100 × 15-mm petri dishes per petunia cultivar containing MS medium supplemented with 2.5 µM BA, 5.0 µM NAA, and 0.5 µM 2,4-dichlorophenoxyacetic acid (2,4-D) (after Binding and Krumbegel-Schroeren, 1984)
- Eight 100 × 15-mm petri dishes per petunia cultivar containing MS medium supplemented with 2.5 µM BA

Follow the instructions in Procedure 18.2 to complete this experiment.

Procedure 18.2
Regeneration of petunia from callus and monitoring somaclonal variation.

Step	Instructions and comments
1	Prepare explants as described in Procedure 18.1 and place on MS medium supplemented with 2.5 µM BA, 5.0 µM NAA, and 0.5 µM 2,4-D. Alternatively, callus tissue can be excised after 3 weeks from the Experiment 1 explants growing on the medium containing both BA and IBA.
2	Regardless of the source, after 3 to 4 weeks excise pieces of callus that are approximately 4 mm^3 and place on MS medium with 2.5 µM BA. At this time prepare new leaf explants and place two per petri dish on MS medium containing 2.5 µM BA.
3	After an additional 4 weeks, excise shoots that formed from callus and from leaf explants, but keep shoots from the treatments separate. Shoots can be rooted in vitro on MS without PGRs or under intermittent mist in vermiculite in a greenhouse.
4	After the shoots are rooted, transplant plantlets as described for the original seedlings. Gradually acclimatize plantlets to the greenhouse environment. Compare the growth of plantlets to that of the original stock plants. After a time, the plants will be sufficiently large to be transplanted into 15 (6-in.) tall pots containing the same soilless medium as before.
5	This experiment can be arranged as a completely randomized design with four replications, each consisting of two explants, in the laboratory. In the greenhouse, a randomized complete block design is best. The number of replications will depend on the number of plants that are acclimatized. Blocking should be done with consideration of potential temperature gradients within the greenhouse.
6	Collect data each week on number and length of shoots that form from callus and leaf explants. Measure initial size of acclimatized plantlets in the greenhouse including width and length of fully expanded leaves. Also note if leaves appear normal or abnormal in any way. Measure differences in flower corolla length, width across the tops of mature flowers, color, and number of stamens and note any abnormalities. Determine plant height and number and length of branches.

Anticipated results

Callus should grow from the original leaf explants placed on the medium with BA, NAA, and 2,4-D. Shoots should form directly from leaf explants and indirectly from the callus when they are cultured on medium containing only BA. Acclimatized plants should be obtained from both the leaf explants and the callus. There is a much greater chance of variability among the plants regenerated from callus compared to the plants regenerated directly from the leaf explants or the original stock plants.

Questions

- What are the disadvantages and/or opportunities that might be presented to the breeder if shoots are regenerated indirectly?
- If a person is interested in clonal micropropagation, how might they avoid somaclonal variation?
- Might somaclonal variation be a problem for genetic transformation?
- How might the variation be increased among the plants regenerated from tissue cultures?

Literature cited

Binding, H. and G. Krumbegel-Schroeren. 1984. Protoplast regeneration. pp.123–132. In: *Petunia*. K.C. Sink (Ed.). Springer-Verlag, Berlin.

Lewis-Smith, A.C., M. Chamberlain, and S.M. Smith. 1990. Genetic and chromosomal variation in *Petunia hybrida* plants regenerated from protoplast and callus cultures. *Biol. Plant*. 32:247–255.

Liberty Hyde Bailey Hortorium. 1976. *Hortus Third: A Concise Dictionary of Plants Cultivated in the United States and Canada*. 3rd ed. Macmillan, New York.

Murashige, T. and F. Skoog. 1962. A revised medium for rapid growth and bioassays with tobacco tissue cultures. *Physiol. Plant*. 15:473–497.

chapter nineteen

Nonzygotic embryogenesis*

Dennis J. Gray

An embryo can be defined as the earliest recognizable multicellular stage of an individual that occurs before it has developed the structures or organs characteristic of a given species. In most organisms, embryos are morphologically distinct entities that function as an intermediate stage in the transition between the gametophytic to sporophytic life cycle (Figure 19.1). For example, in higher plants, we are most familiar with embryos that develop within seeds; such embryos usually arise from gametic fusion products (zygotes) following sexual reproduction and are termed zygotic embryos, although seed-borne embryos also can develop apomictically (i.e., without sexual reproduction). However, plants are unique in that morphologically and functionally correct nonzygotic embryos also can arise from widely disparate cell and tissue types at a number of different points from both the gametophytic and sporophytic phases of the life cycle (Figure 19.2).

The first demonstration that plants could produce nonzygotic embryos in vitro was published in 1958 by Steward et al. Subsequently, Reinert (1959) observed bipolar embryos to differentiate in a culture of carrot roots after transfer from one medium to another. While carrot was the first species in which in vitro nonzygotic embryogenesis was reported, in subsequent years many species of angiosperms and gymnosperms have been added to the list of successes. In fact, demonstrations of nonzygotic embryogenesis are so widespread that it may be regarded to be a universal capability of higher plants.

A plethora of terminology has arisen to designate nonzygotic embryos. Such embryos originally were termed "embryoids" to denote perceived significant differences from zygotic embryos and, unfortunately, this term persists in some literature. However, differences in embryogenic cell origins notwithstanding, distinctions between zygotic and nonzygotic embryos become blurred as our understanding of embryo development increases. Nonzygotic embryos are now shown to be functionally equivalent to zygotic embryos and the suffix "oid" should be dropped. Other terms for nonzygotic embryos are based primarily on differences in their specific sites of origin and often are interchangeable, leading to some inconsistencies in the literature. For example, nonzygotic embryos can arise from plant vegetative cells, reproductive tissues, zygotic embryos, or callus cells derived from any of these. Thus, "somatic embryos" grow from somatic cells; "haploid or pollen embryos" are derived from pollen grains or microspore mother cells; "nucellar embryos" are formed from nonzygotic nucellar seed tissue; "direct secondary embryos" develop from previously formed embryos; etc.

* Florida Agricultural Experiment Station Journal Series No. R-07027.

0-8493-2029-1/00/$0.00+$.50

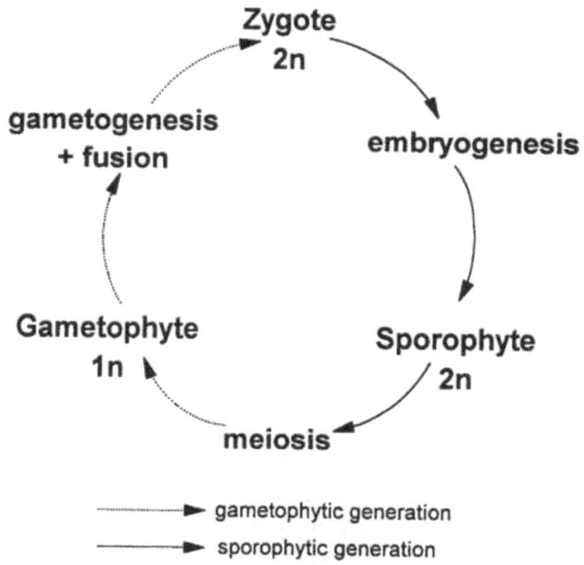

Figure 19.1 Typical angiosperm life cycle showing the natural role of embryogenesis in sporophyte development. (From Trigiano, R. N. and Gray, D. J., Eds., *Plant Tissue Culture Concepts and Laboratory Exercises, First Edition*, CRC Press LLC, Boca Raton, FL, 1996.)

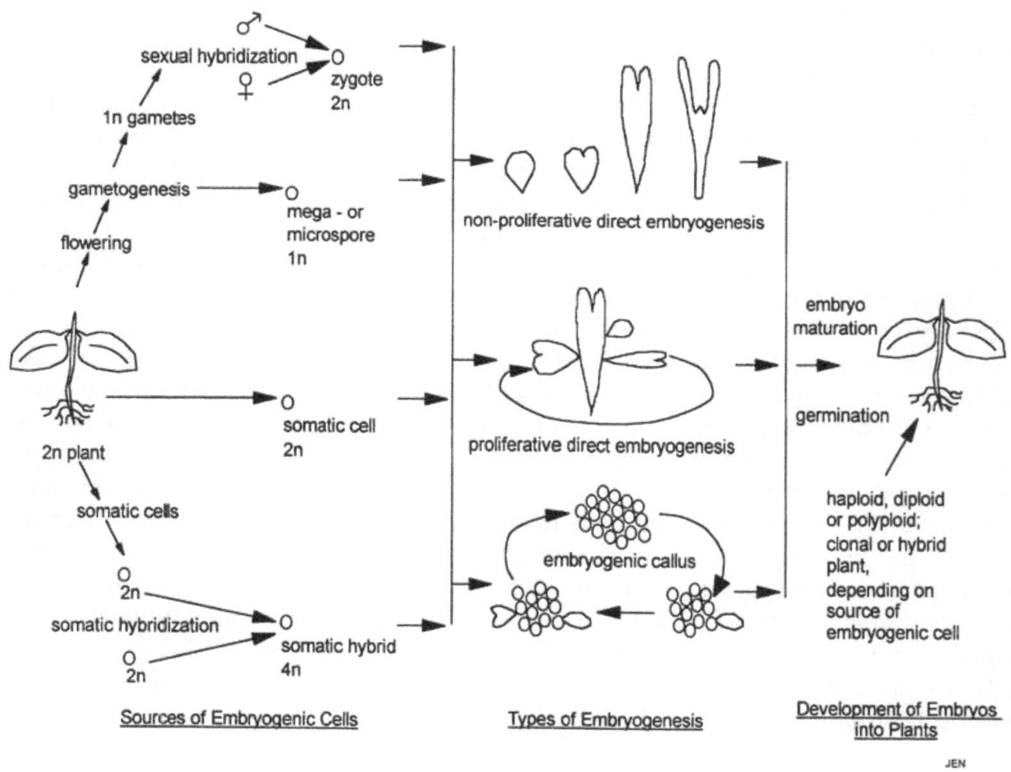

Figure 19.2 Sources of embryogenic cells, types of embryogenesis, and plants obtained via in vitro culture of higher plants. (From Trigiano, R. N. and Gray, D. J., Eds., *Plant Tissue Culture Concepts and Laboratory Exercises, First Edition*, CRC Press LLC, Boca Raton, FL, 1996.)

Most of these embryo types share the common trait of being able to be manipulated via in vitro culture. Such embryogenic culture systems form the basis for many biotechnological approaches to plant improvement, since they allow not only clonal plant propagation, but also specific and directed changes to be introduced into desirable, elite individuals by genetic engineering of somatic cells. Modified individual cells and/or embryos then can be efficiently multiplied in vitro to very high numbers prior to plant development. This approach to genetic improvement bypasses the unwanted consequences of sexual reproduction (mass genetic recombination and required cycles of selection) inherent to conventional breeding technology.

Because of the many commonalities exhibited by these embryo types as well as the various methods available to study and manipulate them, for convenience, this chapter will only utilize the term "nonzygotic embryo" to designate the embryos that develop through in vitro culture, regardless of origin. Thus, the developmental processes and experimental procedures described will tend to be applicable to many of the embryo types mentioned above.

In vivo vs. in vitro growth conditions

In seeds, nutritive tissues directly encase the developing embryo. Early in embryogenesis, nutritive substances enter the zygotic embryo either through the suspensor or the embryo body (Figure 19.3A). The relative reliance of the developing embryo on obtaining nutrition via the suspensor vs. endosperm differs greatly depending on the species. In contrast, nonzygotic embryos develop naked, not encased in seed endosperm, and, as such, are not subjected to specialized and highly regulated nutritional regimes (Gray and Purohit, 1991). Often, a suspensor is the only link between the embryo and growth medium (Figure 19.3B). This demonstrates that the suspensor can serve as the pathway for all needed nutrition and that endosperm is not absolutely necessary for embryogenesis and germination to occur.

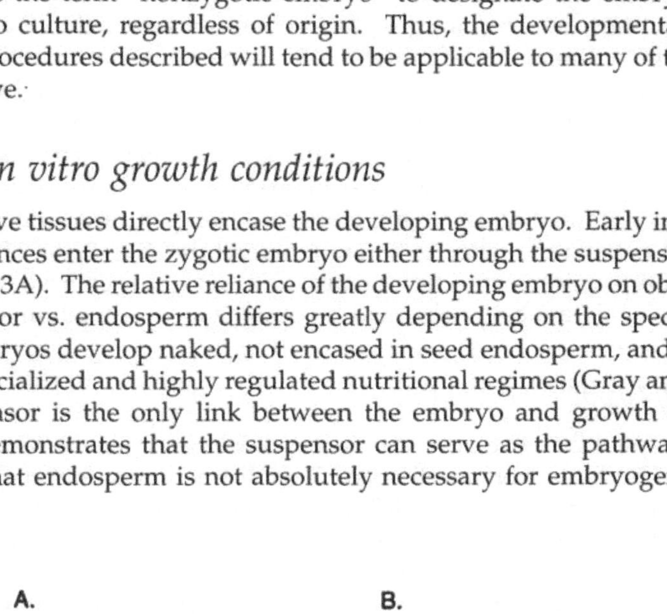

Figure 19.3 Comparison of typical zygotic embryo development in seed (A) with nonzygotic embryo development from callus (B). Zygotic embryos typically develop within nutritive seed tissues (shaded area) and are connected to the mother plant by a suspensor (arrow). In contrast, nonzygotic embryos often develop perched above subtending tissue, the only source of nutrition being the narrowed suspensor. (From Gray, D.J. and A. Purohit. 1991. *CRC Crit. Rev. Plant Sci.* 10:33–61.)

Common attributes of embryogenic culture protocols

Given the range of different explants, culture conditions, and media used to initiate and maintain embryogenic cultures, certain generalities can be made regarding basic methodology. The use of specific plant growth regulators (PGRs), which are required in most instances, will be discussed below with regard to embryogenic cell initiation.

The choice of genotype and explant often is crucial in obtaining an embryogenic response (see Gray, 1990; Gray and Meredith, 1992 for reviews). For example, with corn, only a limited range of genotypes are capable of producing embryogenic cultures and, with few exceptions, only immature embryo explants can be used. With corn, the age of immature embryos also is important, since embryos that are too young do not survive culture and those that are too old do not produce embryogenic callus. Interestingly, most species tested to date can be induced to produce embryogenic cultures from at least some genotypes and tissues. Additionally, the embryogenic response is heritable, such that it can be bred from embryogenic lines into nonembryogenic lines via sexual hybridization, albeit with a certain degree of difficulty and transfer of other, often undesirable, traits via sexual hybridization.

Environmental requirements for optimum culture growth often are quite specific, but not unusual. Cultures may require growth in either dark or light or a combination of both over time, the optimization of which can be experimentally quantified. Culture in dark may be necessary in order to prevent the triggering effect of light on many plant biological processes that may adversely affect the growth of embryogenic cell populations. Culture in dark suppresses unwanted tissue differentiation in explant tissue, for example, by limiting the development of plastids into chloroplasts. Similarly, dark conditions may inhibit precocious germination of young embryos. Temperature requirements also are specific, but tend to be in the range of "room temperature" (i.e., 23 to 27°C). Despite the similarity in culture requirements, improvements in nonzygotic embryo development and maturation increasingly are being obtained by optimization of medium composition.

Origin of nonzygotic embryos

It is generally accepted that, like zygotic embryos, nonzygotic embryos arise from a single cell, in contrast to budding from a cell mass. This distinction is important in considering efficient genetic engineering, since modification of a single embryogenic cell might eventually result in a modified plant, compared to genetic modification of a cell within a bud, which would result in a chimeric plant.

Nonzygotic embryos growing from isolated cells, such as microspores or protoplasts, clearly develop from single cells. However, the origin of nonzygotic embryos that develop from complex intact primary explant tissue or callus is more difficult to resolve, since the action of microscopic single cells cannot be readily followed. Nonzygotic embryos often develop with a well-defined suspensor apparatus identical to that of zygotic embryos, which somewhat suggests a single cell origin, whereas others can develop with a broad basal attachment, suggesting a multicellular budding phenomenon (Figure 19.4).

Initiation of embryogenic cells

In complex explants, nonzygotic embryos typically can be initiated only from the more juvenile or meristematic tissues. For example, immature zygotic embryos or zygotic embryo cotyledons and hypocotyls dissected from ungerminated seeds are commonly used explants. The young leaves, shoot tips, or even roots of established plants sometimes are used to initiate embryogenic cultures. However, the explant response is highly genotype dependent, so that for any given species, only a certain type or range of explant can be used to initiate embryogenic cultures.

There are several pathways by which nonzygotic plant cells become embryo initials (Christensen, 1985). In instances where the explant consists of undifferentiated embryonic tissue, such as an immature zygotic embryo, initiation and maintenance of an embryogenic callus is akin to culturing and propagating a preexisting proembryonal complex. Thus,

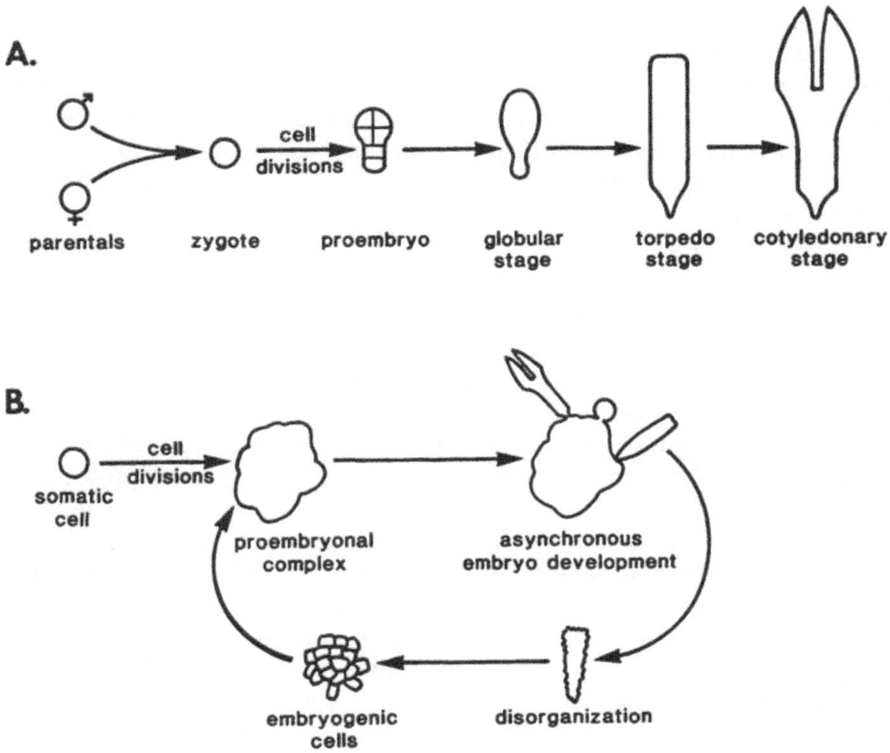

Figure 19.4 Comparison of typical zygotic embryogenesis occurring within seeds (A) with proliferative nonzygotic embryogenesis occurring in vitro (B). Zygotic embryogenesis is characterized as being very regulated, such that embryos synchronously pass through distinct developmental stages, whereas nonzygotic embryogenesis often is nonuniform, with many stages present at a given time. Nonzygotic embryos may bypass maturation and become disorganized, adding to the proembryonic tissue mass. (From Gray, D.J. and A. Purohit. 1991. *CRC Crit. Rev. Plant Sci.* 10:33–61.)

the embryogenic cells present in explant tissue prior to culture initiation are simply propagated and otherwise manipulated in vitro. However, in many instances, embryogenic cells are induced from nonembryogenic cells; this represents a dramatic change in their presumptive fate. The shift in developmental pattern involves a dedifferentiation away from the cell's "normal" fate followed by redetermination toward an embryogenic cell type. For example, cells in leaf explants, which normally would develop into constituents of relatively short-lived parenchymatous tissue, instead become embryogenic under certain conditions. This is a pivotal change in development, since cells that normally would be capable of only a few divisions, at most, before senescence, instead become redirected to become totipotent and capable of possibly unlimited divisions. Such embryogenic cells become immortal in the sense that they reinstate the germ line by being capable of developing into mature, reproductive individuals.

The fact that isolated somatic cells can develop normally into embryos irrevocably demonstrates that the developmental program for embryogenesis is contained within and controlled by the cell itself and not by external factors. However, the exact nature of the triggering mechanism(s) of embryogenesis, whether it be a physical, biochemical, and/or genetic event(s), is unknown. Early attempts to identify genes that became activated as a direct consequence of embryogenic cell induction failed due to inadequate understanding of the basic induction phenomenon, which resulted in faulty experimental designs.

Since the cell cultures used were already induced, any genes critical to the induction step already had been activated. Hence, the genes that were observed in such studies actually were those activated later in embryogenesis such that they were controlling downstream developmental processes and were not related to the triggering event. Despite the technical difficulty of identifying critical induction-controlling genes, a number of other genes and their products have been identified during nonzygotic embryo development (e.g., Zimmerman, 1993). In some instances, the genetic mechanisms resolved for nonzygotic embryos can be related directly to that of the corresponding zygotic embryo.

Inductive plant growth regulators

In practice, the initiation of embryogenic cells requires in vitro culture of the appropriate explant on or in a medium that contains specific PGRs. In fact, a preponderance of reports of embryogenic culture initiation employed a very narrow range of PGRs that typically are added to culture medium. Synthetic auxins, notably 2,4-dichlorophenoxyacetic acid (2,4-D), are added to the medium in the predominance of reported protocols. Similar auxins used include dicamba, indolebutyric acid (IBA), naphthoxyacetic acid (NOA), picloram, and others. In addition, several weaker auxins, such as indoleacetic acid (IAA), a natural plant hormone, and naphthyleneacetic acid (NAA) have been utilized in a few culture systems. Auxins serve to induce the formation of embryogenic cells, possibly by initiating differential gene activation, as noted above, and also appear to promote increase of embryogenic cell populations through repetitive cell division, while simultaneously suppressing cell differentiation and growth into embryos. However, an auxin often is not required in instances where the explant consists of preexisting embryogenic cells, as described above, possibly because a discrete induction step then is not required.

In addition to auxin-like PGRs, cytokinins also are required to induce embryogenesis in many dicotyledonous species. In a few instances, only a cytokinin is required to cause embryogenic cultures to develop. The most commonly used cytokinin is benzyladenine (BA), but others such as thidiazuron (TDZ) and kinetin, and the natural cytokinin, zeatin, also are utilized.

Actual PGR concentration is important for an optimum response, since concentrations that are too low may not trigger the inductive event and concentrations that are too high, particularly when considering phenoxy-auxins, may become toxic. Typically, following the induction period, resulting culture material is transferred to the medium lacking PGRs, which removes the auxin-induced suppression of embryo development and allows embryogenesis to occur; however, not all culture systems require this two-step procedure. For example, as mentioned above, certain species do not require any induction step at all and others, particularly poaceous monocots, become induced and undergo complete embryogenesis in the continuous presence of auxin. Examples of different PGR regimes used in culture systems are illustrated via the laboratory exercises provided after this chapter.

Embryo development

The physical, observable transition from a nonembryogenic to an embryogenic cell may occur when the progenitor cell undergoes an unequal division, resulting in a larger vacuolate cell and a smaller, densely cytoplasmic (embryogenic) cell. Embryogenic cells are readily distinguished by their small size, isodiametric shape, and densely cytoplasmic appearance (Figure 19.5). This type of unequal cell division is identical to that observed in zygotes and may be an early indication of developmental polarity. The embryogenic cell then either continues to divide irregularly to form a proembryonal complex, or divides

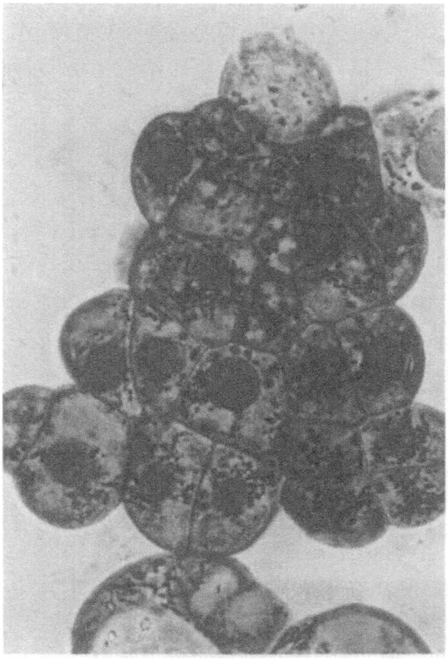

Figure 19.5 Typical embryogenic cells of grape. Note the three-cell stage embryo in the upper center part of the cell mass. (From Gray, D.J., 1995. *Somatic Embryogenesis in Woody Plants*. Reprinted by permission of Kluwer Academic Publishers.)

in a highly organized manner to form a somatic embryo (see Figures 19.2 and 19.4). However, an often undesirable difference exhibited by nonzygotic embryos is that they frequently deviate from the normal pattern of development either by producing callus, undergoing direct secondary embryogenesis, or germinating precociously. This tendency toward erratic development clearly is due to environmental factors as discussed below.

Zygotic and nonzygotic embryos share the same gross pattern of development, with both typically passing through globular, scutellar, and coleoptilar stages for monocots, or globular, heart, torpedo, and cotyledonary stages for dicots and conifers. Generally, the anatomy and morphology of well-developed nonzygotic embryos is faithful to the corresponding zygotic embryotype such that they easily can be identified by eye or with the aid of a stereomicroscope (Figure 19.6). For example, nonzygotic embryos of grass and

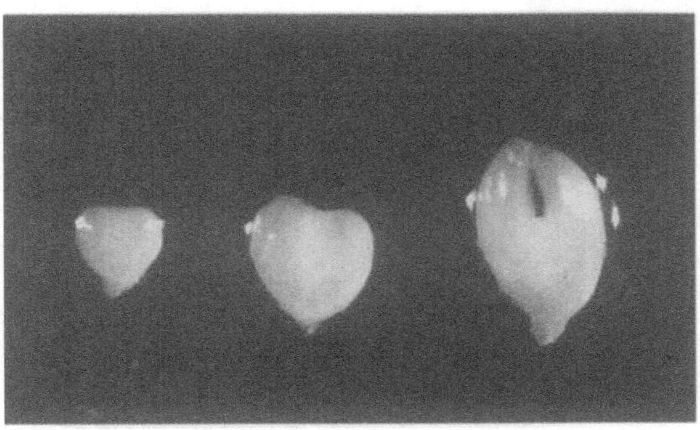

Figure 19.6 Somatic embryos of grape at heart and early cotyledonary stages, illustrating morphologies highly faithful to those of zygotic embryos. (From Trigiano, R. N. and Gray, D. J., Eds., *Plant Tissue Culture Concepts and Laboratory Exercises, First Edition*, CRC Press LLC, Boca Raton, FL, 1996.)

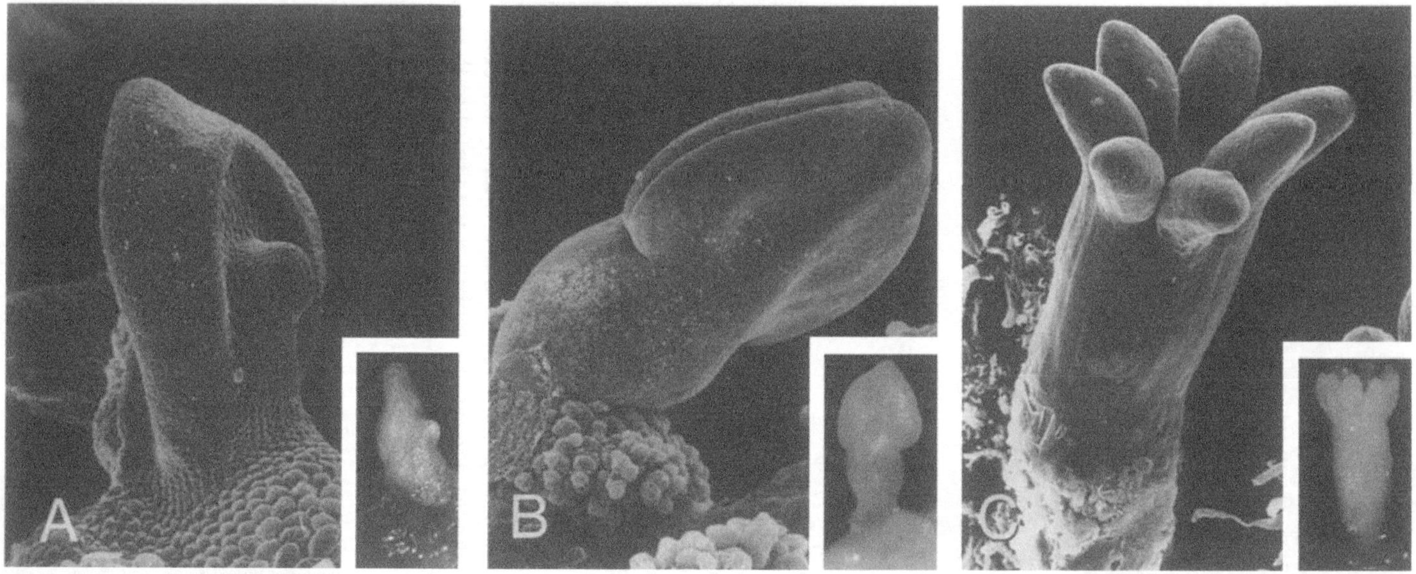

Figure 19.7 Comparison of monocot (A), dicot (B), and gymnosperm (C) nonzygotic embryos. Scanning electron micrographs, with corresponding stereomicrographs (inset). (A) Somatic embryo of orchardgrass growing from an embryogenic callus. Note the protruding coleoptile and notch, through which the first leaves will emerge after germination. The prominent large flattened body of the embryo is the scutellum. (From Gray, D.J., B.V. Conger, and G.E. Hanning. 1984. *Protoplasma* 122:196–202. With permission.) (B) Somatic embryo of grape growing from an embryogenic callus. Note two flattened cotyledons and subtending hypocotyl. (From Gray, D.J. 1995. *Somatic Embryogenesis in Woody Plants*. Reprinted by permission of Kluwer Academic Publishers.) (C) Somatic embryo of Norway Spruce. Note multiple cotyledons and elongated hypocotyl. (From Fowke, L.C. S.M. Altree, and P.J. Rennie. 1994. *Plant Cell Rep*. 13:612–618. With permission.)

cereal species typically possess a scutellum, coleoptile, and embryo axis, which are distinctive embryonic organs of monocots (Figure 19.7A). Embryos of dicotyledonous species have a distinct hypocotyl and cotyledons (usually two) (Figure 19.7B); those of conifers (see Chapter twenty-five) also exhibit a hypocotyl and numerous cotyledons (Figure 19.7C), the number of which is similar to that of the given species.

Embryo development occurs through an exceptionally organized sequence of cell division, enlargement, and differentiation. During early development the embryo assumes a clavate, globular shape and remains essentially an undifferentiated, but organized mass of dividing cells with a well-defined epidermis. Subsequent heart through early torpedo stages are characterized by cell differentiation and polarized growth, notably elongation and initiation of rudimentary cotyledons in dicots (Figure 19.6) and development of the scutellum with initiation of the coleoptilar notch in poaceous monocots. At the same time, obvious tissue differentiation begins with the development of embryonic vasculature (Figure 19.8) and accumulation of intracellular storage substances. Final stages of development toward maturation are distinguished by overall enlargement, increase in cotyledon size in dicots (Figure 19.9), and coleoptilar enlargement in monocots. At the same time, the embryonic axis becomes increasingly developed. In dicots, the root apical meristem becomes well established, embedded in tissue located above the suspensor apparatus and at the base of the hypocotyl, whereas the shoot apical meristem develops externally between the cotyledons. In monocots, the embryo axis develops laterally and parallel to the scutellum. The root apical meristem is embedded and the shoot apical meristem develops externally, but is protected by the coleoptile. All of the events described above occur in concert with each other in a manner that is essentially identical to that of zygotic embryos. However, for a number of reasons, as described below, nonzygotic embryos often differ somewhat from their zygotic counterparts in morphology and/or performance.

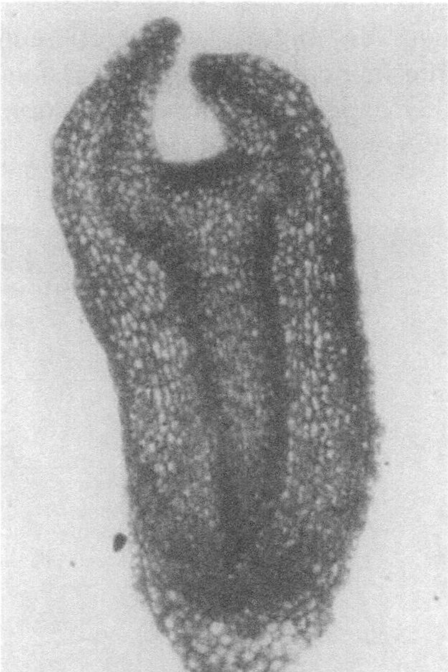

Figure 19.8 Longitudinal section through a grape somatic embryo showing typical vascular system. (From Gray, D.J. 1995. *Somatic Embryogenesis in Woody Plants*. Reprinted with permission of Kluwer Academic Publishers.)

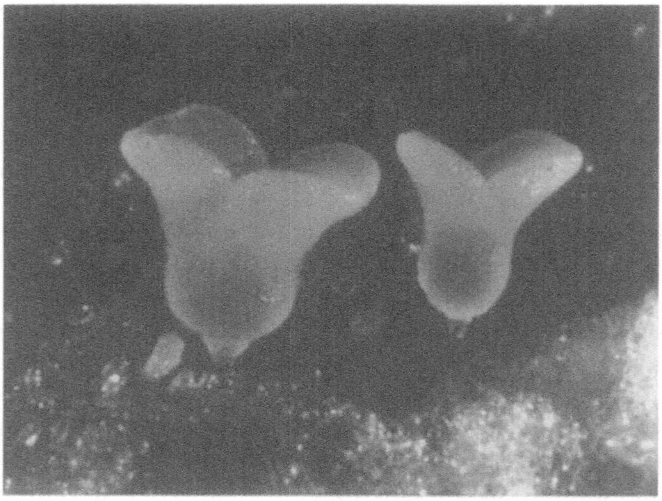

Figure 19.9 Cotyledonary-stage somatic embryos of cantaloupe growing from a cultured cotyledon. Note that the fine suspensor is the only connection to explant. (From Gray, D.J., D.W. McColley, and M.E. Compton. 1993. *J. Amer. Soc. Hort. Sci.* 118:425–432. With permission.)

An obvious difference in gross morphology between nonzygotic embryos growing in vitro and zygotic embryos in seeds is simply caused by the physical constraint on zygotic embryos imposed by the developing seed coat, often causing them to become compressed and/or flattened into a shape and size distinct for a given species or variety. This is made apparent by comparing the morphology of seed-borne zygotic embryos with corresponding nonzygotic embryos of a given species (Figure 19.10). Zygotic embryos excised from seed typically exhibit a highly compressed shape because the embryos become highly flattened during development. In contrast, nonzygotic embryos tend to be larger and have wider hypocotyls and fleshier cotyledons. It is possible that pressure exerted by the seed coat contributes to other aspects of embryo development that are lacking during nonzygotic embryogenesis (Gray and Purohit, 1991).

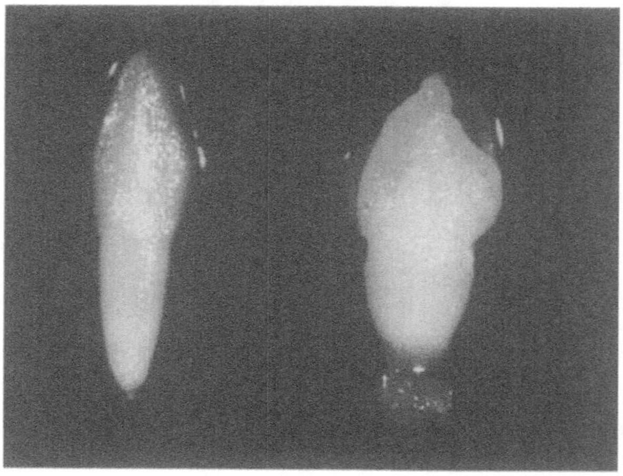

Figure 19.10 Comparison of grape zygotic embryo, compressed and flattened by development within a seed (left), with grape somatic embryo, which is not flattened (right). (From Gray, D.J. and A. Purohit. 1991. *CRC Crit. Rev. Plant Sci.* 10:33–61.)

In addition to differences in development related to simple physical constraints, for several reasons, significantly more instances of abnormal development are known to occur during nonzygotic embryogenesis when compared to zygotic embryogenesis. For example, as mentioned above, species that normally produce zygotic embryos with suspensors often produce clusters of nonzygotic embryos from a proembryonal cell complex (Figure 19.4). This basic change in developmental pattern is likely due to differences between the seed and in vitro environments, since immature zygotic embryos dissected from seeds and cultured also often develop abnormally (Gray and Purohit, 1991).

Nonzygotic embryos growing in mass from proembryonal complexes tend to develop asynchronously so that several stages are present in cultures at any given time (Figures 19.2 and 19.4). Nonzygotic embryos initiated over time are subjected to different nutrient regimes as the medium becomes depleted, then replenished, between and during subcultures. This leads to differences in development even among embryos from a single culture. With such variable and unregulated environmental conditions, nonzygotic embryos often bypass maturation altogether, becoming disorganized, forming new embryogenic cells, and contributing to asynchrony (Figure 19.4). Nonzygotic embryos also often exhibit structural anomalies such as extra cotyledons (Figure 19.11) and poorly developed apical meristems.

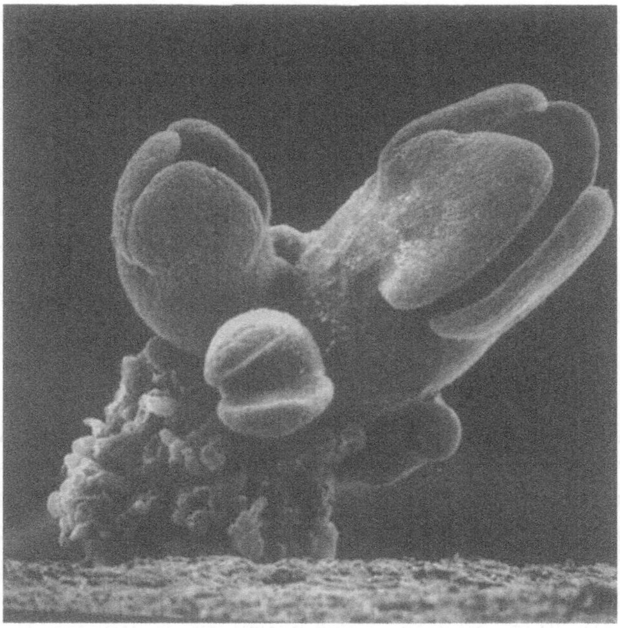

Figure 19.11 Asynchronous and abnormal somatic embryo development from embryogenic callus of grape. Increasing levels of development (clockwise from lower right globular embryo) are evident in a single group of embryos. In addition, the normal cotyledon number of two (lower center embryo) contrasts with supernumery cotyledon development of three and four on the two upper embryos. (From Gray, D.J. and J. A. Mortensen. 1987. *Plant Cell Tissue Organ Cult.* 9:73–80. Reprinted with permission from Kluwer Academic Publishers.)

Embryo maturation

Maturation is the terminal event of embryogenesis and is characterized by the attainment of mature embryo morphology, accumulation of storage carbohydrates, lipids, and proteins (see prior section regarding in vivo vs. in vitro growth conditions), reduction in water content, and often a gradual decline or cessation of metabolism. Nonzygotic embryos typically do not mature properly when compared to zygotic embryos. In fact, in the preponderance of instances, rapid growth continues to occur leading to precocious germination.

Although, complete maturation is not absolutely necessary in order to obtain plants from nonzygotic embryos, it is required to achieve high rates of plant recovery. As such, factors that influence and/or enhance nonzygotic embryo maturation have been explored. A number of culture medium components have been shown to promote maturation. In particular, high-sucrose levels (9 to 12%), timed pulses of abscisic acid (ABA), a naturally occurring plant hormone, and polyethylene glycol (PEG), an osmotically active compound, in conjunction with certain amino acids, notably glutamine, have been remarkably successful. For example, an 8-week treatment with ABA and PEG caused white spruce nonzygotic embryos to accumulate more lipids and to better resemble zygotic embryos morphologically (Attree and Fowke, 1993). Similarly, a short pulse of ABA in conjunction with glutamine resulted in alfalfa nonzygotic embryos that could withstand dehydration as mentioned below (Senaratna et al., 1989).

Quiescence and dormancy

Perhaps the most obvious developmental difference between zygotic and nonzygotic embryos is that the latter lack a quiescent resting phase. By comparison, zygotic embryos of many species begin a resting period during seed maturation. Zygotic embryo quiescence in conjunction with the protective and nutritive tissues that comprise a seed are the major factors allowing seeds to be conveniently stored. Thus, the ability to withstand dehydration appears to be a normal step in embryo development.

However, nonzygotic embryos tend either to grow and germinate without normal maturation, become disorganized into embryogenic tissue, or die. They rarely enter a resting stage. Although quiescence and dormancy have only recently been documented in somatic embryos, it is possible that dormancy occurs in instances where plants cannot be obtained from well-developed embryos (see Gray, 1986; Gray and Purohit, 1991 for reviews).

Embryo germination and plant development

Obtaining plants from nonzygotic embryos often is more difficult than would be expected. The early literature concerning first reports of nonzygotic embryogenesis for a given species or cultivar often did not include information on plant recovery. When plants were obtained, the recovery rate was very low or not reported, suggesting that a majority of nonzygotic embryos were too abnormal to germinate. Typically, plant recovery from nonzygotic embryos ranges from 0 to 50%. This is very low compared to zygotic embryos in commercial seed, for which germination and plant development typically exceed 90% in soil. In all but a few instances, nonzygotic embryos develop very poorly, if at all, when planted in soil like seed. Only recently has research been conducted to raise plant recovery rates.

Progress in culture methodology has resulted in better development and germination of nonzygotic embryos. As mentioned above regarding maturation, pulse treatments with various amino acids, osmotica, and PGRs, particularly ABA, have resulted in nonzygotic embryos with better maturation, including the ability to be dehydrated and stored like seeds, as well as improved germination characteristics (Attree and Fowke, 1993). This demonstrates that careful attention to culture conditions and nutrition, especially with regard to timed pulses of certain factors, results in plant recovery rates from nonzygotic embryos equivalent to those of zygotic embryos. In general, it can be concluded that conditions favoring embryo maturation also favor the recovery of plants.

After the event of germination, the embryo begins development into a plant. Typically, the storage reserves present (lipids, proteins, and/or starch — depending on the species) become depleted concomitant with the start of increased mitotic activity in shoot and root meristems. Eventually, a young photosynthetically competent plant develops (Figure 19.12), which then can be gradually acclimatized to ambient conditions.

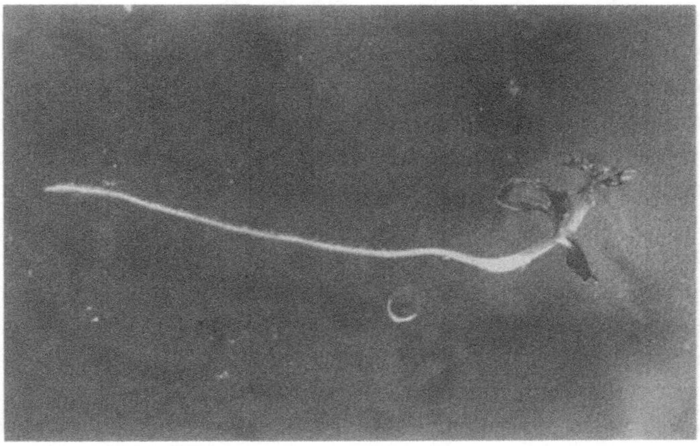

Figure 19.12 Plant development from a germinated melon somatic embryo. Note shoot and root development. (From Gray, D.J., D.W. McColley, and M.E. Compton. 1993. *J. Amer. Soc. Hort. Sci.* 118:425–432. With permission.)

Uses for embryogenic cultures

Embryogenic culture systems are used for a number of purposes.

1. They constitute an important tool for the study of plant development, both due to the unique convenience of in vitro culture over *en planta* growth and the contrasts that can be drawn from differences between nonzygotic embryogenesis in vitro and zygotic embryogenesis in seeds.
2. Embryogenic cultures are an efficient vehicle for genetic engineering, since embryogenic cultures often produce many embryos per volume of cell mass and isolated genes integrated into single embryogenic cells can become incorporated into the genome of the plant that ultimately develops.
3. Many nonzygotic embryos can be produced from a single desirable plant for efficient clonal propagation.

Certain potential uses of embryogenic systems as research tools have been described or suggested throughout this chapter and an in-depth discussion of genetic engineering applications is presented in Chapter thirty-one. The use of nonzygotic embryogenesis as a cloning vehicle is discussed in more detail below.

Synthetic seed technology

An active research front has emerged over the last decade with the goal of developing nonzygotic embryogenesis into a commercially useful method of plant propagation. The technology that has emerged is termed "synthetic (or artificial) seed technology." A synthetic seed is defined as a somatic embryo that is engineered to be of practical use in commercial plant production (Gray and Purohit, 1991). Applications for synthetic seed vary depending on the relative sophistication of existing production systems for a given crop and the opportunities for improvement. Whether or not a cost advantage results from synthetic seed will ultimately determine its commercial use. For seed-propagated agronomic crops, quiescent, nonzygotic embryos produced in bioreactors and encapsulated in synthetic seed coats will be necessary (see Chapter twenty, Experiment 7 for an example). Certain vegetable crops with expensive-to-produce seeds, such as seedless watermelon, are attractive candidates for synthetic seed technology, since the per-plant cost might be reduced.

Similarly, conifers, which are difficult to improve by breeding due to a long life cycle, would benefit from application of synthetic seed technology if elite individuals could be cloned and planted with seed efficiency (Farnum et al., 1983). For vegetatively propagated crops, particularly those with a high per-plant value, naked, hand manipulated, nonquiescent embryos may be cost effective. For example, many ornamental crops are painstakingly commercially micropropagated in tissue culture via adventitious bud proliferation, with the major expense being the labor needed for multiple culture and rooting steps. Substitution of embryogenic culture systems for such crops would greatly reduce labor costs since mass-produced somatic embryos easily could be hand selected and placed directly into planting flats and rooted (Gray and Purohit, 1991).

Ongoing efforts to develop synthetic seed technology constitutes one of the most active areas of nonzygotic embryogenesis research. The potential benefits of using "clonal seeds" in agricultural production is great enough to stimulate continued investigation of factors regulating nonzygotic embryo maturation.

Conclusion

In vitro embryogenesis from nonzygotic cells must be regarded as a universal property of higher plants, albeit one that generally does not occur in nature. When induced from isolated cells, nonzygotic embryogenesis conclusively demonstrates not only cellular totipotency, but the presence of a highly conserved developmental program that is innate to a wide range of cell types. For the student, nonzygotic embryogenesis represents an opportunity to conveniently study in a culture vessel one of the more developmentally complex aspects of the plant life cycle.

Embryogenic cell culture allows the entire genome of a plant to be manipulated with microbiological techniques. Once the cells are proliferated, genetically engineered, and/or otherwise managed, intact plants easily can be reconstituted. From such simple manipulation comes numerous possibilities for better understanding and ultimately improving our use of plants.

Literature cited

Attree, S.M. and L.C. Fowke. 1993. Embryogeny of gymnosperms: advances in synthetic seed technology of conifers. *Plant Cell Tissue Organ Cult.* 35:1–35.

Christensen, M.L. 1985. An embryogenic culture of soybean: towards a general theory of embryogenesis. In: *Tissue Culture in Forestry and Agriculture*, pp. 83–103. R.R. Henke, K.W. Hughes, M.J. Constantine, and A. Hollaender, Eds., Plenum Press, New York.

Farnum, P., R. Timmis, and J.L. Kulp. 1983. Biotechnology of forest yield. *Science* 219:694–702.

Fowke, L. C., S.M. Attree, and P.J. Rennie. 1994. Scanning electron microscopy of hydrated and desiccated mature somatic embryos and zygotic embryos of white spruce (*Picea glauca* (Moench) Voss.). *Plant Cell Rep.* 13:612–618.

Gray, D.J. 1986. Quiescence in monocotyledonous and dicotyledonous somatic embryos induced by dehydration. In: Proc. Symp. Synthetic Seed Technology for the Mass Cloning of Crop Plants: Problems and Perspectives. *HortScience* 22:810–814.

Gray, D.J. 1990. Somatic embryogenesis and cell culture in the Poaceae, In: *Biotechnology in Tall Fescue Improvement*, pp. 25–57. M.J. Kasperbauer, Ed., CRC Press, Boca Raton, FL.

Gray, D.J. 1995. Somatic embryogenesis in grape, In: *Somatic Embryogenesis in Woody Plants*, pp. 191–217. S.M. Jain, P.K. Gupta, and R.J. Newton, Eds., Kluwer Academic Publishers.

Gray, D.J., B.V. Conger, and G.E. Hanning. 1984. Somatic embryogenesis in suspension and suspension-derived callus cultures of *Dactylis glomerata*. *Protoplasma* 122:196–202.

Gray, D.J., D.W. McColley, and M.E. Compton. 1993. High-frequency somatic embryogenesis from quiescent seed cotyledons of *Cucumis melo* cultivars. *J. Amer. Soc. Hort. Sci.* 118:425–432.

Gray, D. J. and C.P. Meredith. 1992. Grape. In: *Biotechnology of Perennial Fruit Crops*. F. A. Hammerschlag and R. E. Litz, Eds., CAB International, Wallingford, U.K.

Gray, D. J. and J. A. Mortensen. 1987. Initiation and maintenance of long term somatic embryogenesis from anthers and ovaries of *Vitis longii* 'Microsperma'. *Plant Cell Tissue Organ. Cult.* 9:73–80.

Gray, D. J. and A. Purohit. 1991. Somatic embryogenesis and the development of synthetic seed technology. *CRC Crit. Rev. Plant Sci.* 10:33–61.

Reinert, J. 1959. Ueber die Kontrolle der Morphogenese und die Induktion von Adventiveembryonen an Gewebekulturen aus Karotten. *Planta* 53:318–333.

Senaratna, T., B.D. McKersie, and S.R. Bowley. 1989. Desiccation tolerance of alfalfa (*Medicago sativa* L.) somatic embryos — influence of abscisic acid, stress pretreatments and drying rates. *Plant Sci.* 65:253–259.

Steward, F.C., M.O. Mapes, and K. Mears. 1958. Growth and organized development of cultured cells. II. Organization in cultures grown from freely suspended cells. *Amer. J. Bot.* 45:705–708.

Zimmerman, L.J. 1993. Somatic embryogenesis: a model for early development in higher plants. *Plant Cell* 5:1411–1423.

chapter twenty

*Embryogenic callus and suspension cultures from leaves of orchardgrass**

Dennis J. Gray, Robert N. Trigiano, and Bob V. Conger

Members of the family Poaceae comprise the single most economically important group of plants. This family includes the nutritious cereal grain crops such as corn (*Zea mays* L.), oats (*Avena sativa* L.), rice (*Oryza sativa* L.), and wheat (*Triticum aestivum* L.), as well as the forage grasses such as annual ryegrass (*Lolium multiflorum* Lam.), tall fescue (*Festuca arundinacea* Schreb.), and the wheatgrasses (*Agropyron* spp.). The cereals are important sources of complex carbohydrates, oils, and proteins, whereas the forages are important in dairy and meat production. Sugarcane (*Saccharum officinarum* L.), another member of this family, is an important source of sucrose.

Much research emphasis in the Poaceae has gone toward developing biotechnological approaches to crop improvement. One area of biotechnology that received early attention by researchers was that of in vitro plant regeneration. The grasses and cereals were initially regarded by researchers to be difficult or impossible to regenerate from tissue and cell cultures. However, continuous effort since the early 1980s led to significant advancements, such that plant regeneration has been obtained for all grass and cereal species that have been attempted (Gray, 1990).

Somatic embryogenesis is the most common mode of regeneration for species in the Poaceae. Embryogenic culture systems are useful in the classroom, not only for illustrating plant regeneration, but also for studying embryo development and morphology. Since the Poaceae is contained within the monocotyledoneae, a taxonomic division based on embryo morphology, their zygotic embryos are distinctly different from dicotyledonous embryos. In many poaceous plants, the induction of somatic embryogenesis is not simple and reliable enough to be used for classroom exercises. Many species require use of immature zygotic embryos as explants, which can only be obtained by careful cultivation and pollination of source plants. Other species tend to produce somatic embryos that have relatively abnormal morphologies when compared to zygotic embryos. However, a few poaceous species, notably orchardgrass (*Dactylis glomerata* L.), are much easier to manipulate.

* Florida Agricultural Experiment Station Journal Series No. R-07028.

0-8493-2029-1/00/$0.00+$.50

Orchardgrass is a perennial cool-season forage species that is grown in temperate regions of the world to produce high-quality hay. It is genetically self-incompatible, which makes breeding of new varieties difficult and time consuming. The potential of using somatic embryos as synthetic seed to efficiently clone outstanding parental lines holds promise in the development of improved varieties (see Gray et al., 1992 for details). The following laboratory exercises cover the methods used to produce somatic embryos from cultures of orchardgrass. Selected clones of this species are highly embryogenic and, as a perennial, can be maintained in the greenhouse as potted plants. These factors account for its convenient use. Leaf tissues are used as explant material for initiating cultures from which morphologically correct somatic embryos are produced. Orchardgrass has "clasping-type" leaves, which are easier to dissect than "whorled-type" leaves. Thus, the orchardgrass system is ideal for demonstrating monocotyledonous somatic embryogenesis.

The following experiments represent a greatly expanded version of exercises that were originally published in *HortTechnology* (Gray et al., 1994) and are arranged to lead students through successful culture establishment, observations of embryogenic callus formation, and somatic embryo development, including the gradient embryogenic response that is typical of this culture system, culture maintenance, plant regeneration, manipulation of liquid suspension cultures, and the production of model "synthetic seeds."

General considerations

Growth of plants

A special clone of orchardgrass, 'Embryogen-P' (Conger and Hanning, 1991), is the most convenient source of leaf explant tissue and is available from Dr. B. V. Conger, Department of Plant and Soil Science, the University of Tennessee, Knoxville, TN 37901-1071. The clone should be maintained in greenhouse pots in a high-quality potting mix, with regular fertilizer applications, and kept in vigorous growing condition. Before becoming root bound, plant number can be increased by splitting entire plants, including the root ball, and transplanting into two or three new pots. Insect and fungal pests should be controlled; however, culture response may be decreased for up to 6 weeks after chemical spraying, especially with systemic fungicides. Therefore, stock plant health and vigor are the most important factors in successful embryogenic culture initiation of orchardgrass.

Culture media

The culture medium used is Schenk and Hildebrandt (SH) basal salt mixture (see Chapter three for composition) and vitamin powder with 6.6 mg/L (30 μM) dicamba (synthetic auxin plant growth regulator [PGR]), 30 g/L (88 mM) sucrose, and 0.7% tissue culture-tested agar. This medium is hereafter referred to as SH30, whereas the medium used for embryo germination studies and lacking dicamba is designated as SH0. The pH is adjusted to 5.4 before autoclaving. The cooling medium then is poured, 25 ml, into each 100 × 15-mm sterile petri dish. SH30-C is a liquid medium (i.e., without agar) that contains 3 g/L casein acid hydrolysate type 1. The casein hydrolysate is dissolved in 25 ml of distilled water, loaded into a 30-ml syringe, and then dispensed through a sterile 0.2-μm syringe filter into cooled, autoclaved, SH30 liquid medium. The resulting SH30-C then is sterilely pipetted (25 ml) into each 125-ml Erlenmeyer flask. Generally at least four petri dishes or flasks of the appropriate medium are required for each treatment and/or transfer for each student or student team.

Exercises

Experiment 1. Culture establishment

Orchardgrass is a tillering-type forage grass in which the plant body consists of dense clumps of tillers. New tillers arise by adventitious budding from the base of previously developing tillers. Each tiller is actually a complete plant in itself, with a root-shoot axis and several flattened, interfolded clasping leaves. Grass leaves grow from a pronounced basal (intercalary) meristem, such that the youngest part of a given leaf is the portion nearest to the shoot apical meristem. Youngest leaves are those nearest to the center. Thus, the youngest, most meristematically active leaf tissues are those nearest the bases of the innermost leaves. Figure 20.1 summarizes the steps of this experiment.

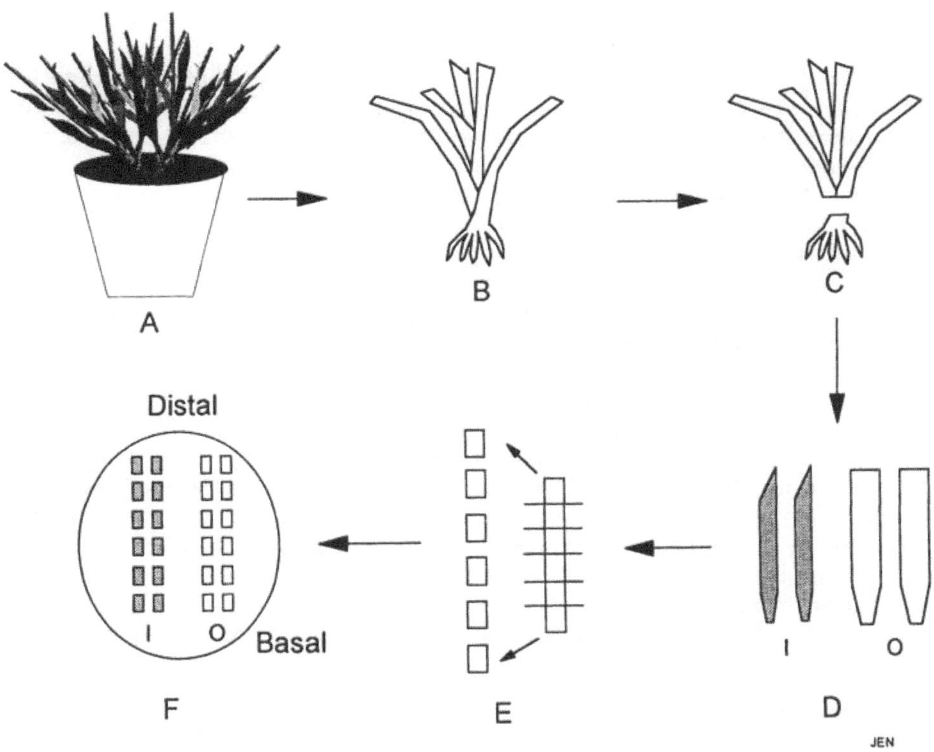

Figure 20.1 Diagrammatic representation of orchardgrass leaf explant preparation for embryogenic culture initiation. (A) Potted plant of orchardgrass variety Embryogen-P. (B) An isolated tiller. (C) Remove and discard basal stem and roots, leaving only the clasped leaves. (D) Carefully separate leaves and retain the innermost and next leaf out. (E) Carefully split each leaf longitudenally into two halves. Cut halves of innermost leaf at an angle to retain its identity, then surface disinfest. (F) Cut each leaf half into sections and plate sections of all four halves on a single petri dish containing SH30 medium. Be careful to retain the identity of each leaf and to keep the sections within each leaf in order of most basal to distal section. (From Trigiano, R. N. and Gray, D. J., Eds., *Plant Tissue Culture Concepts and Laboratory Exercises, First Edition*, CRC Press LLC, Boca Raton, FL, 1996.)

Materials

The following items are needed for each student or team:

- 100-ml of 50% commercial bleach with a drop of surfactant (such as Triton-X100™, Tween 20™, or Ivory Liquid™ soap)
- Twelve (approximately) 100 × 15-mm petri dishes containing SH30 medium
- Stereomicroscope with light source

Follow the protocol provided in Procedure 20.1 to complete this experiment.

Procedure 20.1
Embryogenic culture establishment.

Step	Instructions and comments
1	Remove individual tillers from a rapidly growing plant. Dissect individual leaves carefully so as to prevent damage and identify the innermost two leaves.
2	Remove the basal-most 3 cm of these leaves and separate the two halves of each longitudinally. Each leaf base should yield two strips of blade tissue approximately 3 cm long × 4 mm wide.
3	Place bleach solution into a sterile beaker and prepare three additional beakers with sterile distilled water for rinsing. Surface disinfest the leaf halves by gently agitating in the bleach solution for 2 min. Sterile forceps may be used to accomplish agitation.
4	Gently rinse the leaf halves three times in sterile water.
5	Remove one leaf half at a time and cut on a hard sterile surface (such as a petri dish) with a fresh, sharp scalpel blade. Cut the basal-most part of each leaf half crosswise into six approximately square sections (approximate size = 5 × 5 mm) and discard the remaining portion of each leaf half.
6	Immediately plate the sections onto SH30 medium. All of the dissection steps should be accomplished as rapidly as possible.
7	Arrange the sections from each leaf half in separate rows such that basal sections are at one end of a given row and the most distal sections are at the other end. Sections from four leaf halves should be placed in each petri dish. Seal each petri dish securely with several layers of parafilm.
8	Incubate cultures at 23°C in the dark and evaluate daily for bacterial and fungal contamination. Transfer noncontaminated sections to fresh medium as needed. Forceps should be sterilized between moving of each section when transferring noncontaminated sections in order to avoid spreading of inconspicuous contamination to clean sections.
9	Determine the number of sections from each leaf (i.e., innermost vs. outermost) that became contaminated weekly. Chart the relative contamination rate as a function of leaf position in the tiller.

Anticipated results

A measure of bacterial and fungal contamination growing on the medium or from cultured leaf sections is expected. While up to 50% of sections may become contaminated, a more normal contamination rate is 10%. It is important to follow directions for flaming instruments and manipulating leaf sections during transfer in order to eliminate, rather than spread, contamination.

Questions

- Discuss the importance of proper "transfer etiquette" in the establishment of cultures.
- Are the same procedures required in the maintenance of established cultures? Why?
- Did younger leaves become more or less contaminated than older leaves over time? Why?

Experiment 2. The gradient embryogenic response

The embryogenic response in orchardgrass is related to leaf position within a tiller (i.e., the innermost or youngest leaf vs. the next leaf out or older leaf) as well as the position of cultured leaf sections relative to the shoot apical meristem (i.e., proximal or distal). The following two distinct embryogenic responses are possible (Conger et al., 1983): either (1) an embryogenic callus develops or (2) somatic embryos grow directly from the leaf. Embryogenic callus is white to yellowish in color and friable to wet in consistency. Often, adventitious roots develop from the leaf sections, but never lead to plant regeneration. Somatic embryos, which are described in more detail below, appear as small (approximately 1 mm), white, organized structures, either embedded in callus or growing directly from the leaf section surface.

Materials

As this experiment utilizes the cultures developed previously, additional materials are not required.

Follow the methods outlined in Procedure 20.2 to complete this exercise.

Procedure 20.2
The gradient embryogenic response.

Step	Instructions and comments
1	Plate leaf sections on SH30 medium as described above such that the sections from each leaf half are kept together in a single row and arranged in sequential order from the most proximal to the most distal section.
2	Place the sections from four leaf halves in each dish (i.e., four rows of sections/dish — see Figure 20.1). Mark each leaf from each tiller so that the identity of the innermost vs. the outermost of two leaves per tiller is maintained.
3	Culture and screen for contamination as described in Procedure 20.1.
4	Determine the number of sections from each location within each leaf half that produces embryogenic callus and/or direct embryos at weekly intervals for 8 weeks. Chart the occurrence of embryogenic callus and somatic embryos as a function of time and position of the section in the original tiller.

Anticipated results

The gradient embryogenic response is demonstrated by correlating the relative position in a leaf from which a given section is obtained. A typical gradient response is shown in Figure 20.2. After 4 to 6 weeks, basal sections produce embryogenic callus. The callusing response progressively diminishes in sections from increasingly more distal positions of the leaf, being replaced by a direct embryogenesis response and, eventually, no embryogenic response at all. The differential embryogenic response is related to meristematic activity in the explant tissue. Leaf sections with abundant undifferentiated meristematic cells tend to produce embryogenic callus. As meristematic activity is decreased, embryogenic response is reduced such that only direct somatic embryos develop on sections with limited meristematic activity. Sections in which meristematic activity is too low or has ceased do not exhibit an embryogenic response.

Questions

- Do sections from younger or older leaves or sections from proximal or distal locations within a leaf produce more callus or direct embryos?
- Which sections respond most rapidly?
- How do observed embryogenic responses of individual sections correlate with the meristematic zones in intact orchardgrass leaves?

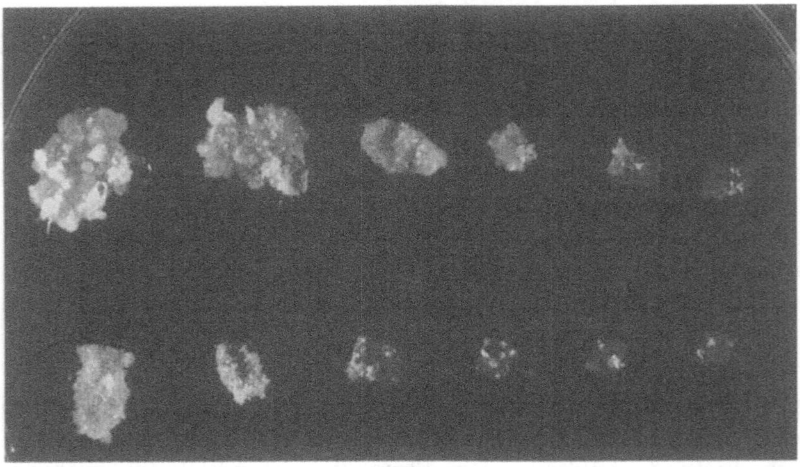

Figure 20.2 Gradient embryogenic response exhibited by the younger innermost (upper row of sections) and older next-out (lower row of sections) leaves of orchardgrass. Note that the basal most-leaf sections (left) produce embryogenic callus, whereas the more-distal-leaf sections (right) produce embryos directly. Also, more of the basal most-leaf sections produce embryogenic callus than the older leaf sections. (From Trigiano, R. N. and Gray, D. J., Eds., *Plant Tissue Culture Concepts and Laboratory Exercises, First Edition*, CRC Press LLC, Boca Raton, FL, 1996.)

Experiment 3. Maintenance of embryogenic callus

Embryogenic callus of orchardgrass can be maintained by monthly transfer of callus and/or somatic embryos to fresh SH30 medium. Embryogenic callus is very heterogeneous and continuously sectors into embryogenic and nonembryogenic portions. For maintenance of long-term embryogenic cultures, it is important to recognize and transfer only embryogenic sectors (Gray et al., 1984).

Materials

The following items are needed for each student or team:

- Ten 100 × 15-mm petri dishes containing SH30

Follow the methods in Procedure 20.3 to complete this experiment.

Procedure 20.3
Embryogenic callus maintenance.

Step	Instructions and comments
1	Isolate embryogenic callus and somatic embryos from leaf cultures. Dissect callus utilizing a stereomicroscope to assist in identifying various types of callus and tissue.
2	Separate dry, friable-type callus from wet, mucilaginous-type callus. Identify and separate sectors with numerous root primordia, as well as clumps of large somatic embryos and masses of small somatic embryos embedded in a watery matrix (Gray et al., 1984). Separate the calluses and tissues by type and plate small clumps of each, five clumps per petri dish, onto SH30 medium (produce five cultures). Be sure to also plate single, isolated somatic embryos (five per petri dish) into five petri dishes in order to test their response.
3	Incubate in the dark for 4 weeks as in Procedure 20.1.

Procedure 20.3 continued Embryogenic callus maintenance.	
Step	Instructions and comments
4	Determine and describe the different callus types obtained based on their morphologies. Determine which of these callus types and/or tissues gave rise to embryogenic callus.

Anticipated results

A typical embryogenic callus is shown in Figure 20.3. Embryogenic calluses or individual embryos are transferred to fresh SH30 medium in order to maintain and increase the embryogenic line. When transferring callus, it is important to isolate only sectors that are producing somatic embryos. The watery matrix material that contains masses of small somatic embryos frequently produces the most vigorous and prolific embryogenic callus. Embryogenic callus continues to sector into embryogenic and nonembryogenic callus types, primarily due to the recallusing of existing somatic embryos. The somatic embryos are anatomically complex and produce a number of different callus types, including embryogenic and rhizogenic (root-forming) calluses, when recultured on SH30.

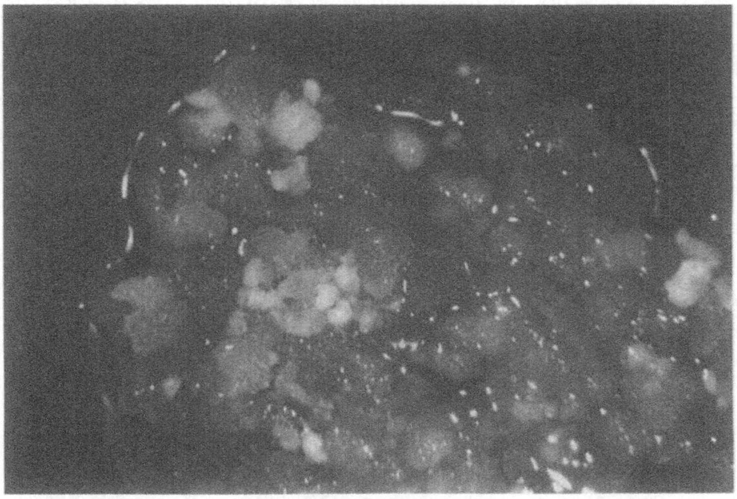

Figure 20.3 Embryogenic callus of orchardgrass. (From Trigiano, R. N. and Gray, D. J., Eds., *Plant Tissue Culture Concepts and Laboratory Exercises, Second Edition*, CRC Press LLC, Boca Raton, FL, 1996.)

Questions

- Which of these callus types produced the most embryogenic callus and somatic embryos?
- Where did embryogenic callus arise from plated somatic embryos?
- What type of tissue isolated from embryogenic callus would be best to use in culture maintenance?
- Why does embryogenic callus continually sector?

Experiment 4. Observations on monocotyledonous somatic embryos

Orchardgrass embryogenic cultures produce monocotyledonous somatic embryos that are of the poaceous type. The typical embryo possesses a scutellum and a coleoptile. While the somatic embryos are small, it is possible to see morphological details using a stereomicroscope. Abnormal embryos are also common. For reference, see Figure 19.7, Conger et al. (1983); Gray et al. (1984); Gray and Conger (1985b); and Trigiano et al. (1989).

Materials

The following items are needed for each student or team:

- Leaf and callus cultures developed through previous experiments

Follow the methods listed in Procedure 20.4 to complete this experiment.

Procedure 20.4
Observations on monocotyledonous somatic embryos.

Step	Instructions and comments
1	Observe somatic embryos developing from callus and directly from leaf cultures. Correlate specific embryos observed with those illustrated in referenced publications.
2	Identify the scutellum and coleoptile.
3	Observe and describe the most commonly occurring types of abnormal embryos.

Anticipated results

When viewed with a stereomicroscope, small, white somatic embryos that are morphologically similar to zygotic embryos either appear in embryogenic callus or emerge directly from uncallused leaf sections. Embryos at all developmental stages can be recovered from the cultures (Figure 20.4). The embryos possess a distinct scutellum and a notch, from which a coleoptile develops and enlarges during germination (see Figure 19.6). Direct embryos are frequently attached to the leaf by a distinct suspensor.

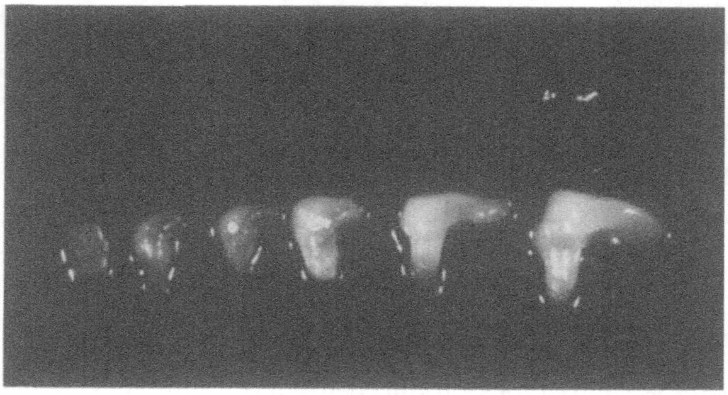

Figure 20.4 Somatic embryos isolated from embryogenic callus of orchardgrass. The youngest recognizable embryonic stages are shown on the left. Well-developed embryos (right) are white and opaque and possess a large, flattened scutellum and a narrowed embryo axis. (From Trigiano, R. N. and Gray, D. J., Eds., *Plant Tissue Culture Concepts and Laboratory Exercises, First Edition*, CRC Press LLC, Boca Raton, FL, 1996.)

Questions

- What types of abnormalities are most common?
- How might such abnormalities have occurred during embryo development?

Experiment 5. Somatic embryo germination and plant recovery

Orchardgrass somatic embryos germinate readily and in a manner similar to that of zygotic embryos when placed on SH0 medium and incubated in light (cool white fluorescent) at 25°C. Small rooted plants are removed from petri dishes, quickly transferred (to avoid dehydration) to moist potting medium in small pots, which are enclosed in plastic bags and cultured in a lighted incubator as above. When plants begin to grow vigorously, often within a week, they are transferred to greenhouse pots and maintained in the same manner as stock plants.

Materials

The following items are needed for each student or team:

- Source of embryos from previous experiments
- Five 100 × 15-mm petri dishes of SH0

Follow the methods listed in Procedure 20.5 to complete this experiment.

Procedure 20.5
Somatic embryo germination and plant recovery.

Step	Instructions and comments
1	Transfer well-developed somatic embryos, five per petri dish, to SH0 medium and incubate as above. Alternatively, transfer small clumps of embryogenic callus to the same medium.
2	Observe each daily for signs of germination. Count the number of shoots and roots present after 2, 4, 6, and 8 days to determine whether shoots or roots emerge first. Identify the developing red-pigmented coleoptiles.
3	Count the number of shoots per callus; this provides an indication of the relative number of well-developed somatic embryos per callus. Determine the percentage of somatic embryos that produce acclimatized plants.

Anticipated results

Individual embryos are induced to germinate into plants by transfer to medium lacking dicamba. Relative germination responses can easily be determined by counting the number of roots and shoots produced per culture. Resulting plants are easily acclimatized to greenhouse conditions.

Questions

- Why does transfer of embryos to medium lacking dicamba cause germination to occur?
- What is the first sign of germination and when does it occur?

Experiment 6. Initiation and manipulation of embryogenic suspension cultures

Embryogenic suspension cultures are produced by culturing embryogenic callus in liquid SH30 medium (i.e., lacking agar). The development of somatic embryos can be either stimulated or stopped by adding or removing various organic sources of nitrogen (Gray et al., 1984; Gray and Conger, 1985a; Trigiano and Conger, 1987).

Materials

The following items are needed for each student or team:

- Rotary shaker
- Four 125-ml Erlenmeyer flasks containing 10 ml liquid SH30
- Twelve 125-ml Erlenmeyer flasks containing 20 ml liquid SH30
- Four 125-ml Erlenmeyer flasks containing 20 ml liquid SH30-C
- Five or more 25-ml pipettes

Follow the methods given in Procedure 20.6 to complete this experiment.

Procedure 20.6
Embryogenic suspension culture.

Step	Instructions and comments
1	Initiate embryogenic suspension cultures by placing approximately 2 g of rapidly growing embryogenic callus into each 125-ml flask containing 10 ml of SH30 medium. Start with four flasks.
2	Cover the neck of each flask with several layers of sterile aluminum foil and seal the edges of the foil to the flask with parafilm. Rotate the cultures at approximately 75 rpm in the dark.
3	After 2 weeks, add 20 ml of SH30 medium to each flask and rotate at 100 rpm for an additional 2 weeks. Observe the increase of culture mass in rapidly growing cultures.
4	Increase and maintain cultures by pouring half of each rapidly growing culture into new flasks (total of eight flasks) containing 20 mL of SH30 medium at 2-week intervals.
5	Observe the remaining culture material with a stereomicroscope and produce fresh-mounted microscope slides to observe cells with a compound microscope. Inspect for somatic embryos or embryogenic cells, which are characterized as being small and densely cytoplasmic (see Chapter nineteen).
6	To induce embryo development, transfer half of each established, proliferating culture into flasks containing fresh SH30 medium and the other half into flasks containing SH30-C medium. Maintain cultures as before and observe twice weekly for the development of small, white somatic embryos, which can be readily observed collecting around the rim of the flask.
7	After 4 weeks growth in SH30 and SH30-C media, decant the liquid from flasks and spill the remaining culture mass into an empty, sterile petri dish.
8	Using a spatula and/or tweezers, plate small amounts of material (clumps about 3 mm in diameter) from each medium treatment, five clumps per petri dish, onto either solidified SH30 or SH0 medium and incubate in the light.
9	Observe the response weekly for 4 weeks.

Anticipated results

A typical suspension culture is shown in Figure 20.5. Suspension cultures will proliferate and can be maintained in liquid SH30 medium but will not produce recognizable somatic embryos unless an organic nitrogen source, such as casein hydrolysate, which contains a complex array of amino acids, is present. This indicates that the nitrogen source is an important controlling factor in somatic embryogenesis. In fact, somatic embryo development can be repeatedly started or stopped by adding or deleting casein hydrolysate in medium during transfer. When culture material from SH30 and SH30-C medium is placed on solidified SH30 medium, both produce embryogenic callus cultures. However, only the tissue from liquid SH30-C medium produces embryos that readily germinate into plants on SH0 medium. This indicates that although both SH30 and SH30-C liquid culture media support the proliferation of embryogenic cells, only medium with casein hydrolysate promoted development of the cells into somatic embryos.

Figure 20.5 An embryogenic suspension culture of orchardgrass. (From Trigiano, R. N. and Gray, D. J., Eds., *Plant Tissue Culture Concepts and Laboratory Exercises, First Edition*, CRC Press LLC, Boca Raton, FL, 1996.)

Questions

- What accounts for the effect of casein hydrolysate on embryo development?
- What do the observed differences between the responses of culture material from the two liquid media indicate when transferred to the two types of solidified media?

Experiment 7. Encapsulation of somatic embryos to produce synthetic seeds

Somatic embryogenesis is an ideal route for vegetative reproduction, since somatic embryos arise from the cells of one "parent." In fact, somatic embryogenesis represents the most efficient vegetative propagation system that can be envisioned, due to the rapid scale-up potential and prolific plant production exhibited by certain systems. In addition, since somatic embryos are nearly identical to zygotic embryos, research had attempted to add desirable seed-like qualities to somatic embryos including a protective coating and ability to become quiescent. Such synthetic seeds would revolutionize certain aspects of agriculture, allowing genetically uniform synthetic seed to be produced indoors at will. The following experiment illustrates an aspect of synthetic seed research, the addition of a protective coating. See Gray and Purohit (1991) for a complete treatment of the subject.

Materials

The following items are needed for each student or team:

- 100 ml of 2% sterile suspension of the sodium salt of alginic acid (Sigma Chemical Co.)
- 100 ml of 25 mM sterile aqueous solution of $CaCl_2$
- 100 ml of liquid SH0 medium
- Ten 100 × 15-mm petri dishes, each containing 25 ml of SH0 medium
- Three 150-ml sterile beakers with magnetic stir bars
- Magnetic stir plate
- Sterile spoonulas
- Sterile 1000-µm nylon screen

Follow the experimental protocol listed in Procedure 20.7. The steps in this procedure are illustrated in Figure 20.6.

Procedure 20.7
Encapsulation of somatic embryos.

Step	Instructions and comments
1	Harvest somatic embryos from suspension cultures by sieving the contents of a flask through a 1000-μm screen. Individual and small clumps of embryos, cells, and cell aggregates will pass through the screen. Discard the material retained by the screen.
2	Using a stereomicroscope, select individual, morphologically mature embryos and place them into the gently stirring 2% alginic acid solution. Alternatively, leaf or callus cultures with mature somatic embryos may be used in place of suspension cultures by plucking embryos directly from the cultures and placing into alginic acid.
3	For a control treatment, place some embryos directly on solidified SH0 medium without alginic acid treatment.
4	Using a wide bore pipette, withdraw a single embryo at a time from the alginic acid solution. Express enough of the embryo/alginic acid suspension to form a small drop at the tip of the pipette, such that the drop contains one embryo.
5	Allow this drop to fall into the gently stirring 25-mM $CaCl_2$ solution. Repeat this step numerous times, trying to perfect the art of forming a uniform drop containing an embryo.
6	Transfer the resulting calcium alginate/embryo beads to a swirling liquid SH0 solution for 15 min and then plate the beads, five per petri dish, to solidified SH0 medium and incubate at 25°C in the light.
7	Compare germination responses with control embryos twice weekly for 2 weeks. Record the number of root and shoot emergences at each time for each treatment.

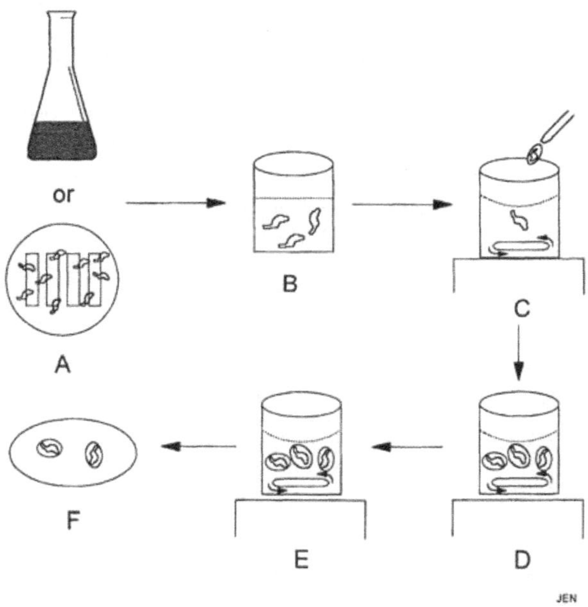

Figure 20.6 Encapsulation of orchardgrass somatic embryos in calcium alginate beads. (A) Somatic embryos harvested from 5-week-old suspension, leaf, or callus culture. (B) Embryos suspended in 2% alginic acid solution. (C) Embryos dropped singly into a beaker containing swirling 25 mM $CaCl_2$ solution. (D) Alginic acid is converted to calcium alginate, forming beads around embryos. (E) Synthetic seeds transferred to liquid SH0 medium to remove excess $CaCl_2$. (F) Plants emerging from synthetic seeds on solidified SH0 medium. (From Trigiano, R. N. and Gray, D. J., Eds., *Plant Tissue Culture Concepts and Laboratory Exercises, First Edition*, CRC Press LLC, Boca Raton, FL, 1996.)

Anticipated results

Immediately after dropping the alginic acid/embryo complex into $CaCl_2$, a cloudy-appearing bead should form around the embryo. This is the salt of alginic acid formed with the divalent cation Ca^{+2}, which is relatively insoluble in water. It may take several minutes for the calcium alginate drop to solidify completely to a soft gel-like consistency. Typically, plants are obtained from encapsulated embryos, but the plant recovery rate often is lower than that of controls.

Encapsulated somatic embryos constitute one type of synthetic seed. The calcium alginate can be regarded as a synthetic seed coat, which, theoretically, offers protection as well as a matrix to hold nutrients, pesticides, etc. See Gray and Purohit (1991) for a detailed discussion.

Questions

- Why did encapsulated embryos differ from controls in plant recovery rate?
- What are some potential advantages of coating somatic embryos?

Literature cited

Conger, B. V. and G. E. Hanning. 1991. Registration of Embryogen-P orchardgrass germplasm with a high capacity for somatic embryogenesis from in vitro cultures. *Crop Sci.* 31:855.

Conger, B. V., G. E. Hanning, D. J. Gray, and J. K. McDaniel. 1983. Direct embryogenesis from mesophyll cells of orchardgrass. *Science* 221:850–851.

Gray, D. J. 1990. Somatic cell culture and embryogenesis in the Poaceae. *Biotechnology in Tall Fescue Improvement.* M. J. Kasperbauer Ed., pp. 25–57, CRC Press, Boca Raton, FL.

Gray, D. J. and B. V. Conger. 1985a. Influence of dicamba and casein hydrolysate on somatic embryo number and quality on cell suspensions of *Dactylis glomerata* (Gramineae). *Plant Cell Tissue Organ Cult.* 4:123–133.

Gray, D. J. and B. V. Conger. 1985b. Time-lapse light photomicrography and scanning electron microscopy of somatic embryo ontogeny from cultured leaves of *Dactylis glomerata* (Gramineae). *Trans. Amer. Microsc. Soc.* 104:395–399.

Gray, D. J., B. V. Conger, and G. E. Hanning. 1984. Somatic embryogenesis in suspension and suspension-derived callus cultures of *Dactylis glomerata. Protoplasma* 122:196–202.

Gray, D. J. and A. Purohit. 1991. Somatic embryogenesis and development of synthetic seed technology. *Crit. Rev. Plant Sci.* 10:33–61.

Gray, D. J., R. N. Trigiano, and B. V. Conger. 1992. Liquid culture of orchardgrass somatic embryos and their use in synthetic cultivar development. SynSeeds. In: *Applications of Synthetic Seeds to Crop Improvement.* K. Redenbaugh, Ed., pp. 351–366. CRC Press, Boca Raton, FL.

Gray, D. J., R. N. Trigiano, and B. V. Conger. 1994. Classroom exercises in the study of orchardgrass somatic embryogenesis. *HortTechnology* 4:322–324.

Trigiano, R. N. and B. V. Conger. 1987. Regulation of growth and somatic embryogenesis by proline and serine in suspension cultures of *Dactylis glomerata. J. Plant Physiol.* 130:49–55.

Trigiano, R. N., D. J. Gray, B. V. Conger, and J. K. McDaniel. 1989. Origin of direct embryos from cultured leaf segments of *Dactylis glomerata. Bot. Gaz.* 150:72–77.

chapter twenty-one

Somatic embryogenesis from seeds of melon*

Dennis J. Gray

Melon, also known as cantaloupe or muskmelon (*Cucumis melo* L.), has been diversified through breeding and selection into hundreds of cultivars. It belongs to the family Cucurbitaceae, the cucurbits, which includes several other important species, notably, cucumber, *Cucumis sativa* L., squash and pumpkin, *Cucurbita pepo* L., and watermelon, *Citrullus lanatus* (Thunb.) Matsum and Nakai. Collectively, cucurbits are one of the world's most important horticultural crops, as can be seen by their abundance year round in nearly all produce markets.

Melon is subject to a number of bacterial, fungal, and viral diseases that severely limit yield and for which adequate levels of native resistance are not available. A solution to this problem may be to genetically engineer resistance into existing germplasm (see Chapters thirty-one through thirty-four for transformation methodologies). Adaptation of this technology to commercial varieties and breeding lines of *C. melo* requires a dependable, high-frequency cell regeneration system. Although plant regeneration through organogenesis has been described for melon (Chee, 1991) and an organogenic system is demonstrated for watermelon (Compton and Gray, 1992), somatic embryogenesis is more efficient (Gray et al., 1993). Melon is an ideal subject for demonstrating somatic embryogenesis, since explants can be obtained from commercially available seeds and the culture system is simple and rapid.

The purpose of this exercise is to demonstrate the development of somatic embryos of a dicotyledonous species and to investigate the effects of differences in explants and genotypes on the frequency of somatic embryogenesis. This exercise will illustrate concepts of nonzygotic embryogenesis discussed in Chapter nineteen. The referenced publication, Gray et al. (1993), is required to illustrate some of the experiments. In addition, somatic embryogenesis in melon is illustrated in Chapters six (Figure 6.2), nineteen (Figures 19.9 and 19.12), and thirty-four (Figure 34.3). These figures will be referenced at appropriate points in the following experiments.

* Florida Agricultural Experiment Station Journal Series No. R-07029.

0-8493-2029-1/00/$0.00+$.50

General considerations

Seed sources

Purchase one pound each of the following seeds: (1) 'Rocky Ford Green Flesh', also known as 'Eden Gem', from D. V. Burrell Seed Growers Co., P. O. Box 150, 405 N. Main, Rocky Ford, CO 81067; (2) 'Super Market' hybrid from D. V. Burrell Seed Company, Rocky Ford, CO 81067; (3) 'Top Mark' also from D. V. Burrell Seed Company. One pound of seed is a sufficient quantity to repeat the described experiments many times. Dry seeds stored at approximately 4°C will retain viability for several years.

Preparation of explants

To facilitate removal of seed integuments and dissection, seeds first are allowed to imbibe water. Rinse seeds briefly in a small kitchen colander or similar screen under running tap water. Place 100 seeds per 100 ml of sterile distilled water in 150-ml beakers for 1 h. Without regard to aseptic conditions, remove the seed coat and underlying papery integument to reveal a white embryo. It is essential not to damage the embryos, since even minute nicks can influence results. A stereomicroscope to view the seeds along with sharp disposable scalpel blades, which are changed frequently, and fine tweezers are used to facilitate clean dissections. Place the embryos immediately into fresh sterile water until all seeds have been processed. Surface disinfest embryos by agitation for 5 min in 100 ml of an aqueous solution of 25% commercial bleach containing a drop of Triton X surfactant, then rinse twice and place for immediate use in sterile distilled water.

Culture media

For simplicity, media for the following experiments have been changed from the formulas given in the first edition. We now recommend a prepackaged powdered medium, which can be ordered from Sigma Chemical Company, PO Box 14508, St. Louis, MO 63178. "Murashige and Skoog Basal Medium with Sucrose and Agar," catalog number 9274 (MS) is used to produce the media needed. The following two culture media are required: embryo initiation (EI) and embryo development (ED) media. The EI medium contains MS with the addition of 5 mg/L (22.7 µM) of 2,4-D and 0.1 mg/L (0.45 µM) of thidiazuron. The latter two substances are a synthetic auxin and cytokinin, respectively. The pH of the media are adjusted to 5.4 and then autoclaved. ED medium is unmodified MS without the added growth regulators. Medium is dispensed, 30 ml, into each 15 × 100-mm sterile petri dish.

Exercises

Experiment 1. Effect of explant type on somatic embryogenesis

In melon, the predominance of embryogenesis occurs only from certain regions of specific embryonic organs. In this exercise, the seed-derived embryos will be dissected into well-defined explant types, cultured, and examined for embryogenesis to determine the site(s) of embryogenically competent cells.

Materials

The following items are needed for each student or student team:

- Twenty surface disinfested zygotic embryos of 'Rocky Ford Green Flesh' (use only this variety for this experiment)

- Twenty petri dishes containing EI medium
- Forty petri dishes containing ED medium

Follow the protocols listed in Procedure 21.1 to complete this experiment.

Procedure 21.1
Effect of explant type on somatic embryogenesis.

Step	Instructions and comments
1	Using fresh, sharp scalpel blades, sterilely dissect zygotic embryos as shown in Figure 21.1. Each embryo should yield one embryo axis explant plus four basal cotyledonary explants and four distal cotyledonary explants. Basal cotyledonary explants will appear similar to those shown in Figures 21.2 and 34.3.
2	Place all nine explants into a single petri dish containing EI medium and arrange in a manner such that the explant type can be identified after callusing occurs. Maintenance of cultures in complete darkness (as in a dark incubator or in a cardboard box covered with aluminum foil) and at 25°C is an absolute requirement for the desired responses to occur.
3	Check cultures every other day under dim light for contamination and move uncontaminated explants away from contaminated explants to fresh medium (see laboratory exercise concerning orchardgrass somatic embryogenesis for procedures for managing contamination).
4	After 2 weeks on EI medium, transfer explants to ED medium, but place the embryo axis explants on a separate dish to provide more growth space. Be sure to maintain the identity of each explant type. Incubate these cultures under light (16-h cool white fluorescent light [60 $\mu mol \cdot m^{-2} \cdot s^{-1}$]/8-h dark cycle).
5	After 3 weeks and again after five weeks on ED medium, determine the total number of each explant type that produced at least one somatic embryo, recognizable through a stereomicroscope. In addition, at 3 weeks estimate the number of somatic embryos on each responding explant without disturbing the cultures and, at 5 weeks, open the petri dishes and count the exact number of embryos per dish.
6	Calculate the standard error of response per explant type for number of responding explants and number of somatic embryos per explant and determine which explant type was most responsive.

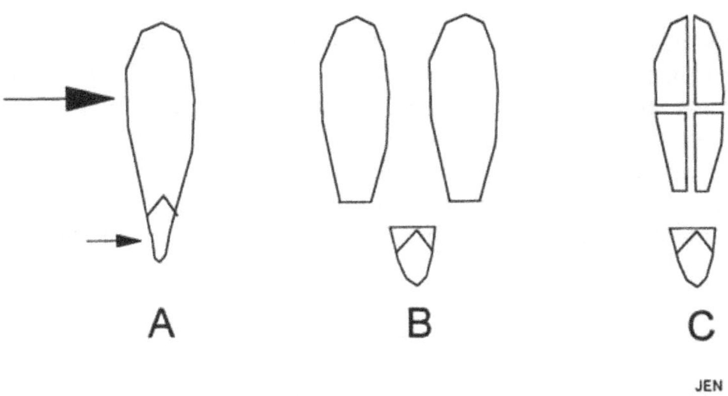

Figure 21.1 Dissection of explants from imbibed melon seed. (A) Embryo from which seed coat and papery integument have been removed. Cotyledons (large arrow) are easily distinguishable from embryo axis (small arrow). (B) Remove embryo axis and separate cotyledons. Be careful to make clean, sharp cuts and to not damage the tissue. (C) Cut each cotyledon into two basal and two distal explants. (From Trigiano, R. N. and Gray, D. J., Eds., *Plant Tissue Culture Concepts and Laboratory Exercises, First Edition*, CRC Press LLC, Boca Raton, FL, 1996.)

Anticipated results

During incubation on EI medium, the explants will enlarge to several times their original size and begin to produce callus. This response is illustrated in Figure 21.2. Explants normally should remain pale yellow in color. Change in pigmentation to a green color on EI medium indicates that adequate dark conditions were not maintained and suggests that a reduced embryogenic response subsequently will occur. Lack of explant expansion and/or onset of browning indicates problems with medium formulation or surface disinfestation. Microscopic examination of explants after 2 weeks on EI medium often will reveal small hyaline clavate structures, some of which are the earliest visible stages of somatic embryogenesis.

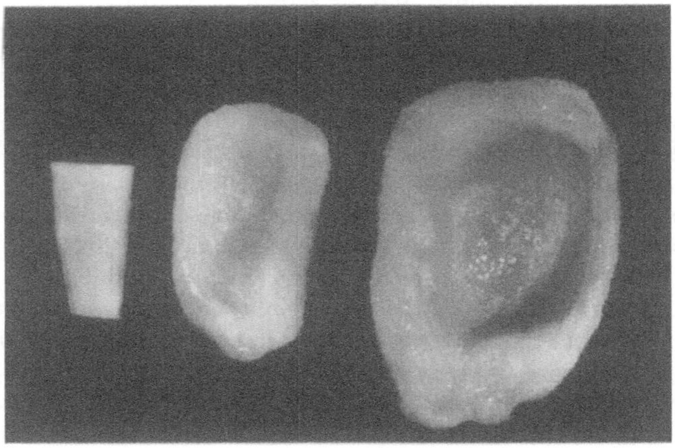

Figure 21.2 Development of basal cotyledon explant on EI and ED media. Newly plated explant (left) is white in coloration. After 2 weeks on EI medium in the dark (center), the explant enlarges, begins to callus along its edges, but remains white to pale yellow in color. Two weeks after transfer to ED medium in the light (right), further enlargement and callusing occurs and the explant becomes green in color. It is at or after this stage that well-developed somatic embryos typically appear. (From Trigiano, R. N. and Gray, D. J., Eds., *Plant Tissue Culture Concepts and Laboratory Exercises, First Edition*, CRC Press LLC, Boca Raton, FL, 1996.)

After transfer to ED medium in the light, explants continue to enlarge and turn green in color (Figure 21.2). White to yellow to green-colored callus also will develop. Somatic embryos in various stages (see the last experiment in this series) will begin to appear as soon as 4 weeks from culture initiation and will continue to develop for at least 8 weeks. Younger embryos initially are white to pale yellow in color, but soon become green. Although some somatic embryos often can be found on all explant types, the great abundance of embryos will develop from the basal cotyledonary explants.

Questions

- What is the rationale behind initiating cultures on medium containing auxin and cytokinin, followed by transfer to basal medium?
- Why is an initial incubation period in dark so important?
- Why do certain explant types respond better than others?

Experiment 2. Effect of genotype on somatic embryogenesis

Embryogenesis in melon is genotype dependent. Relatively high or low embryogenic responses occur depending on the genotype used. In this experiment, three genotypes that differ in embryogenic capacity are compared.

Materials

The following items are needed for each student or student team:

- Fifteen surface-disinfested embryos each of 'Rocky Ford Green Flesh', 'Super Market', and 'Top Mark'
- Thirty petri dishes containing EI medium
- Thirty petri dishes containing ED medium

Follow the protocols outlined in Procedure 21.2 to complete this exercise.

Procedure 21.2
Effect of genotype on somatic embryogenesis.

Step	Instructions and comments
1	Dissect embryos as in Experiment 1, but culture only the basal cotyledon explants.
2	Place the four explants from each seed on a single petri dish containing EI medium. Prepare ten dishes for each explant. Incubate in the dark and manage as described in Procedure 21.1.
3	Determine the relative difference in embryogenesis between the 3 genotypes at 3 and 5 weeks after transfer to ED medium, and process the data as in Procedure 21.1.

Anticipated results

Dramatic differences in embryogenic response will occur among the three genotypes tested. 'Rocky Ford Green Flesh' will exhibit the greatest embryogenic response, whereas 'Super Market' will be intermediate in response and 'Top Mark' will be the worst. Varying rates of contamination may also occur between the genotypes.

Questions

- What is the apparent relationship, if any, between contamination rate and genotype in terms of embryogenic response?
- Speculate as to precise reasons why genetic differences between genotypes may influence their embryogenic response.

Experiment 3. Embryogenic culture maintenance

Materials

The following items are needed for each student or student team:

- Embryogenic explants and callus bearing somatic embryos obtained from the previous experiments
- Fifteen petri dishes of EI medium
- Fifteen petri dishes of ED medium
- Other supplies listed in Experiment 1

Follow the protocol listed in Procedure 21.3 to complete this experiment.

Procedure 21.3
Embryogenic culture maintenance.

Step	Instructions and comments
1	Place five individual basal cotyledon explants or small pieces of callus (approx. 5 mm diameter) or single somatic embryos into each petri dishes of EI and ED medium. Make five replicate plates of each tissue type/medium combination such that 50 inoculated pieces of each tissue are used.
2	After 2 weeks, observe differences between culture growth and embryogenesis, then transfer cultures to fresh medium. Repeat this cycle one more time (i.e., transfer for a total of three cycles).
3	At the end of the experiment, compare the embryogenic responses for each inoculum type on each medium and with the responses that were obtained initially from primary explants.
4	Determine which tissue type/medium combination results in proliferation of the embryogenic culture.

Anticipated results

It is not possible to maintain high-frequency somatic embryogenesis of melon under the conditions described. Cultures placed on ED medium either produce a few plants from germinating embryos or decline and die. Cultures placed on EI medium continue to proliferate as callus, but the number of embryos produced declines rapidly. Additionally, the embryos that are produced become increasingly abnormal in morphology until they barely can be identified. Although the exact reasons for this decline are not known, a relatively high proportion of polyploid plants are known to occur from somatic embryos of melon (Oridate et al., 1992).

Questions

- Why does embryogenesis decline so rapidly in cultures of melon?
- Considering theories relating to induction of embryogenic cells, how can the response of recultured somatic embryos be explained?

Experiment 4. Observation and categorization of embryonic stages

Because of the rapidity and high frequency of the embryogenic response exhibited by melon, it provides a perfect system for studying somatic embryo development. In addition, this system allows all stages of embryogenesis to be followed dynamically, which is not possible with zygotic embryogenesis due to the presence of obscuring seed integuments. Often, all stages of embryogenesis, including instances of abnormal development, can be seen in a single explant. In this experiment, cultured explants are examined and various embryogenic stages are categorized.

Materials

The following items are needed for each student or student team:

- Approximately ten petri dishes containing ED medium and each containing five cultured cotyledonary bases of 'Rocky Ford Green Flesh' that already have been induced on EI medium and transferred to ED medium as described in Experiment 1
- High-quality stereomicroscope and light source

Follow the methods described in Procedure 21.4 to complete this experiment.

Procedure 21.4
Observation and categorization of embryonic stages.

Step	Instructions and comments
1	Use responsive cultures obtained in previous, ongoing experiments.
2	After transfer to ED medium, number each culture and observe at least two times each week. Keep a log of explant response. Identify a few of the explants that begin to produce the most somatic embryos and follow their development.
3	On these explants, identify several young embryo stages and sketch their relative location on the explant, as well as their morphology. Redraw these embryos weekly and determine the changes in morphology.
4	Identify globular, heart-shaped, torpedo, and cotyledonary embryonic stages (see Figures 6.2 and 19.9 as well as Gray et al., 1993 for examples of some stages). Attempt to identify cotyledonary-stage embryos with pronounced shoot apical domes and follow the stages of early shoot development. Identify instances of abnormal embryo development and precocious germination and categorize different types (see Gray et al., 1993 for examples). Identify somatic embryos that have developed into plants (see Figure 19.12 for an example).
5	Determine which embryonic stages result most often in plant development.

Anticipated results

The process of embryogenesis in this system is very rapid. Cultures must be observed twice a week. Once highly responsive explants are identified, the development of a select few embryos will illustrate both normal and abnormal embryogenesis. All typical stages of embryogenesis can be identified and followed.

Questions

- Suggest reasons for the development of fused somatic embryos or those with missing or extra cotyledons.
- When in the development of the embryos must certain abnormalities occur?
- Do normal or precociously germinating embryos ultimately produce more plants?

Literature cited

Chee, P.P. 1991. Plant regeneration from cotyledons of *Cucumis melo* 'Topmark'. *HortScience* 26:908–910.

Compton, M.E. and D.J. Gray. 1992. Shoot organogenesis and plant regeneration from cotyledons of diploid, triploid and tetraploid watermelon. *J. Amer. Soc. Hort. Sci.* 118:151–157.

Gray, D. J., D. W. McColley, and M. E. Compton. 1993. High-frequency somatic embryogenesis from quiescent seed cotyledons of *Cucumis melo* cultivars. *J. Amer. Soc. Hort. Sci.* 118:425-432.

Oridate, T., H. Atsumi, S. Ito, and H. Araki. 1992. Genetic differences in somatic embryogenesis from seeds in melon (*Cucumis melo* L.). *Plant Cell Tissue Organ Cult.* 29:27–30.

chapter twenty-two

Somatic embryogenesis from mature peanut seed

Hazel Y. Wetzstein and Charleen Baker

The cultivated peanut or groundnut (*Arachis hypogaea* L.) is a crop of great economic importance. In the U.S., primary uses of peanut seeds are for confectionery purposes or making peanut butter. However, the peanut is an important worldwide source of oil, protein, food, and feed. A major objective of peanut improvement programs is the production of cultivars that are genetically resistant to pests and diseases. The use of genetic transformation technologies has created the potential to improve agronomic characters and food quality properties. Some valuable traits that could potentially be introduced into the peanut include resistance to insects, viruses, and fungi, herbicide tolerance, increased adaptability to different environments, and improved oil quality.

Before biotechnology methods, such as genetic transformation, can be used for improving the agronomic and culinary characteristics of the peanut, a reliable regeneration system needs to be developed. Somatic embryogenesis is a promising method because potentially high numbers of regenerates can be obtained, each theoretically originating from a single or few cells. High regeneration rates and origin of somatic embryos increase the likelihood of successfully introducing desirable genes and recovering stably transformed plants.

The following laboratory exercises will illustrate a direct somatic embryogenic system using mature peanut embryo axes as the explant or starting materials. Students will be acquainted with the following concepts: (1) the effect that explant part (i.e., epicotyl vs. radicle) has on embryogenesis, (2) how auxin concentration in the primary and secondary media affects embryo induction and development, (3) how environmental factors, such as light, can modify embryo form, and (4) germination of embryos and acclimatization of plants. These exercises are adapted from publications by Baker et al. (1995) and Wetzstein and Baker (1993).

General considerations

Explant material and medium preparation

Most of the media used in the following experiments were adapted for peanut cultivars 'GK7' and 'AT127', which produce embryogenic cultures at high frequencies. However, embryogenic cultures have also been obtained using a number of other cultivars, including 'Florunner', 'Arkansas Valencia', 'NC7', 'Georgia Runner', 'GK17', and 'VC-1', and these are suitable for substitution if the 'GK7' and 'AT127' are not available.

0-8493-2029-1/00/$0.00+$.50

Shelled peanut seed should have a reddish-brown seed coat and well-filled cotyledons. Dissect out the small embryo axes on clean, but not necessarily sterile, paper towels. This can be easily accomplished by using a scalpel blade to separate the two cotyledons, after which the embryo axis can be popped out using the tip of the blade. See Figure 22.1 of an opened peanut seed illustrating the cotyledons and embryo axis. As axes are removed, store them in a petri dish lined with moist filter paper to prevent desiccation. To obtain satisfactory and reproducible results, discard any damaged or malformed axes (i.e., injured during dissection or by pests). For each experiment, some extra axes should be collected, since some may be lost or damaged during disinfestation or subsequent cutting.

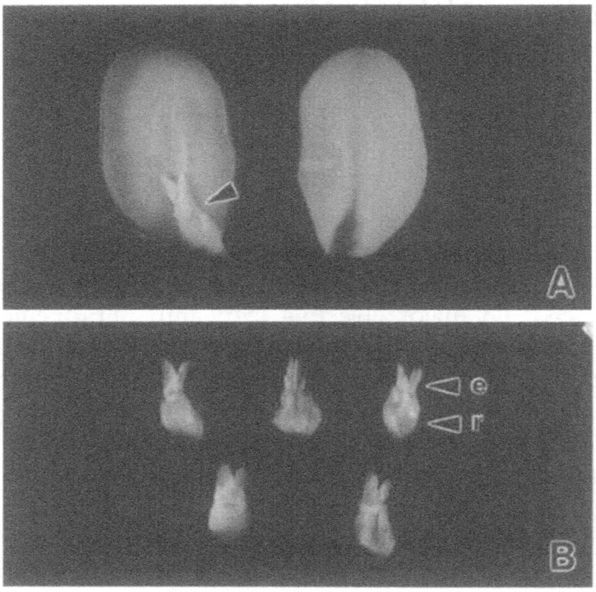

Figure 22.1 (A) A dissected peanut seed showing cotyledons and embryo axis (point); (B) whole, isolated embryo axes. Note epicotyl (e) and radicle (r) ends. (From Trigiano, R. N. and Gray, D. J., Eds., *Plant Tissue Culture Concepts and Laboratory Exercises, First Edition*, CRC Press LLC, Boca Raton, FL, 1996.)

Under a flow hood, transfer the axes to a 250-ml beaker and surface disinfest in 10% commercial bleach (0.26% NaOCl) with two to three drops of Tween-20 (a detergent) with occasional agitation for 5 min. Rinse the axes three times in sterile deionized water.

The basal culture medium is composed of Murashige and Skoog (MS) (1962) salts supplemented with B5 vitamins (Gamborg et al., 1968), 30 g/L (88 mM), and 0.4% (0.4 g/L) Gel-gro (ICN Biochem., Irvine, CA). Plant growth regulators (PGRs) will be used as described in the different experiments. Adjust the pH of the culture media to 5.8 before adding Gel-gro and autoclave. Dispense 25 ml of medium into each 100 × 15-mm petri dish.

Exercises

Experiment 1. Effect of explant tissue on somatic embryo development

The type of explant tissue used is critical for obtaining success in many embryogenic systems. Particularly in difficult-to-work-with species, actively dividing meristematic tissues often gives much better results. The mature cotyledons of peanuts, which are well-differentiated tissues with abundant storage reserves, typically give a poor embryogenic response. In contrast, the embryo axis, which has young, meristematic regions, usually gives a good response. Even parts within the embryo axis react or respond differentially. Having high-frequency embryogenesis and understanding where embryos emerge can be critical if an objective is to apply genetic transformation techniques to an embryogenic system. In this exercise, the whole embryo axis, epicotyl portion, and radicle portion will be used as explants. Two 2,4-dichlorophenoxyacetic acid (2,4-D) concentrations (15 mg/L [68 µM]) and 30 mg/L [136 µM] will be used. Sites of embryo development will be noted. It is recommended that this exercise be completed in groups of two to three students, with four petri dishes used for each explant type and auxin concentration (six treatments in all). This experiment will require about 8 weeks for completion and may be considered as a completely randomized design with four replicates of each treatment. Data should be analyzed, e.g., standard deviation or ANOVA etc. (see Chapter seven).

Materials

The following items are needed for each student or team of students:

- About 150 mature peanut seeds
- Petri dish lined with moistened filter paper
- 100 ml of 10% commercial bleach with two to three drops of Tween 20
- Four 100-mm petri dishes containing MS + 68 µM (15 mg/l) 2,4-D per explant type
- Four 100-mm petri dishes containing MS + 136 µM (30 mg/l) 2,4-D per explant type
- Twenty-four 100-mm petri dishes containing MS without 2,4-D
- Incubator providing darkness at 26°C (or a box for cultures to provide darkness in the growth room)

Follow the protocols outlined in Procedure 22.1 to complete the experiment.

Procedure 22.1
Effect of explant tissue on somatic embryo development.

Step	Instructions and comments
1	Dissect out the embryo axes from the peanut seed as described in the general considerations section and disinfest in 10% commercial bleach +Tween 20 for 5 min with occasional agitation. Rinse three times with sterile water.
2	Retain one third of the explants as whole or entire embryo axes. Using a sharp scalpel, cut the remaining two thirds of the axes into two pieces each; the cut should be made at about one third of the distance from the shoot apex. The apical portion will contain the upper third of the axis, whereas the radicle portion will comprise the lower two thirds of the axis.
3	Place each of the three explant types (whole axis, apical portion, and radicle portion) onto media with either 68 µM (15 mg/L) or 136 µM (30 mg/L) 2,4-D. Refer to Figure 22.2 for a diagrammatic representation of this procedure.
4	Incubate cultures in the dark at 26°C for 30 days, then transfer all explants to fresh basal MS medium without 2,4-D and return to the darkened incubator.
5	Observe cultures every 2 weeks from initiation and note from which tissues somatic embryos are emerging. Record callus production and somatic embryo numbers after 4 weeks on basal medium.

Anticipated results

Explants will exhibit some swelling and enlargement on induction medium containing 2,4-D. Somatic embryos will begin to form after transfer to basal MS medium without 2,4-D. Embryos will originate from apical leaves, with little or no embryo formation on

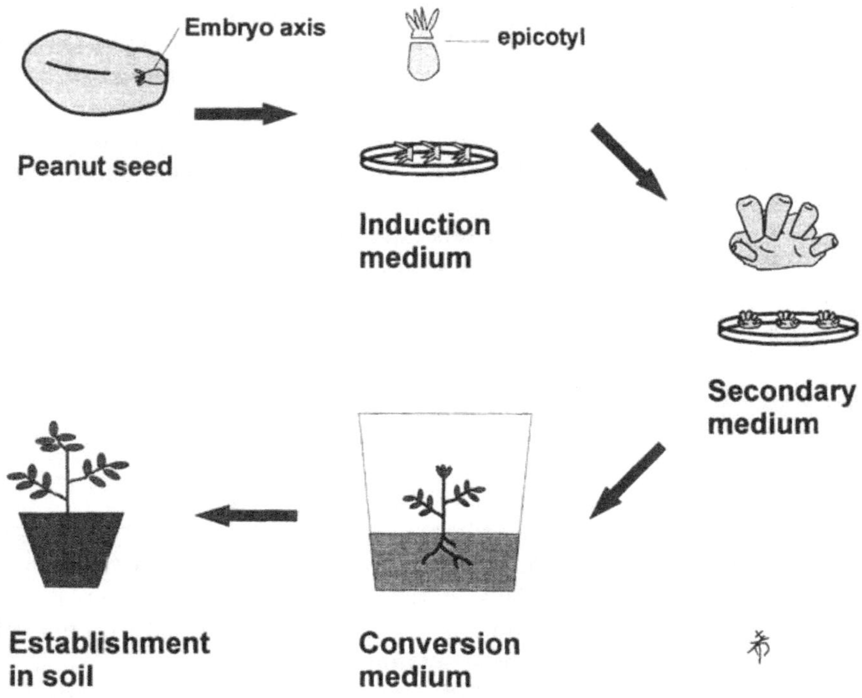

Figure 22.2 Diagrammatic representation of the somatic embryogenic system for the peanut. Shown is an opened mature seed with a cotyledon and embryo axis. Embryo axis explants are dissected out and placed on induction medium with 2,4-D (shown are epicotyl explants). Somatic embryos are evident within 4 weeks after transfer to a secondary medium with no or low 2,4-D concentration. Somatic embryos transferred onto conversion medium germinate and convert into plantlets that can be established into soil.

the radicle portion of whole axes or radicle portion of cut explants. See Figure 22.3 for a representation of developing somatic embryos. Radicle portions may exhibit some browning and callus growth.

Questions

- Why is it necessary to transfer explants after 4 weeks from induction medium to a medium without PGRs?
- Which explant type would be preferable to use in genetic transformation studies and why?

Experiment 2. Effects of auxin concentration and photoperiod on somatic embryo development and morphology

Somatic embryos are usually induced by incorporating an auxin into a primary medium. An important first step in developing any embryogenic protocol is to establish the most appropriate auxin type and concentration for induction of somatic embryogenesis. Also, the proper environmental conditions, such as temperature and light, for cultures should be determined. In this experiment, different concentrations of 2,4-D in the induction medium will be evaluated as well as the effects of growing cultures in dark vs. light conditions.

Figure 22.3 Embryogenic culture from epicotyl explants showing numerous somatic embryos about 4 weeks after transfer to basal secondary medium. (From Trigiano, R. N. and Gray, D. J., Eds., *Plant Tissue Culture Concepts and Laboratory Exercises, First Edition,* CRC Press LLC, Boca Raton, FL, 1996.)

Students should work in teams of two or three with four petri dishes used for each auxin concentration and light/dark treatment (eight treatments in all). If materials or time are limited, omit the light culture treatments, since cultures grown in the dark will produce better embryos. Alternatively, complete all of the dark treatments and include only the 68 (15 mg/L) and 136 µM (30 mg/L) 2,4-D treatments in the light. This experiment will require about 8 weeks to complete and may be considered as a completely randomized design with four replicates of each treatment. Data should be analyzed, e.g., standard deviation or ANOVA etc. (see Chapter seven).

Materials

The following are needed for each student or team of students:

- About 200 peanut seeds
- Four 100-mm petri dishes containing MS without 2,4-D per photoperiod
- Four 100-mm petri dishes containing MS + 68 µM (15 mg/l) 2,4-D per photoperiod
- Four 100-mm petri dishes containing MS + 136 µM (30 mg/l) 2,4-D per photoperiod
- Four 100-mm petri dishes containing MS + 272 µM (60 mg/l) 2,4-D per photoperiod
- Thirty-two 100-mm petri dishes containing MS without PGRs
- Incubators or growth room providing dark or a 16-h photoperiod at 26°C

Follow the protocols listed in Procedure 22.2 to complete the experiment.

Procedure 22.2
Effects of auxin concentration and photoperiod on somatic embryo development and morphology.

Step	Instructions and comments
1	Dissect out the embryo axes from the peanut seed as described in the general considerations section and disinfest in 10% commercial bleach +Tween 20 for 5 min with occasional agitation. Rinse three times in sterile water.

Procedure 22.2 continued
Effects of auxin concentration and photoperiod on somatic embryo development and morphology.

Step	Instructions and comments
2	Cut the embryo axes into two pieces, at about one third of the distance from the shoot apex (see Procedure 22.1 for details). Discard the radicle portion of the axis and use only the epicotyl portions as explants.
3	Place five explants into each 100-mm petri dish containing induction medium with one of the four different concentrations of 2,4-D.
4	Place half the cultures in the dark and half the cultures in the light (under a 16-h photoperiod at about 130 $\mu mol \cdot m^{-2} \cdot sec^{-1}$). Maintain cultures at 26°C.
5	After 30 days on induction medium, transfer the explants to fresh MS medium without growth regulators. Return the cultures to the same respective light or dark conditions.
6	Observe cultures every 2 weeks from initiation and record somatic embryo numbers after 4 weeks on basal medium.

Anticipated results

Somatic embryos will begin to develop after transfer to the secondary medium in cultures grown under both light and dark conditions. Embryos in dark-grown cultures will be white, well-formed, and easily separated for germination. In cultures grown in the light, embryos will be green, less succulent, and more difficult to separate. A white nonembryogenic callus may be prevalent in cultures grown in the light. Embryos will not form in cultures initiated on medium without 2,4-D. Most explants induced on 15 or 30 mg/L 2,4-D will form embryos. On induction media with 60 mg/L 2,4-D, explants will exhibit browning and a lower rate of embryogenesis compared with other treatments in which embryos were formed.

Questions

- What 2,4-D concentration for induction and photoperiod would you recommend using for embryogenesis and why?
- Do you think that embryo morphology will influence embryo germination and development? How?

Experiment 3. Secondary medium effects on repetitive embryogenesis and somatic embryo development

In peanut and many somatic embryogenic systems, explant material is placed sequentially on two types of media. A primary medium, which has an auxin, is used to induce somatic embryos. This is followed by transfer to a secondary medium without auxin, or with a lower auxin concentration. Auxin is proposed to be necessary for somatic embryo induction, but is inhibitory to somatic embryo development. In this experiment, secondary media with varying concentrations of 2,4-D will be used. Some cultures will be retained on induction medium with high 2,4-D (136 µM or 30 mg/L), whereas others will be transferred to medium with low (45 µM or 10 mg/L) or without 2,4-D. Observe the effects that secondary medium has on repetitive embryogenesis (that is, the continued proliferation of somatic embryos from preexisting somatic embryos) and embryo development. This exercise should be completed by teams of two or three students with four petri dishes used per secondary medium (three treatments). The experiment will require about 12 weeks for completion and may be considered as a completely randomized design with four replicates of each treatment. Data should be analyzed, e.g., standard deviation or ANOVA etc. (see Chapter seven).

Materials

The following items are needed for each student or team of students:

- About 100 mature peanut seeds
- Twelve 100-mm petri dishes containing MS + 30 mg/L 2,4-D
- The following media for subculture after 4 weeks and again after 8 weeks:
  Four 100-mm petri dishes containing MS without 2,4-D
  Four 100-mm petri dishes containing MS + 45 μM (10 mg/L) 2,4-D
  Four 100-mm petri dishes containing MS + 136 μM (30 mg/L) 2,4-D

Follow the protocols outlined in Procedure 22.3 to complete the experiment.

Procedure 22.3
Secondary medium effects on repetitive embryogenesis and somatic embryo development.

Step	Instructions and comments
1	Dissect out the embryo axes from the peanut seed as described in the general considerations section and disinfest in 10% commercial bleach +Tween 20 for 5 min with occasional agitation. Rinse three times in sterile water.
2	Cut the embryo axes in two, at about one third of the distance from the shoot apex (see Procedure 22.1 for details). Discard the radicle portion of the axis and use only the epicotyl portions as explants.
3	Place five explants into each 100-mm petri dish, all of which contain MS medium with 136 μM (30 mg/L) 2,4-D. Incubate cultures at 26°C in the dark.
4	After 30 days, transfer the explants to one of three secondary media containing either 0, 45 (10 mg/L), or 136 μM (30 mg/L) 2,4-D. Return the cultures to the same dark conditions.
5	After 4 weeks, subculture onto fresh dishes containing the same secondary medium.
6	Observe cultures after transfer to secondary media and record embryo numbers 8 and 12 weeks after initiation.

Anticipated results

Somatic embryos will begin to form when transferred to secondary media. Some embryos on secondary medium without growth regulators will exhibit precocious or premature germination in culture. Explants on secondary medium with 10 mg/L 2,4-D will produce secondary embryos. Callus production will occur on explants on 30 mg/L 2,4-D with few repetitive embryos formed.

Questions

- What is repetitive or secondary embryogenesis and of what significance is it?
- What are the advantages and disadvantage of somatic embryos germinating in culture?
- What level of 2,4-D would you choose for the secondary medium and why?

Experiment 4. Germination, planting, and acclimatization of somatic embryos

Plants can be obtained from the somatic embryos produced in the previous experiments. Germination is enhanced by transferring the embryos to Magenta boxes in the light. Peanut somatic embryos can have very divergent forms, including those with an elongated axis, trumpet- or horn-shaped embryos, and fan-shaped embryos (Wetzstein and Baker, 1993). In this experiment, students can observe germination and shoot growth of embryos, and acclimatization of plants out of culture. The experiment will illustrate how embryo morphology can affect germination and the ability to produce plants. This experiment will take about 5 to 8 weeks beyond the production of somatic embryos.

Materials

For germination include the following:

- Two to five GA-7 Magenta vessels containing 50 ml basal MS medium (or baby food jars or similar vessels)

For planting in soil and acclimatization, include the following:

- 10-cm-diameter (4-in.) plastic pots
- Equal parts of soilless potting mix/perlite/vermiculite
- MS salts for watering
- Plastic bags

Follow the protocols listed in Procedure 22.4 to complete the experiment.

Procedure 22.4
Germination, planting, and acclimatization of somatic embryos.

Step	Instructions and comments
1	Under sterile conditions, carefully separate individual somatic embryos from the cultures. Insert the radicle end of about ten embryos into MS basal medium contained in each Magenta box. Observe the different types (i.e., funnel-shaped, well-defined axes and cotyledons, etc.) of somatic embryos and note the treatment conditions from which embryos were obtained.
2	Incubate the boxes under a 16-h photoperiod at about 28°C. Periodically observe the embryos for root emergence and shoot growth.
3	Transfer germinated embryos to pots filled with a mixture composed of equal parts of soilless potting mix/perlite/vermiculite, which has been wetted with water. Gently wash agar from roots, place roots in planting mix, and wet each pot with about 100 ml MS salt solution. Cover the plant in the pot with a plastic bag and grow under light conditions.
4	After 2 to 3 days, slowly acclimatize the plants over the next week by cutting progressively larger slits in the plastic bag. After this time remove the bags. To maintain plants, water when dry and fertilize with MS salts every 2 weeks.

Anticipated results

Some somatic embryos should exhibit root and shoot growth within 1 month after transfer into Magenta boxes. Embryo morphology will affect the germination and growth into plants. Horn- or trumpet-shaped embryos will be poor producers of plants. Those with a well-defined axis and cotyledons or foliar structures will produce plants in higher frequencies.

Questions

- Why is it important to wash agar from the roots before transfer to pots?
- Why is acclimatization of plants necessary?

Literature cited

Baker, C.M., R.E. Durham, J.A. Burns, W.A. Parrott, and H.Y. Wetzstein. 1995. High frequency somatic embryogenesis in peanut (*Arachis hypogaea* L.) using mature, dry seed. *Plant Cell Rep*. 15:38–42.

Gamborg, O., R. Miller, and K. Ojima. 1968. Nutrient requirements of suspension cultures of soybean root cells. *Exp. Cell Res.* 50:150–158.

Murashige, T. and F. Skoog. 1962. A revised medium for rapid growth and bioassays with tobacco tissue cultures. *Physiol. Plant.* 15:473–497.

Wetzstein, H.Y. and C.M. Baker. 1993. The relationship between somatic embryo morphology and conversion in peanut (*Arachis hypogaea* L.). *Plant Sci.* 92:81–89.

chapter twenty-three

Direct somatic embryogenesis from leaves and flower receptacles of cineraria

Robert N. Trigiano, Mary Catherine Scott, and Kathleen R. Malueg

The florists' cineraria, *Senecio* × *hybridus* Hyl., is a daisy-like flowering pot plant and a species belonging to the family Asteraceae. The most notable horticultural attributes of the plant are its flowers that are produced in a wide range of colors including white, pink, red, purple, maroon, magenta, and several striking hues of blue. Cineraria is a perennial, but usually grown as an annual, and is very well suited to low-light intensities typical of winter production in the Northern Hemisphere. Cineraria is usually considered to be a minor crop of little economic importance and is underutilized as a floricultural crop. However, cineraria is an excellent subject to demonstrate direct somatic embryogenesis. The following two exercises that are depicted in Figure 23.1 will illustrate the effects of medium composition on the direct formation of somatic embryos from either leaves or flower receptacles of cineraria.

General considerations

Growth of plants

'Hansa' or 'Cindy' mix cineraria seeds (Park Seed Co, Greenwood, SC) are sown sparingly, about four to five per inch, in very shallow trenches of prewetted vermiculite or other suitable seedling mix and then lightly covered with medium. We have found that an 11 × 21-in. plastic flat of seedlings will provide enough material for a class of 10 to 15 students. Cover the trays with plastic wrap and incubate at 21°C with about 50 $\mu mol \cdot m^{-2} \cdot sec^{-1}$ of light provided by cool fluorescent tubes. If a growth chamber is unavailable, a benchtop in a cool laboratory will work well. The plastic wrap should be removed after the seedlings emerge in about 10 days to 2 weeks. The first true leaves will be formed in about 2 weeks and are suitable for explants. Flower formation requires about 26 to 28 weeks of growth, including some time in a cooler. Five weeks after sowing, seedlings are transplanted into bedding plant cell paks. After 4 weeks, plants are moved up to 4-in. pots (any soilless medium will do) and, after an additional 4 weeks of growth, plants are placed in a cooler at 15°C. The plants are moved after 6 weeks to a cool greenhouse (about 18°C) and should flower between 8 and 10 weeks later (see Larsen, 1985 for scheduling). For receptacle explants, choose flower buds that are showing slight color, but are more or less tightly closed.

0-8493-2029-1/00/$0.00+$.50

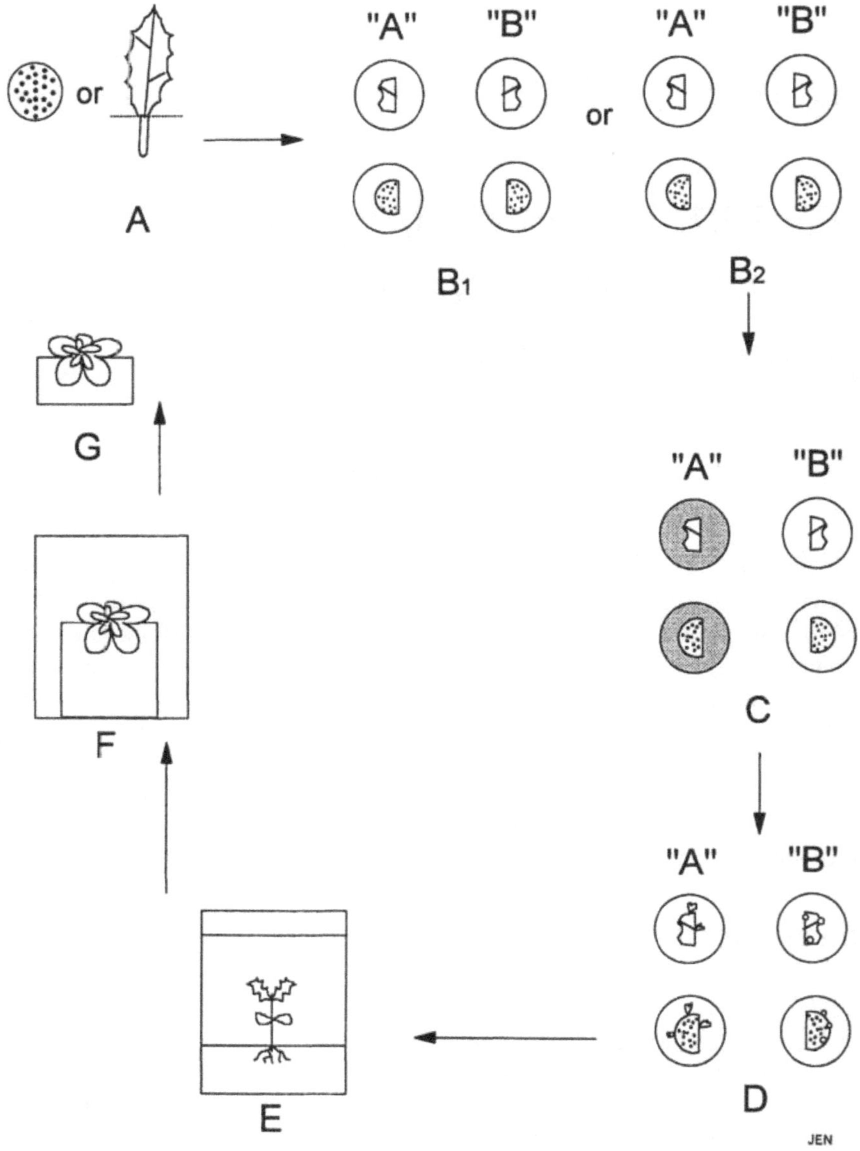

Figure 23.1 Diagrammatic representation of exercises for direct somatic embryogenesis from leaves and receptacles of cineraria. (A) Leaf and receptacle tissue before surface disinfestation; (B_1) tissue cut into equal halves and placed on SH medium "A" and MS medium "B" (Experiment 1); (B_2) tissue cut into equal halves and both placed on the MS medium containing PGRs (Experiment 2); (C) tissue cultured on "A" transferred to charcoal containing medium while "B" remains on the original medium; (D) tissue cultured on "A" transferred to medium without PGRs, whereas tissue on "B" is transferred to fresh MS medium containing PGRs; (E) somatic embryo formed on explants in "A" germinated on medium without PGRs in a Magenta vessel; (F) plantlet grown in a peat pellet and acclimatized to ambient growing conditions. Plantlet potted in soilless medium and transferred to a greenhouse or cooler. (From Trigiano, R. N. and Gray, D. J., Eds., *Plant Tissue Culture Concepts and Laboratory Exercises, First Edition*, CRC Press LLC, Boca Raton, FL, 1996.)

Explant preparation and basal culture media

The first true leaves, including the petiole, may be harvested for explants about 2 weeks after the seedlings emerge. Surface disinfest the leaves first by immersing in 70% ethanol for 30 sec, followed by gently agitating in 10% bleach solution for 10 min, and finally rinsing three times with sterile distilled water. Excise and discard the petiole and bisect the leaves (lamina) longitudinally (lengthwise) into two equal halves (Figure 23.1). If receptacle tissue is to be used as explants, flower buds with about 2 cm of the subtending pedicle (stem) should be immersed in 70% ethanol for 1 min and then the alcohol ignited by quickly passing the bud through a flame. The buds are soaked in 20% Clorox for 10 min followed by three rinses in sterile distilled water. Using an aseptic technique, remove the subtending pedicle and strip the calyx (green portion) and all the florets from the receptacle. Cut the receptacle into two equal halves. An optional disinfecting treatment is to immerse the naked receptacle in 10% Clorox solution for 5 min followed by three rinses with sterile water before bisection.

There are two treatments in both of the experiments included in this chapter, therefore the exercises may be set up simply as paired variate designs. In this experimental design, we assume that an individual leaf or flower receptacle is composed of relatively homogenous tissue in which any portion (e.g., a leaf half) is capable of responding to any treatment in a similar manner compared to the corresponding sister half. Therefore, if the sister halves are placed on different treatments, variations in responses will be due to treatment and not to differences in the ability of the tissue to respond or to other extraneous factors. This is a very powerful experimental design and eliminates the need for extensive replication. Regardless of the type of tissue used in these experiments, place one half of the explant onto one medium (treatment) and place the corresponding sister half onto the other (this is the "pair" in the paired variate design). Label the first petri dish A-1 and the second B-1, where the letter indicates the treatment and the numeral designates the replicate or explant (see Figure 23.1). Data (e.g., the number of somatic embryos) should be analyzed using a t-test for paired variates (Snedecor and Cochran, 1967 or other comparable statistics textbook).

The culture media in the following two experiments are composed of either Murashige and Skoog (MS) (1962) or Schenk and Hildebrandt (SH) (1972) basal salts (see Chapter three for composition) supplemented with 30 g (88 mM) sucrose, 1 g (0.55 mM) myo-inositol, 5 mg (2.9 μM) thiamine-HCl, 3 mg (13.5 μM) 2,4-dichlorophenoxyacetic acid (2,4-D), 1 mg (4.5 μM) benzyladenine (BA), and 8 g of agar per liter. The pH of the medium is adjusted to 5.8 before sterilization and approximately 10 ml of medium poured into each 60 × 15-mm plastic petri dish.

Exercises

Experiment 1. The effect of basal medium on initiation of somatic embryos

The composition of the basal medium can have dramatic effects on the response of an explant in culture. This simple experiment is designed to test the ability of cineraria explants to form somatic embryos on two commonly used tissue culture media, SH and MS. The experiment will require about 4 weeks to complete.

Materials

The following items are needed for each student or team of students:

- Six seedlings or six flower receptacles
- Twelve 60 × 15-mm petri dishes, six each containing SH (treatment "A") and MS medium (treatment "B") and both media supplemented with 13.5 μM 2,4-D and 4.5 μM BA
- 10 and 20% clorox solutions + 0.1% Triton X-100

Follow the experimental outline provided in Procedure 23.1.

Procedure 23.1
The effect of basal medium on initiation of somatic embryos.

Step	Instructions and comments
1	Surface disinfest leaves or flower receptacles with ethanol and bleach as described under explant preparation.
2	Cut the leaves or receptacles into identical "sister halves" and place one half onto each of the experimental media. Label each of the dishes with the treatment "A" or "B" and be **absolutely** certain that the dishes containing the "sister halves" are labeled with the same number.
3	Incubate the cultures in the dark between 22 and 25°C. Examine weekly for embryo development. If one culture becomes contaminated (e.g., A-2), then the corresponding culture (B-2) should also be discarded.
4	After 2 and 4 weeks, count the number of somatic embryos formed on explants cultured on each of the treatments (media). We suggest that the class share and analyze the data using a t-test for paired variates (see Table 23.1) for an example.

Anticipated results

Both leaf explants should begin to curl and become somewhat contorted after 4 to 5 days in culture. Late in the second week of culture, explants grown on SH medium or treatment "A" (Figure 23.2a) will produce some crystalline-looking callus and occasionally a somatic embryo, whereas numerous globular-stage yellow somatic embryos will begin to form directly from explants cultured on MS medium or treatment "B" (Figure 23.2b). Four weeks after culture initiation, leaf portions cultured on SH medium will have produced abundant callus, whereas receptacle explants will be curled with little evidence of callus growth (Figure 23.2c). Explants cultured on MS medium will usually be covered with yellow-to-green globular-to-heart-shaped somatic embryos that may be beginning callus (Figure 23.2d). Very few, if any, morphologically mature embryos with well-developed cotyledons will be present.

Questions

- Why did MS and not SH medium support the development of somatic embryos? Hint: consider the nitrogen sources and total salt concentration of each medium.
- Why was the ontogeny of somatic embryos on MS medium arrested in the globular or heart-shaped stage of development? What are some culture strategies that would permit continued morphological development?

Experiment 2. Maturation of somatic embryos and regeneration of plants

Somatic embryo initiation and development are often affected by nutrition, as demonstrated in the previous exercise, and also by plant growth regulators (PGRs). Sequential media transfer systems are often necessary to promote induction, maturation, and germination of somatic embryos (see Chapter nineteen). Typically, an induction or primary

Table 23.1 Example of Data Analysis Using a t-test for Paired Variates

Treatments	Mathematical procedures				
Pair Number	"A"[a]	"B"[a]	D = A-B	$D - D_m$[b]	$(D-D_m)^2$
1	1	9	–8	–2	4
2	0	9	–9	–3	9
3	1	12	–11	–5	25
4	1	6	–5	1	1
5	0	0	0	6	36
6	2	5	–3	3	9
			–36	0	84

Note: Calculations: (1) $D_{mean} = -36/6 = -6$; (2) standard deviation $(84/5)^{1/2} = (16.8)^{1/2}$; (3) standard error $(16.8)^{1/2}/(6)^{1/2} = (2.8)^{1/2} = 1.65$; (4) $t = (-6 - 0)/1.65 = -3.64$: Note "0" is used to include the null hypothesis of no difference between treatments; (5) for 5 degrees of freedom and $p = 0.05 = -2.57$ (see Snedecor and Cochran, 1967); (6) Since $t = -2.57$ is greater than –3.64, the result is significant — explants cultured on treatment "B" produced significantly more embryos than were produced on treatment "A."

[a] Number of somatic embryos.

[b] Mean difference between treatments.

From Trigiano, R. N. and Gray, D. J., Eds., *Plant Tissue Culture Concepts and Laboratory Exercises, First Edition*, CRC Press LLC, Boca Raton, FL, 1996.

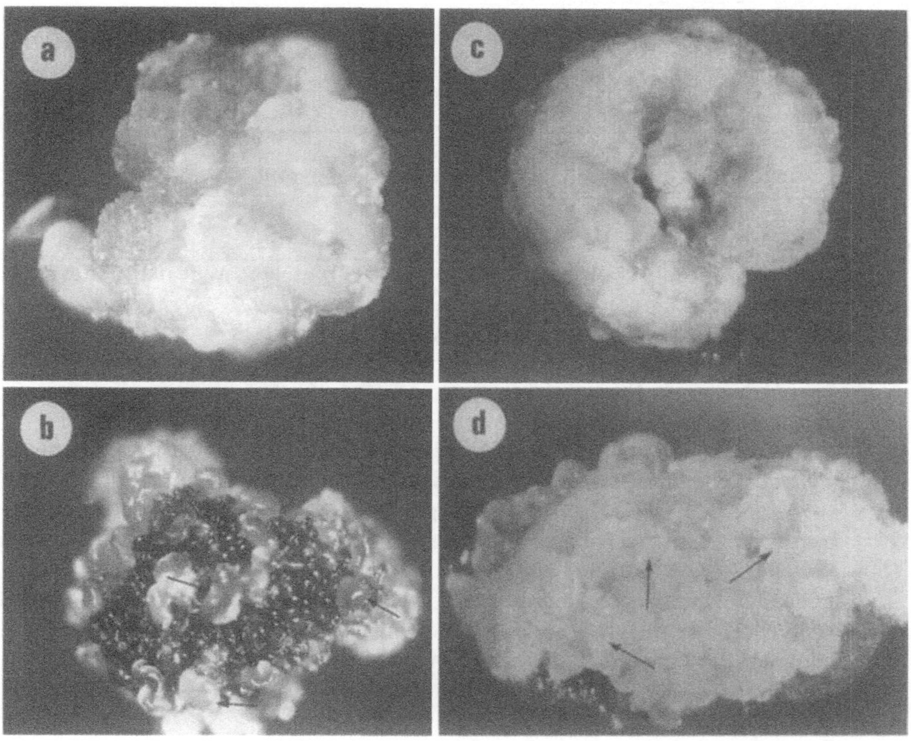

Figure 23.2 Somatic embryogenesis from leaf and flower receptacle tissues of cineraria. (a) A leaf explant cultured on SH medium (treatment "A") does not produce somatic embryos; (b) the sister leaf half-cultured on MS medium (treatment "B") produced numerous somatic embryos (arrows); (c) one half of a receptacle cultured on SH medium becomes contorted, but does not produce somatic embryos; (d) the other half of the receptacle formed many embryos (arrows), which in this case are beginning to callus. (From Trigiano, R. N. and Gray, D. J., Eds., *Plant Tissue Culture Concepts and Laboratory Exercises, First Edition*, CRC Press LLC, Boca Raton, FL, 1996.)

medium for somatic embryogenesis contains relatively high concentrations of PGRs, especially auxin or auxin-like compounds, whereas the levels of PGRs in a maturation medium are usually low, and in some, may even be excluded. A germination medium may be the same as a maturation medium, may have reduced concentrations of basal salts and sugar, or may have incorporated into it low levels of some PGRs. The following exercise illustrates a three-medium scheme to produce somatic embryos that can germinate and form plants and is adapted from a publication by Malueg et al. (1994). The experiment will require about 5 weeks for somatic embryo production or if acclimatized regenerated plants are desired, about 11 weeks.

Materials

Each student or team of students will need the following materials:

- Eighteen 60 × 15-mm petri dishes containing MS medium amended with 13.5 µM 2,4-D and 4.5 µM BA
- Six 60 × 15-mm petri dishes containing MS medium without PGRs, but supplemented with 0.5% (5 g/L) activated charcoal (AC)
- Six 60 × 15-mm petri dishes containing MS medium without PGRs
- Three GA-7 vessels containing MS medium without PGRs
- Jiffy-7 Peat Pellets (Jiffy Products of America, Inc., Chicago, IL)
- Fafard no. 2 potting mix (Conrad Fafard, Inc., Agawam, MA)
- Small plastic bags

Follow the experimental protocol outlined in Procedure 23.2.

Procedure 23.2
Maturation of somatic embryos and regeneration of plants.

Step	Instructions and comments
1	Surface disinfest explants with ethanol and/or bleach as indicated under explant preparation.
2	Bisect explants and place one half on MS medium with PGRs and label "A-1" and the sister half into another dish containing the same medium, but label "B-1." The letters designate different treatments. We suggest that each team prepare six replications and incubate them in the dark at 22 to 25°C for 2 weeks.
3	Transfer the explants from the petri dishes labeled "A" to medium containing activated charcoal (AC) only and label with "A" and the explant number. Incubate as before for 3 days, then transfer explants to MS medium without PGRs.
4	Transfer the explants from "B" treatment to fresh MS medium with PGRs. Be sure to retain the original lids with labels on all the new petri dishes.
5	Incubate all cultures for an additional 3 weeks in the same conditions described above. Count and categorize somatic embryos by developmental stage (i.e., globular, cotyledonary) for each of the treatments. Analyze data by embryo development stage and total number of embryos using a t-test for paired variates (see Table 23.1).
6	Remove cotyledonary-stage somatic embryos from the explants and transfer to MS medium without PGRs contained in GA-7 vessels. Incubate in 25 to 75 $\mu mol \cdot m^{-2} \cdot sec^{-1}$ at 22°C. When plants are about 2 cm tall, transfer to moist Jiffy 7 peat pellets and cover with plastic bags.
7	After 2 or 3 weeks, plants can be acclimatized to ambient conditions by opening the bags for increasingly longer intervals over a 2-week period. Plants can then be potted in Fafard No. 2 potting medium and grown as discussed under General considerations, growth of plants.

Anticipated results

Somatic embryos should form directly on all explants capable of responding to the treatments. However, there should be more developmentally advanced (torpedo and cotyledonary-stage) embryos on explants that were transferred through the series of three media. Most embryos formed on explants that remained on the initial medium with PGRs should be callused as shown in Figure 23.2d.

About one third of the cotyledonary-stage embryos transferred to MS without PGRs should produce a radicle and then a shoot in less than 10 days; the others will only form a radicle or will never germinate. Most of the plants derived from somatic embryos will have multiple stems, should acclimatize easily, and will produce normal colored and shaped — but smaller — flowers.

Questions

- Why do some or most germinating embryos develop multiple shoot apices?
- Why do less than half of the embryos placed on a germination medium form plants?
- How would you determine if somatic embryos were formed directly or indirectly?

Literature cited

Larsen, R. 1985. Temperature and flower induction of *Senico* × *hybridus* Hyl. *Swed. J. Agric. Res.* 15:87–92.

Malueg, K. R., G. L. McDaniel, E. T. Graham, and R. N. Trigiano. 1994. A three media transfer system for direct somatic embryogenesis from leaves of *Senico* × *hybridus* Hyl. (Asteraceae). *Plant Cell Tissue Organ Cult.* 36:249–253.

Murashige, T. and F. Skoog. 1962. A revised medium for rapid growth and bioassays with tobacco tissue cultures. *Physiol. Plant.* 15:473–497.

Schenk, R. U. and A. C. Hildebrandt. 1972. Medium and techniques for induction and growth of monocotyledonous and dicotyledonous plant cell cultures. *Can. J. Bot.* 50:199–204.

Snedecor, G. W. and W. G. Cochran. 1967. *Statistical Methods*, 6th ed. Iowa State University Press, Ames, 593p.

chapter twenty-four

Somatic embryogenesis from immature seeds of yellow poplar

Scott A. Merkle

Liriodendron tulipifera L. (yellow poplar, tulip poplar, tuliptree) is one of the most distinctive and valuable hardwoods in the eastern U.S. The tree is characterized by rapid height growth on good sites, reaching up to 40 m in 50 years, and by its straight trunk and conical crown. Large volumes of yellow poplar wood are used for furniture, plywood, corestock, millwork, siding, and other light construction lumber. It is also used for pulp and for products manufactured from chips or flakes (Russell, 1977). Yellow poplar is known as an excellent honey species and as a highly desirable ornamental, for which a number of horticultural cultivars have been described (Santamour and McArdle, 1984).

As with other forest trees, vegetative propagation of yellow poplar is a desirable goal as a means of capturing genetic superiority. It is also true that usually less than 10% of open-pollinated yellow poplar seeds are filled (Wilcox and Taft, 1969), making them an unreliable method for propagation. A number of vegetative propagation methods have been reported for yellow poplar, including rooted stem cuttings, root cuttings, and grafting. The ability to root stem cuttings of yellow poplar was very poor until it was discovered that over 70% success could be obtained by using cuttings taken from stump sprouts or epicormic shoots (McAlpine, 1964; Kormanick and Porterfield, 1966). However, in order to obtain stump sprouts or epicormic shoots, the tree must be either partially girdled or sacrificed entirely. The most successful method reported to date for in vitro propagation of yellow poplar is via somatic embryogenesis (Merkle and Sommer, 1986; Merkle et al., 1990). As with most embryogenic systems reported for forest trees, the major drawback of the yellow poplar system with regard to application to tree improvement is the fact that initiation of the cultures relies on the use of immature zygotic embryos, which are of unknown genetic value. However, even with this limitation, the cultures have proven to be valuable tools for such applications as protoplast culture (Merkle and Sommer, 1987) and microprojectile-mediated gene transfer (Wilde et al., 1992; Rugh et al., 1998).

Traditionally, woody plants, in general, and forest trees, in particular, have been challenging subjects for in vitro propagation compared to agronomic and other herbaceous species, due to a number of properties intrinsic to their biology. The following laboratory exercises will illustrate some of these problems and how they can be handled. In addition, yellow poplar embryogenic cultures have the rare property among forest tree embryogenic cultures of proliferating as proembryogenic masses (PEMs; Halperin, 1966). PEMs may be thought of as very early-stage proembryos caught in a repetitive cycle of proliferation.

0-8493-2029-1/00/$0.00+$.50

This property is one feature that makes yellow poplar cultures excellent material for mass production of somatic embryos and microprojectile-mediated gene transfer. The objectives of the exercises are to give students experience with the following techniques: (1) initiation of repetitive embryogenic cultures from immature seeds and (2) production of somatic embryos from PEMs.

Exercises

Experiment 1. Effect of seed developmental stage on embryogenic culture initiation

As stated in the introduction, most embryogenic cultures of forest trees are initiated from seed tissues or seedlings. One factor that has been repeatedly demonstrated to be critical for the successful initiation of embryogenic cultures of many species, including most woody plants, is the developmental stage of the zygotic embryo or seed when it is placed in culture. In many cases, it has been shown that immature seeds have a much higher potential to produce embryogenic cultures than do mature seeds or seedlings. This experiment was designed to illustrate this concept and will also demonstrate the special challenges that forest trees present with regard to procuring the optimal explant. In fact, it should be noted that precisely because this culture initiation experiment deals with obtaining material that is only available during a certain time of year, it may not fit easily into a class schedule, unless the class is offered during the summer months. If the class schedule does not fit in with the requirements of the experiment, it will be up to the instructor to collect the explant material and initiate the cultures so that the embryogenic material will be available for the other experiment in the chapter. Refer to Figure 24.1, A to F, for a diagrammatic representation of this exercise. Experiment 1 will require approximately 15 weeks to complete.

Materials

The following materials will be needed for each student or team of students:

- 80 ml of 5% commercial bleach
- 10 ml Roccal (National Laboratories, Montvale, NJ)
- 50 ml 70% ethanol
- 80 ml of 0.01 N HCl
- Eight sterile 150-ml beakers
- Grafting knife
- Scissors
- Two spoonulas (Fisher Scientific, Pittsburgh, PA)
- One stand magnifier or dissecting microscope
- Ten 60-mm plastic petri dishes containing yellow poplar induction medium (Table 24.1)

Follow the protocols listed in Procedure 24.1 to complete this experiment.

Anticipated results

At the end of the first month on induction medium, some of the explants may have produced a small amount of callus, which is usually nonmorphogenic. By the end of the second month, however, a yellow, nodular cluster of PEMs should appear at one end of some of the explants (Figure 24.2). At that time, since each of these PEM-producing explants represents an independent clonal line, each should be transferred to its own plate of induction medium. PEM proliferation can be maintained by monthly transfer to fresh

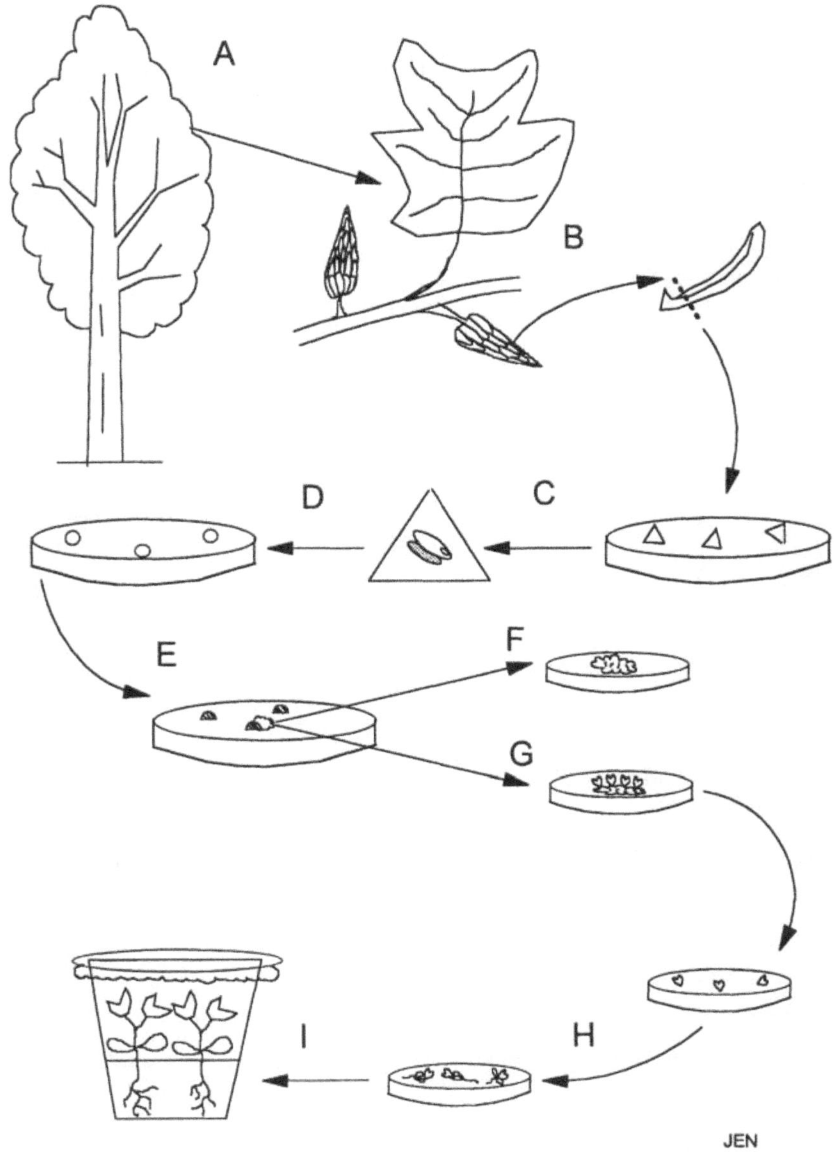

Figure 24.1 Diagrammatic representation of yellow poplar embryogenic culture initiation and somatic embryo and plantlet production. (A) Immature fruit are collected from yellow poplar tree; (B) aggregates are dissected to individual samaras and dewinged; (C) dewinged samaras are surface sterilized; (D) dewinged samaras are sliced open with a scalpel to reveal seed contents (embryo and endosperm); (E) seed contents are scraped out and cultured on plates of induction medium; (F) explants are transferred to fresh medium after 1 month, and within another month, PEMs proliferate from some of the explants; (G) PEMs transferred to basal medium continue to proliferate as PEMs; (H) PEMs transferred to basal medium begin to produce somatic embryos within 1 month; (I) cotyledon-stage somatic embryos transferred to basal medium lacking casein hydrolysate and cultured in the light germinate and produce green leaves within 2 weeks; (J) germinated somatic embryos grow into yellow poplar plantlets following transfer to potting mix. (From Trigiano, R. N. and Gray, D. J., Eds., *Plant Tissue Culture Concepts and Laboratory Exercises, First Edition,* CRC Press LLC, Boca Raton, FL, 1996.)

Table 24.1 Composition of Yellow Poplar Basal and Induction Media

Component	Concentration (mg/L)	Ref.
Major salts		Witham et al. (1971)
KNO_3	1000	
$Ca(NO_3)_2 \cdot 4H_2O$	500	
$MgSO_4 \cdot 7H_2O$	72	
KCl	65	
NH_4NO_3	1000	
Minor salts		Sommer and Brown (1980)
$MnSO_4 \cdot 4H_2O$	8	
$ZnSO_4 \cdot 7H_2O$	4	
H_3BO_3	2	
KI	0.6	
$Na_2MoO_4 \cdot 2H_2O$	0.025	
$CuSO_4 \cdot 5H_2O$	0.025	
$CoCl_2 \cdot 6H_2O$	0.025	
Iron		Murashige and Skoog (1962)
$FeSO_4 \cdot 7H_2O$	27.85	
Na_2 - EDTA	41.25	
Vitamins		Gresshoff and Doy (1972)
Thiamine-HCl	1	
Nicotinic acid	0.1	
Pyridoxine-HCl	0.1	
Myo-inositol	0.1	
Others		
Myo-inositol (additional)	100	
Casein hydrolysate	1000	
Sucrose	40000	
Phytagar	8000	
2,4-D[a]	2	
BAP[a]	0.25	

[a] PGRs are included only in induction medium.

Procedure 24.1
Effect of seed developmental stage on embryogenic culture initiation.

Step	Instructions and comments
1	Collect yellow poplar fruit. Bright yellow-green, conical aggregates of samaras should be collected for the first time approximately 8 weeks following flowering (which occurs between April and June, depending on latitude and elevation), and, if possible, again 4 weeks or more later. Sufficient aggregates should be collected so that each student or team of students has at least two fruit from each collection date. For larger trees, reaching the fruit may require pruning poles. Fruit should be kept at 4°C in sealed plastic bags until used.
2	Cut each aggregate in half lengthwise and pry samaras apart from each other using the grafting knife and fingers. Cut wings off samaras and discard. If samaras will not be used immediately, store in a sealed plastic bag at 4°C.
3	Surface disinfest the dewinged samaras using eight beakers in the following sequence: 70% ethanol for 20 sec, 10% Roccal (with sterile water) for 3 min; repeat the first two steps using the same solutions over again, 5% bleach for 5 min, sterile water for 3 min, 0.01 N HCl for 3 min, and three more sterile water rinses for 3 min each. Samaras may remain in the last rinse beaker for a few hours without damage.

Procedure 24.1 continued Effect of seed developmental stage on embryogenic culture initiation.	
Step	Instructions and comments
4	Remove three or four samaras from the last rinse with sterile forceps and place them on sterile filter paper in a sterile glass petri dish. Use the scalpel with #11 blade to cut completely through the seed-bearing portion of each samara, slicing down through the vertical ridge, which should bisect both seeds inside. Most seeds will probably be empty, but viable seeds will be filled with white endosperm. Using the fixed blade scalpel, scoop as much of the contents out of each seed as possible and place on induction medium. Embryos are only about 1 mm long and are difficult to distinguish from endosperm, but if the whole endosperm is removed, chances are good that the embryo is embedded in it.
5	Culture the contents of two or three seeds per petri dish of induction medium. Seal the dishes with Parafilm and incubate in the dark at about 25°C.
6	After 1 month, transfer the explants and any callus that they have produced to fresh dishes of induction medium and incubate for another month.

medium. If fruit was collected for culture initiation on both of the dates suggested above, most if not all of the cultures producing PEMs will probably be derived from seeds collected on the first date. This is due to the fact that in yellow poplar, as in a number of other species, immature zygotic embryos have a higher potential for production of embryogenic material than do more mature embryos (Sotak et al., 1991).

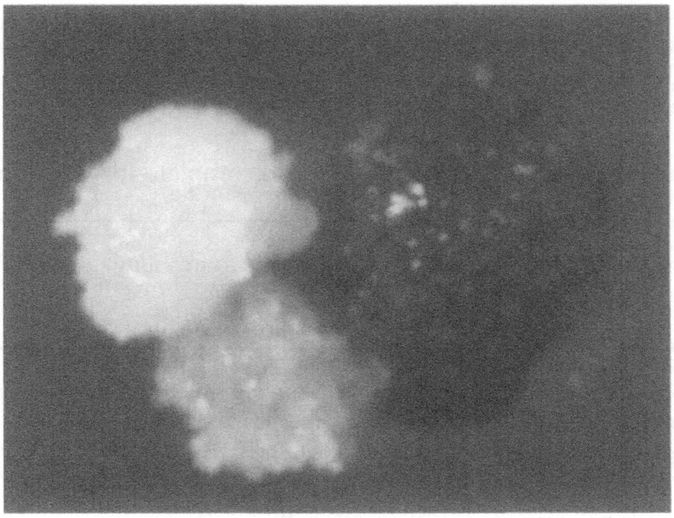

Figure 24.2 Proliferation of PEMs from explanted yellow poplar embryo and endosperm following 7 weeks of culture on induction medium. Bar = 500 μm. (From Trigiano, R. N. and Gray, D. J., Eds., *Plant Tissue Culture Concepts and Laboratory Exercises, First Edition*, CRC Press LLC, Boca Raton, FL, 1996.)

Questions

- Why is it that the more immature zygotic embryos seem to have a higher potential for production of embryogenic material than the more mature embryos?
- Does an embryogenic culture derived from the seed embryo of a given tree effectively "clone" the genotype of that tree? If not, why not? Is it possible for a seed explant to produce an embryogenic culture that does "clone" the source tree? How?

Experiment 2. Production of somatic embryos and plants from yellow poplar PEMs

Once proliferating PEM cultures are obtained, it is not difficult to produce somatic embryos and plantlets from them. Like many embryogenic systems, including the model carrot system, yellow poplar PEMs are maintained in a proliferative state by some minimal level of exogenous auxin, and once the concentration of auxin is reduced below this level or removed entirely, clusters of cells on the surface of the PEMs are "released" to develop into somatic embryos. This experiment is designed to demonstrate how PEMs are "released" by removal of 2,4-D from the medium, and how germination of these embryos is enhanced by transfer to a medium without casein hydrolysate. Refer to Figure 24.1, parts F to J, for a diagrammatic representation of this exercise. Experiment 2 will require about 8 weeks to complete.

Materials

The following materials will be needed for each student or team of students:

- One dish each of five embryogenic yellow poplar lines producing PEMs (or as many lines as are available). PEMs should have been transferred to fresh induction medium 2 to 3 weeks previous to the experiment
- Ten 60-mm plastic petri dishes containing yellow poplar induction medium (see recipe — Table 24.1)
- Twenty 60-mm plastic petri dishes containing yellow poplar basal medium (see recipe — Table 24.1)
- Ten 60-mm plastic petri dishes containing yellow poplar basal medium without CH
- Two spoonulas (Fisher Scientific)
- Two or three GA-7 magenta vessels
- Potting mix (e.g., Fafard Peat-Lite)

Follow the instructions listed in Procedure 24.2 to complete this experiment.

Procedure 24.2
Production of somatic embryos and plants from yellow poplar PEMs.

Step	Instructions and comments
1	For each embryogenic line, transfer about 1/4-spoonula-full of PEMs to two dishes of basal medium and two dishes of induction medium. Use the back of the spoonula to spread the PEMs into a thin layer on the surface of the medium.
2	Seal plates with Parafilm and incubate in the dark at about 25°C. Check the plates weekly, starting about 3 weeks after transfer, and score the number of somatic embryos that arise from each by the end of 6 weeks.
3	After 6 weeks, from each of the cultures that produces somatic embryos, use forceps to collect individual cotyledon-stage embryos and transfer them to either a fresh petri dish of basal medium or basal medium lacking casein hydrolysate. Lay the embryos horizontally on the surface of the medium, spacing them at least 0.5 cm apart. Try to harvest enough embryos from each culture to place five to ten embryos on each of two dishes of each medium.
4	Seal dishes with Parafilm and incubate under cool white fluorescent lights for 16 h per day, at about 30°C.
5	After 2 weeks, score each of the cultures for the number of embryos that have produced roots and/or have green cotyledons or other leaves.

Procedure 24.2 continued
Production of somatic embryos and plants from yellow poplar PEMs.

Step	Instructions and comments
6	Plant germinated embryos in GA-7 magenta vessels 1/3 filled with moistened potting mix, four germinants per box. Cover GA-7 vessels with plastic wrap (e.g., Saran Wrap) and incubate under cool white fluorescent lights with 16-h photoperiod at about 25°C.
7	After two or three new leaves are produced, punch four to six small holes in the plastic wrap so the humidity can begin to decrease inside the vessel. As new leaves appear, more holes can be punched and finally the plantlets can be repotted and moved to the greenhouse. While the plantlets are in the GA-7s, fertilize weekly with Hoagland's solution or other liquid plant food.

Anticipated results

Following 6 weeks of culture, PEMs transferred to basal medium should have produced a number of somatic embryos (Figure 24.3). Globular-stage somatic embryos can be distinguished from PEMs by their smooth, shiny appearance (as opposed to the bumpy surfaces of the PEMs), and later stages are characterized by the emergence of cotyledons and radicle. PEMs transferred to induction medium should continue to proliferate as PEMs. After 2 weeks, some of the somatic embryos transferred to basal medium should have germinated, with elongating radicles and greening cotyledons. A significantly higher percentage of the somatic embryos transferred to basal medium lacking casein hydrolysate should have germinated, and those that have germinated should have longer radicles and more apical development than those cultured on basal medium with CH. Germinants planted in potting mix and fertilized weekly should produce at least one new leaf within a few weeks, and a branching root system 5 to 10 cm long should develop within 6 weeks.

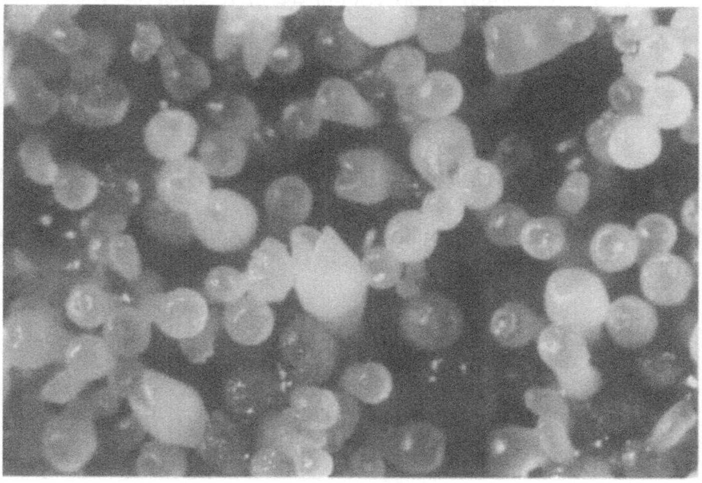

Figure 24.3 Developing yellow poplar somatic embryos 6 weeks following transfer of PEMs from induction medium to basal medium. (From Trigiano, R. N. and Gray, D. J., Eds., *Plant Tissue Culture Concepts and Laboratory Exercises, First Edition,* CRC Press LLC, Boca Raton, FL, 1996.)

Questions

- Why do PEMs transferred to basal medium begin to produce somatic embryos while those transferred to induction medium continue to produce PEMs?
- What are some possible explanations for why the removal of casein hydrolysate from the basal medium enhances germination of somatic embryos?
- Why is a gradual lowering of relative humidity required in order for the plantlets to survive transfer to ex vitro conditions?

Literature cited

Gresshoff, P.M. and C.H. Doy. 1972. Development and differentiation of haploid *Lycopersicon esculentum* (tomato). *Planta* 107:161–170.

Halperin, W. 1966. Alternative morphogenic events in cell suspensions. *Amer. J. Bot.* 53:443–453.

Kormanick, P.P. and E.J. Porterfield. 1966. Rooting yellow poplar cuttings. *Forest Farmer* 26:24, 41–42.

McAlpine, R.G. 1964. A method for producing clones of yellow poplar. *J. For.* 62:115–116.

Merkle, S.A. and H.E. Sommer. 1986. Somatic embryogenesis in tissue cultures of *Liriodendron tulipifera* L. *Can. J. For. Res.* 16:420–422.

Merkle, S.A. and H.E. Sommer. 1987. Regeneration of *Liriodendron tulipifera* (family Magnoliaceae) from protoplast culture. *Amer. J. Bot.* 74:1317–1321.

Merkle, S.A., R.J. Sotak, A.T. Wiecko, and H.E. Sommer. 1990. Maturation and conversion of *Liriodendron tulipifera* somatic embryos. *In Vitro Cell. Dev. Biol.* 26:1086–1093.

Murashige, T. and F. Skoog. 1962. A revised medium for rapid growth and bioassays with tobacco tissue culture. *Physiol. Plant.* 15:473–497.

Rugh, C.L., R.B. Meagher, J.F. Senecoff, and S.A. Merkle. 1998. Development of transgenic yellow poplar for mercury phytoremediation. *Nature Biotechnol.* 10:925–928.

Russell, T.E. 1977. Planting yellow poplar — where we stand today. USDA Forest Service Gen. Tech. Rep. SO-17. South. For. Exp. Stn., New Orleans, LA. 8 p.

Santamour, F.S., Jr. and A.J. McArdle. 1984. Cultivar checklist for *Liquidambar* and *Liriodendron*. *J. Arboriculture* 10:309–312.

Sommer, H.E. and C.L. Brown. 1980. Embryogenesis in tissue cultures of sweetgum. *For. Sci.* 26:257–260.

Sotak, R.J., H.E. Sommer, and S.A. Merkle. 1991. Relation of the developmental stage of zygotic embryos of yellow poplar to their somatic embryogenic potential. *Plant Cell Rep.* 10:175–178.

Wilcox, J.R. and K.A. Taft, Jr. 1969. Genetics of yellow poplar. USDA Forest Service Research Paper WO-6. 12 p.

Wilde, H.D., R.B. Meagher, and S.A. Merkle. 1992. Expression of foreign genes in transgenic yellow poplar plants. *Plant Physiol.* 98:114–120.

Witham, F.H., D.F. Blaydes, and R.M. Devlin. 1971. *Experiments in Plant Physiology.* Van Nostrand-Reinhold Co., New York. 245 pp.

chapter twenty-five

Induction of somatic embryogenesis in conifers

Stephen M. Attree, Patricia J. Rennie, and Larry C. Fowke

The genus *Picea*, the spruces, is a member of the Pinacea, the largest family of conifers. There are about 40 species of spruce restricted mostly to the Northern Hemisphere. Due to whiteness of the wood, low resin content, and fiber qualities, the main use of spruce is as pulpwood for the manufacture of paper.

After logging, forests are often left to regenerate naturally, which is slow and unpredictable, as a succession of different plant species will follow. However, increased forest productivity is needed to meet an increasing world demand for wood and wood products. Improved forest productivity can be achieved by more intensively managing forests by, for example, using improved silvicultural practices and replanting with nursery-grown trees following logging. The nursery trees are usually grown from seed collected from the wild; however, forest productivity may be further improved by replanting with genetically superior trees grown from seed obtained from tree-breeding programs. Characteristics that can be selected for in-tree breeding programs include faster growth rate, resistance to pests and diseases, and tolerance to environmental stresses such as frost or drought. Also important is the lignin content of the wood. Wood with a low lignin content is more suitable for pulping, whereas wood with a high lignin content is more desirable for lumber. Elite seed from breeding programs is expensive and available only in low quantity; therefore, it is used to generate clonal seed-production orchards that mass produce seed for reforestation. The clonal orchards are established using various conventional means of vegetative propagation, such as rooting cuttings, or grafting scions (shootstocks) onto seedling rootstocks. Mature trees in such orchards are pollinated by wind, and the resulting seed is then collected. This improved seed is then grown in nurseries and the young plants are replanted at the reforestation sites.

Conventional breeding of conifers is a very long-term endeavor. Vegetative propagation would alleviate time constraints, but has a low success rate and is especially difficult with mature trees. However, it is the mature trees that are of most interest for establishing clonal orchards, as desirable traits can be readily identified. Due to these difficulties, clonal propagation and breeding procedures are not extensively used to supply improved trees to forest nurseries.

In vitro culture techniques, particularly somatic embryogenesis, provide an alternative means to rapidly clone large numbers of plants with improved characteristics cost effectively for direct use by forest nurseries. Somatic embryogenesis may be induced from elite seed of known parents from tree improvement programs. Somatic embryogenesis

0-8493-2029-1/00/$0.00+$.50

also provides a regeneration system suitable for producing genetically modified plants. The steps to produce improved plants for direct reforestation via conventional means or somatic embryogenesis are outlined in Figure 25.1. While somatic embryogenesis in angiosperms has been studied for almost 30 years, somatic embryogenesis in conifers was only reported relatively recently (reviewed by Attree and Fowke, 1993). Since the initial reports of conifer somatic embryogenesis, our understanding of factors controlling induction and development of somatic embryos has made remarkable progress. For conifers, somatic embryogenesis usually involves the culture of zygotic embryos on medium containing auxin and cytokinin. From a proportion of the cultured zygotic embryos arise somatic embryos that undergo continued proliferation in the presence of auxin and cytokinin with rapid proliferation occurring in liquid suspension culture. Thus, from a single zygotic embryo, unlimited numbers of immature somatic embryos can be derived. These may be induced to undergo synchronous development to mature embryo forms (Chapter nineteen) by removing the auxin and cytokinin and culturing in bioreactors in the presence of abscisic acid and osmoticum (Attree et al., 1994).

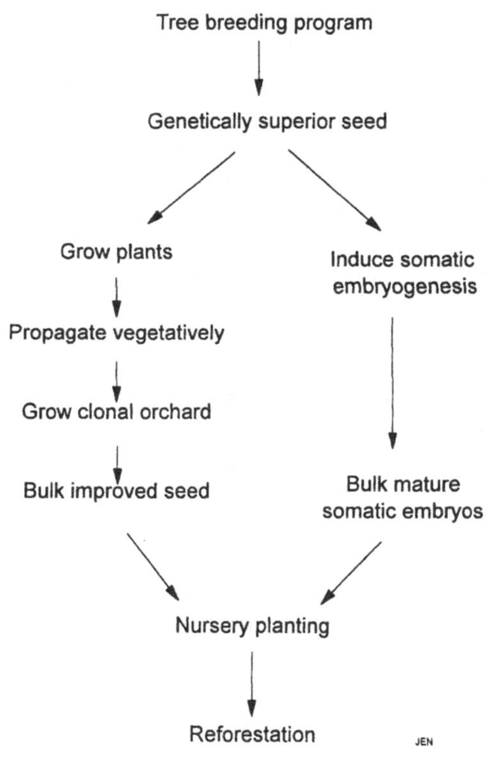

Figure 25.1 Steps for improving forest productivity by conventional or in vitro means. (From Trigiano, R. N. and Gray, D. J., Eds., *Plant Tissue Culture Concepts and Laboratory Exercises, First Edition*, CRC Press LLC, Boca Raton, FL, 1996.)

The simple methods provided in this chapter allow students to establish and maintain embryogenic cultures of spruce on gelled medium. They can then become familiar with the appearance of conifer embryogenic cultures and the cultures can be compared with embryogenic cultures of angiosperms, as described in other chapters of this volume. While methods in this chapter may be applied to a wide variety of conifers, spruces are most suitable for study, as somatic embryos may be induced from immature developing seed and mature stored seed (Tautorus et al., 1990), or cotyledonary explants excised from in vitro germinated seedlings (Attree et al., 1990). Induction of somatic embryogenesis in pines (*Pinus* spp.) is more difficult than the spruces, often requiring precotyledonary stage zygotic embryos to be co-cultured with the megagametophyte. Even then, induction occurs at only about a 5% frequency. Also, firs (*Abies* spp.) may require cytokinin alone for inducion. Not all conifer species are, at present, capable of embryogenic induction; a recent list is given by Attree and Fowke (1993).

Exercises

Experiment 1. Induction and maintenance of somatic embryos

Seed sources and basal medium

Seed may be obtained from a seed orchard; however, the ubiquitous nature of spruces should allow cones containing seed to be collected from local trees. For spruce, mature seed is suitable, but induction frequencies may be higher if immature developing seed is collected every 1 to 2 weeks in early summer. Seed collected shortly after pollination will contain early stages of zygotic embryos at a stage comparable to the embryos within the embryogenic cultures. A variety of culture media are suitable for culturing conifer somatic embryos; however, half-strength Litvay (1/2 LV) medium (Litvay et al., 1981) has been found to work well for many spruces and is available commercially (Sigma product no. L4272). The 1/2 LV medium should also contain 1% sucrose, 9 mM 2,4-D, 4.5 mM benzyladenine, 500 mg/L casein hydrolysate, and 0.6% agar. The growth regulators may be added prior to autoclaving at 121°C for 20 min.

Materials

For inducing and culturing somatic embryos, the following items are needed for each student or team of students:

- 50 to 150 seeds, preferably of a spruce
- 50 ml of 15% hydrogen peroxide
- A nonsterile flask capped with muslin cloth secured with a rubber band for soaking the seeds
- Binocular dissecting microscope
- 100 × 15-mm petri dishes containing culture medium
- Darkened incubator at 25°C

To complete this experiment, follow the protocols outlined in Procedure 25.1.

Procedure 25.1
Induction and maintenance of somatic embryos from spruce.

Step	Instructions and comments
1	The procedure for induction and maintenance of somatic embryos is summarized diagrammatically in Figure 25.2. Soak seed overnight in nonsterile tap water to imbibe seeds and soften the seed coat for ease of dissection. Discard the empty nonviable seeds that float.
2	Drain the water and transfer the seeds to a sterile 100-mm-diameter petri dish and surface disinfest the seed for 15 to 20 min in about 20 ml of a 15% solution of hydrogen peroxide. Remove the hydrogen peroxide by pipetting, and rinse the seeds three times with the sterile distilled water.

Procedure 25.1 continued
Induction and maintenance of somatic embryos from spruce.

Step	Instructions and comments
3	Using a dissecting binocular microscope and aseptic techniques, excise the zygotic embryos from within the white megagametophyte and place 10 to 15 embryos on the surface of the culture medium in each petri dish. Excision is best perfomed by holding the seed in the forceps and gently cutting the seed coat lengthwise, then turning the seed over and cutting the seedcoat down the other side. Remove the white megagametophyte, incise lengthwise, and pry open with the forceps to reveal the contents. The embryo is situated in the central cavity running lengthwise down the megagametophyte (Figure 25.3a). Conifers undergo single fertilization to produce the diploid zygote and the megagametophyte is haploid. The megagametophyte carries out a nutritive function similiar to that of the endosperm of angiosperms, which results from double fertilization. Zygotic embryos early in development may be barely detectable as a small, thin, hair-like strand of translucent tissue consisting largely of suspensor cells. As zygotic embryo development proceeds, the embryos become easier to distinguish and manipulate as they pass through larger globular and then cotyledonary stages.
4	Seal the dishes containing the plated zygotic embryo explants and incubate in the dark at 20 to 25°C. Transfer to fresh medium after 2 to 4 weeks. Within 3 to 6 weeks a translucent white mucilaginous embryogenic tissue, consisting of hair-like immature somatic embryos, should be evident proliferating from a proportion of the explants (Figure 25.3b). Using a pair of fine forceps, separate this tissue from the nonembryogenic tissues and culture in a separate petri dish. Nonembryogenic tissue is readily distinguishable as a white or green undifferentiated hard callus.
5	Subculture every 2 to 4 weeks to fresh maintenance medium. During subculture remove a small portion (about 5 mm diameter) of the white tissue from the edge of the embryogenic tissue and transfer to fresh medium. About seven clumps can be maintained per dish and several dishes should be kept to guard against contamination.

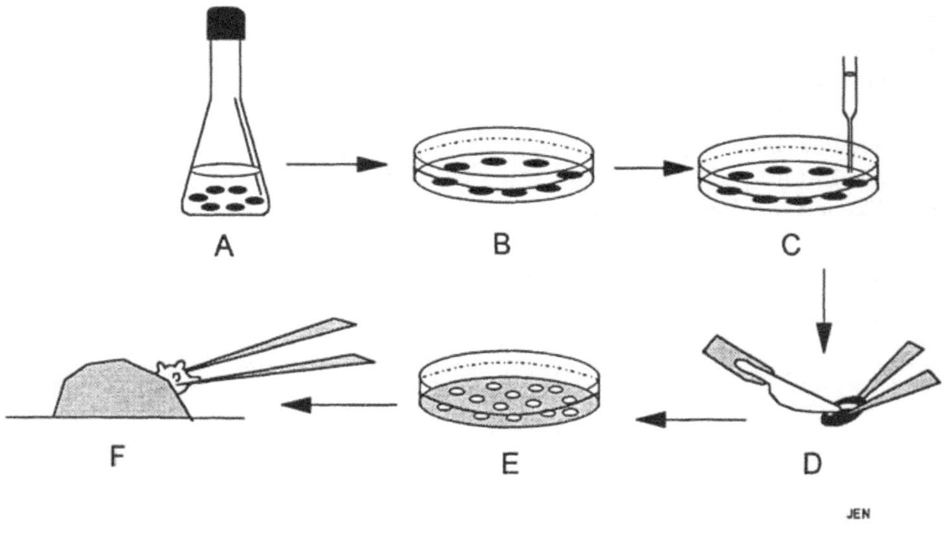

Figure 25.2 Diagrammatic representation of the methods used for inducing somatic embryos in conifers. (A) Soak seeds overnight in water; (B) surface disinfest seeds in 15% hydrogen peroxide; (C) rinse seeds in sterile, distilled water; (D) using a scalpel, excise zygotic embryos from seeds; (E) incubate embryos on culture medium; (F) subculture embryogenic tissues on medium. (From Trigiano, R. N. and Gray, D. J., Eds., *Plant Tissue Culture Concepts and Laboratory Exercises, First Edition,* CRC Press LLC, Boca Raton, FL, 1996.)

Anticipated results

The frequency of induction varies according to conifer species and maturity of seed, or age of seedling if the latter is used as explant material. However, induction frequencies of 30% can be expected and may go as high as 70% if immature spruce seeds are used. Spruce somatic embryos tend to arise usually within 4 to 6 weeks of plating; however, they sometimes appear as early as 2 weeks after plating if immature zygotic embryos are used as explants. Also, the embryogenic tissue may arise over a large proportion of the explant when immature zygotic embryos are used (Figure 25.3b). It generally takes longer (e.g., up to 12 weeks) for mature zygotic embryo explants to give rise to embryogenic tissue. Mature embryo explants more often undergo some nonembryogenic callusing

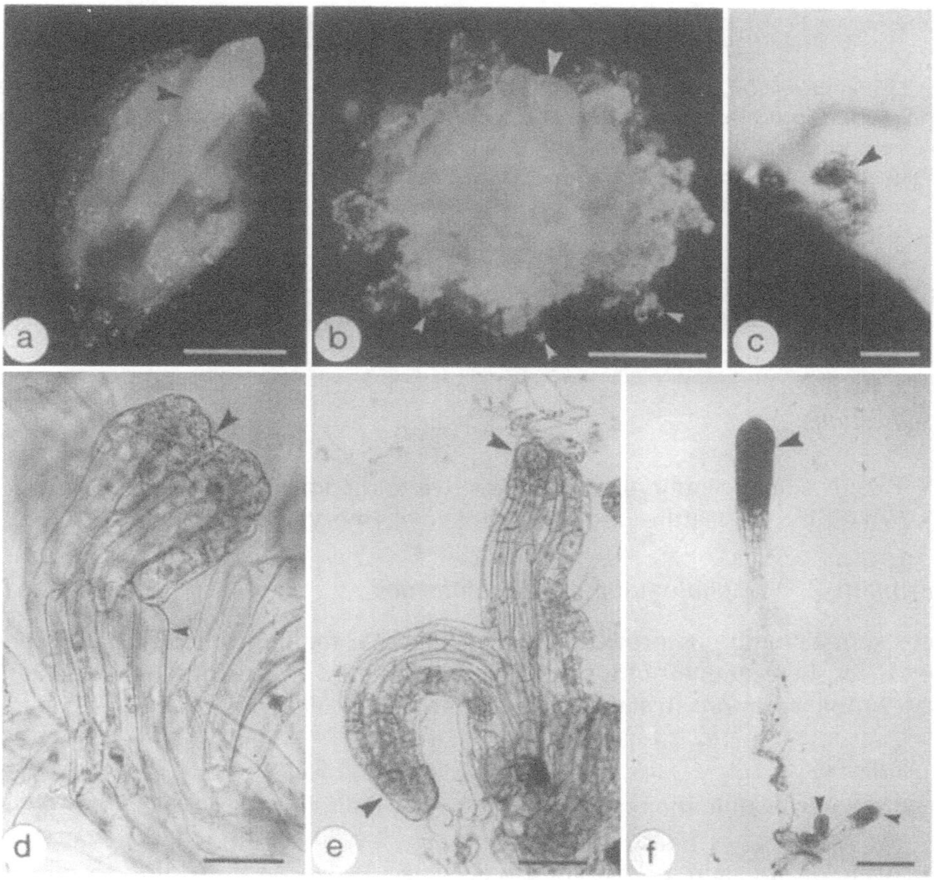

Figure 25.3 Light micrographs showing zygotic and somatic embryos of conifers. (a) Dissected seed of white spruce (*Picea glauca* [Moench.] Voss.) to show mature zygotic embryo (arrow); bar = 1 mm. (b) Translucent embryogenic tissue arising from a cultured zygotic embryo (large arrow) of white spruce. Note the immature somatic embryos (small arrows) protruding from the surface of the embryogenic tissue; bar = 1 mm. (c) Single white spruce somatic embryo (arrow) growing from the surface of a cultured white spruce shoot explant; bar = 0.25 mm. (From Attree, S.M. et al. 1990. *Can. J. Bot.* 68:30–34. With permission.) (d) Somatic embryo of white spruce showing early stage of embryo head cleavage (large arrow). The suspensor (small arrow) consists of elongate highly vacuolated cells; bar = 0.1 mm. (e) Late stage of cleavage showing two somatic embryos (arrows) of white spruce; bar = 0.2 mm. (f) Early jack pine (*Pinus banksiana* Lamb.) zygotic embryos undergoing cleavage. One of the embryos matures (large arrow) while the others (small arrows) degenerate; bar = 0.2 mm. (Reprinted with permission from Attree, S.M. and L.C. Fowke. 1993. *Plant Cell Tissue Organ Cult.* 35:1-35. With permission.)

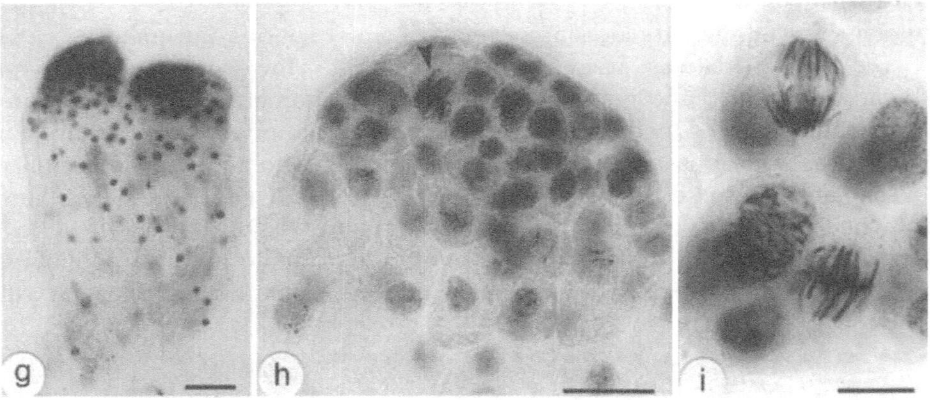

Figure 25.3 (continued) Light micrographs showing zygotic and somatic embryos of conifers. (g) Cleaving white spruce somatic embryo at stage similar to that shown in Figure 25.3d stained with aceto-orcein to reveal the distribution of nuclei; bar = 0.1 mm. (h) White spruce somatic embryo head stained with aceto-orcein to show nuclei and mitotic figures. Note the cell at metaphase (arrow); bar = 50 mm. (i) Cells of white spruce somatic embryo head stained to show nuclei and mitotic figures; bar = 20 mm.

prior to appearance of embryogenic tissue, and somatic embryos may arise from just one point on the explant (Figure 25.3c). A lack of any growth of the explants is probably due to lack of viability of the embryo explant, which is not uncommon.

Questions

- Why is somatic embryogenesis attractive to the forest industry?
- What physical features are characteristic of embryogenic tissue?

Experiment 2. Morphology of somatic embryos

Conifer somatic embryos provide excellent material for light microscopic observation. Embryos can be examined in a bright field compound microscope either unstained or using a simple aceto-orcein stain, which colors the nuclei and chromosomes red.

Materials

For microscopic examination of somatic embryos, the following items are needed:

- Aceto-orcein stain: prepare a 45% acetic acid solution by adding 45 ml of glacial acetic acid to 55 ml distilled water; add 0.15 g orcein powdered stain to 15 ml of the 45% acetic acid solution.
- 1.5-ml eppendorf tubes
- Pasteur pipettes and bulbs
- Forceps and dissecting needles
- Slides and coverslips
- Alcohol lamp or other heat source
- Paper tissues for blotting
- Compound microscope
- Small beaker or water glass

Follow the protocols in Procedure 25.2 to complete this experiment.

Procedure 25.2
Morphology of spruce somatic embryos.

Step	Instructions and comments
1	Place a small mass of embryogenic tissue in a drop of water on a microscope slide. Tease tissue apart with dissecting needles, add a coverslip, and examine using a light microscope. Close the condensor diaphram slightly to increase contrast. Locate individual embryos and try to determine which part of the embryo contains actively dividing cells and whether or not cytoplasmic streaming is occurring in any of the cells.
2	To stain nuclei and chromosomes, place a small, solid clump or mass of cells into a few drops of aceto-orcein stain in an eppendorf tube for 30 min or longer depending on the size of tissue mass (1 h in order for stain to diffuse into large compact embryo heads). Now, heat the tube in a beaker of water to 60 to 80°C for 10 min.
3	Remove a small sample of stained tissue from the tube and spread on a slide in a drop of stain. Observe under the microscope, being careful to keep stain solution away from the microscope lenses (acetic acid will damage lens mount) and skin. The nuclei of the embryos should stain an intense red against the background of stain on the slide. To intensify the stain gently warm the slide plus stain over the alcohol lamp, being careful not to boil the stain.
4	When the nuclei are adequately stained place a coverslip over the sample and remove residual stain by placing folded tissues at one side of coverslip and adding 45% acetic acid solution to the opposite side. When most of the stain has been replaced with acetic acid, blot slide and observe. Nuclei and chromosomes should remain red while other cellular materials are colorless.

Anticipated results

With light microscopy, the proliferating embryogenic tissue consists of differentiated immature somatic embryos, which resemble immature zygotic embryos. The somatic embryos are polarized structures that are organized into a meristematic embryonic head region subtended by elongate suspensor cells (Figures 25.3d and e and Hakman et al., 1987). With unstained living preparations, it is sometimes possible to observe cytoplasmic streaming in the suspensor cells closest to the embryonic region.

Under the influence of auxin and cytokinin, the immature somatic embryos continue to proliferate by splitting into two without further development (Figures 25.3d and e) — a process termed cleavage polyembryogenesis. This type of cleavage occurs naturally within seeds of certain conifer species when the early developing zygotic embryo separates into four independently developing embryos (Figure 25.3f). Competition among these embryos results in only the most aggressive surviving and maturing. Conifer embryogenic tissues differ in appearance from those of angiosperms. The early stages of angiosperm somatic embryogenesis generally consist of proliferating cell clusters called proembryonic cell complexes (Haccius, 1978), that are disorganized and so do not form recognizable structures unless auxin is reduced to encourage the later stages of embryo development. These differences may reflect a long period of evolutionary separation of these two important plant groups (Attree and Fowke, 1993).

With stained preparations at low magnification, areas of meristematic cells should stain strongly in contrast to suspensor regions which remain unstained (Figure 25.3g). At higher magnification mitotic figures may be observed in embryo head regions (Figures 25.3h and i). Mitotic frequencies may be calculated by dividing the number of cells with mitotic figures by total number of cells in a sample.

Questions

- Describe the general morphology of a conifer somatic embryo. Is it similar to the corresponding zygotic embryo or angiosperm zygotic and somatic embryos?
- How do conifer somatic embryos proliferate in culture?
- Where do cell divisions occur in the somatic embryos?

Literature cited

Attree, S. M., S. Budimir, and L. C. Fowke. 1990. Somatic embryogenesis and plantlet regeneration from cultured shoots and cotyledons of seedlings germinated from stored seed of black and white spruce (*Picea mariana* and *Picea glauca*). *Can. J. Bot.* 68: 30–34.

Attree, S.M. and L. C. Fowke. 1993. Embryogeny of gymnosperms: advances in synthetic seed technology of conifers. *Plant Cell Tissue Organ Cult.* 35: 1–35.

Attree, S. M., M. K. Pomeroy, and L. C. Fowke. 1994. Production of vigorous, desiccation tolerant white spruce (*Picea glauca* [Moench.] Voss.) synthetic seeds in a bioreactor. *Plant Cell Rep.* 13: 601–606.

Hakman, I., P. Rennie, and L. C. Fowke. 1987. A light and electron microscope study of *Picea glauca* (White spruce) somatic embryos. *Protoplasma* 140: 100–109.

Haccius, B. 1978. Question of unicellular origin of nonzygotic embryos in callus cultures. *Phytomorphology* 28: 74–81.

Litvay, J. D., M. A. Johnson, D. Verma, D. Einspahr, and K. Weyrauch. 1981. Conifer suspension culture medium development using analytical data from developing seeds. *Tech. Pap. Ser. Inst. Paper Chem.* 115: 1–17.

Tautorus, T.E., S. M. Attree, L. C. Fowke, and D. I. Dunstan. 1990. Somatic embryogenesis from immature and mature zygotic embryos, and embryo regeneration from protoplasts in black spruce (*Picea mariana* Mill.). *Plant Sci.* 67: 115–124.

Part V

Crop improvement techniques

chapter twenty-six

Use of protoplasts for plant improvement

Michael E. Compton, James A. Saunders, and Richard E. Veilleux

A fundamental difference between plant and animal cells is that certain plant cells are totipotent and can be induced to divide and develop into complete plants. Another significant difference is the presence of a cell wall in plants. The plant cell wall, although playing a key role in processes involving plant structure and function, is a major impediment in exploiting direct DNA transfer to individual cells and the production of somatic hybrids by cell fusion. Fortunately, the cell wall can be temporarily removed from plant cells without a significant loss of viability. Plant cells from which the cell wall has been removed are termed protoplasts. Protoplasts are somewhat unique in plant cell culture in that they exist as separate cells without cytoplasmic continuity with neighboring cells.

Plant protoplasts were first isolated by Klercker in 1892 by slicing onion bulb scales with a fine knife in a plasmolyzing solution, resulting in the release of protoplasts when cells were cut through the wall (Bhojwani and Razdan, 1983). Protoplast yield was low and the procedure was restricted to highly vacuolated, nonmeristematic cells. Isolation of plant protoplasts became more widely practiced in the 1960s with the extraction and purification of enzymes that could be used to degrade the plant cell wall. Cocking (1960) discovered that intact protoplasts could easily be obtained by incubating plant tissues in a concentrated cellulase solution prepared from the fungus *Myrothecium verrucaria*. By 1968, commercial preparations of purified cell wall degrading enzymes, such as macerozyme and cellulase, were available that allowed easy isolation of protoplasts from plant leaves and nonvacuolated cells.

Isolation of plant protoplasts

Protoplast isolation by enzymatic cell wall digestion involves the use of cellulase, hemicellulase, and/or pectinase. These hydrolytic enzymes are extracted from various sources including fungi, snail, and termite gut and are available commercially in differing formulations that vary in purity. Digestion by a combination of these three enzymes is generally conducted at a pH of 5.5 to 5.8 over a period of 3 to 8 h. No matter which procedure is used, protoplasts can be collected and purified by centrifugation techniques designed to separate broken and damaged cells from intact protoplasts by taking advantage of their differing buoyant densities.

0-8493-2029-1/00/$0.00+$.50

Uses for plant protoplasts

Plant breeders have used sexual hybridization to improve cultivated crops for centuries. This process is generally limited to plants within a species or to wild species that are closely related to the cultivated crop. When possible, sexual hybrids between distantly related species have been useful for the incorporation of single gene traits such as insect and disease resistance. However, intraspecific and interspecific incompatibility barriers limit the use of sexual hybridization for crop improvement.

Protoplast culture has been used to develop plants with improved agronomic and horticultural characteristics and improved disease resistance through the recovery of culture-induced variants, parasexual hybrids, and genetically engineered plants. By isolating protoplasts and regenerating plants from them, variants (somaclones) with improved characteristics have been obtained. Parasexual or somatic hybrids can be obtained by fusing protoplasts of unrelated or distantly related species. Through genetic engineering, foreign genes can be transferred to plant cells and improved plants with the new genes regenerated from transformed cells.

Improved plants obtained by protoplast manipulations can be used in breeding programs to develop new cultivars. This chapter discusses the use of protoplasts in plant improvement and is accompanied by chapters with laboratory exercises demonstrating protoplast isolation from leaves of chrysanthemum (*Dendrathema grandiflorum*), orchardgrass (*Dactylis glomerata* L.), and tobacco (*Nicotiana tabacum* L.) and fusion of potato (*Solanum tuberosum* L.) protoplasts.

Procedures for regenerating plants from protoplasts

Leaf mesophyll tissue from plants grown in a greenhouse or growth chamber with controlled light (16:8 or 18:6 h [light/dark] photoperiod, 60 to 300 $\mu mol \cdot m^{-2} \cdot s^{-1}$ light intensity) and temperature (20 to 25°C) is commonly used as a source of plant protoplasts. Uncontrolled greenhouse environment may hinder the repeatability of experiments, making the environment undesirable for growing plants for protoplast isolation.

Cells within the leaf mesophyll are loosely packed, which allows penetration of digestive enzymes that facilitate protoplast release. Typically 10^6 to 5×10^7 protoplasts can be obtained per gram of leaf tissue. Generally, the youngest, fully expanded leaves from young plants or seedlings are used. Preconditioning plants in darkness and/or cold (4 to 10°C) for 24 to 72 h before protoplast isolation usually improves protoplast yield.

Leaves from greenhouse- or growth chamber-grown plants require surface disinfestation prior to protoplast isolation. Therefore, in vitro plantlets are often preferred as a source of protoplasts because of the advantages conferred by aseptic conditions and a controlled environment. In vitro sources of protoplasts include callus, cotyledons, hypocotyls, embryogenic suspension cultures, leaves, shoots, or somatic embryos.

Before protoplast isolation, plant tissues are cut into small pieces and floated in an osmotically adjusted solution at 20 to 25°C for 1 to 24 h. During this step water moves out of the cells, which causes their contents to shrink and draw away from the cell wall (plasmolysis). This allows cells to retain their integrity after the cell wall is removed. Sometimes it is beneficial to conduct this step at low temperature (4 to 10°C). Next, leaf pieces are incubated in digestive enzymes for removal of the cell walls and middle lamellae. Incubation is conducted in darkness on a shaker (30 to 50 rpm) for 3 to 18 h at 25 to 30°C. Incubation time and temperature vary with species and tissue. Plasmolysis and enzyme incubation steps may be conducted simultaneously.

Following enzyme treatment, the digest is gently swirled to release protoplasts and filtered (50- to 100-µm mesh size) to separate protoplasts from large debris. The protoplast suspension is centrifuged at low-speed (50 × *g*) for about 10 min. Protoplasts and debris collect in the pellet at the bottom of the tube. The supernatant is discarded and protoplasts are resuspended in a high sucrose (flotation) medium. The sucrose layer is overlaid with 1 mL of rinse medium containing mannitol. During centrifugation, viable protoplasts collect at the interface of the two media while debris concentrates in the pellet. Protoplasts are removed from the interface of the two media with a Pasteur pipet and washed two to three times before transfer to culture medium. There are several variations for purification of protoplasts that combine or separate the steps outlined in this general procedure.

After the final wash, protoplasts are suspended in a small volume (~1 ml) of liquid culture medium. A sample is removed and transferred to a hemacytometer to determine protoplast density. Enough culture medium should be added to adjust the protoplast density to 10^4 to 10^6 cells per milliliter. Protoplasts may be cultured in a thin layer of liquid medium embedded in agarose, cultured in liquid medium on top of solid medium, or cultured in agarose droplets suspended in liquid. The type of culture method depends on the species and/or purpose of the experiment. Liquid medium is often used because cell density and osmotic pressure can be efficiently adjusted. In addition, cells of some species are unable to divide in agar-solidified medium. The development of individual cells can be observed if protoplasts are embedded in agarose because they remain stationary.

Protoplast viability may be estimated by staining a sample with fluorescein diacetate (FDA). The stain is mixed with protoplasts and fluorescence observed under UV light. Viable protoplasts actively absorb the stain and exhibit a green fluorescence, whereas nonviable protoplasts do not. Although nonviable mesophyll protoplasts fluoresce red due to autofluorescence of chlorophyll, they are not metabolically active. Another viability stain, Evans blue, does not require fluorescence microscopy to detect protoplast viability.

Freshly isolated protoplasts are spherical (Figure 26.1) because they are unrestricted by a cell wall. Under suitable conditions, viable protoplasts regenerate a new cell wall within 48 to 96 h after isolation. Presence of the cell wall can be determined by a change in shape from spherical to ovoid or by staining cells with Calcofluor White, which detects the presence of wall materials. Stained cells or protoplasts with remnant cell wall material or protoplasts that have resynthesized a new cell wall fluoresce bluish-white under UV

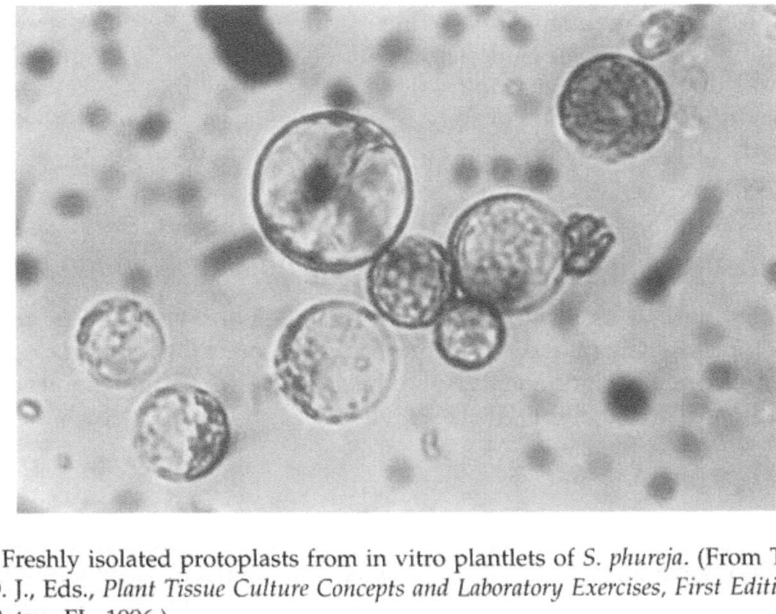

Figure 26.1 Freshly isolated protoplasts from in vitro plantlets of *S. phureja*. (From Trigiano, R. N. and Gray, D. J., Eds., *Plant Tissue Culture Concepts and Laboratory Exercises, First Edition*, CRC Press LLC, Boca Raton, FL, 1996.)

light. Protoplasts without a wall do not fluoresce. Protoplasts that fail to regenerate a wall generally will not divide and eventually die.

Not all healthy protoplasts divide. Therefore, the plating efficiency (PE = number of dividing protoplasts/total number of protoplasts) is used to estimate cell vigor. PE is usually calculated 1 week after protoplast isolation and varies from 0.1 to 80%. However, a PE of 20 to 30% is most common. The first cell division (Figure 26.2) often occurs within 2 to 10 days after isolation. Division of protoplasts is affected by culture medium, osmoticum, plating density, storage conditions, and source tissue.

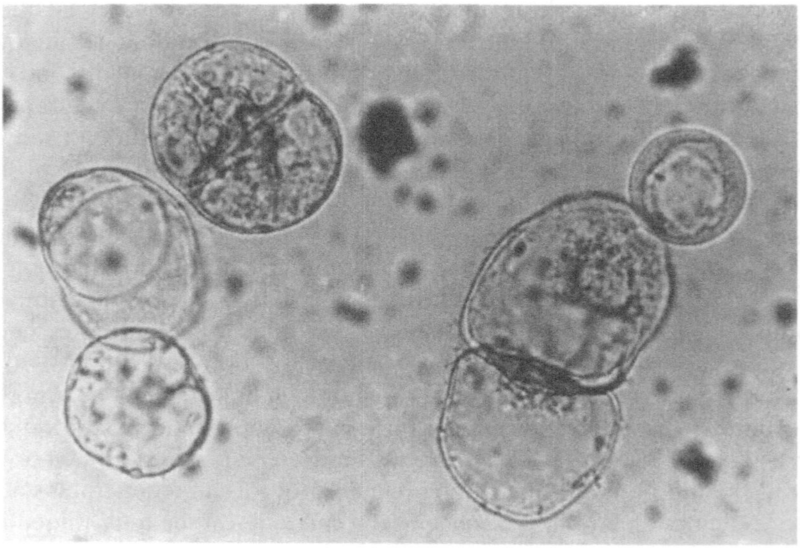

Figure 26.2 First mitotic division after protoplast isolation of *S. phureja*. Notice that the cells are no longer spherical and are more opaque than the freshly isolated protoplasts. (From Cheng, J. and R.E. Veilleux, 1991. *Plant Sci.* 75:257–265.)

After protoplasts have regenerated a new wall and divided, they are diluted with culture medium that has lower osmoticum. Multicellular colonies (Figure 26.3) develop within 14 to 21 days and macroscopic calli (Figure 26.4) within 4 weeks. Plant regeneration occurs by shoot formation or somatic embryogenesis. The first shoots or embryos may be seen as early as 1 month, but may require 6 months to 1 year after protoplast isolation to form.

Factors that influence protoplast growth and development

Modifications to normal tissue culture procedures often are required due to the delicate nature of protoplasts. Changes often include adjustment of the inorganic salt concentration, addition of organic components, vitamins, sugars to control osmolarity, and plant growth regulators (PGR) to stimulate cell division.

Culture media

The best culture medium for protoplasts is often similar to, or slightly modified from, that used for organ regeneration from other explants. Adjustment of ammonium nitrate and calcium is often required to stimulate cell division. Ammonium nitrate is essential to promote

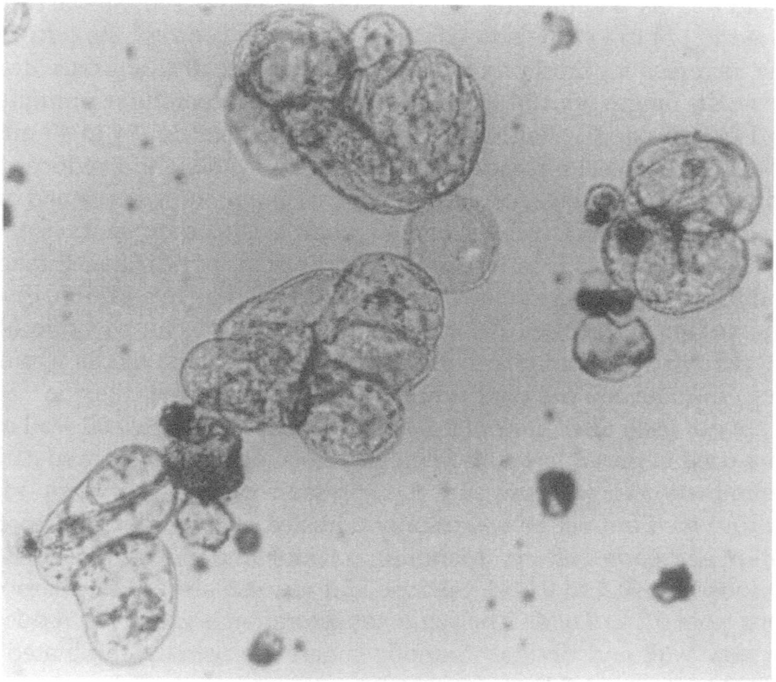

Figure 26.3 Multicellular colonies 7 days after isolation of protoplasts of *S. phureja*. (From Cheng, J. and R.E.Veilleux, 1991. *Plant Sci.* 75:257–265. With permission from Elsevier Science.)

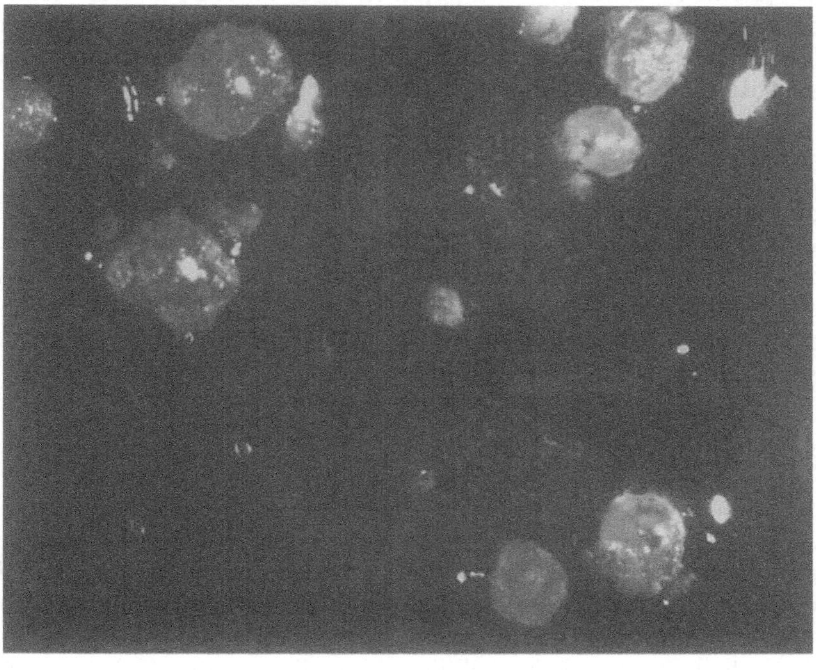

Figure 26.4 Macroscopic p-calli several weeks after isolation of protoplasts of *S. phureja*. (From Trigiano, R. N. and Gray, D. J., Eds., *Plant Tissue Culture Concepts and Laboratory Exercises, First Edition*, CRC Press LLC, Boca Raton, FL, 1996.)

cell division but is toxic to protoplasts at concentrations (20 mM) used in most tissue culture media. To avoid toxicity and promote cell division, the ammonium nitrate concentration is usually lowered to 1/4 to 1/2 the normal concentration. In contrast, the calcium concentration is usually increased for most protoplast procedures. The calcium concentration of most tissue culture media ranges from 0.5 to 3 mM. At this low concentration, protoplasts typically aggregate and brown rapidly. Raising the calcium concentration (14 to 40 mM) promotes early cell division, cell synchronization, decreases aggregation, and reduces browning of protoplasts during the early stages of culture. Normal ammonium nitrate and calcium levels are restored after protoplasts have regenerated a new wall and divided.

Organic components such as inositol, nicotinic acid, pyridoxine, thiamine, glycine, folic acid, and biotin are found in many tissue culture media. In addition, casein hydrolysate, D-Ca-pantothenate, choline chloride, cysteine, malic acid, ascorbic acid, adenine sulfate, riboflavin, and glutamine are often added to protoplast media in small amounts (0.01 to 10 mg/L) to hasten cell wall synthesis and promote cell division. Most of these compounds are not used after protoplasts have regenerated a new cell wall and divided.

Sugars are used in protoplast culture media as osmotic stabilizers and carbon sources. Due to the elimination of wall pressure (i.e., pressure potential), a component of water potential, culture media must be osmotically adjusted to prevent rupture of protoplasts during isolation and early culture. Mannitol, sorbitol, and glucose are used as osmotica at concentrations from 0.3 to 0.7 M. Sucrose and glucose are used as carbon sources at concentrations from 0.2 to 0.6 M. The sugar concentration is gradually reduced after cells synthesize a new wall and divide. Osmotic sugars are usually eliminated by the time macroscopic colonies are visible.

Exogenous PGRs are necessary to promote cell division. Both auxins and cytokinins typically are used. The type and concentration differ with the species and mode of regeneration. Naphthaleneacetic acid (NAA) and 2,4-dichlorophenoxyacetic acid (2,4-D) at 0.45 to 10.7 μM are typical auxins and concentrations. BA and zeatin are usual cytokinins used at concentrations of 2 to 5 μM. Kinetin and zeatin riboside are sometimes used instead of, or in combination with, BA or zeatin at similar concentrations. Coconut milk has been substituted for cytokinin at concentrations of 20 to 40 ml/L.

Culture environment

Following isolation, protoplasts are sensitive to light and should be incubated in darkness or exposed only to filtered light (5 to 30 $\mu mol \cdot m^{-2} \cdot s^{-1}$) until they synthesize a new wall and divide. At this point, they can be exposed to elevated light levels (30 to 300 $\mu mol \cdot m^{-2} \cdot s^{-1}$). White light is often best as blue or red light alone may inhibit shoot formation. Protoplasts are generally incubated at 20 to 25°C. Optimum environmental conditions vary and should be tested for each species.

Somaclonal variation

Spontaneous mutation rates in plants vary depending on the species and DNA sequence. When mutations occur in somatic cells, they generally will not be observed because somatic cells within mature plant tissue do not divide to give rise to mutated sectors of tissue. Protoplast culture can be used as a means of recovering somatic mutations because the culture environment encourages the division of individual somatic cells. About 10^4 to 10^6 protoplasts are cultured per milliliter of medium, with 40,000 to 1,000,000 cells present in a typical 15 × 60-mm petri dish with 4 to 10 ml of medium. Given that each cell is viable and has a chance of regenerating into a plant, the possibility of recovering somatic variants is improved.

Somaclonal variants have been recovered from protoplast cultures of various species. Some variations include changes in leaf and flower morphology, fertility, improved disease resistance, and variation in secondary product production. For example, protoclones, i.e., protoplast-derived plants of an asexually (clonally) propagated crop, have been examined in potato with the hope of obtaining plants with improved horticultural characteristics. Protoclones that exhibited compact growth habit; variation in tuber characteristics such as maturity date, color, russeting, protein content, size, and number of tubers per plant; changes in flowering response to photoperiod; and improved resistance to early blight (*Alternaria solani*) and late blight (*Phytophthora infestans*) were regenerated from protoplasts of potato, *S. tuberosum* L. 'Russet Burbank' (Shepard et al., 1980) and 'Bintje' (Burg et al., 1989). Plants with improved resistance to fusarium wilt (*Fusarium oxysporum* f. sp. *lycoperici* race 2) were obtained by challenging tomato (*Lycopersicon esculentum* Mill. cv. UC 82) protoplasts and protoplast-derived calli (p-calli) with fusaric acid (Shahin and Spivey, 1986), and plants with elevated lanatoside C (a glycoside that yields digoxin, glucose, and acetic acid on hydrolysis) content were regenerated from *Digitalis lanata* J.F. Ehrh. protoplasts (Diettrich et al., 1991).

The most frequent form of somaclonal variation is the regeneration of polyploid plants. This occurs through endopolyploidization (doubling of the chromosome number during mitosis without cytokineses), spontaneous cell fusion, or regeneration from polyploid cells already present in the plant tissue from which protoplasts were isolated. Recovery of aneuploids (plants having a chromosome number that is not a multiple of the monoploid number) is also a problem. The problem of polyploidy or aneuploidy of regenerants may be so great that none of the regenerants contain the original chromosome number. However, more often somaclonal variants of this type comprise only a small portion of the total number of regenerants.

Protoplast fusion and somatic hybridization

Once purified protoplasts have been obtained from two different plant or tissue sources, various treatments can be applied to induce them to fuse together to form hybrid cell lines. The simplest way to do this is to use spontaneously fusogenic protoplasts. Such cell lines have been described in detail in carrot and other systems (Boss, 1987), but they are not common enough for widespread use in plant protoplast research. Generally, chemical agents or electrical manipulation are necessary to induce membrane instability that leads to protoplast fusion.

Polyethylene glycol-mediated protoplast fusion

A number of aqueous solutions have been used to induce chemical fusion of plant protoplasts (see Saunders and Bates, 1987 for review). These include various salt solutions (NaCl, KCl, $NaNO_3$, KNO_3), dextran sulfate, polyvinyl alcohol, lysolethicin, and polyethylene glycol (PEG). Of these fusogenic agents, PEG is used most frequently in conjunction with alkaline pH and high calcium concentrations.

There are a number of steps in the fusion of plant protoplasts using PEG as a chemical facilitator (Figure 26.5). Initially, the cell membranes must be brought into close physical contact (agglutination). As most cell membranes possess a net negative surface charge, adjacent cells with similar charges tend to repel each other. Chemical fusogenic facilitators, such as concanavalin A and immune antisera, promote cell fusion by overcoming the repellent effect of similar net negative surface charges on the protoplast membranes. PEG is a potent cell agglutinator, but also functions as a membrane modifier. There is evidence to suggest that once cell membranes are in close contact, their surface proteins migrate to create areas of lipid-rich regions. It is during this period that the dehydrating effect of PEG on the cell membrane, and the ability of PEG to bind to phospholipids within the membrane, become important. Continuities between adjacent cells begin to form during this cell adhesion period. Subsequently, successive washings with high concentrations of calcium in a buffer with an alkaline pH are done. Experiments involving chemical fusion are presented in Chapter twenty-seven.

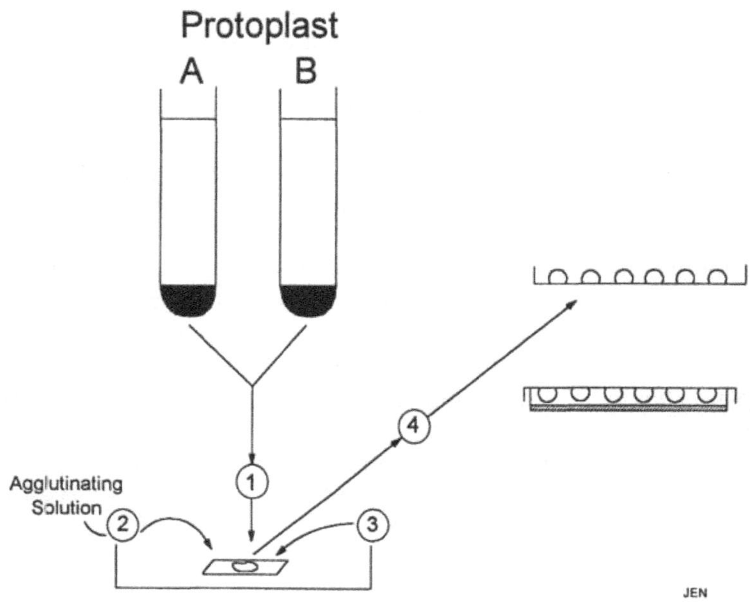

Figure 26.5 Typical protocol for protoplast fusion of potato using PEG. (1) Dilute protoplast mixture to 10^4 cells per milliliter in 0.5 M glucose, 3 mM $CaCl_2$, 0.7 mM KH_2PO_4, pH 5.7; (2) add 3 to 200 µl volumes of agglutinating solution (0.09 mM PEG 6000, 10 mM $CaCl_2$, 0.7 mM KH_2PO_4, pH 5.5); (3) add five 100-µl volumes of diluting solution (50 mM $CaCl_2$, 0.5 M glucose, 50 mM glycine buffer, pH 10.0); (4) wash cells with culture medium and culture in hanging drop suspensions. (From Trigiano, R. N. and Gray, D. J., Eds., *Plant Tissue Culture Concepts and Laboratory Exercises, First Edition*, CRC Press LLC, Boca Raton, FL, 1996.)

Electrofusion-mediated protoplast fusion

Another type of cell fusion that has emerged in recent years involves the manipulation of cell membranes by electrical currents. Electrofusion of plant protoplasts is often preferred over PEG fusion because it does not employ reagents that are toxic to the cells being fused. As with all other procedures, conditions for electrofusion must be optimized for specific cell types to achieve maximum effectiveness. Typically, a yield of 20% or greater fusion products can be obtained by electrofusion of protoplasts compared to less than 1% fusion products with PEG.

Protoplasts must be brought closely together (agglutination) and their cell membranes perturbed by an electrical shock to promote membrane fusion between the two cells. With electrofusion, dielectrophoresis is used to bring the protoplasts into close physical contact. The protoplasts are subjected to a low-voltage (50 to 100 V/cm electrode gap) oscillating sine wave current of approximately 1 MHz frequency that polarizes the cells so that each has a positive and negative area. The positive end of one protoplast is drawn toward the negative area of an adjacent protoplast such that "pearl chains" of protoplasts are formed prior to fusion. The fusion itself may be accomplished by a relatively high voltage (0.5-1.5 kV/cm electrode gap) pulse that destabilizes the cell membranes at specific sites on each protoplast. The electrical pulse does not affect the organelles inside the cell because the current only passes along the cell membranes that are generally more conductive than the cytoplasm. It is necessary to supply the voltage of the fusion pulse for some critical duration at a minimal threshold level in order to achieve successful fusions.

There are several reports of useful somatic hybrid plants being produced by electrofusion. For example, the Colorado potato beetle voraciously eats the leaves of cultivated potato, but will not eat the foliage of a closely related, wild species, *Solanum chacoense* Bitt. Protoplasts from somatic leaf tissue of *S. tuberosum* have been successfully fused with those of *S. chacoense* by electrofusion. The fused cells grew as p-calli for several weeks in tissue culture and were subsequently regenerated into hybrid plants with traits of both "parents." When leaves of the hybrid plants were tested in bioassays with Colorado potato beetles, the insects showed a three-fold greater preference for the cultivated potato than for either *S. chacoense* or the somatic hybrid (Figure 26.6). Although the tubers of these hybrids were unsuitably small for potato production, the somatic hybrids could be used in potato breeding programs to incorporate insect resistance into cultivated lines.

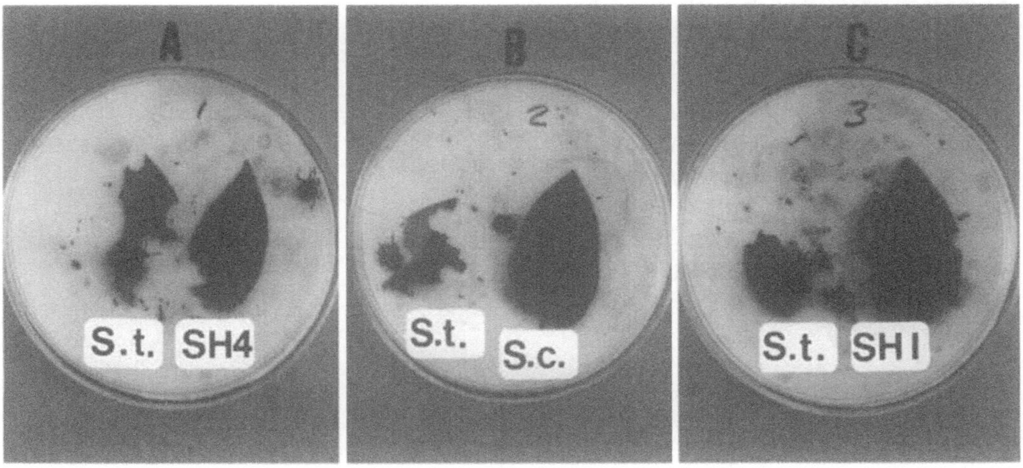

Figure 26.6 An in vitro feeding perference assay using adult Colorado potato beetles feeding on freshly excised leaves of *S. tuberosum* (S.t.), *S. chacoense* (S.c.), or somatic hybrids produced from these two parents. A (somatic hybrid SH4) and C (somatic hybrid SH1) depict typical assays of feeding trials of leaves from two somatic hybrids. Pictures were taken 4 h after initiation of the feeding. In all cases, a preference was shown to the leaf of the control parent, *S. tuberosum* cv. Kennebec, which was included as a choice in each assay. (From Cheng et al., 1995. *In Vitro Cell. Dev. Biol.* 31:90–95.)

Selection of hybrid cell lines

An important feature of any somatic hybridization procedure is the identification and selection of fused cells. Typically, it is best to separate the hybrid cell lines from all other cells in the mixture because cell lines that result from fusion of two or more cells from the same parent or from cells that did not undergo fusion may greatly outnumber the heterokaryotic fusion products. Furthermore, there is no guarantee that just because a cell fusion occurred, nuclear fusion will follow. Fusion of the two-membrane bound nuclei within a single cell and coordination of mitosis involving the chromosomes from what were formerly two independent nuclei are required to form a stable hybrid. Unless there is some identification and selection system incorporated into each of the parental cell lines prior to fusion, collection of hybrid cell lines can be very difficult.

A number of selection systems have been developed based on the specific requirements of prospective parent cell lines. These selective systems include antibiotic-resistant cell lines, herbicide-resistant cell lines, and cell lines with the ability to grow on specific amino acid analogs, among others. It is most convenient to use two parental cell lines with differing requirements as selective screens. By invoking these selective screens on the populations of cells subjected to a fusion treatment, only those fused cells that possess the complementary traits of both parents will thrive.

Unfortunately, most commercially important cultivars of crop plants that are candidates for cell fusion research do not have such sophisticated selection systems. Therefore, a manual selection system based on the internal traits of each parent must be used. In the case of the insect-resistant potato hybrids cited above, the *S. chacoense* cell line contained a visible red-colored anthocyanin pigment, but grew poorly in tissue culture and did not regenerate shoots (Figure 26.7). The *S. tuberosum* cell line did not contain visible anthocyanins but regenerated shoots readily. In addition, callus tissue from hybrid fusions grew more vigorously than calli from either parent. As a result putative somatic hybrids were identified by selecting the most rapidly growing p-calli that regenerated shoots with anthocyanins.

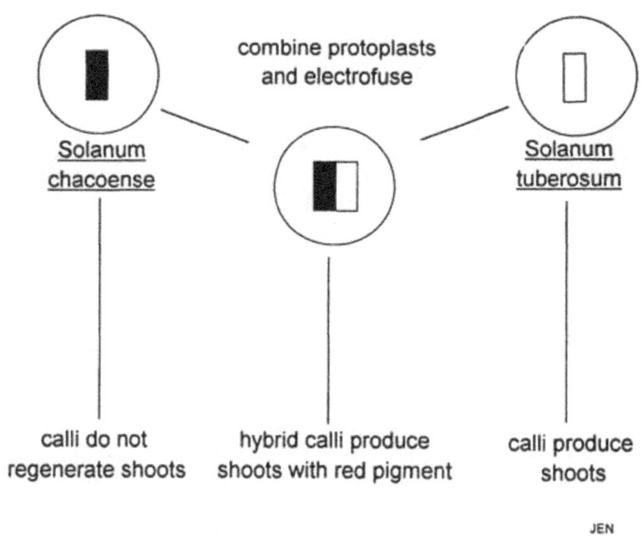

Figure 26.7 Schematic diagram of selection of calli and shoots produced from electrofused protoplasts from *S. chacoense* and *S. tuberosum*. *S. chacoense* protoplasts regenerate calli with red pigment but do not regenerate shoots from p-calli. *S. tuberosum* protoplasts regenerate calli that produce shoots without red pigmentation. Calli from fused cells contain anthocyanin and produce shoots with red pigment. (From Trigiano, R. N. and Gray, D. J., Eds., *Plant Tissue Culture Concepts and Laboratory Exercises, First Edition*, CRC Press LLC, Boca Raton, FL, 1996.)

Even when such obvious differences are lacking, parental cell lines can be given different properties by the use of vital stains (stains that do not adversely affect the viability of cells) that can be recognized by fluorescence detectors. Two such stains, fluorescein isothiocyanate (FITC) and rhodamine isothiocyanate (RITC), have been used successfully on plant protoplasts. FITC stains cells green and RITC stains cells red when viewed using a fluorescence microscope. Fusion products that have both FITC and RITC fluoresce yellow. The fused (yellow) cells can be isolated mechanically using a micromanipulator and cultured individually on an enriched tissue culture medium supplemented either with nurse cultures or "conditioned" medium. Large numbers of fused cells marked with fluorescent dyes can be isolated more efficiently using a laser-activated flow cytometer, also known as a cell sorter. Cells are sorted according to the wavelength of their fluorescent emission, collected, and cultured in enriched medium. Although these procedures have been used successfully, it may be difficult to maintain sterile conditions throughout the lengthy process.

Putative somatic hybrid plants can also be identified by counting the number of chloroplasts present in the guard cells of leaves. Hybrid plants typically contain double the number of chloroplasts per guard cell as unfused parents. Although this is a rapid method of obtaining data about the hybrid nature of regenerated plants, occasionally true hybrids lose the characteristic of doubled chloroplasts.

Genetic transformation

Genetic transformation of single cells has been accomplished by co-cultivating protoplasts with *Agrobacterium tumefaciens* or by direct DNA transfer using polycationic chemicals, electroporation, liposomes, microinjection, or sonication (Sawahel and Cove 1992; Fish and Dandekar, 1993). Transformation of plant protoplasts using polycationic chemicals and/or electroporation has been most often reported.

Agrobacterium-*mediated transformation*

As discussed in Chapter thirty-one, *A. tumefaciens*, the causative agent of the crown gall disease, stimulates gall formation by inserting a portion of the tumor-inducing (Ti) plasmid into plant cells. For *Agrobacterium*-mediated gene transfer, protoplasts are isolated and cultured for 2 to 3 days prior to co-cultivation with *Agrobacterium* cells. Protoplasts are mixed with *Agrobacterium* at a rate of 10^4 to 10^6 plant cells to 10^6 to 10^8 bacterial cells per milliliter and incubated together for 24 to 30 h. After co-cultivation, antibiotics that inhibit growth of the bacterium are added to the culture medium. Treated plant cells are allowed to grow to the multicellular stage before the addition of selective antibiotics (kanamycin, hygromycin, etc.) that inhibit growth of nontransformed cells. Transformed cells grown on selective medium are able to regenerate plants with the introduced gene. Transformation frequencies usually range from 0.26 to 15.3%. Use of *Agrobacterium* to transform plant protoplasts has lost favor in recent years because not all species, especially cereals and grasses, are amenable to *Agrobacterium*-mediated transformation and it is sometimes difficult to inhibit the growth of *Agrobacterium* following co-cultivation. In addition, use of simpler regenerative culture systems have eliminated some of the need to use protoplasts for transformation (see Chapters thirty-two and thirty-four).

Polycationic chemical-mediated transformation

The most commonly used polycationic chemical for direct DNA transformation is polyethylene glycol (PEG). It has been used to transform protoplasts from a wide variety of plant species. Transformation frequencies vary from 0.8×10^{-6} to 6×10^{-3} depending on the plant species.

To transform plant cells using PEG, protoplasts are isolated and adjusted to the proper cell density before adding transforming DNA and PEG. The mixture is incubated for 10 to 30 min before dilution with culture medium. With some species it is beneficial to subject protoplasts to a heat shock treatment (~45°C) prior to the addition of DNA. It is not completely known how PEG promotes direct DNA uptake. It is thought that the polycationic charge of PEG interacts with negatively charged DNA to form a positively charged complex that binds to the anionic plasmalemma. Absorption of DNA into cells occurs through endocytosis or active uptake. PEG-mediated transformation is a popular method for transforming plant protoplasts because it has low cell toxicity, promotes high transformation frequencies, increases membrane permeability, protects DNA against degradation by nucleases, is independent of DNA structure (supercoiled, circular, or linear), allows integration of DNA into the host genome without a carrier, and does not have a limited host range.

Hexademethrine bromide (Polybrene) acts similarly to PEG in that it promotes gene transfer by interacting with the anionic plasmalemma to develop regions of positive charge that react with negatively charged DNA. The polybrene method has been used to transform maize (*Zea mays* L. 'Black Mexican Sweet' [BMS]) protoplasts with genomic DNA from BMS kan® cell line 28 (Antonelli and Stadler, 1990). Polybrene-mediated direct DNA uptake is a promising method for transforming plant protoplasts because it has low cell toxicity, promotes high transformation frequencies, and is not limited by DNA size or structure.

Electroporation-mediated transformation

In electroporation, an electric pulse is used to induce DNA uptake by plant protoplasts. It is thought that DNA uptake occurs through pores in the cell membrane generated in response to the electric pulse. In general, transforming DNA is mixed with plant protoplasts prior to one or more electric pulses of 200 to 1000 V/cm and 16 to 1000 µF capacitance. Short pulses (typically microseconds) are typically less lethal to protoplasts than long (millisecond) pulses. Optimum electroporation parameters vary and should be tested for each new species.

Transformation frequencies and cell viability are generally lower for electroporation than polycationic chemical methods but can be improved by adding PEG (4 to 8%) to the electroporation medium. Stable transformation using electroporation either with or without PEG has been obtained in protoplasts of many plant species. Genes incorporated into plant protoplasts using electroporation have included antibiotic resistance, *GUS A*, or phosphoenolpyruvate from *Sorghum bicolor* (L.) Moench. Electroporation is a fast and efficient means of delivering DNA to plant cells and protoplasts, is not limited by the size or structure of transforming DNA, and does not have a limited host range. However, electroporation is often lethal to protoplasts, resulting in low viability.

Sonication-, liposome-, microinjection-, optoporation-, and laser-mediated transformation

Other methods of incorporating foreign DNA into plant protoplasts show considerable promise. During sonication, protoplasts are mixed with DNA and given sonic pulses (500 to 900 msec at 0.65 to 1.6 V/cm) that disrupt the cell membrane and allow DNA to enter cells through pores created in response to the electric pulse. Liposomes are hydrogenated phospholipids in which DNA or RNA is encapsulated and delivered directly to plant cells through endocytosis or active uptake (Antonelli and Stadler, 1990). Microinjection uses a fine capillary needle to deliver DNA, chromosomes, or nuclei directly into plant cells in a manner that preserves cell integrity and viability. Optoporation works similar to microinjection. A laser microbeam (355 nm) can be used to puncture holes in the cell wall and membrane and the difference in osmotic potential between the interior and exterior of osmotically treated cells facilitates movement of plasmid DNA through the membrane openings (Guo et al., 1995). The benefit of optoporation over microinjection is that holes made by the laser are smaller and possess less of a risk to cell survival. Lasers can also be used to cut, splice, and move large fragments of chromosomal DNA into cells (Berns, 1998). Laser scissors have been used to make specific cuts on chromosomes to excise large fragments of cell DNA. Laser tweezers have been developed that remove DNA fragments from cells and splice the remaining chromosome pieces together. This technology has potential applications for transformation of plant protoplasts since laser scissors and tweezers could be used to insert large DNA fragments into specific regions on plant chromosomes.

Conclusions and future prospects

Plants with improved characteristics have been obtained through somaclonal variation and somatic hybridization. Due to the limited host range of *Agrobacterium*, direct gene transfer utilizing protoplasts can be used to insert beneficial genes into crops, especially cereals and grasses, that are not amenable to *Agrobacterium*-mediated transformation. These methods can be used to create plants with improved agronomic and horticultural characteristics as well as increased insect and disease resistance for use by plant breeders to develop improved cultivars.

Literature cited

Antonelli, N. M. and J. Stadler. 1990. Genomic DNA can be used with cationic methods for highly efficient transformation of maize protoplasts. *Theor. Appl. Genet.* 80:395–401.

Berns, M.W. 1998. Cell micromanipulation with laser scissors and tweezers. *In Vitro Cell. Dev. Biol. Plant* 34:24A (abstract).

Bhojwani, S.S. and M.K. Razdan. 1983. *Plant Tissue Culture: Theory and Practice*. Elsevier, Amsterdam, pp. 237–260.

Boss, W. 1987. Fusion-permissive protoplasts. A plant system for studying cell fusion. pp. 145–166. In: *Cell Fusion*. A.E. Sowers, Ed. Plenum Press, New York.

Burg, H. C. J., K. S. Ramulu, G. M. M. Bredemeijer, S. Roest, P. Dijkhuis, J. J. van Hoogen, and A. Houwing. 1989. Patterns of phenotypic and tuber protein variation in plants derived from protoplasts of potato (*Solanum tuberosum* L. cv. Bintje). *Plant Sci.* 64:113–124.

Cheng, J. and R.E. Veilleux. 1991. Genetic analysis of protoplast culturability in *Solanum phureja*. *Plant Sci.* 75:257–265.

Cocking, E.C. 1960. A method for the isolation of plant protoplasts and vacuoles. *Nature* 187:927–929.

Diettrich, B., V. Schneider, and M. Luckner. 1991. High variation in cardenolide content of plants regenerated from protoplasts of the embryogenic cell strain VII of *Digitalis lanata*. *J. Plant Physiol.* 139:199–204.

Fisk, H.J. and A.M. Dandekar. 1993. The introduction and expression of transgenes in plants. *Scientia Hort.* 55:5–36.

Guo, Y., H. Liang, and M.W. Berns. 1995. Laser-mediated gene transfer in rice. *Physiol. Plant.* 93:19–24.

Saunders, J. A. and G. W. Bates. 1987. Chemically induced fusion of plant protoplasts. pp. 497–520 In: *Cell Fusion*. A.E. Sowers, Ed. Plenum Press, New York.

Sawahel, W.A. and D.J. Cove. 1992. Gene transfer strategies in plants. *Biotech. Adv.* 10:393–412.

Shahin, E.A. and R. Spivey. 1986. A single dominant gene for *Fusarium* wilt resistance in protoplast-derived tomato plants. *Theor. Appl. Genet.* 73:164–169.

Shepard, J. F., D. Bidney, and E. Shahin. 1980. Potato protoplasts in crop improvement. *Science* 208:17–24.

chapter twenty-seven

Isolation, culture, and fusion of tobacco and potato protoplasts

Richard E. Veilleux and Michael E. Compton

Isolation of plant protoplasts, followed by fusion and subsequent regeneration of somatic hybrids, are remarkable techniques that allow exchange of genetic information between sexually incompatible species. Success in application of these techniques was first achieved with tobacco (*Nicotiana tabacum* L.) for which Nagata and Takebe (1970) developed a protocol for cell wall regrowth and cell division, followed eventually by plant regeneration (Nagata and Takebe, 1970). Carlson et al. (1972) reported the first interspecific somatic hybrids. The list of species amenable to these techniques has grown to include many of the major crops. However, the source of explants for protoplast extraction, the choice of genotype, and the specific conditions of isolation and medium changes must be customized for each species.

The objectives of somatic hybridization may be to develop hybrids with the complete genome of the two "parents" (somatic or parasexual hybrids), partial genome transfer between species (asymmetric hybrids), or cytoplasmic transfer to develop "cybrids" with the nucleus of one parent and the cytoplasm of the other. Somatic hybrids between closely related species are generally polyploid and have the potential of fertility. Potato has been the object of somatic hybridization with a range of related species from sexually compatible tuber-bearing *Solanum* relatives to sexually incompatible nontuberous *Solanum*, and even to tomato (*Lycopersicon esculentum* Mill.), nightshade (*Solanum nigrum* L.), and tobacco. Both its plasticity in tissue culture and vegetative propagation have made it attractive to geneticists wishing to examine the potential of somatic hybridization in an important economic crop. Somatic hybrids between potato (*S. tuberosum* L.) and a weedy, nontuber-bearing relative, *S. brevidens* Phil., contained the additive number of chromosomes of the two parents (i.e., 72 with 48 from potato and 24 from *S. brevidens*). These somatic hybrids could be backcrossed to potato, resulting in fertile tuber-producing plants with considerable breeding potential as sources of disease resistance (Helgeson et al., 1986).

As the taxonomic divergence between species increases, the possibility of obtaining fertile somatic hybrids decreases and the likelihood of chromosome elimination, i.e., partial or complete spontaneous loss of chromosomes of one or the other parent, increases. Such chromosome elimination results in aneuploid regenerants or asymmetric hybrids. A recent example has been the regeneration of plants following protoplast fusion between tobacco and carrot (*Daucus carota* L.) reported by Kisaka and Kameya (1994). The somatic hybrids contained between 23 and 32 chromosomes instead of the additive number of 66 (48 from tobacco and 18 from carrot).

0-8493-2029-1/00/$0.00+$.50

As opposed to somatic hybrids in which nuclear fusion has occurred, cybrids between even distantly related parents are more likely to be fertile (Rose et al., 1990). Nuclei from protoplasts of one fusion partner can be inactivated by irradiation or mutagenic treatment before fusion to prevent transmission of the nuclear genome. Protoplasts of the other fusion partner may be metabolically inhibited by iodoacetate treatment. In this way, unfused protoplasts of either parent will not regenerate and the preponderance of regenerants will be cybrids. Such a scheme has recently been successful to develop cybrids between potato and either of the two sexually incompatible wild species, *S. bulbocastanum* Dun. or *S. pinnatisectum* Dun. (Sidorov et al., 1994). Complete lack of transmission of nuclear material of the irradiated protoplasts cannot be guaranteed, such that occasional asymmetric hybrids may result (Matibiri and Mantell, 1994).

The following exercises illustrate the basic procedures of protoplast isolation and fusion. The more elaborate schemes mentioned above may be built upon these exercises. The conditions under which donor plants are grown can greatly affect the success with protoplast yield and culture. This is especially true of potato. Therefore, the first exercise has been designed for tobacco protoplast isolation because it is much more tolerant of a range of conditions, making it easier to handle and more likely to succeed as a student lab. Protoplast fusion can be induced in many different ways (see Chapter twenty-six). The exercise described in Experiment 2 is a small-scale chemical fusion of protoplasts using polyethylene glycol and high Ca^{2+} to facilitate membrane fusion between adjacent protoplasts.

General considerations

Tobacco

Most cultivars of tobacco will respond well to protoplast isolation and culture. However, Nagata and Takebe's (1970) initial protoplast culture of tobacco was hindered by inadvertent choice of a cultivar prone to seasonal fluctuations that prevented repeatability of experiments. We have often used 'Samsun,' an older cultivar that has been used extensively in tissue culture research. Seeds can be obtained from: Dr. Verne Sisson, Oxford Tobacco Research Station, P.O. Box 1555, Oxford, NC 27565-1555 (Tel. 919 693-5151 ext. 228).

Potato

Response of potato to protoplast manipulation is under genetic control (Cheng and Veilleux, 1991) so care must be taken in selection of genotypes; some are completely recalcitrant in protoplast culture, whereas others respond easily. In a comparison of 36 cultivars or breeding lines for ease of manipulation in protoplast culture, Haberlach et al. (1985) listed *S. tuberosum* cv. Russet Burbank, US-W 5328.4, US-W 9546.46, PI 423654, *S. etuberosum* Lindl. PIs 245924 and 245939, and *S. brevidens* Phil. 245763 as highly regenerable. (Of these, Russet Burbank is tetraploid [2n = 2x = 48] and the rest are diploid [2n = 2x = 24].) Many other genotypes yielded few or no p-calli or would not regenerate from p-calli. In vitro copies of regenerable cultivars or species may be obtained from the NRSP-6 Potato Introduction Project, Sturgeon Bay, WI 54235. Several genotypes should be used such that each student or student team will extract protoplasts from one genotype and then fuse his/her protoplasts with those extracted by another student or student team. It may be desirable to use one highly regenerable and one recalcitrant genotype for protoplast fusion. In this way, unfused protoplasts of one parent would not be expected to regenerate in the fusion experiments. Haberlach et al. (1985) list *S. tuberosum* PI 256977 (tetraploid), Kennebec haploid 731.3 (diploid), US-W 9587.24 (diploid), and 77-16 (diploid) as unable to form p-calli after protoplast isolation and culture.

Growth of plants

Tobacco

The plants should be approximately 50 to 60 days from planting at the time of protoplast isolation. The seedlings are slow to start so seeds can be planted in a single pot for the first month and then transplanted individually to 3.7-L nursery pots. One plant per student should be sufficient. Any of the soilless potting mixes is suitable. Tobacco is a heavy feeder so weekly fertilization with a 20:20:20 (N:P:K) liquid fertilizer is recommended. Light and temperature conditions in the greenhouse are not especially critical. Tobacco is susceptible to infestation with whitefly under greenhouse conditions, so treatment with a systemic insecticide (e.g., Marathon, Olympic Horticultural Products, PO Box 230, Mainland, PA 19451) shortly after transplanting is recommended. Otherwise, microbial contamination is very likely in protoplast cultures.

Potato

The best source of plant material for protoplast manipulation of potato is in vitro plantlets. Therefore, an exercise in making approximately 10 to 15 copies using single nodes of previously established in vitro plantlets of the selected genotypes on propagation medium for each student should be planned approximately 3 weeks before the protoplast fusion lab. Obviously, adequate in vitro multiplication will have been necessary 3 to 4 weeks prior to this round of multiplication. In vitro potato can harbor bacterial contaminants that are not apparent during routine subculture on basal medium. However, such contaminants often become obvious during protoplast manipulations, resulting in failed experiments. A precaution against such "cryptic" contamination is to use the antibiotic, cefotaxime, in the propagation medium (see Table 28.4) used just prior to the transfer when students will multiply their genotypes.

Explant preparation

Tobacco

Fully expanded young leaves yield the best protoplasts. Leaves should be harvested and brought to the lab on the day of the exercise.

Potato

In vitro potato cultures should be placed in the dark in a refrigerator for 48 h prior to protoplast extraction. Examine cultures carefully and remove any obviously contaminated ones.

Culture medium

For tobacco protoplast isolation, five media are needed (Tables 27.1 to 27.3). Three of these are simple variations of protoplast isolation (PI) medium (Table 27.2) and can be made from the same preparation of PI (preplasmolysis = PI + 13% mannitol; flotation = PI + 20% sucrose; and rinse = PI + 10% mannitol). The other two media (enzyme solution and culture medium) have a different basal composition so must be prepared separately. Only the enzyme solution requires filter sterilization; the other four media can be autoclaved. It is advisable to dispense each medium prior to autoclaving into properly labeled 25 × 150-ml culture tubes with Magenta two-way caps in aliquots sufficient for each student or

Table 27.1 Enzyme Medium for Tobacco Laboratory

Component	%
Onozuka R10 cellulase	0.5
Onozuka R10 macerozyme	0.1
Mannitol	13.0
pH 5.8	

From Trigiano, R. N. and Gray, D. J., Eds., *Plant Tissue Culture Concepts and Laboratory Exercises, First Edition*, CRC Press LLC, Boca Raton, FL, 1996.

Table 27.2 Protoplast Isolation (PI) Medium for Tobacco Laboratory

Component	mg/L
$CaCl_2 \cdot H_2O$	1480.0
KH_2PO_4	27.2
KNO_3	101.0
$MgSO_4 \cdot 7H_2O$	246.0
$CuSO_45 \cdot H_2O$	0.025
KI	0.16
pH 5.8	

From Trigiano, R. N. and Gray, D. J., Eds., *Plant Tissue Culture Concepts and Laboratory Exercises, First Edition*, CRC Press LLC, Boca Raton, FL, 1996.

Table 27.3 Protoplast culture (PC) Medium for Tobacco Laboratory

Component	mg/L	Component	mg/L
$Ca(H_2PO_4) \cdot H_2O$	100.0	Sequestrene 330	28.0
$CaCl_2 \cdot 2H_2O$	450.0	Sucrose	10,000
KNO_3	2500.0	Glucose	18,000
$MgSO_4 \cdot 7H_2O$	250.0	Mannitol	100,000
$NaH_2PO_4 \cdot 2H_2O$	170.0	Inositol	100.0
$(NH_4)_2SO_4$	134.0	Nicotinic acid	1.0
$CoCl_2 \cdot 6H_2O$	0.025	Pyridoxine-HCl	1.0
$CuSO_4 \cdot 5H_2O$	0.025	Thiamine-HCl	10.0
H_3BO_3	3.0	2,4-D	0.1
KI	0.75	NAA	1.0
$MnSO_4 \cdot 4H_2O$	13.2	BA	1.0
$Na_2MoO_4 \cdot 2H_2O$	0.25	pH	5.8
$ZnSO_4 \cdot 7H_2O$	2.0		

From Trigiano, R. N. and Gray, D. J., Eds., *Plant Tissue Culture Concepts and Laboratory Exercises, First Edition*, CRC Press LLC, Boca Raton, FL, 1996.

student team. If different students or student teams share a common supply of medium, contamination is much more likely.

For potato protoplast isolation and fusion, nine different media (Tables 27.4 to 27.12) are required, including the protoplast propagation medium for the micropropagation step conducted 3 weeks before isolation. Four of these (propagation, preplasmolysis, enzyme, and culture) are variations of Murashige and Skoog (1962) basal medium, such that stocks 2, 3, 4, and 5 for MS medium preparation can be used. A modified MS1 stock consisting of 9.5 g KNO_3, 0.85 g KH_2PO_4, and 0.903 g $MgSO_4$ per 100 ml is necessary; this stock differs from routine MS1 by elimination of NH_4NO_3 because of the ammonium sensitivity of protoplasts. Four of these media (preplasmolysis, enzyme, culture, and Ca^{2+}) require filter sterilization. The other five (propagation, flotation, rinse, CPW 13M, and PEG 22.5) can be autoclaved. If the propagation medium contains cefotaxime, this component must be filter sterilized separately and added to the autoclaved propagation medium as it cools. As with the tobacco experiment, it is advisable to dispense these media (prior to autoclaving) into aliquots sufficient for each student or student team. For filter-sterilized media, dispensing into autoclaved culture tubes after filter sterilization is recommended.

Table 27.4 Propagation Medium for In Vitro Potato Plantlets

Component	Amount/L	Component	Amount/L
MS1 and MS2	20 ml	Casein hydrolysate	500 mg
S3, MS4, MS5	10 ml	Cefotaxime[a]	250 mg
Sucrose	20 g	Agar	7 g
Myo-inositol	100 mg	pH 5.7	
KH_2PO_4	170 mg		

[a] Cefotaxime is optional but is especially recommended for propagation of stock plantlets one or more clonal generations before propagation for protoplast extraction to reduce bacterial contamination. It is not autoclavable so must be filter sterilized and added to the autoclaved medium as it cools.

From Trigiano, R. N. and Gray, D. J., Eds., *Plant Tissue Culture Concepts and Laboratory Exercises, First Edition*, CRC Press LLC, Boca Raton, FL, 1996.

Table 27.5 Preplasmolysis Medium for Potato Laboratory

Component	Amount/100 ml of Medium	Component	Amount/100 ml of Medium
MS2	6.7 ml	Mannitol	9 g
MS5	1 ml	MES	58.6 mg
KH_2PO_4	3.3 mg	pH 5.8	
KNO_3	10.1 mg	Filter sterilize	
$MgSO_4$	600 mg		

From Trigiano, R. N. and Gray, D. J., Eds., *Plant Tissue Culture Concepts and Laboratory Exercises, First Edition*, CRC Press LLC, Boca Raton, FL, 1996.

Table 27.6 Enzyme Medium for Potato Laboratory

Component	Amount/100 ml of Medium	Component	Amount/100 ml of Medium
Mannitol	7.3 g	Cellulase Onozuka R-10	1 g
Glucose	1.8 g	Macerozyme R-10	0.1 g
Modified MS1[a]	1 mL	pH 5.8	
MS2	3 mL	Filter sterilize, make fresh on the day of isolation	
MS3, MS4, MS5	0.5 ml each		

[a] MMS1 has 9.5 g KNO_3, 0.85 g KH_2PO_4, and 0.903 g $MgSO_4$ per 100 ml.

From Trigiano, R. N. and Gray, D. J., Eds., *Plant Tissue Culture Concepts and Laboratory Exercises, First Edition*, CRC Press LLC, Boca Raton, FL, 1996.

Table 27.7 Rinse Medium for Potato Laboratory

Component	Amount/200 ml of Medium
KCl	4.4 g
MS2	6 ml
pH 5.8	
Autoclave	

From Trigiano, R. N. and Gray, D. J., Eds., *Plant Tissue Culture Concepts and Laboratory Exercises, First Edition*, CRC Press LLC, Boca Raton, FL, 1996.

Table 27.8 Flotation Medium for Potato Laboratory

Component	Amount/100 ml of Medium
Sucrose	17.1 g
MS2	3 ml
pH 5.8	
Autoclave	

From Trigiano, R. N. and Gray, D. J., Eds., *Plant Tissue Culture Concepts and Laboratory Exercises, First Edition*, CRC Press LLC, Boca Raton, FL, 1996.

Table 27.9 Culture Medium for Potato Laboratory

Component	Amount/100 ml of Medium
Modified MS1	1 ml
MS3, MS4, MS5	0.5 ml each
MS2	3 ml
Glucose	3 g
Sucrose	1 g
Sorbitol	5 g
Casein hydrolysate	50 mg
Glutamine	10 mg
Serine	1 mg
Myo-inositol	10 mg
Thiamine	1 mg
NAA	0.125 mg
2,4-D	0.025 mg
Zeatin	0.1 mg
pH 5.8	
Filter sterilize	

From Trigiano, R. N. and Gray, D. J., Eds., *Plant Tissue Culture Concepts and Laboratory Exercises, First Edition*, CRC Press LLC, Boca Raton, FL, 1996.

Table 27.10 CPW 13M Medium for Potato Laboratory

Component	Amount/L
KH_2PO_4	27.2 mg
KNO_3	0.101 g
$CaCl_2 \cdot 2H_2O$	1.48 g
$MgSO_4 \cdot 7H_2O$	246 mg
KI	0.16 mg[a]
$CuSO_4 \cdot 5H_2O$	0.025 mg[b]
Mannitol	130 g
pH 5.8	
Autoclave	

[a] 8 mg/50 ml stock — use 1 ml.

[b] 2.5 mg/50 ml stock — use 0.5 ml.

From Trigiano, R. N. and Gray, D. J., Eds., *Plant Tissue Culture Concepts and Laboratory Exercises, First Edition,* CRC Press LLC, Boca Raton, FL, 1996.

Table 27.11 Fusogen PEG 22.5 Medium for Potato Laboratory

Component	Amount/100 ml	%
PEG 8000	22.5 g	22.5
Sucrose	1.8 g	1.8
$CaCl_2 \cdot 2H_2O$	0.150 g	0.15
KH_2PO_4	0.01 g	0.01
pH 5.8[a]		
Autoclave		

[a] Use 1 N KOH buffer.

From Trigiano, R. N. and Gray, D. J., Eds., *Plant Tissue Culture Concepts and Laboratory Exercises, First Edition,* CRC Press LLC, Boca Raton, FL, 1996.

Table 27.12 Ca^{2+} Washing for Potato Laboratory

Component	g/100 ml	%
$CaCl_2 \cdot 2H_2O$	0.74 g	0.74
Glycine	0.38 g	0.38
Sucrose	11.0 g	11.0
pH 5.8		
Filter sterilize		

From Trigiano, R. N. and Gray, D. J., Eds., *Plant Tissue Culture Concepts and Laboratory Exercises, First Edition,* CRC Press LLC, Boca Raton, FL, 1996.

Exercises

Experiment 1. Protoplast isolation and culture of tobacco

Materials

The following are needed for each student or student team:

- Five sterile 300-ml fleakers to surface sterilize leaves (one for EtOH, one for bleach, and three for sterile water)
- 100 ml 20% commercial bleach
- Hemacytometer
- Centrifuge
- Fluorescein diacetate (FDA) stain (5 mg FDA [product no. F 7378, Sigma Chemical Co., St. Louis, MO] in 1 ml acetone)
- A 250-ml beaker containing a 63-µm filter. (Such a filter can be constructed by ordering 63-µm wire mesh from the Newark Wire Mesh Co., Newark, NJ, cutting the wire mesh into squares just large enough to cover the cap end of a Nalgene 125-ml autoclavable animal watering bottle, and fusing the wire mesh to the plastic on a hotplate covered with aluminum foil. After the mesh has embedded into the melted plastic, the excess mesh can be trimmed with scissors and the bottom of the bottle can be cut off with a scalpel.) The filter can be autoclaved in the beaker covered with foil and used repeatedly.) Alternatively, nylon mesh of approximately the same size can be used but most nylon cannot be autoclaved so must be sterilized by an ethanol soak.
- 15-ml centrifuge tubes
- 10 ml enzyme solution
- 20 ml preplasmolysis medium
- 10 ml flotation medium
- 30 ml rinse medium
- 25 ml protoplast culture medium
- Calcofluor White stain (0.1% Fluorescent Brightener 28 [product no. F 6259, Sigma Chemical Co., St. Louis, MO] in 0.7 M mannitol)

Follow the protocols outlined in Procedure 27.1 to complete this experiment.

Procedure 27.1
Tobacco protoplast extraction and culture.

Step	Instructions and comments
1	**Day 1:** Surface disinfest one tobacco leaf by immersion in 70% ethanol for 30 sec and transfer to 20% bleach/Tween for 15 min. Rinse three times in sterile, distilled water.
2	Place the tobacco leaf in a sterile petri dish and peel sections of the lower epidermis from the leaf and/or cut the leaf into 1-mm strips. Even though the peeling is tedious, it is more effective than cutting. The epidermis can be stripped by lifting at a leaf vein with fine forceps and pulling gently across the lamina. This should be repeated until at least one half of the epidermis has been removed.
3	Add 20 ml preplasmolysis medium and incubate the peeled leaf for 1 to 8 h.
4	Remove preplasmolysis solution using a sterile Pasteur pipet and replace with 10 ml enzyme solution. Wrap the petri dish with parafilm, label, and place in the dark overnight at low-speed (40 rpm) on a shaker.

Procedure 27.1 continued
Tobacco protoplast extraction and culture.

Step	Instructions and comments
5	**Day 2:** Gently agitate the plate of protoplasts in enzyme solution to facilitate the release of protoplasts. Transfer protoplast suspension to the 63-μm filter in a 250-ml beaker using a Pasteur pipet.
6	Add 3 ml rinse medium to the plate with leaf debris, shake vigorously, and combine with the protoplast/enzyme mixture in the 250-ml beaker.
7	Use an additional 2 ml rinse medium to rinse protoplasts off the filter into the beaker by holding the filter with forceps as you rinse. Transfer the protoplast/enzyme mixture to a 15-ml sterile centrifuge tube and spin at $50 \times g$ for 10 min.
8	Remove the supernatant and resuspend the pellet in 10 ml flotation medium, add 1 ml rinse medium to the top of the protoplasts in flotation medium dropwise so as not to mix the two media but rather to have distinct layers of medium (dripping the rinse medium down the side of the tube works better than splashing it on the surface of the flotation medium), and recentrifuge for 10 min at $50 \times g$. Viable protoplasts should float to the surface.
9	Remove the band of green protoplasts with a Pasteur pipet, transfer it to a fresh centrifuge tube, add 10 ml rinse medium, and recentrifuge. Protoplasts should sink to the bottom and form a pellet.
10	Rinse the protoplasts again in 10 ml rinse medium. (This second rinse can be omitted if the protoplast pellet is small, i.e., barely visible.) Remove the supernatant and resuspend the protoplasts in approximately 1 ml culture medium.
11	Place a drop of protoplast on a hemacytometer and estimate protoplast density (number of cells/grid × 10,000). To culture the protoplasts, dilute to 50,000 cells/ml with protoplast culture (PC) medium and transfer to a sterile petri dish.
12	Stain a protoplast sample with FDA stain by diluting one drop of FDA stain with 25 drops of culture medium. Add one drop of protoplast to one drop of diluted FDA stain on a microscope slide. Observe cell viability under a fluorescence microscope. Viable protoplasts fluoresce green; dead cells fluoresce red due to autofluorescence of chlorophyll. Record the frequency of viable protoplasts in a sample of 100 to 200 cells. Seal, label, and culture in the dark at 28°C overnight.
13	**Day 3:** Move culture to low light (10 to 20 $\mu mol \cdot sec^{-1} \cdot m^{-2}$) at a 16-h photoperiod (e.g., under cool white fluorescent light covered by a layer of cheesecloth) for 2 days.
14	**Day 5:** Move culture to higher light intensity (50 to 75 $\mu mol \cdot sec^{-1} \cdot m^{-2}$) by removing the layer of cheesecloth.
15	**Day 7:** Examine cultures for cell wall regrowth by staining a sample with Calcofluor White stain and observing under a fluorescence microscope. Is there any evidence of cell wall regrowth or cell division evident by blue fluorescence? If so, record the plating efficiency of the protoplast cultures, i.e., the number of dividing cells divided by the total number of cultured cells, in a sample of 100 to 200 cultured protoplasts. Is there evidence of contamination?

Anticipated results

By day 7, the protoplasts should fluoresce blue after staining with Calcofluor White. Protoplasts that have died prior to wall resynthesis do not fluoresce. Some protoplasts should have divided already such that two- or four-celled colonies are visible. If cell colonies can be found, the protoplasts can be diluted with fresh culture medium with lower mannitol to observe the development of p-calli. After approximately 8 to 10 weeks, p-calli can be transferred to shoot regeneration medium.

Questions

- Why do the protoplasts float to the interface of the flotation and rinse solutions while the debris sinks to the bottom of the sucrose solution during centrifugation?
- Why do viable protoplasts fluoresce green when stained with FDA stain?
- Report the viability of freshly isolated protoplasts and the plating efficiency after 7 days. Explain why there is a difference between frequency of viable protoplasts and plating efficiency.
- Would you expect the plants regenerated from individual protoplasts derived from a single leaf to be genetically identical? Explain.

Experiment 2. Protoplast fusion of potato

Materials

For each student or student team the following materials are needed:

- All the materials as described in Experiment 1 except solutions
- Mineral oil (autoclaved)
- Microscope cover slips (autoclaved in a glass petri dish wrapped in foil)
- 10 ml preplasmolysis medium
- 10 ml enzyme medium containing either RITC or FITC (10 µl of a 3-mg/mL stock in ethanol)
- 30 ml rinse medium
- 10 ml flotation medium
- 10 ml CPW 13M
- 1 ml fusogen (PEG 22.5)
- 5 ml Ca^{2+} wash medium
- 10 ml culture medium

Procedures for the isolation and fusion of potato protoplasts are outlined in Figure 27.1. Follow the protocol provided in Procedure 27.2 to complete this experiment.

Anticipated results

After the Ca^{2+} washing medium has been added, protoplast fusion can be observed under an inverted microscope. If the microscope is equipped with fluorescence, FITC-stained protoplasts should fluoresce green, RITC-stained protoplasts should fluoresce red, and fusion products should fluoresce yellow. Without a selection scheme for growing the fusion products exclusively, it would be difficult to recover somatic hybrids from these cultures. Using iodoacetate or irradiation to prevent unfused protoplasts from regenerating complicates the experiment unnecessarily for a student lab. However, it is possible that the hybrid p-calli will differ morphologically from the control p-calli, such that putative hybrids can be identified.

Questions

- How would you select for fusion products?
- Why is mannitol used in protoplast culture?
- What frequency of fused protoplasts was observed?
- How did the plating efficiency compare between the fusion plates and the control plates?
- Was there a difference in p-calli frequency, morphology, or growth rate between the control plates and the fusion plates?

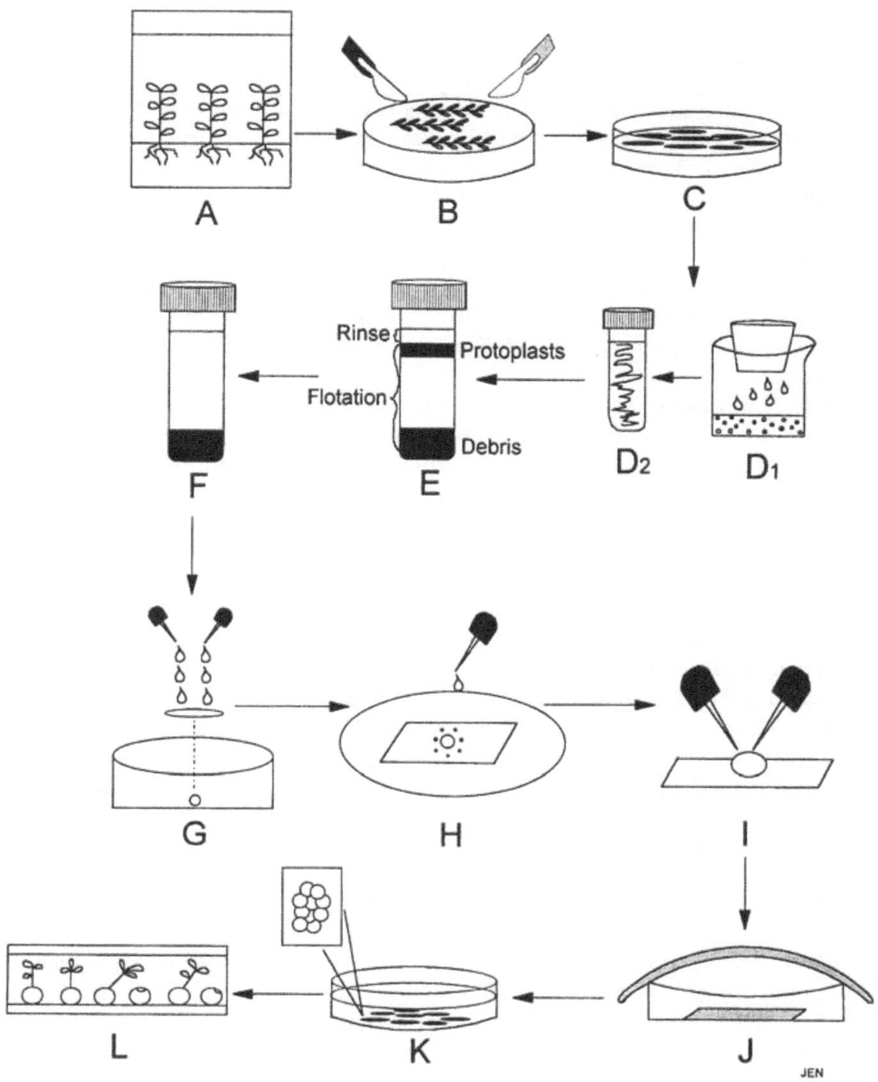

Figure 27.1 Experimental procedures for fusing potato protoplasts. (A) Propagate potato plantlets in vitro until 15 copies of each genotype are obtained. (B) Cut shoots and leaves into fine pieces using two scalpels. (C) Incubate cut pieces in plasmolysis medium for 1 to 8 h followed by 4 to 16 h in enzyme medium. (D_1) Pour protoplast suspension through a sterile 63-µm filter into a 250-ml beaker. (D_2) Transfer protoplast suspension into a sterile centrifuge tube and centrifuge. (E) Resuspend pellet in 10 mL flotation medium overlaid with 1 ml rinse medium and centrifuge (protoplasts collect at the interface of the two media during centrifugation). (F) Rinse protoplasts and resuspend in CPW 13M at a density of 1×10^6 per milliliter. (G) (1) In the center of a sterile petri dish add one small drop of sterile mineral oil; (2) position a sterile cover slip over the droplet; (3) add three drops of each of the two protoplast suspensions. (H) Fuse protoplasts by gradually adding six drops of Fusogen PEG 22.5 medium. (I) Wash protoplasts with Ca^{2+} washing medium. Replace washing medium with culture medium. (J) Incubate protoplasts in dim light at 25°C. (K) After 2 to 3 weeks embed multicellular colonies in medium with agarose. (L) Regenerate shoots and transfer to MS basal medium for rooting. (From Trigiano, R. N. and Gray, D. J., Eds., *Plant Tissue Culture Concepts and Laboratory Exercises, First Edition*, CRC Press LLC, Boca Raton, FL, 1996.)

Procedure 27.2
Protoplast extraction and PEG-mediated fusion of potato.

Step	Instructions and comments
1	**Day 1 — Preparation:** In the late afternoon, cut in vitro shoots and leaves of potato plantlets into fine pieces using two scalpels in a 10-cm petri dish and add preplasmolysis medium for 1 to 8 h.
2	Remove preplasmolysis medium with a Pasteur pipet and add the enzyme medium with either FITC or RITC (one for one fusion partner and one for the other) if a fluorescence microscope is available. Incubate in the dark at room temperature on a shaker (60 rpm) for 4 to 16 h.
3	**Day 2 — Isolation:** Gently swirl the petri dishes to loosen protoplasts from debris; then transfer protoplasts in enzyme media to a sterile 63-µm filter in a 250-ml beaker.
4	Add 3 ml rinse solution to the petri dish, swirl vigorously, and add this to the enzyme solution. Rinse the filter with an additional 2 ml rinse medium. Remove the filter and pour the protoplast solution into a sterile centrifuge tube. Centrifuge at $50 \times g$ for 5 min.
5	Remove supernatant with a Pasteur pipet. Resuspend the pellet in 10 ml flotation medium. Add 1 ml (25 drops) rinse medium dropwise to the top of the sucrose solution by dripping it gently down the side of the centrifuge tube so that the two layers remain intact. Centrifuge at $50 \times g$ for 10 min.
6	Collect protoplast band at the interface between rinse and sucrose media with a Pasteur pipet. Transfer protoplasts to a sterile 15-ml centrifuge tube, add rinse medium to 10 ml, and spin at $50 \times g$ for 5 min.
7	Remove rinse medium with a Pasteur pipet, being careful not to disturb the pellet. Adjust the protoplast density to 1×10^6 per ml in CPW 13M medium. If insufficient protoplasts have been obtained to reach this density, adjust both fusion partners to the same density as close to this as possible.
8	**Day 2 — Fusion:** Students should be paired such that each student in a pair has isolated protoplasts from different parents with different stains. The students can then share their protoplasts with each other and retain the remainder for control plates.
9	Place a small drop of sterile mineral oil in the center of a plastic petri dish and gently lower a sterile cover slip onto the oil.
10	Pipet three drops of each of the two protoplast suspensions (one incubated with RITC and isolated by one student in the pair and the other incubated with FITC and isolated by the other student in the pair) onto the cover slip. Leave undisturbed for 10 min to allow the protoplasts to settle and stick to the cover slip.
11	Add six drops fusogen PEG 22.5, drop by drop, surrounding the protoplasts, leaving the last drop for the center. Incubate at room temperature for 20 to 25 min.
12	Gently add three drops of Ca^{2+} washing medium every 5 min for 20 min to one side of the protoplast culture while removing three drops of fusion medium from the opposite side using a second Pasteur pipet. Agglutination begins during this washing procedure. It is important that the protoplasts are disturbed as little as possible during this process.
13	Replace the Ca^{2+} washing medium with culture medium. Place three drops of culture medium at one side of the protoplast mixture. Then, using a second Pasteur pipet, remove some of the washing medium from the opposite side; repeat two more times. Then flood the cover slip with culture medium and place a few drops of culture medium on the plate surrounding the cover slip to maintain humidity. Seal the plate and incubate in dim light at 25°C.

Procedure 27.2 continued
Protoplast extraction and PEG-mediated fusion of potato.

Step	Instructions and comments
14	Prepare control dishes of the remaining unfused protoplasts by centrifuging for 5 min, removing the supernatant, and replacing it with sufficient culture medium to adjust the density to 2.5×10^5 per ml. Dispense to the appropriate size petri dish, depending upon the amount of culture obtained — 4 to 6 ml in a 100 × 20-mm dish; 2 to 4 ml in a 60 × 15-mm dish; 1 to 2 ml in a 35 × 10-mm dish. Seal the dish and incubate in dim light at 25°C, as above.
15	**Postfusion handling of protoplasts:** After 2 and 7 days, the cultures should be checked for cell wall regrowth and cell division.
16	At 7 to 10 days after isolation, plating efficiency of the fusions and control plates of each parent should be recorded.
17	If the protoplasts are dividing and callus colonies are growing, the cultures should be diluted with fresh culture medium after approximately 2 weeks. This can be done by preparing double-strength filter-sterilized culture medium and autoclaving an equal volume of double-strength (0.6%) agarose in distilled water. The two can then be mixed to provide a volume of medium equal to what is in the plates of protoplasts. Then the protoplasts in liquid culture medium are added to the fresh medium to embed the protoplasts in agarose while replenishing the medium simultaneously.
18	After two more weeks, the number of visible callus colonies should be recorded for the fusion plates and the controls. Callus morphology often differs among clones for obvious morphological traits such as color, texture, and size. The morphology of callus in each plate should be noted. Fast growing or obviously different callus in the fusion plates may be indicative of putative somatic hybrids.
19	The number of p-calli per plate should be recorded for fusions and controls after 4 to 6 weeks of culture. If there are too many to count, a sample region of equal size representative of each plate can be scored.

Literature cited

Carlson, P.S., H.H. Smith, and R.D. Dearing. 1972. Parasexual interspecific plant hybridisation. *Proc. Natl. Acad. Sci.* 69:2292–2294.

Cheng, J. and R.E. Veilleux. 1991. Genetic analysis of protoplast culturability in *Solanum phureja*. *Plant Sci.* 75:257–265.

Haberlach, G.T., B.A. Cohen, N.A. Reichert, M.A. Baer, L.E. Towill, and J.P. Helgeson. 1985. Isolation, culture and regeneration of protoplasts from potato and several related *Solanum* species. *Plant Sci.* 39:67–74.

Helgeson, J.P., G.J. Hunt, G.T. Haberlach, and S. Austin. 1986. Somatic hybrids between *Solanum brevidens* and *Solanum tuberosum*: expression of a late blight resistance gene and potato leaf roll resistance. *Plant Cell Rep.* 3:212–214.

Kisaka, H. and T. Kameya. 1994. Production of somatic hybrids between *Daucus carota* L. and *Nicotiana tabacum*. *Theor. Appl. Genet.* 88:75–80.

Matibiri, E.A. and S.H. Mantell. 1994. Cybridization in *Nicotiana tabacum* L. using double inactivation of parental protoplasts and post-fusion selection based on nuclear-encoded and chloroplast-encoded marker genes. *Theor. Appl. Genet.* 88:1017–1022.

Murashige, T. and F. Skoog. 1962. A revised medium for rapid growth and bioassays with tobacco tissue cultures. *Physiol. Plant.* 15:473–497.

Nagata, T. and I. Takebe. 1970. Cell wall regeneration and cell division in isolated tobacco mesophyll protoplasts. *Planta* 92:301–308.

Rose, R.J., M.R. Thomas, and J.T. Fitter. 1990. The transfer of cytoplasmic and nuclear genomes by somatic hybridisation. *Aust. J. Plant Physiol.* 17:303–321.

Sidorov, V.A., D.P. Yevtushenko, A.M. Shakhovsky, and Y.Y. Gleba. 1994. Cybrid production based on mutagenic inactivation of protoplasts and rescuing of mutant plastids in fusion products: potato with a plastome from *S. bulbocastanum* and *S. pinnatisectum*. *Theor. Appl. Genet.* 88:525–529.

chapter twenty-eight

Isolation of protoplasts from leaves of chrysanthemum and orchardgrass

Robert N. Trigiano

The following laboratory exercises use leaves of either chrysanthemum (*Dendrathema grandiflora* Tzvelev) or orchardgrass (*Dactylis glomerata* L.) to illustrate some aspects of protoplast technology. We have found that, although protoplasts are relatively easy to isolate from leaf tissue of these two species, it is more difficult to regenerate plants from chrysanthemum (Sauvadet et al., 1990) and nearly impossible from orchardgrass. As such this laboratory is devised only to demonstrate some of the general principles of protoplast methodologies, i.e., donor tissue preparation, preculture, tissue digestion, purification, and assessment of purity and viability. We suggest that the laboratory exercise be completed without regard to aseptic conditions, since callus and plants will not be regenerated, and that some of the complex protoplast culture media discussed in Chapter twenty-six are not required. The mechanics of the experiment may be completed in one day if some students can help out "off and on" for the entire day or if the instructor prepares some of the initial steps of the procedure in advance of the class meeting. The laboratory may also be conducted as a demonstration.

Exercise

Experiment 1. Isolation of protoplasts from leaves

Materials

- 30 to 60 tillers of orchardgrass or about 15 young, expanding, light green leaves of chrysanthemum — 'Iridon' or any other cultivar
- Pectolyase Y-23 (Seishin Pharm. Co. Ltd., 4-13, Koamicho, Nihonbashi, Tokyo, Japan)
- RS "Onozuka" Cellulase (Yakult Honsha Co., Ltd. 1.1.19 Higashi Shinabashi, Minato-ku, Tokyo, 105, Japan)
- The following four isolation media:
  1. One-half strength SH salts (see Chapter three) amended with 0.38 M (*circa* 7%) mannitol, 3 mM MES, 2 mM $CaCl_2$, pH = 5.7
  2. Same medium as above except use 0.49 M (*circa* 9%) mannitol

0-8493-2029-1/00/$0.00+$.50

3. Same medium as above except use 0.60 M (*circa* 11%) mannitol
4. Same medium as above except use 0.71 M (*circa* 13%) mannitol

- The following digestion media: same as the four isolation media listed above except include 1% (w/v) RS cellulase and 0.4% (w/v) Y-23 pectolyase
- 1% calcofluor in water
- 0.05% fluorescein diacetate (FDA) dissolved in acetone
- Hemocytometers
- Compound microscope equipped for epifluorescence and filters
- Nylon mesh sieves with 40 to 100 μm cross-sectional openings in beakers
- Low-speed table-top centrifuge
- 25- or 50-ml glass or plastic centrifuge tubes
- Any solidified agar medium in petri dishes

General procedure

The general procedure is outlined diagrammatically in Figure 28.1. The experiment, as listed below, is designed to investigate only the influence of isolation medium, specifically

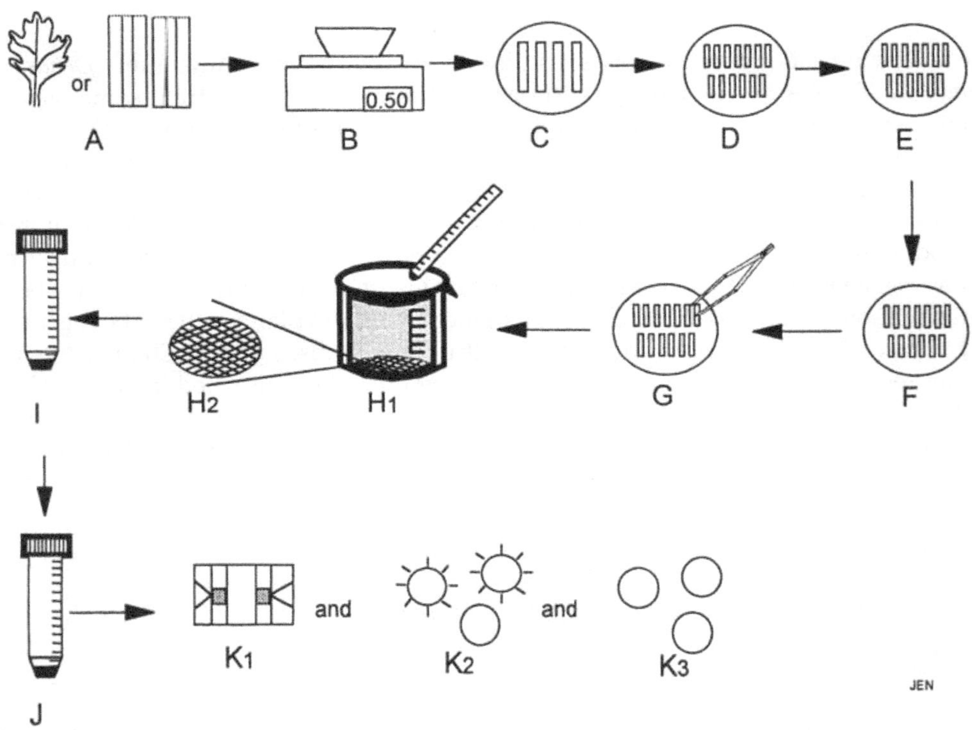

Figure 28.1 Diagrammatic representation of the protocol for making protoplasts from chrysanthemum or orchardgrass leaves. (A) Leaves of chrysanthemum and orchardgrass. (B) Weigh leaves. (C) Preculture in cold isolation medium. (D) Cut tissue into small linear strips. (E) Rinse cut tissue several times with isolation medium. (F) Expose tissue to digestion medium for 3 to 4 h. (G) Remove large pieces of undigested tissue with fine-tipped forceps. (H_1 and H_2) Pipette digestion medium and small debris into sieve apparatus (H_1) and express protoplasts through 40- to 100-μm sieve (H_2). (I) Collect protoplasts by centrifugation. (J) Rinse protoplasts with isolation medium. (K_1 to K_3) Determine the following: concentration of protoplasts with a hemocytometer (K_1), viability (K_2), and purity (K_3) of protoplasts with fluorescence. (From Trigiano, R. N. and Gray, D. J., Eds., *Plant Tissue Culture Concepts and Laboratory Exercises, First Edition*, CRC Press LLC, Boca Raton, FL, 1996.)

the osmotic pressure, on successful procurement of viable protoplasts. As alternative treatments, the concentration of hydrolytic enzymes may be varied. This is a little more difficult since the appropriate isolation medium must first be determined. We suggest that each student or student team work with either one or two of the isolation media listed above. Students should then share and analyze data, i.e., number of protoplasts etc. The experiment may be considered as a completely randomized design for statistical analyses. Alternatively, if desired, only one replication of each treatment may be prepared and the class divided into four groups or teams. In this scenario, the experiment can be regarded as a demonstration. Follow the instuctions in Procedure 28.1 to complete this laboratory exercise.

Procedure 28.1
Preculture of plant material and enzymatic digestion of plant cell walls.

Step	Instructions and comments
1	Collect ten tillers of orchardgrass for each replication of isolation media and dissect the innermost two leaves for culture as outlined in Chapter twenty. Only use the bottom 2 cm from each of the leaves. For mums, three to five leaves of any cultivar for each replication of a treatment will be sufficient.
2	Record the fresh weight of the leaf tissue and place in a 60 × 15-mm petri dish.
3	Add 15 ml of ice cold (4°C) isolation medium and place in the refrigerator for 30 min.
4	Using fine forceps and a scalpel with a #10 blade, cut the leaves into very thin (<1-mm) strips. Orchardgrass leaves are easily "feathered" by cutting parallel to the veins; mum leaves should be cut along the long axis. Care should be taken not to tear the leaves — a sharp cut works best for protoplast formation and thinner strips ensure a higher yield of protoplasts.
5	After all the leaves are cut, use a pasteur pipette to remove most of the isolation medium and replace it with more of the appropriate cold isolation medium. Repeat this step until the fluid is free of most small debris and green color. For mums, the preparation should be odor free.
6	Prepare digestion media about 1 h before use. Dissolve 200 mg (1%) RS cellulase and 80 mg (0.4%) Y-23 Pectolyase in 20 ml of each of the four isolation solutions — use 100-ml beakers and vigorously stir for at least 30 min. Remove the predigestion media and add the digestion solutions to the cut tissues.
7	Wrap the petri dishes with several layers of Parafilm and incubate at room temperature (22 to 25°C) in the dark. Cultures should remain stationary — shaking the petri dishes at this time is detrimental to protoplast formation.
8	Using a compound microscope, observe the tissue preparation for release of protoplasts. Focus on the scalpel blade marks on the bottom of the petri dish — often the smaller, more cytoplasmic dense protoplasts will sink to the bottom. Also observe the cut edges of the leaves for plasmolyzed cells and the release of protoplasts (Figure 28.2). Repeat this operation and record your observations every 1/2 h for a total of 3 or 4 h.

Protoplasts are now isolated from cells, and they must be collected and concentrated by centrifugation. Note the following two cautionary items before centrifugation: (1) be sure that the "centrifuge or rotor is balanced." Insert a tube of equal weight in the rotor station exactly opposite your tube containing protoplasts. (2) Do not confuse $100 \times g$ with 100 rpm (revolutions per minute) — they are not the same. Check the specifications of the centrifuge (the manufacturer usually provides a conversion chart or a mathematical formula to calculate relative force [*g*] from rpm) and adjust rpm to achieve $100 \times g$. Follow the instructions provided in Procedure 28.2 to concentrate the protoplasts.

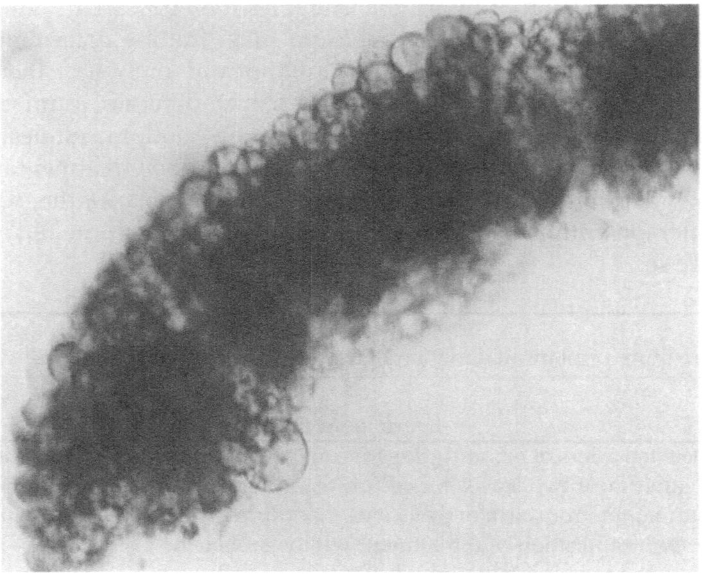

Figure 28.2 Protoplasts formed from a strip of a chrysanthemum leaf exposed to release medium with cell wall-degrading enzymes for 2 h. (From Trigiano, R. N. and Gray, D. J., Eds., *Plant Tissue Culture Concepts and Laboratory Exercises, First Edition*, CRC Press LLC, Boca Raton, FL, 1996.)

Procedure 28.2
Isolation of protoplasts.

Step	Instructions and comments
1	After the plant material is digested, place the petri dishes on a rotary shaker set for the slowest possible speed for 5 min; the liquid should barely move.
2	Remove as much of the undigested debris as possible using a fine pair of forceps. The debris will be very soft and adhere to the forcep blades, but can easily be removed by running the forcep blades through some agar medium in a petri dish.
3	Gently swirl the contents of the petri dish and using a wide-bore glass pipette equipped with a bulb or pump, carefully remove the liquid from the petri dish.
4	Hold the tip of the pipette against the inside of the nylon mesh sieve (40- to 100-μm cross-sectional openings) housing and very gently express the liquid medium down the side — protoplasts are very fragile and may be damaged if not handled carefully. If the surface tension between the liquid and sieve prevents passage of the liquid, gently "scratch" the nylon sieve surface with the forcep blades and the liquid will flow into the collection beaker.
5	Using a wide-bore pipette, gently transfer the sieved medium to either a 25- or 50-ml plastic centrifuge tube. The medium may appear "cloudy" — no problem, this is great! This is a sign that many protoplasts were formed during digestion. Centrifuge the suspension at 100 × *g* for 5 min following the precautions listed above.
6	Decant or pipette (long pasteur pipette) the digestion medium from the tubes without disturbing the green pellets of protoplasts on the bottom. Carefully pipette 20 ml of isolation medium down the inside of the centrifuge tube. Centrifuge as in Step 5 and repeat this washing step twice more. All hydrolytic enzymes must be removed if the protoplasts are to be cultured. Cell regeneration will not occur in the presence of these enzymes.

Determining the yield of protoplasts

Protoplasts are now concentrated at the bottom of the tube, and we can estimate the number of protoplasts formed per gram of leaf tissue. This information is necessary to compare treatment effects on protoplast isolation. Follow the outline in Procedure 28.3 to calculate the yield of protoplasts.

Procedure 28.3
Determining the yield of protoplasts.

Step	Instructions and comments
1	Resuspend the pelleted protoplasts in 5 ml of fresh isolation medium.
2	Estimate the number of protoplasts in the suspension using a hemacytometer. Sum the number of protoplasts in five squares (Figure 28.3). Only count those protoplasts that lie entirely within the boundaries of the counting squares; those touching a line should not be counted. Calculate the mean from at least three samples.
3	Calculate the number of protoplasts per milliliter using the following equation: X × 2000 = P, where X is the mean number of protoplasts and P = number of protoplasts/milliliter; e.g., if X = 50, then P = 10^5 protoplast ±10%/milliliter. Since the total volume of the suspension is 5 ml, the estimate of total number of protoplasts is 5 ml × 10^5 protoplasts/milliliter = 5 × 10^5 or a total of 500,000 ±10% protoplasts in the tube.
4	Now estimate the yield of protoplasts per gram of tissue, e.g., 0.5 g of tissue was used in the isolation, then 5 × 10^5 protoplasts divided by 0.5 g = 10^6 or 1,000,000 protoplast per gram of tissue. The yield of protoplasts from the different treatments now can be compared since the data are all expressed on a per-gram basis.

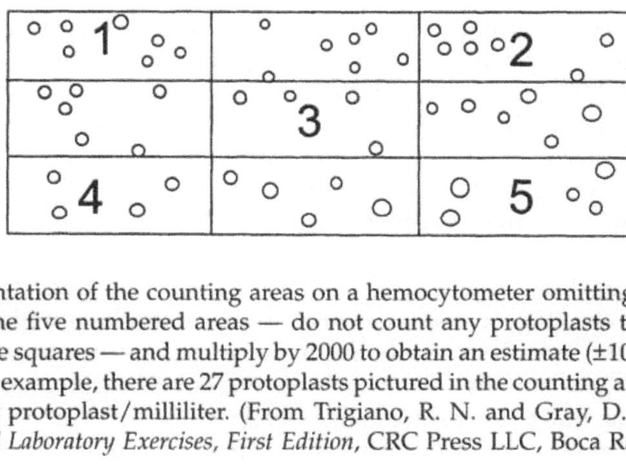

Figure 28.3 Representation of the counting areas on a hemocytometer omitting the fine lines. Sum the protoplasts in the five numbered areas — do not count any protoplasts that touch any of the boundary lines of the squares — and multiply by 2000 to obtain an estimate (±10%) of the protoplasts per milliliter. In this example, there are 27 protoplasts pictured in the counting areas (1 to 5), therefore an estimated 54,000 protoplast/milliliter. (From Trigiano, R. N. and Gray, D. J., Eds., *Plant Tissue Culture Concepts and Laboratory Exercises, First Edition*, CRC Press LLC, Boca Raton, FL, 1996.)

Methods for evaluating the quality of protoplasts

When counting protoplasts you may have noticed that they were almost perfectly spherical; normally this would indicate that all of the cell wall material was digested by the enzymes. However, "patches" of cell wall may still remain. The best way to detect wall materials is with calcofluor, a chemical that binds with cellulose, a major plant cell wall polymer. The stain is made by mixing 1 ml of 1% aqueous calcofluor with 20 ml of the appropriate isolation medium. Pipette 0.25 ml of protoplast suspension into 1 ml of the stain solution. Swirl to mix, mount a few drops of the mixture on a glass slide, and cover with a #1 or 1 1/2 coverslip. First, observe protoplasts with transmitted visible light using a 10 or 20 × objective lens. Next, examine the protoplasts using an ultraviolet light (UV) source with the following filters and mirror: 400- to 440-nm excitation; 470-nm barrier; and 455-nm dichroic mirror (equivalent to Nikon filter assembly BV-2A). If the wall is not completely digested, the calcofluor bound to cellulose will fluoresce a brilliant blue that will not fade with time. Some protoplasts may still retain wall fragments, which are visible as irregularly shaped blue glowing patches. If true protoplasts (i.e., the cell minus the wall), the field will appear dark and the protoplast may not be discernable except for perhaps some red autofloresence of chloroplasts. Observe at least 50 protoplasts and calculate the percentage that are free of cell wall. Incidentally, this is the same optical brightener that makes white clothing "bright" in natural light or fluoresce with black light.

The protoplast isolation procedure often damages or kills a percentage of cells; therefore, it is advisable to assess the condition of the protoplasts. Mount a small drop of the protoplast suspension on a glass slide, coverslip, and observe using 20 and 40 objective lens. The cytoplasm of some or a few may appear to be streaming or moving in a circle — look for the small "particles" (organelles) in the cytoplasm. This phenomenon is termed "cyclosis" and only occurs in living cells. However, a cell that does not exhibit cyclosis is not necessarily dead. Also, look for "bulges" or "budding" of the plasmalemma (cell membrane) that may indicate damaged, dying, or fused protoplasts.

There are a number of presumptive tests to determine viability of cells. The tests are presumptive because they do not actually measure which cells are "living" and those cells that are "dead," but usually assess some physiological or physical properties associated with living cells. Most of these tests are based on enzyme activity or permeability of the plasmalemma to certain indicator chemicals, such as dyes or stains.

Fluorescein diacetate often has been used to estimate the percentage of viable protoplasts in a preparation (Widholm, 1972). This technique uses fluorescence and is very efficient — many protoplasts can be examined in a short time. Fluorescein diacetate normally does not fluoresce, but once across the cell membrane, an esterase in the cytoplasm cleaves the acetate groups from the main portion of the molecule. The resulting moiety will fluoresce green when exposed to the proper wavelengths of UV light.

Mix 0.25 ml of 0.05% fluorescein diacetate (dissolved in acetone) with 20 ml of the appropriate isolation medium to make the "stain" solution. Pipette 0.25 ml of the protoplast suspension into 1.0 ml of the stain solution. Swirl gently to mix and mount some of the protoplasts on a glass slide. First, look at the protoplasts using a 10 or 20 × objective lens. Now view the protoplasts using the UV source and the following filters and mirror: 420- to 490-nm excitation; 520-nm barrier; and a 510-nm dichroic mirror (equivalent to Nikon filter assembly — B-3A). The cytoplasm of "living" protoplasts should fluoresce green (Figure 28.4); vacuoles in the protoplasts will appear dark. Protoplasts in which the cytoplasmic esterase is not active are presumably dead and will appear to be dark. Occasionally, the autofluorescence of chlorophyll or other compounds will impart a reddish hue to the protoplasts. Observe at least 50 protoplasts and calculate the percentage of living cells. Multiply the estimate of number of protoplasts by this percentage to obtain an estimate of the number of living protoplasts. From the example above, there are 5×10^5 protoplasts per milliliter multiplied by 0.95 viable protoplasts, equals 4.75×10^5 living protoplasts per milliliter. A good isolation will yield between 90 and 95% viable protoplasts.

As an alternative to epifluorescence microscopy to determine viability, Evan's blue stain may be used with visible light (Gaff and Okong 'O-Ogola, 1971). Mix 2 ml of 1% aqueous stain with 10 ml of the appropriate isolation medium. Pipette 0.25 ml of the protoplast suspension into 1.0 ml of the stain mixture and allow to stand for about 15 min. The dye will be excluded from the protoplasts with intact and functional plasmalemma (presumed living cells), whereas those protoplasts whose membrane are dysfunctional will turn blue and are considered nonviable.

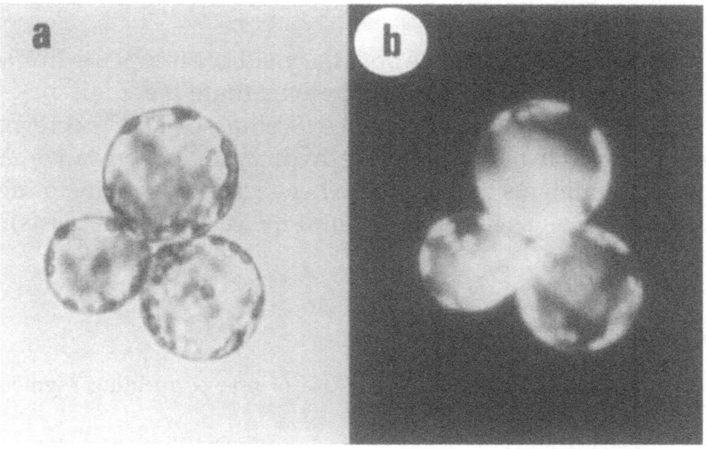

Figure 28.4 Determining viability of protoplasts using fluorescein diacetate. (a) Protoplasts viewed with transmitted visible light; (b) protoplasts "stained" with fluorescein diacetate and viewed with UV light. The fluorescence light is green. (From Trigiano, R. N. and Gray, D. J., Eds., *Plant Tissue Culture Concepts and Laboratory Exercises, First Edition*, CRC Press LLC, Boca Raton, FL, 1996.)

Anticipated results

Preculturing and cutting the donor tissue in isolation medium should minimize damage to cells. Rinsing the donor tissue before digestion will remove cellular debris and chemical compounds released from broken cells during the cutting operation. Some of these chemicals are thought to be detrimental to the remaining cells.

A very few protoplasts should be released within the first hour and will be difficult to find. However, focusing on a cut edge of the leaf tissue (see Figure 28.2) should show cells with various degrees of cell wall degradation. Also, plasmolysis of cells should be easily observed. As the experiment progresses, more and more protoplasts will be released and can easily be seen on the bottom of the petri dish or floating in the medium. The brief shaking at the conclusion of the experiment helps to maximize the release of protoplasts. Because so many factors influence the yield of protoplasts, there is not a way to accurately foretell which release medium will produce the best results. Typically, this protocol can be optimized for both orchardgrass and chrysanthemum leaves to yield between 3×10^5 and 5×10^6 protoplasts per gram of tissue. Although some protoplasts should be produced throughout the range of osmotic pressures developed in the four media, protoplasts in some of the media may be very fragile and may not have survived the centrifugation steps. It is not unusual to observe in the lowest osmotic pressure medium that intact, viable protoplasts will not be produced. Therefore, when approaching the question of how to isolate protoplasts from a "new species," it is often useful to initially try a wide range of media and enzyme concentrations.

Questions

- Why would preculturing the donor tissue in isolation medium at 4°C facilitate production of protoplasts?
- Why are protoplast isolation and release media made with relatively high concentrations (%) of mannitol? Could other osmotically active sugar alcohols be used?
- Why are protoplasts isolated from relatively young, meristematic tissues? Hint: consider cell wall properties, also.

- Why is it important to remove all of the enzyme solution before trying to culture protoplasts?
- Why is it important to determine both the yield of protoplasts from the donor tissue and the concentration of protoplasts in the culture medium?
- Are the cell wall degradation enzymes ("cellulase" and "pectinase") used in this experiment pure? If not, what other hydrolytic and oxidative enzymes may be present as contaminants and are these beneficial (aids in liberating protoplasts) or detrimental (may cause harm or death to the cells or protoplasts)?
- What is fluorescence?

Literature cited

Gaff, D. F. and O. Okong 'O-Ogola. 1971. The use of non-permeating pigments for testing the survival of cells. *J. Exp. Bot.* 22: 756–758.

Sauvadet, M.-A., P. Brochard, and J. B. Gibod. 1990. A protoplast-to-plant system in chrysanthemum: differential response among several commercial clones. *Plant Cell Rep.* 8: 692–695.

Widholm, J. 1972. The use of fluorescein diacetate and phenosafranine for determining viability of cultured plant cells. *Stain Technol.* 47: 189–194.

chapter twenty-nine

Haploid cultures

Sandra M. Reed

Haploid plants are a tremendous asset in plant breeding and genetic studies. Due to the presence of only one set of chromosomes, recessive mutations are easily identified. In addition, doubling the chromosome number of a haploid to produce a dihaploid results in a completely homozygous plant. Theoretically, the genotypes present among a large group of dihaploids derived from an F_1 hybrid represent, in a fixed form, the genotypes expected from an F_2 population. Use of dihaploids in breeding programs can thus greatly reduce the time required for development of improved cultivars. To be most useful, a large number of haploids from many different genotypes are required.

Haploids have been available for genetic studies for many years. Prior to the 1960s, they were mostly obtained spontaneously following interspecific hybridization or through the use of irradiated pollen, but usually only infrequently and in very small numbers. Haploid methodology took a giant step forward 30 years ago when Guha and Maheshwari (1964) found that haploid plants could be obtained on a regular basis and in relatively large numbers by placing immature anthers (male) of *Datura innoxia* Mill. into culture. This work was rapidly expanded using tobacco (*Nicotiana tabacum* L.), which became the "model species" for anther culture experiments. To date, haploids have been produced via anther culture in over 170 species; several good reviews provide lists of these species (Maheshwari et al., 1982; Bajaj, 1983; Heberle-Bors, 1985). While the work has been much more limited, haploids have also been obtained from in vitro culture of the female gametophyte.

Although much of the terminology used in this chapter has been discussed in previous chapters, the in vitro induction of haploids involves a few specialized terms that need to be defined before further consideration of the topic. A haploid is a plant with the gametic or "n" number of chromosomes. Dihaploids, or doubled haploids, are chromosome-doubled haploids or "2n" plants. When anthers are cultured intact, the procedure is called anther culture; in contrast, microspore culture involves isolating the microspores from the anthers before culture and is sometimes referred to in the literature as pollen culture. The development of haploid plants via anther or pollen culture occurs through a process commonly known as androgenesis.

As with the other in vitro techniques discussed in this book, many variables can and do affect the success of anther culture. This chapter will begin with a discussion of these variables and be followed by a consideration of androgenesis and the different pathways by which haploid plants arise. Last, some of the basics of anther culture methodology will be summarized.

0-8493-2029-1/00/$0.00+$.50

Factors affecting anther culture

Genotype

The choice of starting material for an anther culture project is of the utmost importance. In particular, genotype plays a major role in determining the success or failure of an experiment. Haploid plant production through anther culture has been very limited or nonexistent in many plant species. Furthermore, within a species, differences exist for ability to produce haploid plants. For example, some corn (*Zea mays* L.) cultivars are completely unresponsive in anther culture, whereas a few haploids can be obtained from others (Wan and Widholm, 1993). Even within an amenable species, such as tobacco, some genotypes produce haploids at a much higher rate than do others. Because of this genotypic effect, it is important to include as much genetic diversity as possible when developing protocols for producing haploid plants via anther culture.

Condition of donor plants

The age and physiological condition of donor plants often affect the outcome of anther culture experiments. In most species, the best response comes from the first set of flowers produced by a plant. As a general rule, anthers should be cultured from buds collected as early as possible during the course of flowering. Various environmental factors that the donor plants are exposed to may also affect haploid plant production. Light intensity, photoperiod, and temperature have been investigated and, at least for some species, found to influence the number of plants produced from anther cultures. Specific optimum growing conditions differ from species to species and are reviewed by Powell (1990). In general, the best results are obtained from healthy, vigorously growing plants.

Stage of microspore development

The most critical factor affecting haploid production from anther culture is the stage of microspore development; for many species, anthers are only responsive during the uninucleate stage of pollen development. In contrast, optimum response is obtained in tobacco from anthers cultured just before, during, and just after the first pollen mitosis (late uninucleate to early binucleate microspores).

In developing a protocol for anther culture, one anther from each bud is usually set aside and later cytologically observed to determine the stage of microspore development. Generally, anthers within a bud are sufficiently synchronized to allow this one anther to represent the remaining cultured anthers. Measurements of physical characteristics of the flower, such as calyx and corolla length and anther color, shape, and size, are also recorded. Results of the experiments are analyzed to determine which microspore stage was the most responsive. The physical descriptions of the buds and anthers are then examined to determine if this microspore stage correlates to any easily identified influorescence, flower, or anther characteristic(s). For example, in tobacco, buds in which the calyx and corolla are almost identical in length usually contain anthers having microspores at or near the first pollen mitosis. A researcher wishing to produce a maximum number of haploid plants of tobacco would collect only buds fitting this physical description.

Pretreatment and preincubation

For some species, a pretreatment following collection of buds, but before surface disinfestation and excision of anthers, has been found to be beneficial. Yields of tobacco haploids are often increased by storing excised buds at 7 to 8°C for 12 days prior to anther excision and culture (Sunderland and Roberts, 1979). For other species, temperatures from 4 to 10°C and durations from 3 days to 3 weeks have been utilized. For any one species, there may be more than one optimum temperature–length of treatment combination. In general, lower temperatures require shorter durations, whereas a longer pretreatment time is dictated for temperatures at the upper end of the cold pretreatment range mentioned. High temperature pre-incubations have been effective in some species. For example, haploid plant production was increased in rape (*Brassica campestris* L.) by culturing the anthers at 35°C for 1 to 3 days prior to culture at 25°C (Keller and Armstrong, 1979).

Media

Androgenesis can be induced on a simple medium such as that developed by Nitsch and Nitsch (1969) for tobacco and a few other species. For most other species, the most commonly used media for anther culture are MS (Murashige and Skoog, 1962) and N6 (Chu, 1978) or variations on these media. In some cases, complex organic compounds, such as potato extract, coconut milk, and casein hydrolysate, have been added to the media. For many species, 58 to 88 mM (2 to 3%) sucrose is added to the media, whereas other species, particularly the cereals, have responded better to higher (up to 435 mM or about 15%) concentrations of sucrose. The higher levels of sucrose fulfill an osmotic, rather than a nutritional requirement. Other sugars, such as ribose, maltose, and glucose, have been found to be superior to sucrose for some species.

For a few species, such as tobacco, it is not necessary to add plant growth regulators (PGRs) to the anther culture media. Most species, however, require a low concentration of some form of auxin in the media. Cytokinin is sometimes used in combination with auxin, especially in species in which a callus phase is intermediate in the production of haploid plants.

Anther culture media are often solidified using agar. Because agar may contain compounds inhibitory to the androgenic process in some species, the use of alternative gelling agents has been investigated. Agarose has been reported to be a superior gelling agent for anther culture of many of the cereal species. Alternatively, the use of liquid medium has been advocated by some researchers as a way to avoid the potentially inhibitory substances in gelling agents. In this technique, the anthers are placed on the surface of the medium, forming a so-called "float culture."

Androgenesis

Development of haploids

Haploid plants develop from anther culture either directly or indirectly through a callus phase. Direct androgenesis mimics zygotic embryogenesis with the exception that a true suspensor is not observed in androgenic embryos (see Chapter nineteen). At the globular stage of development, most of the embryos are released from the pollen cell wall (exine). They continue to develop, and after 4 to 8 weeks, the cotyledons unfold and plantlets emerge from the anthers. Direct androgenesis is primarily found among members of the tobacco (Solanaceae) and mustard (Cruciferae) families.

During indirect androgenesis, the early cell division pattern is similar to that found in the zygotic embryogenic and direct androgenic pathways. After the globular stage, irregular and asynchronous divisions occur and callus is formed. This callus must then undergo organogenesis for haploid plants to be recovered. The cereals are among the species that undergo indirect androgenesis.

The early cell divisions that occur in cultured anthers have been studied (for review, see Reynolds, 1990). For species cultured during the uninucleate stage, the microspore either undergoes a normal mitosis and forms a vegetative and a generative nucleus or divides to form two "similar looking" nuclei. In those cases where vegetative and generative nuclei are formed in culture, or where binucleate microspores are placed into culture, it is usually the vegetative nucleus that participates in androgenesis. The only species in which the generative nucleus has been found to be actively involved in androgenesis is black henbane (*Hyoscyamus niger* L.). When similar-looking nuclei are formed, one or both nuclei may undergo further divisions. In some cases, the two nuclei will fuse, producing homozygous diploid plants or callus. Since diploid callus may also arise from somatic tissue associated with the anther, diploids produced from anther culture cannot be assumed to be homozygous. To verify that plants produced from anther culture are haploid, chromosome counts should be made from root tips or other meristematic somatic tissues. Because haploids derived from diploid species are expected to be sterile or have greatly reduced fertility, pollen staining, which is much quicker and requires less skill than chromosome counting, can also be used to identify potential diploids. Pollen staining, however, will not identify plants that have reduced fertility because they have a few extra or missing chromosomes, i.e., aneuploids.

Problems associated with anther culture

Problems encountered in plants during or as a result of anther culture range from low yields to genetic instability. Many of the major horticultural and agronomic crops do not yield sufficient haploids to allow them to be useful in breeding programs. In other species, genetic instability has often been observed from plants recovered from anther and microspore cultures.

The term 'gametoclonal variation' has been coined to refer to the variation observed among plants regenerated from cultured gametic cells. Gametoclonal variation has been observed in many species. While often negative in nature, some useful traits have been observed among plants recovered from anther and microspore culture. Gametoclonal variation may arise from changes in chromosome number (i.e., polyploidy or aneuploidy) or chromosome structure (e.g., duplications, deletion, translocations, inversions, etc.). In tobacco, gametoclonal variation is associated with an increase in amount of nuclear DNA without a concomitant increase in chromosome number (DNA amplification). Changes in cytoplasmic DNA have been observed among albino cereal haploids. Good discussions and reviews of gametoclonal variation can be found in several chapters of Jain et al. (1996).

General anther culture procedures

Collection of floral buds

Floral buds may be collected from plants grown in the field, greenhouse, or growth chamber. Entire influorescences or individual buds are harvested and kept moist until ready for culturing. If buds are to be pretreated (i.e., low temperature), they should be wrapped in a moistened paper tissue and placed into a small zipper-type plastic bag.

Disinfestation, excision, and culture

Disinfestation typically uses a 5% sodium or calcium hypochlorite solution for 5 to 10 min and then a thorough rinse in sterile distilled water. Anthers are aseptically excised in a laminar flow hood, taking care not to cause injury. If the anther is still attached to the filament, the filament is carefully removed.

If a solid medium is used, the anthers are gently pressed onto the surface of the medium (just enough to adhere to the medium), but should not be deeply embedded. When using a liquid medium, the anthers are floated on the surface. Care must be taken when moving liquid cultures so as not to cause the anthers to sink below the surface. For most species, disposable petri dishes are utilized for anther cultures. For a species with large anthers, like tobacco, the anthers from four to five buds (20 to 25 anthers) may be cultured together on one 100 × 15-mm diameter petri dish. For species with smaller anthers, or for certain experimental designs, smaller petri dishes or other containers may be more useful. Petri dishes are usually sealed and placed into an incubator; the specific temperature and light requirements of the incubator may depend on the species being cultured.

Determining stage of microspore development

For most species, stage of microspore development can be determined by "squashing" an entire anther in aceto-carmine or propiono-carmine and then observing the preparation under a low-power objective of a light microscope. The early uninucleate microspore is lightly staining with a centrally located nucleus. As the uninucleate microspore develops, its size increases and a large central vacuole is formed. As the microspore nears the first pollen mitosis, the nucleus is pressed up near the periphery of the microspore. Staining will still be fairly light. Pollen mitosis is of short duration, but it may sometimes be observed; it is recognized by the presence of condensed chromosomes. The product of the first pollen mitosis is a binucleate microspore containing a large vegetative and a small generative nucleus. The vegetative nucleus is often difficult to recognize because it is so diffuse and lightly staining. However, this stage may be definitively identified by the presence of the small densely staining generative nucleus. As the binucleate microspore ages, the intensity of the staining increases and starch granules begin to accumulate. Eventually, both nuclei may be hidden by the dark-staining starch granules.

Handling of haploid plantlets

For species undergoing direct androgenesis, small plantlets can usually be seen emerging from the anthers 4 to 8 weeks after culture. When these get large enough to handle, they should be teased apart using fine-pointed forceps and can then either be placed on a rooting medium (usually low salt, with small amount of auxin) or transplanted directly into a small pot filled with soilless potting mixture. The callus produced in species that undergo indirect androgenesis must be removed from the anther and placed onto a regeneration medium containing the appropriate ratio of cytokinin to auxin.

To produce dihaploid plants, it is necessary to double the chromosome number of the haploids, and for many species, a colchicine treatment is used. Published procedures for producing polyploids from diploids can be modified for use with anther culture-derived haploids. For example, it may be possible to use a colchicine treatment designed for small seedlings with haploid plants directly out of anther culture. Alternatively, established procedures using larger plants may be used. For tobacco, midveins of fully expanded leaves are excised and cultured on a regeneration-promoting medium (Kasperbauer and Collins, 1972). On the average, depending primarily on genotype, 50% of the plants regenerated from this medium will be dihaploids.

In summary, haploids of many plant species can be produced using anther culture. Yields of haploids differ greatly depending on species, and are also affected by cultural conditions, such as media formulation, stage of microspore development at time of culture, and use of a low- or high-temperature pretreatment. Haploids may either arise directly from the anthers or may be produced indirectly through a callus intermediate.

The primary reason for producing haploids is for use in plant breeding, where they can hasten the development of superior cultivars. While anther culture is not effective in all species, for many it is the only or most efficient method of producing haploids. Continued research into cultural conditions and media requirements should eventually lead to improved anther culture protocols for many important crop species.

Literature cited

Bajaj, Y.P.S. 1983. In vitro production of haploids, pp. 228-287. In: *Handbook of Plant Cell Culture,* Vol. 1, Techniques for Propagation and Breeding. Evans, D.A., W.R. Sharp, P.V. Ammirato, and Y. Yamada, Eds. Macmillan, New York.

Chu, C. 1978. The N6 medium and its applications to anther culture of cereal crops, pp. 51–56. In: *Proceedings of Symposium on Plant Tissue Culture.* Science Press, Peking.

Guha, S. and S.C. Maheshwari. 1964. In vitro production of embryos from anthers of *Datura. Nature* 204:497.

Heberle-Bors, E. 1985. In vitro haploid formation from pollen: a critical review. *Theor. Appl. Genet.* 71:361–374.

Jain, S.M., S.K. Sopory, and R.E. Veilleux, Eds. 1996. *In Vitro Haploid Production in Higher Plants,* Vol. 2, Applications. Kluwer Academic Publishers, Dordrecht.

Kasperbauer, M.J. and G.B. Collins. 1972. Reconstitution of diploids from leaf tissue of anther-derived haploids in tobacco. *Crop Sci.* 12:98–101.

Keller, W.A. and K.C. Armstrong. 1979. Stimulation of embryogenesis and haploid production in *Brassica campestris* anther cultures by elevated temperature treatments. *Theor. Appl. Genet.* 55:65–67.

Maheshwari, S.C., A. Rashid, and A.K. Tyagi. 1982. Haploids from pollen grains — retrospect and prospect. *Amer. J. Bot.* 69:865–879.

Murashige, T. and F. Skoog. 1962. A revised medium for rapid growth and bioassays with tobacco tissue cultures. *Physiol. Plant.* 15:473–497.

Nitsch, J.P. and C. Nitsch. 1969. Haploid plants from pollen grains. *Science.* 163:85–87.

Powell, W. 1990. Environmental and genetical aspects of pollen embryogenesis, pp. 45–65. In: *Biotechnology in Agriculture and Forestry,* Vol. 12, Haploids in Crop Improvement. Bajaj, Y.P.S., Ed. Springer-Verlag, Berlin.

Reynolds, T.L. 1990. Ultrastructure of pollen embryogenesis, pp. 66–82. In: *Biotechnology in Agriculture and Forestry,* Vol. 12, Haploids in Crop Improvement. Bajaj, Y.P.S, Ed. Springer-Verlag, Berlin.

Sunderland, N. and M. Roberts. 1979. Cold-treatment of excised flower buds in float culture of tobacco anthers. *Ann. Bot.* 43:405–414.

Wan, Y. and J.M. Widholm. 1993. Anther culture of maize, pp. 199–224. In: *Plant Breeding Review,* Vol. 11. Janick, J., Ed. John Wiley & Sons, New York.

chapter thirty

Production of haploid tobacco plants using anther culture

Sandra M. Reed

Haploids are plants that have the gametic, or n, number of chromosomes. They are very useful in genetic and breeding studies because recessive genotypes are easily identified. Moreover, completely homozygous plants can be obtained quickly by doubling the chromosome number of the haploids to produce dihaploids ($2 \times n$ number of chromosomes); these dihaploids are very useful in plant breeding programs (see Chapter twenty-nine).

For many years, few haploids were available to geneticists and breeders. This situation began to change in the 1960s with the discovery that haploids of *Datura* species and tobacco (*Nicotiana tabacum* L.) could be produced by culturing anthers. Today, haploids of over 170 species have been produced via anther culture. In no species has anther culture been more successful than tobacco, where large numbers of haploids can be obtained from many different genotypes. Because it is so easy to obtain tobacco haploids, it is the ideal species to demonstrate and learn anther culture methodology.

The following laboratory exercises illustrate the technique of anther culture using tobacco. Students will become familiar with the following techniques and concepts: (1) basic anther culture procedures, (2) determining the optimum stage of microspore development for anther culture, and (3) the use of cold pretreatments for enhancing haploid production from anther culture. The general procedure used in both experiments is illustrated in Figure 31.1.

Directions are given for student teams. It should take 30 to 60 min of laminar flow hood use for each team to complete the culturing stage of each of the experiments. Preparation of media, staging of microspores, and evaluation of cultures will require additional laboratory time, but minimum usage of the flow hood. Depending on class size, facilities, and time available, two to four students per team are recommended.

General considerations

Growth of plants

A small quantity of tobacco seeds can be obtained from Dr. Verne Sisson, Department of Crop Science, North Carolina State University, Box 7620, Raleigh, NC 27695 and any cultivar may be used for the experiments. After completing the anther culture experiments, allow a few plants to continue to flower. Approximately 1 month after the flowers

0-8493-2029-1/00/$0.00+$.50

open, open-pollinated seed capsules will turn brown and dry. Seed can be collected and stored for several years under cool, dry conditions. A single capsule will contain 1 to 2000 seed that can be used as a source of plants for future experiments.

Fill a 10-cm-diameter clay pot with vermiculite and water from the bottom by placing the pot in a pan of water. Lightly sow seed on the surface of vermiculite and keep the pot in a saucer of water until the seeds germinate in about 1 week. Continue to water seedlings from the bottom of the pot, but do not allow the vermiculite to be continually soaked nor to dry out completely. Seedlings may be grown in the greenhouse or in a growth chamber with 22 to 26°C and 16/8-h light/dark cycle.

After about 6 weeks, the seedlings will be large enough to transplant to individual 10- or 15-cm-diameter clay pots and any of the soilless potting mixtures may be used. A slow release all-purpose fertilizer may be incorporated into the medium at this time or pots can be watered twice weekly with an all-purpose liquid fertilizer. Plants should be grown in a greenhouse. Supplemental light is not required during the winter, but, if provided, may accelerate growth of the plants and hasten flowering. Five plants per team of students should be sufficient for both experiments included in this chapter. Plants should flower 2 to 4 months after transplant to 15-cm-diameter pots, depending on growing conditions, cultivar, and size of pot (plants can be grown in a smaller pot to encourage more rapid flowering).

Explant preparation and culture handling

Collect buds soon after the plants begin to flower and transport to the laboratory in plastic bags containing moistened paper tissues. Surface disinfest the buds in groups of 20 to 25. Place a single bud in a sterile 100 × 15-mm plastic petri dish. Holding the bud gently with one set of forceps, remove the calyx (green) and corolla (colored) using another set of sharp fine-pointed forceps. Using the fine-pointed forceps, push the anthers away from the filaments (they should separate easily). Taking care not to squeeze the anthers, lift them gently with the forceps and place on the surface of the agar-solidified medium (Figure 30.1). Seal each petri dish and place in a 25°C incubator or growth chamber under constant light or 16/8-h light regime.

Cultures can be scored for number of haploids produced approximately 8 weeks after culture initiation. Using fine-tipped forceps, gently separate plants in order to get accurate counts. Unless these plants are to be used for other experiments, cultures can be scored outside of aseptic conditions.

Culture medium

The medium that will be used in these two experiments is that of Nitsch and Nitsch (1969; see Chapter 3 for composition). Each liter of media will be supplemented with 20 g (58 mM) sucrose and 8 g of agar. Plant growth regulators (PGRs) will not be added. Pour the medium into a number of 100 × 15-mm sterile plastic petri dishes.

Exercises

Experiment 1. Effect of stage of microspore development on yield of haploid plants

As described earlier in Chapter twenty-nine, the stage of microspore development at the time of culture is a critical factor affecting androgenesis. Students will be culturing anthers from buds of different developmental stages, identifying stage of microspore development,

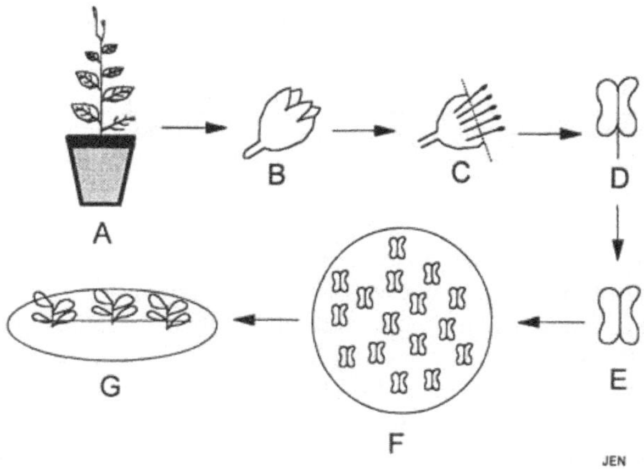

Figure 30.1 Diagrammatic representation of the anther culture protocol for tobacco. (A) Flowering diploid plant. (B) Floral bud excised from plant. Calyx and corolla are of approximately equal lengths. (C) Floral bud with calyx and corolla removed. Buds normally contain five anthers, which are shown still attached to filaments. (D) Isolated stamen with anther and filament. (E) Anther removed from filament. (F) Petri dish with cultured anthers. (G) Plantlets emerging from anthers. (From Trigiano, R. N. and Gray, D. J., Eds., *Plant Tissue Culture Concepts and Laboratory Exercises, First Edition*, CRC Press LLC, Boca Raton, FL, 1996.

and determining whether physical characteristics of the tobacco buds and anthers are correlated to stage of microspore development and the ability to produce haploid plants.

Materials

The following items are needed for each team of students:

- 100 mL of 20% commercial bleach
- Nine 100 × 15-mm petri dishes containing Nitsch and Nitsch medium
- Lighted incubator providing approximately 25°C
- Acetocarmine stain (combine 45 ml glacial acetic acid, 55 ml of distilled water, 0.5 g carmine); place in a beaker; cover with aluminum foil and gently boil solution for 5 min under a fume hood. Filter the solution (may require several days and filter papers) and store in a refrigerator. The solution should be dark red and it may need to be filtered again before use.

Follow the protocols listed in Procedure 30.1 to complete this experiment.

Procedure 30.1
Effect of stage of microspore development on yield of tobacco haploids.

Step	Instructions and comments
1	Collect 36 tobacco flowers buds. The calyx and corolla on 12 of the buds should be of approximately equal length (medium-length buds). Twelve of the buds should have a corolla that extends 5 to 7 mm beyond the calyx (long-length buds). The remaining 12 buds should be the largest of the buds in which the corolla is not visible at all beneath the calyx (short-length buds).
2	Using an indelible marker, divide the bottom of each of the nine media dishes into four quadrants and label from 1 to 36, which will correspond to the culture numbers.

Procedure 30.1 continued
Effect of stage of microspore development on yield of tobacco haploids.

Step	Instructions and comments
3	Surface disinfect medium length buds in 70% ethanol for 1 min, followed by 20% commercial bleach. Rinse twice in sterile distilled water.
4	Follow explant preparation and culture handling instructions given in general directions. Place four of the anthers onto culture medium in one of the numbered quadrants. Place the remaining anther onto a microscope slide which should be marked with the culture number. Repeat with remaining 11 medium-length buds.
5	Repeat steps 3 and 4 with long-length buds, then repeat with short-length buds.
6	Squash anthers on microscope slides in a small drop of acetocarmine stain, remove debris, and cover with cover slip. Heat slide over steam for a few seconds and observe under the microscope. Using photographs in Figures 30.2a to d and the descriptions presented in Chapter twenty-nine, determine stage of microspore development for each anther. If there is not enough time to examine the anthers when cultures are initiated, place them in fixative (3 parts 95% ethanol and 1 part glacial acetic acid; mix under fume hood). After 2 days, replace fixative with 70% ethanol. Leave at room temperature for up to a week, then place in refrigerator. Anthers can be removed from fixative or alcohol at any time, squashed, and examined following the previous directions.
7	For each culture, record bud length (short, medium, or long) and stage of microspore development.
8	Count and record number of plants in each culture approximately 8 weeks after culture initiation.

Anticipated results

Four to six weeks after initiating the anther cultures, small plantlets will be seen emerging from some of the anthers (Figure 30.3). Even under the best of conditions, not all tobacco anthers will produce haploids. Numbers of plantlets emerging from each anther will vary from none to a few to many — there is no way to predict the results of any given experiment.

Anthers from short-length buds will produce few, if any, haploid plantlets. Uninucleate microspores will have been observed from the anthers of this size bud. Anthers from the long-length buds will also produce few, if any, plantlets. This size bud will correspond to a mid-to-late binucleate microspore stage. All or almost all of the plantlets produced in this experiment will come from the anthers in which the calyx and corolla were equal in length. This size bud corresponds to the late-uninucleate to early-binucleate microspore stage, which produces the most haploid plants in tobacco.

Questions

- What is the purpose of trying to correlate a physical characteristic of buds and/or anthers to stage of microspore development?
- How would plantlets have been handled if you had been trying to produce dihaploid plants?
- How can you be sure that the plants you obtained are haploids?

Experiment 2. Effect of cold pretreatment on yield of haploids from anther cultures of tobacco

Subjecting buds to several days of low temperatures after collection, but before culture, increases yields of haploid plants in tobacco and several other species. This experiment

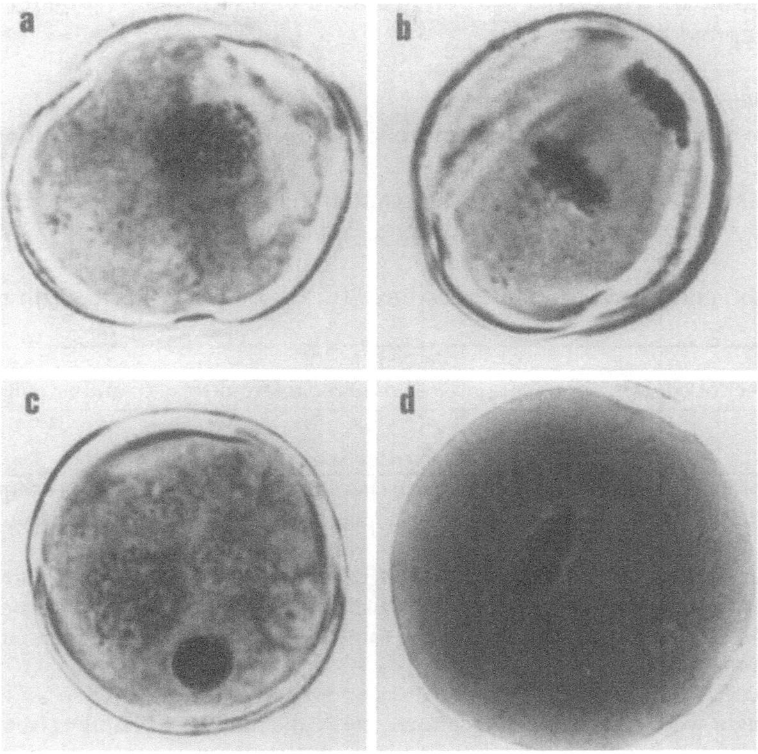

Figure 30.2 Microspores of tobacco. (a) Late uninucleate microspores. Vacuole can be seen on right side of cell. (b) Anaphase of first pollen mitosis. (c) Early binucleate microspore. Large diffuse vegetative nucleus and smaller, more darkly staining generative nucleus can be distinguished. (d) Mid-binucleate microspore. Staining is darker, obscuring vegetative nucleus, but the generative nucleus is still visible. (From Trigiano, R. N. and Gray, D. J., Eds., *Plant Tissue Culture Concepts and Laboratory Exercises, First Edition*, CRC Press LLC, Boca Raton, FL, 1996.)

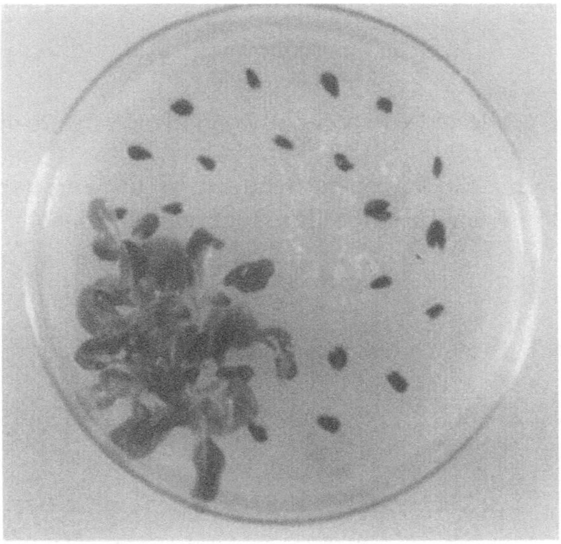

Figure 30.3 Haploid tobacco plantlets emerging from an anther. (From Trigiano, R. N. and Gray, D. J., Eds., *Plant Tissue Culture Concepts and Laboratory Exercises, First Edition*, CRC Press LLC, Boca Raton, FL, 1996.

will demonstrate the effect that cold pretreatment of flower buds (anthers) has on yields of tobacco haploid plants.

Materials

Each student team will require the materials listed in Experiment 1 and the following:

- Ten 100-mm-diameter petri dishes containing Nitsch and Nitsch (1969) medium
- Refrigerator or incubator providing temperature of 7 to 8°C

Follow the directions provided in Procedure 30.2 to complete this experiment.

Procedure 30.2
Effect of cold pretreatment on yield of haploids from tobacco anther cultures.

Step	Instructions and comments
1	Collect 40 tobacco buds, all of which should have a calyx and corolla of equal length.
2	Place 20 buds in a zipper-type plastic bag with a water-moistened tissue. Seal bag and place in a 7 to 8°C incubator or refrigerator.
3	Sterilize remaining 20 buds as directed in step 3, Procedure 30.1. Follow explant preparation and culture handling instructions given in general directions. Place the anthers from four buds in the same dish of medium (20 anthers/dish of medium).
4	After 10 to 12 days, repeat step 3 with cold pretreated buds.
5	Eight weeks after initiating cultures, count the number of plantlets from both the untreated and cold pretreated buds. Calculate the mean number and standard deviation (see Chapter seven) of plants/anther for both groups of buds.

Anticipated results

The mean number of plants/anther should be significantly greater for the buds that received a cold pretreatment; however, differences between treatments will not be as great as observed in Experiment 1. It may be advisable to pool results from all the teams in the class in order to see the effect of the cold pretreatment and statistical analysis.

Questions

- Why were buds collected when the calyx and corolla were approximately equal in length?
- What temperatures and lengths of treatment would you test if you were trying to work out an anther culture procedure for a new species?

Literature cited

Nitsch, J. P. and C. Nitsch. 1969. Haploid plants from pollen grains. *Science*. 163:85–87.

chapter thirty-one

Transformation

Melina López-Meyer, Ignacio E. Maldonado-Mendoza, and Craig L. Nessler

Transformation can be broadly defined as the process of introducing DNA into cells. The importance of this technique comes from the fact that DNA, the molecule that carries the blueprints for life, can be cut, spliced, and otherwise altered in vitro. Genetic manipulation techniques allow scientists to isolate a single DNA sequence from one organism and join it to that of a completely different organism to form what is now known as recombinant DNA.

The first recombinant DNA molecules were generated at Stanford University in 1972 by utilizing the cleavage properties of restriction enzymes and the ability of DNA ligase to join DNA strands together. The isolation of these two kinds of enzymes in the late 1960s was a major milestone in the development of genetic engineering. In 1973 DNA fragments were joined to the plasmid pSC101. These recombinant molecules behaved as replicons, i.e., they could replicate when introduced (transformed) into *Escherichia coli* cells.

Typically, when a particular DNA sequence is to be transformed into a living cell, it is first inserted in a vector. Vectors are used to assist in the transfer, replication, and sometimes expression of a specific DNA sequence in a target cell.

There are certain essential features that vectors should possess. Ideally, they should be fairly small DNA molecules to facilitate their isolation and handling. They must also have an origin of replication, so that their DNA can be copied and thus maintained in the cell population as the host organism grows and divides. It is also important for a vector to have some type of selectable marker, usually resistance to an antibiotic, that will enable the vector to be selected from a large population of cells that have not taken up the foreign DNA. Finally, the vector must also have at least one unique restriction endonuclease recognition site to serve as the site of DNA insertion during the production of recombinants.

Bacteria and some other organisms contain relatively small, circular DNA molecules that are separate from the bacterial chromosome and replicate independently. These molecules are called plasmids. Although plasmids are generally dispensable, i.e., not essential for cell growth and division, they often provide a selective advantage to the host organism such as antibiotic resistance. Because of these features, plasmids are extensively used as vectors, particularly in the "construction" of complex recombinant molecules.

In addition to plasmids, bacteriophages have also been used as vectors. Bacteriophages are viruses of bacteria that are useful as vectors because they can act as carrier molecules for relatively large pieces of DNA.

0-8493-2029-1/00/$0.00+$.50

Not all bacteria can be easily transformed, and it was not until the early 1970s that transformation was demonstrated in *E. coli*. Today, the protocol for transforming *E. coli* is very simple. It is carried out by mixing plasmid DNA with spheroplasts (prepared from cells that were pretreated with calcium chloride), incubating the mixture on ice, and then giving a brief heat shock to facilitate the uptake of DNA into the cells.

Although the initial steps of gene manipulation and transformation methodologies were developed for prokaryotic systems (which still represent a major part of this technology), their potential application to eukaryotic systems such as plants was soon realized. The problems of introducing recombinant DNA into plants, however, are much different than those encountered with bacterial systems, and the methods available for introducing DNA into bacterial cells are not easily transferred to other cell types. It was not until 1983 that the first report of transformed plant cells was published (Fraley et al., 1983).

Several different approaches since developed for the transformation of vascular plants are discussed in this chapter. However, the most widely used system, which is based on the *Agrobacterium tumefaciens* tumor-inducing (Ti) plasmid, is described in the greatest detail. Laboratory exercises that follow this chapter illustrate *Agrobacterium*-mediated transformation systems.

Agrobacterium *transformation system*

Agrobacterium tumefaciens is a soil-inhabiting bacterium that is responsible for the crown gall disease of many plants. The bacterium infects through a wound in the stem of the plant, and a tumor develops at the junction of root and stem (crown). The agent responsible for crown gall formation is not the bacterium itself, but a plasmid known as the Ti plasmid. Ti plasmids are large, ranging in size from 140 to 235 kb (1 kb = 1000 base pairs). During infection, a small portion of Ti plasmid DNA (15 to 30 kb), called T-DNA, is transferred to the plant cell nucleus, where it becomes covalently inserted into the nuclear DNA. In this manner, the T-DNA becomes stably maintained in the genome of transformed cells.

T-DNA carries the genes responsible for tumor formation and for the synthesis of unusual amino acid derivatives known as opines. The biological relevance of opines is that they can be used as the sole carbon and/or nitrogen source for the inducing *Agrobacterium* strain. The genes responsible for opines catabolism reside in the Ti plasmid. Strains of *Agrobacterium* can be classified according to the type of opines that are synthesized by the crown gall tumors that they induce. The most common type of opines are octopine and nopaline.

The genes responsible for transfer of T-DNA are also contained on the Ti plasmid and are called virulence genes (vir genes). *Agrobacterium* infection requires wounded plant tissue because vir genes are induced by phenolic compounds released by the injured plant cells. Induction of the vir genes results in the transfer of T-DNA (Figure 31.1).

The only regions of T-DNA that are absolutely required for its transfer and integration into the plant genome are the border regions. These are short repeat sequences of 25 base pairs. Any DNA sequence inserted between the border repeats will be transferred to and integrated into the plant genome. Therefore, Ti-based plasmids are excellent vectors for introducing foreign genes into plants.

In order to use Ti-based plasmids as vectors, the genes responsible for tumor formation must be removed. Ti-based plasmids lacking tumorigenic functions are known as disarmed vectors. Cells infected with such plasmids do not produce tumors and are consequently much easier to regenerate into normal, fertile plants by standard tissue culture techniques (see Chapter thirty-three).

Even though Ti plasmid can be disarmed, they are too large to be conveniently used as vectors. Thus, smaller vectors (based on the features of the Ti plasmid) have been constructed that are suitable for manipulation in vitro.

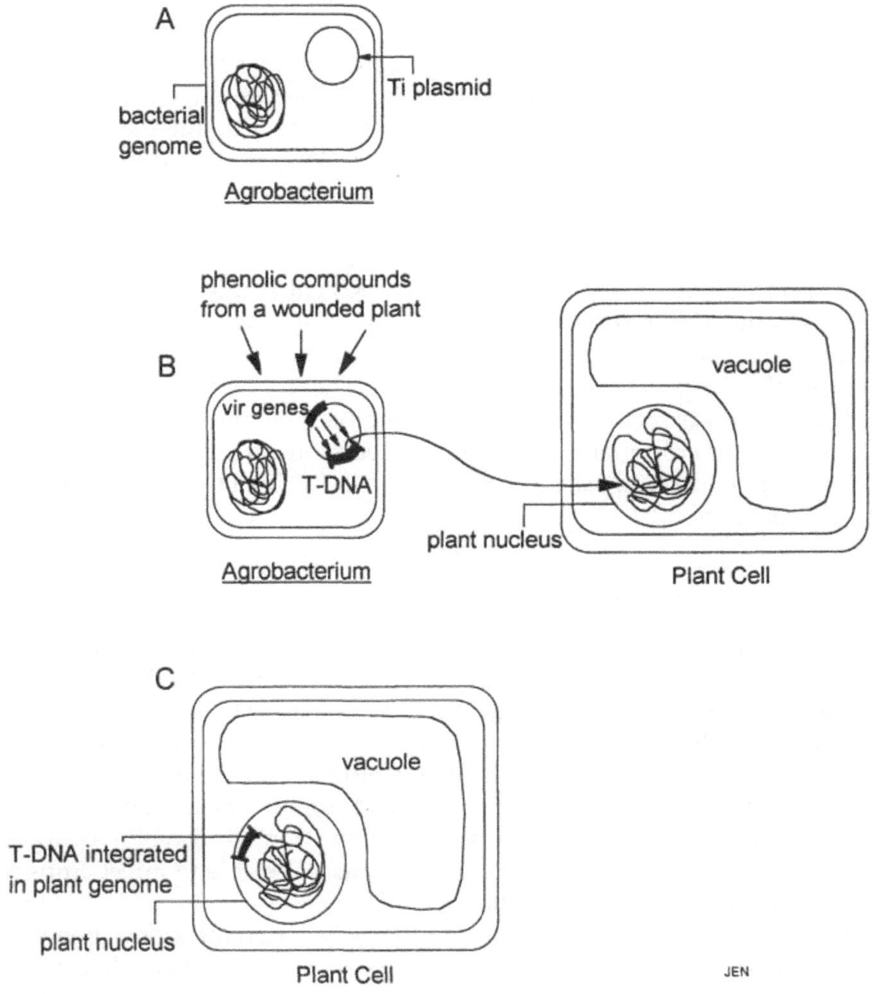

Figure 31.1 *Agrobacterium*-based plant cell transformation. (A) *A. tumefaciens* is a Gram-negative soil bacterium. (B) Wounded plant tissues release phenolic compounds, which induce the vir genes. Vir gene products mobilize and transfer T-DNA into the plant cell. (C) T-DNA is incorporated into chromosomal DNA of the plant cell. Genes in the T-DNA encode proteins, which make opines that feed other bacteria near the wound as well as compounds that induce tumor formation. (From Trigiano, R. N. and Gray, D. J., Eds., *Plant Tissue Culture Concepts and Laboratory Exercises, First Edition*, CRC Press LLC, Boca Raton, FL, 1996.)

Cointegration and binary vector system

Two strategies have been developed for using Ti-based plasmids: a methodology called cointegration and the binary vector system. The cointegration technique is based on in vivo recombination of two plasmids. One plasmid carries desirable DNA sequence; the other plasmid contains vir genes and the border repeats of the T-DNA. After recombination, a very large recombinant Ti plasmid is obtained, which now can be used to transform plants.

The most frequently used approach for *Agrobacterium*-based plant transformation is the binary vector system, which uses two separate plasmids: one to supply the disarmed

T-DNA and a second Ti plasmid, which includes the virulence functions (encoded by vir genes). The plasmid carrying the disarmed T-DNA is called mini-Ti plasmid. This plasmid bears the gene construct that will be inserted into the plant genome, along with an eukaryotic selectable marker between T-DNA border sequences, so that both genes will be inserted as a unit. When a mini-Ti plasmid is then placed into an *Agrobacterium* strain containing a plasmid with virulence functions, the vir gene products are able to drive the transfer of T-DNA into plant cells, even though T-DNA is located on a separate DNA molecule. This approach has been used extensively because mini-Ti plasmids are very easy to manipulate using standard recombinant DNA techniques.

For a long time it was thought that *Agrobacterium*-based transformation systems were only applicable to dicotyledonous plants. Recent studies have shown that many monocots can indeed be transformed with *Agrobacterium;* however, certain "tricks" must be used depending on the target species. These include using special strains of *Agrobacterium* and choosing just the explant to infect. The use of *Agrobacterium* in monocots represents a major advance in plant genetic engineering since many of the most economically important crop plants are monocotyledonous cereals including wheat (*Triticum aestivum* L.), maize (*Zea mays* L.), rice (*Oriza sativa* L.), barley (*Hordeum vulgare* L.), and oats (*Avena sativa* L.).

Plant viruses as vectors

Plant viruses have several special features as potential vectors for plant genetic engineering. For example, purified viral RNA/DNA is directly infectious to plants. Thus, simply rubbing a solution of viral nucleic acid on the leaf of a susceptible plant is often enough to obtain infection. In addition, some viruses can spread to every cell in the plant, which eliminates the regeneration step required in the *Agrobacterium* system. Also, relatively large amounts of virus can be produced from infected plants. This can result in the accumulation of very high levels of viral and foreign proteins being expressed from recombinant viruses. Furthermore, some viruses are able to infect crops which are difficult to transform with *Agrobacterium.*

Regardless of these potential advantages, the virus-based transformation systems also have some major disadvantages. In particular, viral nucleic acids do not become integrated into the plant genome like the *Agrobacterium* T-DNA. Thus, recombinant DNA will not be inherited by the next generation through the seed. In addition, viral infections usually debilitate plants to varying degrees, although mild virus strains might be used in this kind of approach. Another major drawback of using plant viruses as vectors is the limited amount of recombinant DNA that they can carry. Despite these disadvantages, plant viruses still represent a potentially useful way to introduce genetic information into plants.

Direct transformation systems

In addition to natural transformation systems, such as *Agrobacterium* and plant viruses, alternative methods to deliver DNA directly into plant cells have been developed in recent years. Plant cells have a rigid wall, which is a barrier to DNA uptake. All direct methods of DNA transfer must somehow penetrate this barrier, either by removing the wall or by penetrating through it. For example, cellulolytic and pectolytic enzymes can be used to remove the cell wall. This produces protoplasts, which can be made to take up DNA as described below and then cultured to regenerate their walls and eventually form intact plants.

Direct DNA uptake into plant cells is possible by adsorbing DNA molecules onto the surface of protoplasts using calcium phosphate or polyethylene glycol (PEG). The frequency of transformation using this technique is very low. However, because protoplasts can be produced in great numbers, sufficient transformants can be generated to make this approach practical. Protoplast transformation is a useful technique that is not limited by viral or *Agrobacterium* host range and, thus, can be used on species which are recalcitrant to these methods.

Another direct method of introducing DNA into plant cells is a technique called electroporation. Electroporation is based on the fact that high electrical pulses create transient holes in the cell membrane through which large molecules, including DNA, can enter the cell. This technique has several drawbacks, which include the following: (1) the percentage of stable transformed cells is low, (2) cell viability drops dramatically after the electrical shock, and (3) for some plants, regeneration is difficult after electroporation. Nevertheless, this technique has been useful for the transformation of some monocotyledonous species such as maize and rice.

Microinjection is an alternative to using protoplasts for the direct transformation of plant cells. In this technique DNA is directly injected into the plant cell nucleus. This is accomplished by holding the cell on a glass tube by mild suction and injecting the DNA through a very finely drawn glass micropipette. Microinjection requires a mechanical micromanipulator and a high-quality microscope. Although it has been shown to work well for plant cells, microinjection requires a great deal of patience, as well as technical ability and training, in order to carry out the injections in the proper way.

Particle bombardment is a relatively recent development that has proved to be extremely useful in transforming plant cells. The technique involves literally shooting DNA into cells, using specialized "particle guns" or "gene guns." To accomplish particle bombardment, DNA first is used to coat microscopic tungsten or gold particles known as microprojectiles, which are then accelerated toward the cells either by firing a gunpowder charge, releasing a burst of compressed gas, or accelerating the particles through an electric field. Upon contact with plant target tissue, DNA-coated particles penetrate through the cell walls and, in some cases, stable transformation events occur. As with protoplast-based transformation, successful particle-mediated transformation depends both on generating a large number of target cells as well as their efficient selection from among an even greater number of untransformed cells. In the past few years, particle bombardment has become a commonly used approach for transformation of monocotyledonous crop plants such as maize, rice, wheat, barley, and sugarcane (*Saccharum* species). The continued development of *Agrobacterium*-based systems for these crops may soon replace particle bombardment as the method of choice for transforming these important crops.

Reasons for plant transformation

There are two important reasons for producing transgenic plants: (1) to insert specific DNA sequences that confer novel desired characteristics to recipient plants (i.e., pest and herbicide resistance, improved yield, etc.) and (2) to answer basic questions concerning the regulation of plant gene expression.

Creating transgenic plants with useful characteristics

Genetic manipulation of plants has actually been practiced by humans for thousands of years by the process of selective breeding. This approach has been extremely successful, and it will continue to play a major role in modern agriculture. However, classic plant-breeding programs rely on being able to carry out genetic crosses between individual plants. The recipient and donor plants must be sexually compatible, and, thus, it is generally not possible to combine genetic traits from widely divergent species. The advent of recombinant DNA techniques and transformation methods for plants has removed this constraint and has given agricultural scientists a powerful new way of incorporating defined genetic changes into plants.

An example of the application of these techniques is the production of plants resistant to certain insect predators. For instance, the bacterium *Bacillus thuringiensis* synthesizes a glycoprotein that is highly toxic to some Lepidopterans (moths and butterflies). Other varieties *of B. thuringiensis* produce toxins that act specifically on Dipterans, such as mosquitoes and blackflies, and on Coleopterans (beetles). The genes for these so-called BT proteins have been characterized, and their introduction into plants has had a dramatic effect on predators, such as caterpillars of the cabbage white butterfly in brassicas. The gene for *B. thuringiensis* var. *berliner* toxin has been introduced in tobacco (*Nicotiana tabacum* L.) and confers resistance to tobacco hornworm infestations (Vaeck et al., 1987).

This approach depends not only on the introduction of proper protein coding sequences into the plant genome, but also the expression of that sequence to produce the toxic protein. Thus, the "transgene" has to be correctly transcribed and translated.

To express a protein coding sequence conferring a desirable characteristic in a transgenic organism, it has to be coupled to a promoter to direct its transcription. A promoter is a specific DNA sequence to which an RNA polymerase complex binds to initiate mRNA synthesis. The mRNA is subsequently translated in the cytoplasm to produce a protein that is actually the molecule that confers a specific characteristic or phenotype.

Promoters can be constitutive or inducible. When DNA sequences are introduced into plants to produce a high amount of a specific protein, constitutive promoters are usually preferred because they drive expression in all plant tissues and are independent of the development state of the transformed plant. Inducible promoters are able to induce expression only under certain conditions; consequently, they can be viewed as "molecular switches" that will promote transcription under a specific set of circumstances.

Analysis of gene regulation

Many plant genes are highly regulated; this means that they are only transcribed in specific tissues and/or at certain stages of development. Still other genes are only expressed when a plant encounters a particular environmental stimulus. Differential expression of genes is controlled by specific regulatory sequences (promoters) that are generally outside the DNA protein coding sequence located upstream (i.e., 5' to the transcription start site).

In order to study how the expression of a specific gene is controlled, its regulatory sequences can be fused to a reporter gene. The main feature of a reporter gene is that its expression can be easily monitored. Once plants are transformed with a plasmid containing a specific regulatory sequence and a reporter gene, the expression of the reporter gene is analyzed in different tissues to learn about the tissue-specific expression, or through different developmental stages of the plant to know how the gene is expressed during the development of the plant. The effects of different environmental stimuli on gene expression also can be analyzed by exposing the transformed plants to the stimuli and then monitoring the expression and activity of the reporter gene.

Conclusion

Since the first report of plant cell transformation was published in 1983, many genes have been introduced into different plant species and additional transformation techniques have been developed. Considerable success has been achieved in transforming monocotyledonous species through the modification of *Agrobacterium* systems and direct DNA transfer, and further work will undoubtedly extend the number of species capable of being transformed and regenerated. The special emphasis on transforming agronomically important crop plants will certainly continue because of the great impact that genetically engineered crops will likely have on agriculture.

In addition, detailed studies of plant gene regulation have been facilitated by the study of regulatory sequences in transgenic plants. This approach continues to be very useful for understanding how specific regulatory motifs control specific aspects of gene expression. Gene manipulation and transformation techniques will surely continue to give new insights into the processes by which plants grow and develop.

Literature cited

Fraley, R.T., S.G. Rogers, R.B. Horsch, P.R. Sanders, J.S. Flick, S.P. Adams, M.L. Bittner, L.A. Brand, C.L. Fink, J.S. Fry, G.R. Galluppi, S.B. Goldberg, N.L. Hoffman, and S.C. Woo. 1983. Expression of bacterial genes in plant cells. *Proc. Natl. Acad. Sci. U.S.A.* 80:4804–4807.

Nicholl, D.S.T. 1994. *An Introduction to Genetic Engineering*. Cambridge University Press, Cambridge.

Old, R.W. and S.B. Primrose. 1989. *Principles of Gene Manipulation*. pp. 222–254. Blackwell Scientific Publications, Oxford.

Vaeck, M., A. Reynaerts, H. Hofte, S. Jansens, M. De Beuckeller, C. Dean, M. Zabeau, M. Van Montagu, and J. Leemans. 1987. Transgenic plants protected from insects attack. *Nature*. 328:33–37.

chapter thirty-two

Transformation of tobacco and carrot using Agrobacterium tumefaciens *and expression of the β-glucuronidase (GUS) reporter gene*

Ignacio E. Maldonado-Mendoza, Melina López-Meyer, and Craig L. Nessler

Agrobacterium tumefaciens, a Gram-negative soil bacterium, has become the most commonly used organism to achieve plant transformation by mimicking the naturally occurring process in which DNA is transferred from the bacterium to plant cells. This transformation protocol uses a binary vector system and, thus, requires two plasmids, as described in Chapter thirty-one.

The first plasmid is the tumor-inducing (Ti) plasmid normally present in *Agrobacterium* cells. This plasmid encodes the virulence (vir) genes responsible for the excision, transport, and insertion of DNA sequences contained within the borders of the transfer-DNA (T-DNA) into the plant genome. These functions are carried out by the gene products (proteins) encoded by the vir genes that can act in "trans" to transfer a T-DNA. Acting in trans simply means that the T-DNA region does not have to be on the same plasmid as the vir genes. Therefore, the piece of DNA that we wish to transfer into a plant can be easily manipulated in *Escherichia coli* and then placed back into *Agrobacterium*.

The second plasmid used in this exercise is pBIN19 developed by Michael Bevan (1984). The pBin19 plasmid has the direct repeats found in the borders of the T-DNA, which define the region of DNA that will be transferred into the plant genome. But, unlike the wild-type plasmid, pBIN19 has had its Ti genes removed so that transformed cells will not undergo uncontrolled cell division. Our gene of interest (promoter-GUS fusion) has been inserted between the T-DNA borders of pBIN19 along with a gene conferring kanamycin resistance, which will be used as a selectable marker. This means that only plant cells that were successfully transformed with the T-DNA will be able to survive in the presence of the selective agent (kanamycin), while untransformed cells will not. Thus, the pBin19 plasmid acts as a carrier for the genetic information that we want to insert into the plant genome. In the *Agrobacterium*-based system, both plasmids work together to

0-8493-2029-1/00/$0.00+$.50

permit the transfer of the desired piece of genetic material and its integration in a stable, heritable form into the plant genome.

In this chapter, we will make use of a reporter gene system to study gene expression in transgenic plants. The term reporter gene is applied to a gene whose activity is not found in the species in which it will be used. Thus, it can "report" gene activity without background activity from the organism itself. Reporter genes generally are isolated from a different species than the one in which it will be used and can be employed to monitor regulatory elements of genes (i.e., promoters). During the past decade the bacterial β-glucuronidase gene (uidA), originally isolated from *E. coli*, has become the major reporter for the analysis of plant gene expression (Jefferson et al., 1987). The activity of this gene, affectionately known as GUS, cleaves the β-glucuronide bond in a variety of useful substrates, some of which produce colored or fluorescent products. To use the GUS gene fusion system, typically the GUS-coding region (the region that encodes a functional enzyme) is coupled to a gene regulatory element (i.e., promoter region) (see Figure 32.2 for more detailed information on the vectors used in these exercises). Once this fusion is introduced into plant cells, GUS activity mimics the expression of the original gene and reports its activity. With few exceptions, plants normally lack appreciable GUS activity so that detection of the GUS enzyme either in whole tissue or tissue homogenates can be used as a sensitive background-free measure of gene expression.

The wide acceptance of GUS as a reporter gene is mainly due to the fact that it can be quickly assayed without using radioactive substrates. GUS assays are extremely sensitive, and it is possible to obtain both quantitative (i.e., level of expression) and qualitative (i.e., localization of expression in tissues and organs) data with this reporter gene (Martin et al., 1992). GUS is used for a wide range of studies in transgenic plants. We and others have found this system to be very useful for plant promoter analyses (Burnett et al., 1993; Stockhaus et al., 1989; Liu et al., 1991) and in the development and optimization of transformation/regeneration protocols for different plant species (Maldonado-Mendoza and Nessler, 1993).

Design of successful transformation protocols has been made possible by advances in molecular biology, combined with the development of in vitro culture methods for regeneration of different important crop species. The experiments described in this chapter were designed not only to show how transformation protocols work, but also to illustrate how we can use these powerful tools to study gene regulation and alter the expression of any gene by using specific regulatory elements (promoters).

The following laboratory exercises will help students become familiar with the following concepts: (1) *Agrobacterium*-based transformation, (2) the use of reporter and marker genes as a way to confirm plant transformation, (3) the expression of a gene under the regulation of an inducible and a constitutive promoter, and (4) different patterns of plant gene expression.

General considerations

Growth of tobacco plants

Nicotiana tabacum L. 'Xanthi' has been chosen for these experiments, although other cultivars may be suitable. All plant material used for the *Agrobacterium* transformation experiment should be grown under sterile conditions. Tobacco seeds are very small, but easy to handle. They can be surface disinfected in a 1.7-ml microcentrifuge tube. One or two plants will be enough for each student. Surface disinfest the seeds by soaking them in 5% triton X-100 for 3 min, then rinse repeatedly with sterile water until foam is no longer seen. Soak the "wetted" seeds in 70% ethanol for 1 min and transfer to a dilute commercial bleach solution (1:10 with sterile water) for 3 min. Rinse the seeds three to five times with sterile water to remove all traces of chlorine. Resuspend the seeds in 0.5 ml of water and pipette them into a Magenta GA-7 culture vessel, each containing 75 ml of half-strength MS medium (see Chapter three) supplemented with 20 g/L (59 mM) sucrose and 2.5 g/L of Phytagel. Place the Magenta vessels in a growth chamber illuminated with 50 to 100 $\mu mol \cdot m^{-2} \cdot sec^{-1}$ fluorescent light at 25 to 28°C. Because 'Xanthi' is day-neutral, the day length in the chamber is not important — continuous illumination works well. When the plants have reached approximately 5 cm in height (about 1 month), transfer individual plants into Magenta vessels to avoid competition for space and nutrients. Plants will be ready when their leaves reach the lid of the Magenta boxes (1 month after the transfer).

Germination of carrot seeds

Carrot (*Daucus carota* L.) seeds Denver's 126 (Northrup King Seed Co.) or Imperator S8 (Southland Garden Seed) can be used and purchased in any local supermarket. Seeds can be surface sterilized using a 10% solution of commercial bleach (0.5% NaOCl) with one drop of Tween 20 per each 500 ml. After 10 min, the seeds should be washed four times in sterile deionized water and then placed on MSr medium (half-strength MS medium macrosalts, complete strength of microsalts, 2 g/L (5.8 mM) of sucrose, 8 g/L (22.2 mM) of glucose, MS vitamins with an additional 0.4 mg/L thiamine, and 8 g/L Difco Bactoagar). Place seeds in an incubator with a day cycle of 16 h light (25 to 50 $\mu mol \cdot m^{-2} \cdot sec^{-1}$) and 8 h darkness at 25 to 28°C.

Preparation of A. tumefaciens

The following two different plasmids will be introduced in *A. tumefaciens* (strain LBA4404): (1) pBI121 and (2) pCAH-1678 (Burnett et al., 1993). Both plasmids may be purchased from Clontech Laboratories, Palo Alto, CA. Introduction of the two plasmids into *A. tumefaciens* strain LBA4404 can be done by electroporation or triparental mating, as described below. For this procedure prepare a culture of the bacteria in a 500-ml Erlenmeyer flask containing 250 ml of Luria Broth (LB — see below for composition). Grow the culture at room temperature for 2 to 3 days with continuous shaking at 150 to 200 rpm. Harvest the cells by low-speed centrifugation (3000 rpm for 10 min) and wash five times by resuspending the pellet in sterile water and recentrifuging. Resuspend the washed cells in about 3 ml of a 10% glycerol solution and aliquot 100 µl into 0.5-mL microcentrifuge tubes. Freeze the tubes in liquid nitrogen immediately, and store in –80°C freezer. These cells are now ready to be used for electroporation, but can be used for this purpose for up to 1 year if stored at –80°C.

Electroporation

To electroporate the *Agrobacterium* cells, slowly thaw a tube of the LBA4404 electrocompetent cells and immediately place on ice. Add 0.5 µl of the plasmid solution (diluted to 1 µg/µl) to the chilled cells and transfer to a prechilled electroporation chamber (2- or 5-mm width). Set the electroporation apparatus to the following conditions: 2.5 kV, 200 Ω of resistance, 25 µF capacitance, and set the capacitance extender to 960 µF. Discharge the apparatus and transfer the bacteria into a disposable culture tube (13 × 100 mm) containing 1 ml of LB medium using a sterile pasteur pipette. Incubate the culture tube for 1 h in an orbital shaker at 150 rpm at room temperature. After this "recovery" period spread 250 µl of the bacterial suspension on LB plates containing 50 mg/L of kanamycin. Place the plates in an incubator at 28°C or in a cabinet at room temperature and let them grow until colonies easily are seen (2 to 3 days). *Agrobacterium* cannot grow at temperatures much above 30°C and should not be placed in a 37°C incubator used to grow *E. coli.*

Because the pBI121 and pCAH-1678 plasmids contain a kanamycin resistance gene, only those bacterial cells carrying the plasmid will survive and grow into colonies. It is important not to skip the recovery step in liquid LB medium because the bacteria need time to synthesize neomycin phosphotransferase, the protein conferring kanamycin resistance, before they come in contact with the antibiotic. Otherwise, the bacteria may not grow on LB + kanamycin plates. Pick a single colony and streak it out on a fresh LB + kanamycin plate. Incubate the plate for 2 to 3 days at 28°C. Take a single colony from this plate and place it into a flask of liquid LB medium with 50 mg/L kanamycin. A 125-ml Erlenmeyer flask with 50 ml of medium is enough to transform two to three plants. Start the liquid culture 2 to 3 days before the bacteria are needed for plant transformation.

LB is composed of 5 g sodium chloride (85.5 mM), 5 g yeast extract, 10 g bactotryptone, 10 ml of a 1 M magnesium sulfate solution (10 mM) per liter; the pH is adjusted to 7.5 with 1 N NaOH. If the LB is used in petri dishes, 16 g/L agar is added prior to autoclaving. Antibiotics are filter sterilized and added after autoclaving when the media has cooled enough to be handled with bare hands (40 to 50°C).

Triparental mating

In case an electroporation machine is not available, an alternative way to obtain the *Agrobacterium* constructions is by using a method called triparental or tripartite cross. It involves the use of an *A. tumefaciens* strain, which contains the Ti plasmid harboring the virulence genes responsible for plant transformation. The recombinant plasmid pBIN19 is present in one *E. coli* strain, and a conjugation-proficient plasmid in another. When the three strains are mixed, the conjugation-proficient "helper" plasmid transfers to the strain carrying the recombinant plasmid, which is then mobilized and transferred to the *Agrobacterium* (*A. tumefaciens* strain LBA4404, pBIN19 in MC1022 *E. coli* strain, and the conjugative plasmid pRK2013 in HB101 *E. coli* cells can be purchased from Clontech).

Start a liquid culture of LBA4404 *Agrobacterium* cells in M9 sucrose with 250 μg/ml streptomycin and grow at 28 to 30°C. The culture should be 2 days old, and the cells grow in a fine suspension in this medium. One day before the mating experiment, both *E. coli* cultures should be started. Grow HB101 cells (harboring helper plasmid pRK2013) in LB medium, and start the pBIN19/MC1022 culture in LB medium with 50 μg/ml kanamycin. Grow at 37°C.

To perform the mating, combine one drop of each of the three cultures on an LB or M9 plate without antibiotics. Grow overnight at 28 to 30°C. Scrape off the resultant culture and place into 2 ml of LB medium with 50 μg/ml kanamycin and 250 μg/ml streptomycin. The amount should be the size of an inoculation loop. Grow overnight at 28 to 30°C.

Remove 50 μl from the overnight culture and dilute into 1 ml of LB liquid medium and continue dilutions to 10^{-6} and plate on M9 sucrose medium with 50 to 100 μg/ml kanamycin and 250 μg/ml streptomycin. Grow overnight at 28 to 30°C. The colonies should contain your recombinant plasmid in *Agrobacterium* by now!

M9 medium is very good at selecting against the growth of *E. coli*. To make 1 L of 10 × M9 salts, in water dissolve 60 g of Na_2HPO_4, 30 g of KH_2PO_4, 5 g of NaCl, and 10 g NH_4Cl. Adjust the pH to 7.4 with 1 N HCl or NaOH, bring to 1 L, and autoclave. Also prepare individual stock solutions by dissolving the following in water: 20% sucrose, 1 M $MgSO_4$, 1 M $CaCl_2$, 1 mg/ml thiamine-HCl. Autoclave all solutions.

Finally, to make 1 L M9 for pouring into petri dishes, add 15 g of Bactoagar to 890 ml of MilliQ water, autoclave 20 min, cool to 50 to 60°C (the flask should be warm, but not hot to the touch). Add each component taking care to swirl after each addition. First, add 100 ml

10 × M9 salts, then add the following components in this order: 2 ml $MgSO_4$, 10 ml 20% sucrose, 1 ml thiamine-HCl, 0.1 ml $CaCl_2$; swirl fast at this step as some precipitate will form. Finally, add the antibiotics. If the medium looks dark brown the glucose has carmelized and should be discarded. The medium should be light brown with a slight blue/clear color.

Exercises

Experiment 1. Transformation and regeneration of tobacco plants expressing the GUS reporter gene under the regulation of constitutive and inducible promoters

Tobacco products are prepared from the dried leaves of *N. tabacum* L., an annual herb of the potato family (Solanaceae), native to tropical America. Tobacco contains nicotine, which is a health-threatening, addictive drug. However, it is used as the active ingredient in certain insecticides and to kill intestinal worms in farm animals.

Tobacco provides one of the most easy-to-use systems for plant transformation because of its high transformation and regeneration frequencies and relatively short life cycle. Since the first reports of *Agrobacterium*-based transformation, tobacco has been a model system for such studies. In spite of the fact that many important crops like corn and soybeans can now be transformed and regenerated, tobacco is still widely used in a variety of studies dealing with the expression of plant genes. This is particularly true with genes isolated from species with very long life cycles, such as trees, and for genes isolated from species that cannot yet be regenerated.

This experiment introduces concepts underlying the methodology of plant transformation. The basic concepts introduced in this experiment are also applicable to the second experiment in this chapter. Refer to Figure 32.1 for a diagram outlining this exercise. Experiment 1 will require approximately 10 to 12 weeks for completion.

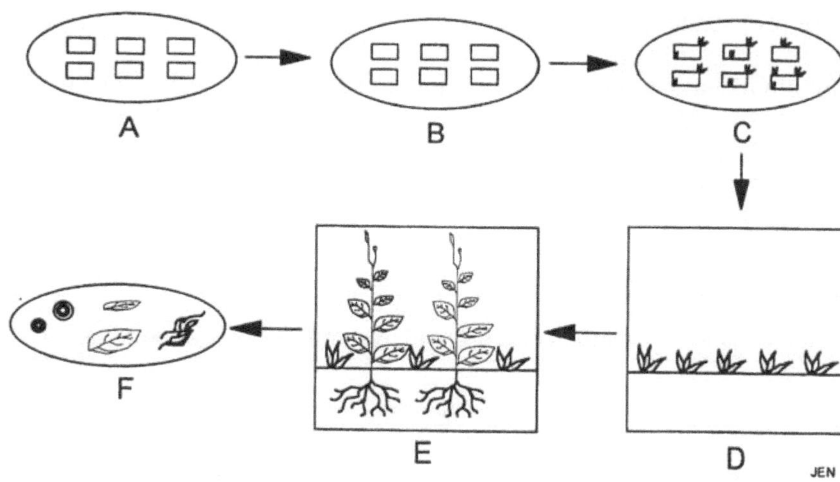

Figure 32.1 Outline of the tobacco transformation protocol. (A) Leaf sections incubated with *A. tumefaciens* strain LBA4404; (B) 2-day incubation in 1/2 MS medium; (C) callus proliferation and shoot formation after 4 to 6 weeks in TR medium with kanamycin and cefotaxime; (D) shoots (0.5 to 1 cm) transferred to a Magenta box containing 1/2 MS medium with kanamycin and cefotaxime; (E) after 5 to 6 weeks kanamycin-resistant shoots are rooted and ready to analyze; (F) harvest kanamycin-resistant plants and test for transformation. (From Trigiano, R. N. and Gray, D. J., Eds., *Plant Tissue Culture Concepts and Laboratory Exercises, First Edition*, CRC Press LLC, Boca Raton, FL, 1996.)

Note: The following experiments must be conducted in strict accordance with APHIS guidelines. All facilities, including laboratory, growth rooms, and greenhouses, must be inspected and approved by an APHIS representative as well as the appropriate office of your institution. The instructor and students must follow all procedures. *Agrobacterium tumefaciens* is a plant pathogen. Even though you will be using a "disarmed" strain (LBA4404) that cannot form plant tumors, it should be handled as if it could "escape" into the environment. All containers should be sterilized after use. Disposable tubes, petri plates, etc. should be placed in biohazard bags and autoclaved before discarding.

Materials

The following items are needed for each student or team of two to four students for the first part of this experiment.

- Two 125-ml Erlenmeyer flasks containing 50 ml of sterile half-strength MS medium
- Two 50-ml sterile Falcon™ tubes or sterile-capped centrifuge tubes for concentration of bacteria
- One to two sterilely grown tobacco plants in GA-7 Magenta vessels per each *A. tumefaciens* strain (two to four in total)
- Six 100 × 15-mm petri dishes containing one half-strength MS (1/2 MS) solid medium
- Six 100 × 15-mm petri dishes containing TR medium (MS basal medium containing 0.1 mg/L (0.5 μM) NAA, 0.2 mg/L (0.88 μM) BA, 30 g/L (87.6 mM) sucrose, 2.5 g/L Phytagel, pH 5.8, amended with 100 mg/L kanamycin and 200 mg/L cefotaxime)
- Ten GA-7 Magenta vessels or other similar-size vessels containing 75 ml of 1/2 MS medium supplemented with 100 mg/L kanamycin and 200 mg/L cefotaxime

The final part of the experiment (test of transformation) requires the following items per each group of students:

- 40 ml of GUS-assay buffer containing 0.1 M phosphate buffer pH 7.0, 5 mM potassium ferricyanide, 5 mM potassium ferrocyanide, and 10 mM EDTA
- 40 mg of X-gluc (5-bromo-4-chloro-3-indolyl β-*d*-glucuronic acid)
- 400 μ*l* of *N,N*-dimethylformamide

Follow the protocol described in Procedure 32.1 to complete this experiment.

Procedure 32.1
Transformation and regeneration of tobacco plants to study expression of reporter genes.

Step	Instructions and comments
1	Slice 1-week-old carrot hypocotyls into 10- to 15-mm sections and place about 20 hypocotyl sections into each petri dish containing 2B5 medium (B5 medium of Gamborg et al. [1968] — see Chapter three) with 2 mg/L (9 μM) 2,4-D and 8 g/L Difco Bactoagar.
2	Keep the tissue in the dark at 20 to 22°C for 2 to 3 days before the experiment.
3	Place the 2-day precultured hypocotyls in *Agrobacterium* solution for 5 min (prepared as described in Experiment 1).
4	Blot the excess liquid from the tissue, and place the hypocotyls onto 2B5 solid medium at 20 to 22°C for 2 to 3 days to allow for T-DNA transfer and the development of kanamycin resistance.

Procedure 32.1 continued
Transformation and regeneration of tobacco plants to study expression of reporter genes.

Step	Instructions and comments
5	Once the *Agrobacterium* cells are evident upon the explants, they should be transferred to 2B5 medium plates containing kanamycin and carbenicillin. Use two petri dishes per bacterial strain.
6	Place the cultures in the dark at 20 to 22°C, and after approximately 6 weeks, kanamycin-resistant calli will develop at the cut edges and inside the hypocotyls.
7	Transfer the callus tissue, trying to avoid taking the original explant tissues, to fresh 2B5 medium plates with kanamycin and carbenicillin and grow them for 6 more weeks under the same conditions. You need only use two petri dishes per bacterial strain. To shorten the time for this exercise, the kanamycin-resistant calli from the first transfer can be placed directly into suspension culture medium (step 8), although the number of truly transformed calli may be reduced.
8	Following the second transfer to solid media, calli are inoculated into the liquid suspension culture medium (2B5 medium with kanamycin and carbenicillin). No more than 1 g fresh weight of callus per flask should be used as the initial inoculum. Initial cell density should be about 1×10^5 cells/milliliter. Start at least three flasks per each bacterial strain.
9	Grow the cultures at 20 to 22°C in the dark with continuous shaking at 150 rpm. Subcultures should be made every 8 weeks by diluting 1:40 in fresh suspension culture medium, which results in a good approximation to the optimum initial cell density.
10	The number of cells at any time point of growth can be estimated by measuring the turbidity of the cell cultures with a Klett-Summerson spectrocolorimeter. Readings in this apparatus are expressed in arbitrary units. For example, Klett 100 (K100) corresponds to about 2×10^6 cells. The relationship is linear up to K150 using this apparatus. If this instrument is not available, use the estimate for cell density described in step 9.
11	After 8 weeks of growth in suspension culture, transfer cells to fresh medium.
12	Seven days after this transfer, collect the cells in a sterile 50-ml Falcon tube and centrifuge at low-speed (3000 rpm for 5 min).
13	Decant the supernatant and wash the pelleted cells by resuspending in fresh B5 medium. Repeat this operation three more times.
14	Take an aliquot of cells and incubate them in the X-gluc reagent as described in the previous exercise.
15	Observe the stained cells under a high-power dissection microscope, and compare the results obtained between different transformed cultures and between the two kinds of promoter-GUS fusions.
16	To initiate embryogenesis, cell suspension cultures should be transferred to B5 medium without 2,4-D. Start two flasks per bacterial strain with an initial cell density of 8×10^5 cells/milliliter.
17	Incubate the cultures for 3 days under the same conditions and then dilute them to low density — i.e., 2 to 5×10^4 cells/milliliter in 20 ml of liquid B5 medium.
18	Grow the cultures in PGR-free B5 media for 4 weeks at 20 to 22°C without shaking.
19	Remove a sample from each flask and stain with X-gluc. Focus your observations on the developmental differences found in different stages of embryo development. Also note any variations in GUS expression in lines transformed with pBI121 vs. pCAH-1678.
20	To recover whole carrot plants from these cultures, allow them to grow in B5 medium until they reach the torpedo stage (see Chapter nineteen).
21	They can then be transferred to solid MSr medium containing kanamycin and carbenicillin and placed in the light to germinate.
22	Later, when green, healthy plantlets are seen, transfer them to soil.

Anticipated results

After the initial 2 to 3 days of incubation with *Agrobacterium*, you may find bacteria overgrowing the tissue explants. If this happens, start over. The next time be sure that you blot the tissue very well before placing it into the 1/2 MS medium. It is also possible that your cefotaxime was added when the medium was too hot, or that it was added at too low a concentration. You may want to prepare a new batch of plates to eliminate this possibility. Once the tissue is covered by bacteria, it is almost impossible to remove, even after washing and rinsing the tissues in antibiotic solution.

Four weeks after transfer to regeneration medium (TR) the leaf tissue may assume a "puckered" shape, and callus may have formed around the cut edges of the leaf sections. Small buds may also be apparent by the last 2 weeks of this period.

After buds are transferred to plant growth regulator (PGR)-free medium, roots will probably start emerging from the base of the shoots within 2 to 3 weeks. Nevertheless, you should wait until the primary root branches before continuing to the next stage. Sometimes some untransformed plants (escapes) are able to form a primary root, but this root will not develop extensive root hairs nor will it branch. Besides showing good root growth, kanamycin-resistant plants will grow taller, healthier, and greener than nonresistant plants. At the end of this period (5 to 6 weeks), you should be able to distinguish which plants are actually transformed and which are not. Untransformed shoots will look sick, pale-green to almost white in color, and be very stunted. You should find up to 50% of the plants growing in the rooting medium to be transformed.

During the analysis of all transformed plants, you will notice that each one can exhibit a different level of GUS expression. Low or high expression levels will be indicated by pale or dark blue staining, respectively. This phenomenon is generally ascribed to the "position effect," which is thought to result from the random positioning of the inserted T-DNA in the plant genome. Some of these random insertion sites appear to favor the expression of inserted genes, whereas others seem to suppress their expression.

You should also notice differences in the histological staining of both types of plants. The pCAH-1678 contains the promoter region of a 3-hydroxy-3-methylglutaryl coenzyme A reductase (HMGR) gene from *Camptotheca acuminata* (Decaisne), a Chinese medicinal tree (Figure 32.2). This gene is involved in the production of mevalonate, which is a precursor for a large number of the plant terpenoids. One site of terpenoid synthesis, and thus of GUS staining, should be the glandular hairs on the surface of the leaves and stem. Normally, tobacco accumulates defense compounds in these glands, which require large amounts of mevalonate for their synthesis. To check the pattern of GUS expression in these plants, see Burnett et al. (1993). Plants transformed with the pBI121 vector have GUS expressed under the control of plant viral promoter, the 35S promoter from the cauliflower mosaic virus. This is considered to be a more constitutive promoter, so you should see widespread expression in virtually every part of the plant.

Questions

- What characteristics should a gene have to be used as a reporter? Can a highly expressed gene in a species be used as a reporter gene in the same species? Why or why not?
- How can you explain the differences seen in GUS staining between individual transformed plants and plants transformed with different promoter-GUS fusions?
- List three desirable characteristics that could be given to a transgenic plant using these transformation techniques and explain how they would work.

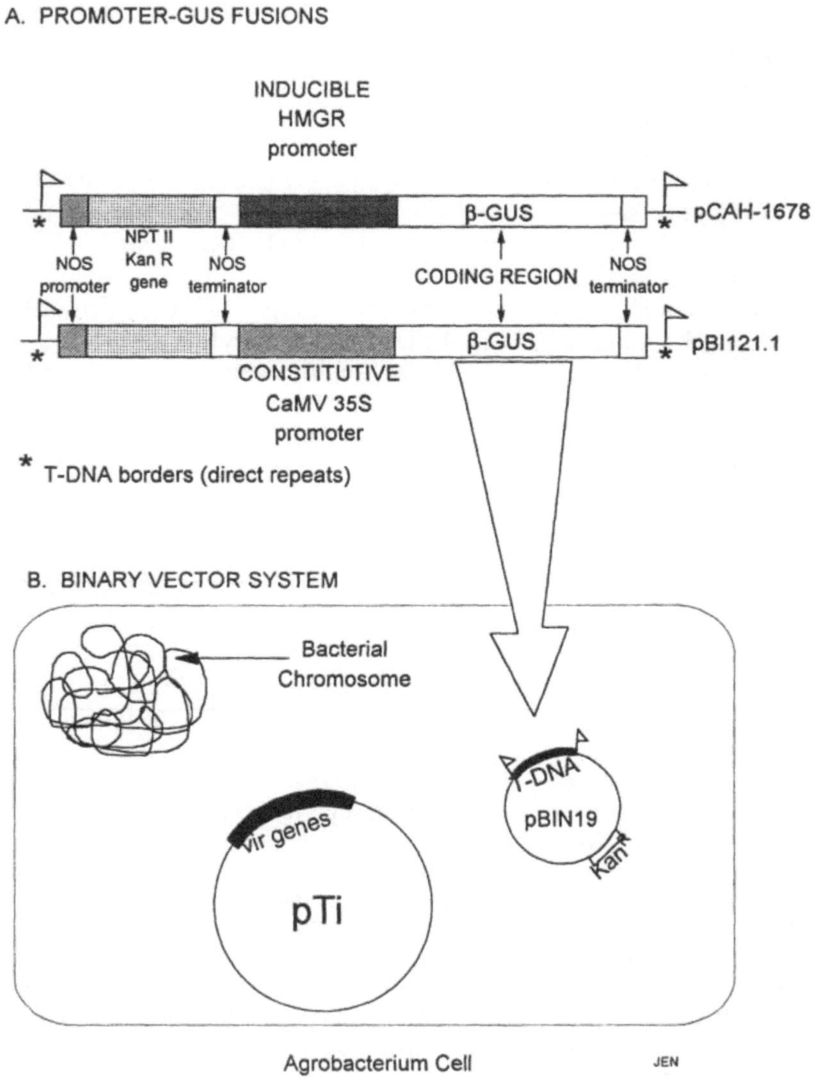

Figure 32.2 (A) Physical map of the promoter-GUS fusions used in these exercises; (B) diagram of the binary vector system. (From Trigiano, R. N. and Gray, D. J., Eds., *Plant Tissue Culture Concepts and Laboratory Exercises, First Edition*, CRC Press LLC, Boca Raton, FL, 1996.)

- How could a promoter conferring tissue specificity (i.e., high expression in the leaf hairs) be used to genetically modify a crop plant?

Experiment 2. Effect of wounding on expression of the GUS gene under the regulation of different promoters

Although this part of the exercise is optional, it illustrates not only an additional way to test for transformation, but it also represents an easy way to analyze quantitatively differences between specifically promoter-driven activity. It can be done in parallel to the test for transformation; however, it may be preferable to identify plants that express high levels of GUS activity and use them in this experiment.

Materials

- Flask containing 50 ml 1/2 MS medium
- One fully expanded transformed tobacco leaf per each plasmid used
- Cork borer (#3 or 4) or paper hole punch
- 1.7-ml microcentrifuge tubes
- Liquid nitrogen or ultrafreezer at –80°C (if a –80°C freezer is not available, a freezer at –20°C can be used)
- Pellet-pestle grinder or any other type of tissue homogenizer
- Microcentrifuge
- 20 ml of lysis buffer containing 50 mM $NaPO_4$ (pH 7.0), 0.1% Triton X-100, 10 mM β-mercaptoethanol (BME), 10 mM EDTA, 0.1% lauryl sarkosyl
- 1 ml substrate buffer containing 2 mM 4-methylumbilliferone glucuronide (MUG); prepared by dissolving 7.6 mg of MUG in 10 ml of lysis buffer
- 20 to 40 ml stop buffer solution (0.2 M Na_2CO_3)
- Fluorimeter

Follow the methodology outlined in Procedure 32.2 to complete this experiment.

Procedure 32.2
Effect of wounding on the regulation of different promoters.

Step	Instructions and comments
1	Set up two petri dishes containing 10 to 15 ml of 1/2 MS medium (one per each promoter-GUS fusion). The experiment will require prolonged incubation times (up to 72 h), which is enough time for bacterial or fungal contamination to occur. Since contamination would affect the results, it is advisable to set this experiment up in a laminar flow hood using sterile techniques.
2	Starting from the apex, note the size of leaves as you near the base. Select the largest leaf nearest the apex that represents the youngest fully expanded leaf.
3	Cut 5- to 7-mm disks from the leaf using a hole punch or cork borer. Avoid cutting disks from large veins. Try to obtain all of the disks from the same leaf because basal GUS expression varies depending on the developmental stage of the leaf.
4	Float 18 disks (obtained from one leaf) in the 1/2 MS medium in each petri dish.
5	Incubate the leaf disks at room temperature under continuous light 25 to 50 ($\mu mol \cdot m^{-2} \cdot sec^{-1}$).
6	Remove three leaf disks, place them into a 1.7-ml microcentrifuge tube labeled as "zero" and freeze them in liquid nitrogen or in –80°C freezer. Do this at the beginning of the experiment and this will be your zero time control.
7	Collect, label, and freeze three disks at 12, 24, 48, and 72 h after wounding — this means after you cut them. Once all disks have been collected you are ready to quantitate their GUS activity. GUS activity can be quantitated by fluorimetry. The protocol described in this exercise involves the use of MUG, a fluorochrome substrate. MUG produces the fluorescent compound 4-methylumbelliferone after GUS enzymatic cleavage. The amount of fluorescence detected in this assay is a measure of relative GUS activity and, thus, the level of GUS expressed by the promoter used in the construct.
8	Take the microcentrifuge tubes containing the frozen tissues, add 100 µl of lysis buffer to each tube, and then grind the tissue using a pellet pestle grinder or any available tissue homogenizer. The tissue should be thoroughly homogenized.
9	Spin the tubes for 3 to 5 min in a microcentrifuge at full speed (10000 to 14000 rpm) to pellet the cuticle and other debris.

Procedure 32.2 continued
Effect of wounding on the regulation of different promoters.

Step	Instructions and comments
10	Transfer the supernatant to another labeled tube without disturbing the pellet (particulate material will reflect light and produce artificially high readings). Discard the tube containing the pellet.
11	Pipette 25 µl from the supernatant tube to two additional tubes.
12	Add 1 ml of stop buffer to one tube which should be labeled "time __ blank" where __ is 0, 12, 24, 48, or 72 h. This sample will be used to calibrate the fluorimeter before reading each experimental sample.
13	Add 25 µl of substrate buffer containing 2 mM MUG (10 ml lysis buffer, 7.6 mg MUG) to each tube.
14	Mix well by vortexing, centrifuge briefly, and incubate at 37°C for 30 min.
15	After incubation, add 1 ml of stop buffer to the experimental tubes and mix well.
16	Assay samples in fluorimeter according to the instrument instructions. If the sample exceeds the linear range of the fluorimeter, it may be diluted with reaction stop buffer and read again.
17	Draw a graph for each transformed plant assayed plotting the time of induction (Y axis) vs. units of fluorescence per hour (X axis).

Anticipated results

This experiment involves the induction of GUS expression by wounding, which stimulates the HMGR promoter. Thus, there should be an increase in GUS activity with elapsed time (Burnett et al., 1993). In contrast, the constitutive 35S promoter should not show any wound induction, but should show the same levels of expression throughout the experiment.

Questions

- Why does GUS activity increase over time during the wounding experiment in the pCAH-1678 transgenic plants and not in the pBI121 transgenic plants?
- When do you think it would be desirable to use a constitutive promoter and when would you use an inducible promoter to engineer a plant?
- How can you explain the fact that the same construct produces transformed tobacco plants with different levels of gene expression?

Experiment 3. Somatic embryogenesis from carrot hypocotyls transformed using Agrobacterium tumefaciens

Carrot is an edible biennial plant which produces a modified, fleshy storage root. This storage organ develops from the root and hypocotyl during the first year of growth, and the stored food is mobilized the following year when the carrot plant "bolts" to make flowers and set seeds. Carrots are hardy plants that can tolerate a wide range of climate and soils. They belong to the celery family (Apiaceae), which is abundantly represented in the Northern Hemisphere.

Somatic embryogenesis in carrot has been used as a model system for the study of plant development for over 30 years (Steward et al., 1964). The isolation of large quantities of staged carrot somatic embryos is relatively simple, which is not always true of seed-derived embryos.

This procedure was designed to generate transformed carrot plants by coupling the *Agrobacterium*-based transformation system to the process of somatic embryogenesis. The exercise also illustrates the application of transformation technology as a tool to study a specific biological process, in this case plant development. Development involves temporal and spatial changes in cellular phenotype. In addition to morphological changes, specific enzyme activities appear and disappear during morphogenesis. These biochemical changes are specifically regulated as sets of gene functions that are simultaneously activated and inactivated.

This experiment will require approximately 26 weeks for completion, although it can be shortened to 20 weeks (see diagram Figure 32.3). This exercise is adapted from a paper by Thomas et al. (1989).

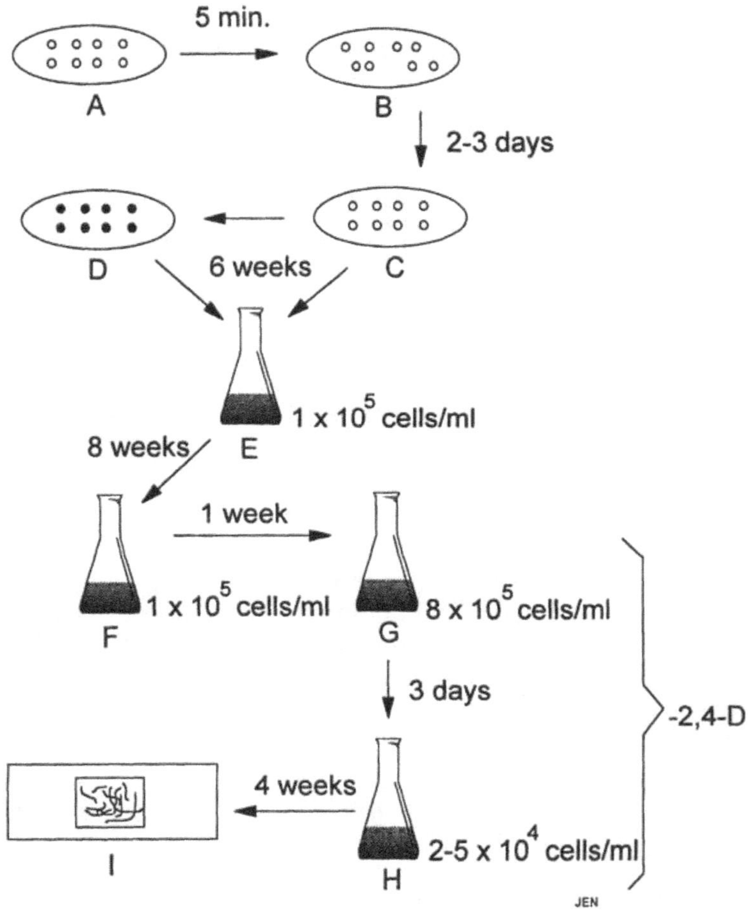

Figure 32.3 Outline of the carrot transformation protocol. (A) Hypocotyl sections incubated with *A. tumefaciens* strain LBA4404; (B) incubation for 2 to 3 days in 2B5 solid medium; (C) transfer of the sections into 2B5 medium containing kanamycin and carbenicillin and growth for 6 weeks; (D) second transfer to fresh plates and six more weeks of growth or (E) direct transfer of callus tissue into suspension culture medium; (F) culture in liquid medium (initial cell density 1×10^5 cells/milliliter) for 8 weeks; (G) transfer to fresh suspension culture medium and incubation for 7 days followed by removal of 2,4-D by washing four times in B5 medium — transfer and incubate for 3 days at high cell density (8×10^5 cells/milliliter) in B5 medium; (H) dilution of the cells to low density (2 to 5×10^4 cells/milliliter) and incubation for 4 weeks without shaking; (I) harvest embryos and examine for transgene expression. (From Trigiano, R. N. and Gray, D. J., Eds., *Plant Tissue Culture Concepts and Laboratory Exercises, First Edition*, CRC Press LLC, Boca Raton, FL, 1996.)

Materials

- Four GA-7 Magenta vessels containing 75 ml of MSr medium
- Bacterial suspensions of pBI121 and pCAH-1678 *Agrobacterium* strains prepared as in Experiment 1 (resuspend in PGR-free B5 medium)
- Eight 100 × 15-mm petri dishes containing 2B5 medium (B5 medium containing 2 mg/L (9 μM) 2,4-D and 8 g/L of Difco Bactoagar)
- Eight 100 × 15-mm petri dishes containing 2B5 medium amended with 100 mg/L kanamycin and 400 mg/L carbenicillin
- Twelve 125-ml Erlenmeyer flasks containing suspension culture medium (40 ml of 2B5 liquid medium supplemented with 100 mg/L kanamycin and 400 mg/L carbenicillin)
- Six 50-ml sterile Falcon tubes or any similar size sterile-capped centrifuge tubes
- One 1-L Erlenmeyer flask containing 500 ml of sterile PGR-free B5 medium
- Six 125-ml Erlenmeyer flasks containing 40 ml of plain B5 medium
- Klett-Summerson spectrocolorimeter (optional)

Follow the protocols listed in Procedure 32.3 to complete this experiment.

Procedure 32.3
Transformation of carrot hypocotyls to study a developmental process: somatic embryogenesis.

Step	Instructions and comments
1	Slice 1-week-old carrot hypocotyls into 10- to 15-mm sections and place about 20 hypocotyl sections into each petri dish containing 2B5 medium (B5 medium of Gamborg et al. [1968] — see Chapter three) with 2 mg/L (9 μM) 2,4-D and 8 g/L Difco Bactoagar.
2	Keep the tissue in the dark at 20 to 22°C for 2 to 3 days before the experiment.
3	Place the 2-day precultured hypocotyls in *Agrobacterium* solution for 5 min (prepared as described in Experiment 1).
4	Blot the excess liquid from the tissue, and place the hypocotyls onto 2B5 solid medium at 20 to 22°C for 2 to 3 days to allow for T-DNA transfer and the development of kanamycin resistance.
5	Once the *Agrobacterium* cells are evident upon the explants, they should be transferred to 2B5 medium plates containing kanamycin and carbenicillin. Use two petri dishes per bacterial strain.
6	Place the cultures in the dark at 20 to 22°C, and after approximately 6 weeks, kanamycin-resistant calli will develop at the cut edges and inside the hypocotyls.
7	Transfer the callus tissue, trying to avoid taking the original explant tissues, to fresh 2B5 medium plates with kanamycin and carbenicillin and grow them for 6 more weeks under the same conditions. You need only use two petri dishes per bacterial strain. To shorten the time for this exercise, the kanamycin-resistant calli from the first transfer can be placed directly into suspension culture medium (step 8), although the number of truly transformed calli may be reduced.
8	Following the second transfer to solid media, calli are inoculated into the liquid suspension culture medium (2B5 medium with kanamycin and carbenicillin). No more than 1 g fresh weight of callus per flask should be used as the initial inoculum. Initial cell density should be about 1×10^5 cells/milliliter. Start at least three flasks per each bacterial strain.
9	Grow the cultures at 20 to 22°C in the dark with continuous shaking at 150 rpm. Subcultures should be made every 8 weeks by diluting 1:40 in fresh suspension culture medium, which results in a good approximation to the optimum initial cell density.

Procedure 32.3 continued
Transformation of carrot hypocotyls to study a developmental process: somatic embryogenesis.

Step	Instructions and comments
10	The number of cells at any time point of growth can be estimated by measuring the turbidity of the cell cultures with a Klett-Summerson spectrocolorimeter. Readings in this apparatus are expressed in arbitrary units. For example, Klett 100 (K100) corresponds to about 2×10^6 cells. The relationship is linear up to K150 using this apparatus. If this instrument is not available, use the estimate for cell density described in step 9.
11	After 8 weeks of growth in suspension culture, transfer cells to fresh medium.
12	Seven days after this transfer, collect the cells in a sterile 50-ml Falcon tube and centrifuge at low-speed (3000 rpm for 5 min).
13	Decant the supernatant and wash the pelleted cells by resuspending in fresh B5 medium. Repeat this operation three more times.
14	Take an aliquot of cells and incubate them in the X-gluc reagent as described in the previous exercise.
15	Observe the stained cells under a high-power dissection microscope, and compare the results obtained between different transformed cultures and between the two kinds of promoter-GUS fusions.
16	To initiate embryogenesis, cell suspension cultures should be transferred to B5 medium without 2,4-D. Start two flasks per bacterial strain with an initial cell density of 8×10^5 cells/milliliter.
17	Incubate the cultures for 3 days under the same conditions and then dilute them to low density — i.e., 2 to 5×10^4 cells/milliliter in 20 ml of liquid B5 medium.
18	Grow the cultures in PGR-free B5 media for 4 weeks at 20 to 22°C without shaking.
19	Remove a sample from each flask and stain with X-gluc. Focus your observations on the developmental differences found in different stages of embryo development. Also note any variations in GUS expression in lines transformed with pBI121 vs. pCAH-1678.
20	To recover whole carrot plants from these cultures, allow them to grow in B5 medium until they reach the torpedo stage (see Chapter nineteen).
21	They can then be transferred to solid MSr medium containing kanamycin and carbenicillin and placed in the light to germinate.
22	Later, when green, healthy plantlets are seen, transfer them to soil.

Anticipated Results

Four weeks after incubating *Agrobacterium*, infected callus tissue will start forming around the inoculated hypocotyls. By 6 weeks the kanamycin-resistant calli will be up to 1 cm in diameter. A second transfer of calli to fresh medium is recommended because it reduces the possibility of continued growth of untransformed calli. We have found that all of the calli growing after the second subculture were resistant to kanamycin and truly transformed. Nevertheless, as mentioned earlier, the second transfer may be skipped to shorten the exercise.

Four weeks after transferring the cell suspensions into embryogenic conditions (low cell density, absence of 2,4-D) you should get a mixed population of somatic embryos. Globular embryos will appear as spherical, multicellular structures 250 to 350 μm in diameter. Heart- and torpedo-staged embryos can reach sizes up to 1 mm.

GUS expression should be detectable throughout the developmental stages of somatic embryogenesis in the cell lines transformed with pBI121, which contains the constitutive promoter-GUS fusion. On the other hand, transformed plants containing the pCAH-1678, the HMGR promoter, will show variability in GUS staining during different developmental stages.

Questions

- What are some possible reasons for any differences in GUS expression seen in cell lines transformed with each *Agrobacterium* line? Are these differences correlated with embryo development?
- How could the somatic embryogenesis system be used to study gene regulation?
- How does the wound inducibility of the pCAH-1678 promoter relate to developmental expression of the promoter in carrot? Compare these results to the experiment in tobacco.

Literature cited

Bevan, M.W. 1984. Binary *Agrobacterium* vectors for plant cell transformation. *Nucleic Acids Res.* 12:8711–8721.

Burnett, R.J., I.E. Maldonado-Mendoza, T.D. McKnight, and C.L. Nessler. 1993. Expression of a 3-hydroxy-3-methylglutaryl coenzyme A reductase gene from *Camptotheca acuminata* is differentially regulated by wounding and methyl jasmonate. *Plant Physiol.* 103:41–48.

Gamborg, O.L., R.A. Miller, and K. Ojima. 1968. Nutrient requirements of suspension cultures of soybean root cells. *Exp. Cell. Res.* 50:151–158.

Jefferson, R.A., T.A. Kavanagh, and M.W. Bevan. 1987. GUS fusions β-glucuronidase as a sensitive and versatile gene fusion marker in higher plants. *EMBO J.* 6:3901–3907.

Liu, X.J., M. Rocha-Sosa, S. Rosahl, L. Willmitzer, and W. Frommer. 1991. A detailed study of regulation and evolution of the two classes of patatin genes. *Plant Mol. Biol.* 17:1139–1154.

Maldonado-Mendoza, I.E. and C.L. Nessler. 1993. Biolistic transient and stable expression of β-glucuronidase (GUS) in embryogenic cell suspension cultures of *Papaver somniferum. Plant Physiol.* (Suppl.) 102:152.

Martin, T., R.V. Wöhner, S. Hummel, L. Willmitzer, and W.B. Frommer. 1992. The GUS reporter system as a tool to study plant gene expression. In: *GUS Protocols. Using the GUS Gene as a Reporter of Gene Expression*. Gallagher, S.R., Ed. Academic Press, San Diego.

Murashige, T. and F. Skoog. 1962. A revised medium for the rapid growth and bioassay with tobacco tissue cultures. *Physiol. Plant.* 15:473–479.

Stewart, F.C., M.O. Mapes, A.E. Kent, and R.D. Holsten. 1964. Growth and development of cultured plant cells. *Science.* 14:320–327.

Stockhaus, J., J. Schell, and L. Willmitzer. 1989. Correlation of the expression of the photosynthetic gene ST-LS1 with the presence of chloroplasts. *EMBO J.* 8:2445–2451.

Thomas, J.C., M.J. Guiltinan, S. Bustos, T. Thomas, and C.L. Nessler. 1989. Carrot (*Daucus carota*) hypocotyl transformation using *Agrobacterium tumefaciens. Plant Cell Rep.* 8:354–357.

chapter thirty-three

Transformation of chrysanthemum leaf explants using Agrobacterium tumefaciens

Robert N. Trigiano

In the previous decade, many methodologies were developed to introduce or move a wide variety of genes from one species into another. Many of these techniques have been discussed in other chapters and will not be considered further in this chapter. However, in the following experiments, we will use *Agrobacterium tumefaciens*, a soil inhabiting, pathogenic (disease-causing) bacterium occurring naturally in many areas of the world, as the agent/vector or shuttle that transfers gene(s) between the two species. This bacterium can be aptly described as nature's own and the original genetic engineer since during pathogenesis (disease development) several genes located on a plasmid (a small circular piece of DNA) within the bacterium are delivered and incorporated into the genome of host plant cells. These genes cause an unusually high number of host cell divisions (hyperplasia) to occur and, as a result, a tumor or gall is formed. Hence the common name of the disease is crown gall. We can take advantage of this natural gene delivery system. Via molecular biological techniques, genes that incite disease can be removed from the T-DNA region of the plasmid and replaced with gene(s) and promoters of our choice.

In recent years there has been increased interest in the transformation of ornamental plants, and chrysanthemum (*Dendranthema grandiflora* Tzvelev.) is no exception (e.g., de Jong et al., 1993; Lowe et al., 1993; Urban et al., 1994). Most of the interest in genetically transforming chrysanthemum thus far has been in the areas of creating new flower colors and conferring resistance to tomato spotted wilt virus. Before chrysanthemum could be genetically engineered, a reliable regeneration system had to be developed. The plant regeneration system described in Chapter fifteen is ideally suited in this regard for the following reasons: chrysanthemums are very susceptible to infection by certain strains of *A. tumefaciens*; shoots are formed directly (without an intervening callus phase); shoots develop from the cut edges of the explant, which are easily infiltrated with a bacterial vector; and the plant tissue has high regenerative capacity. For a more complete discussion of plant transformation with *A. tumefaciens*, refer to Chapters thirty-one and thirty-two.

A note of caution before beginning the experiments in this chapter: when working with any plant pathogen, in this case *A. tumefaciens*, precautions and care must be taken to limit contamination of the work area and "bystander plants" both in the laboratory and in the environment. Inoculated plants and contaminated instruments and cultures must never

0-8493-2029-1/00/$0.00+$.50

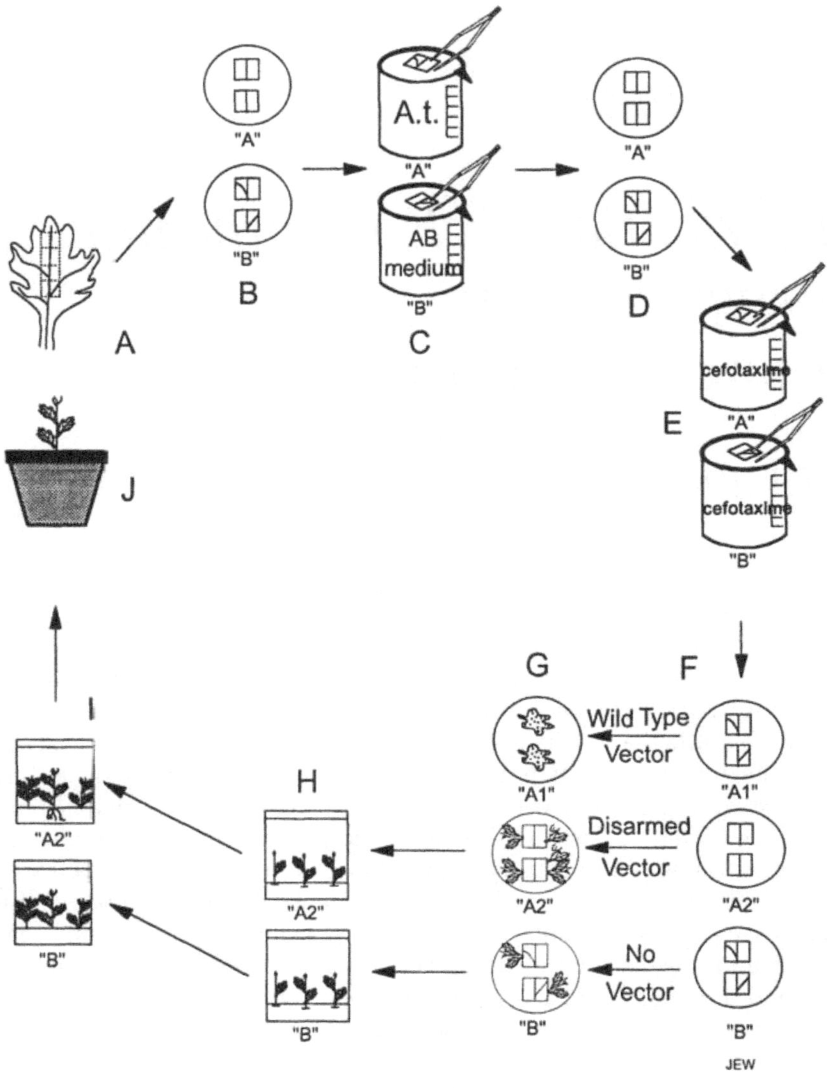

Figure 33.1 Diagrammatic representation of transformation experiments using chrysanthemum leaves and *Agrobacterium tumefaciens*. (A) Leaves used for midrib explants. (B) Preincubation of leaf sections for 2 days before inoculation with treatment. (C) Treatments: "A" explants will be inoculated with bacterium (A.t.); "B" explants will be a control-dish treated only with culture medium (AB). (D) Co-culture with bacteria ("A") for 2 days. (E) Transfer of explants from both treatments to medium containing cefotaxime to kill bacteria. (F) Explants transferred to medium lacking PGRs, but amended with cefotaxime and kanamycin ("A2" and "B"). "A1" = inoculated with "wild type" bacteria and "A2" = disarmed bacteria with NPTII gene; "B" without bacteria. (G) "A1" transformed explants producing callus; "A2" transformed shoots; "B" shoots from control treatment. (H) Rooting of putative transformed and "escape" shoots from "A2" and "B" on kanamycin-containing medium. (I) Some of the shoots produced in "A2" are able to form roots (transformed), whereas others cannot (escapes) in the presence of kanamycin; none of the shoots of "B" produce shoots (all escapes). (J) Growth of plant harboring the NPTII gene. (From Trigiano, R. N. and Gray, D. J., Eds., *Plant Tissue Culture Concepts and Laboratory Exercises, First Edition*, CRC Press LLC, Boca Raton, FL, 1996.)

leave the laboratory. All cultures of bacteria (except stock cultures), plant tissues, and plants exposed to the bacterium must be destroyed, by autoclaving, at the conclusion of the experiments. All instruments and glassware also should be decontaminated by autoclaving.

The experiments outlined in this chapter (Figure 33.1) are designed for smaller, more advanced classes (less than ten students) or students participating in special topics or advanced individual study problems. Generally, they require more developed skills, careful attention to detail, and closer supervision by the instructor. These experiments are adapted primarily from those published by Trigiano and May (1994).

Exercises

Experiment 1. Transformation of cells using "wild-type" Agrobacterium tumefaciens

This experiment is designed as a paired variate (Chapter twenty-three) and uses the mum cultivars 'Goldmine' and 'Iridon', although others may be substituted. We suggest that one half of the students or team of students work with one cultivar, while the other half works with the remaining cultivar.

Materials

The following items are required for each student or team of students:

- Liquid and solidified AB medium (Table 33.1)
- *Agrobacterium tumefaciens* strain 281 (Hood et al., 1986) or any other virulent strain may be secured from any number of sources, including a Plant Pathology or Plant Sciences Department, but first ascertain whether or not a USDA permit for pathogens is needed.
- 125-ml erlenmeyer flasks
- 0.22-µm filters and syringes
- Autoclavable biological hazard bags
- Several pieces of sterile 9-cm filter paper
- Four sterile 50-ml beakers
- Six 60 × 15-mm petri dishes containing MS medium amended with 0.22 mg/L (1.0 µM) BA and 2 mg/L (11.5 µM) IAA (shoot initiation medium)
- Six 60 × 15-mm petri dishes containing shoot initiation medium and 100 mg/L cefotaxime
- Six 60 × 15-mm petri dishes containing MS medium without plant growth regulators (PGRs), but amended with 100 µg/ml cefotaxime
- Liquid MS medium with 100 µg/ml cefotaxime

Note: Gloves and a particle mask should be worn when working with antibiotics and PGRs.

These antibiotic solutions should be sterilized using a syringe and a 0.22-µm filter and added after the basal medium has been autoclaved and cooled, but not hardened.

Follow the protocols outlined in Procedure 33.1 to complete this experiment.

Anticipated results

Leaf sections of 'Goldmine' not treated with bacteria ("B" petri dishes) should not have formed shoots, but may have a few adventitious roots and, perhaps, some scanty callus along the cut edges. Leaf sections of 'Iridon' not treated with bacteria ("B" dishes) should

Table 33.1 Composition of AB Medium for Growth of *Agrobacterium tumefaciens*

20× Salts (A) (500 ml)	20× Buffer (B) (500 ml)	Vitamin (100 ml)
10 g NH_4Cl	30 g K_2HPO_4	20 mg biotin (filter sterilized)
12.5 g $MgSO_4 \cdot 7H_2O$	11.5 g NaH_2PO_4	
1.5 g KCl		
0.1 g $CaCl_2$		
0.025 g $FeSO_4$		

Note: Start with 400 ml of water and add 25 ml each of the 20× salts (A) and buffer (B). Dissolve 2.5 g of D-glucose, bring volume to 500 ml, and add 7.5 g of agar. Autoclave, allow to cool (but not harden) and aseptically pipette 0.1 ml of biotin stock into the medium. Swirl to mix biotin and disperse agar and pour into vessels.

Derived from White, F.F. and E.W. Nester, *J. Bateriol.* 141, 1134, 1980; Trigiano, R. N. and Gray, D. J., Eds., *Plant Tissue Culture Concepts and Laboratory Exercises, First Edition,* CRC Press LLC, Boca Raton, FL, 1996.

Procedure 33.1
Transformation of cells using "wild type" *Agrobacterium tumefaciens.*

Step	Instructions and comments
1	Surface disinfest mum leaves with 10% bleach for 5 min and then rinse three times with sterile distilled water.
2	Excise four midrib sections from each leaf and place two sections per petri dish containing MS medium with 1.0 µM BA and 11.5 µM IAA. Label the dishes "1A" and "1B" etc. The number indicates the leaf number and the "A" and "B" constitute the "pair" in the design of the experiment. Using this design, specific results may be attributed to a treatment and not to other variables, such as physiological status of the explant. Maintain the identity of the pairs and incubate the cultures at 25°C with 25 $\mu mol \cdot m^{-2} \cdot sec^{-1}$ for 3 days.
3	Twenty-four hours before leaf sections are to be inoculated with bacteria, start liquid cultures of *A. tumefaciens* A281 — supervirulent (Hood et al., 1986) or other. Select a plate with a "well-grown lawn" of bacterial colonies and pipette about 1.0 ml of sterile AB medium onto the surface of the agar. Rub gently with a sterile glass rod or an inoculation loop to dislodge bacteria. Aseptically transfer about 0.5 ml of the bacterial suspension to 50 ml of AB medium in each of two 125-ml erlenmeyer flasks. Incubate on a rotary shaker at 75 to 100 rpm at 25°C for 24 to 36 h when the suspension appears cloudy or almost opaque. Do not use suspension older than 48 h. *Note:* many procedures will require that the bacterial suspension be adjusted to a certain colony-forming unit (cfu) concentration or optical density at a specified wavelength, usually between 500 and 600 nm (see Urban et al., 1994). Although in most experiments cfu should be standardized, in our experience this step is not necessary for the successful completion of this laboratory exercise.
4	Three days after the initiation of leaf section cultures, aseptically transfer about 10 ml of the bacterial suspension to sterile 50-ml beakers. Immerse all leaf sections from the petri dishes labeled "A," one at a time, in the suspension of bacteria for 10 sec. Remove excess fluid and bacteria by blotting the leaf sections on sterile, dry filter paper in petri dishes. Return the sections to their original petri dishes. Using different sterile forceps, submerge leaf sections from petri dishes labeled "B" in sterile AB medium contained in sterile 50-ml beakers for 10 sec, blot dry, and return to their original dishes. Incubate all petri dishes as before for a maximum of 2 days. The suspension of bacteria and filter papers should be autoclaved before discarding.

Procedure 33.1 continued Transformation of cells using "wild type" *Agrobacterium tumefaciens*	
Step	Instructions and comments
5	Filter sterilize 25 ml of MS liquid medium containing 100 μg/ml cefotaxime (Sigma Chemical Co, St. Louis, MO) into each of two sterile 50-ml beakers. Vigorously agitate the leaf sections that have been co-cultivated with bacteria in the solution and blot dry with sterile filter paper. Use different forceps and solutions for washing bacteria-treated and bacteria-free leaf sections. Incubate all sections on MS medium amended with PGRs and 100 μg/ml cefotaxime for 9 days in the conditions described in Step 2.
6	Transfer all leaf sections to MS medium without PGRs, but containing 100 μg/ml. Incubate all petri dishes for an additional 21 days or for a total of 35 days from the initiation of cultures.

have produced plentiful shoots, a few roots, and little or no callus at the cut edges (Figure 33.2a). Successful transformation of cells within the leaf explants of the corresponding pairs of both 'Goldmine' and 'Iridon' ("A" petri dishes) is indicated by the production of copious amounts of yellow-green callus that is typically not produced during shoot formation (Figure 33.2b). Callus growth may be patchy on some explants. Shoots and roots should not be present on explants of either cultivar in this treatment.

Callus and/or the explants treated with bacteria may be transferred to MS medium without PGRs supplemented with 100 μg/ml cefotaxime every 3 to 4 weeks. Callus should continue to grow (autotrophic for PGRs) and can be maintained for indefinite periods. At

Figure 33.2 Transformation of mum leaf explants of 'Iridon' using *A. tumefaciens* wild type A281. (a) Explant treated only with AB medium (without bacteria) produced shoots, whereas (b) the explant treated with bacteria formed copious amounts of callus. From Trigiano, R. N. and Gray, D. J., Eds., *Plant Tissue Culture Concepts and Laboratory Exercises, First Edition,* CRC Press LLC, Boca Raton, FL, 1996.)

the conclusion of the experiment, autoclave all cultures including the petri dishes containing used medium.

Questions

- Why is the treatment of dipping leaf sections into sterile AB medium without bacteria included in the experiment?
- Why is it necessary to maintain cefotaxime in the medium after co-cultivation with the bacteria?
- Why is the callus produced on the leaf sections autotrophic with regard to PGRs?

Experiment 2. Regeneration of transgenic chrysanthemum plants

The protocol used in this experiment is similar to the protocol used in Experiment 1 except that *A. tumefaciens* "wild-type" A281 is replaced with EHA105 (Urban et al., 1994), a disarmed A281 lacking genes encoding PGRs but harboring a gene (and CMV-35s promoter) for neomycin phosphotransferase (NPT II). This gene, when incorporated in the plant genome, will confer resistance to kanamycin, an antibiotic. The experiment is designed as a paired variate (see Experiment 1); two leaf sections co-cultivated with bacteria and two leaf sections not treated with bacteria from the same leaf will serve as the pairs. Only the cultivar 'Iridon' is used in this exercise.

Materials

In addition to the items listed under Experiment 1, each student or team of students will require the following:

- Six 60 × 15-mm petri dishes containing MS medium without PGRs and supplemented with 100 µg/ml cefotaxime and 50 µg/ml kanamycin (Sigma)
- Disarmed strain of A281 harboring the NPTII gene (EHA105) (see suggested procurement procedures listed in Experiment 1)

Note: Gloves and a particle mask should be worn when working with antibiotics. These solutions should be filter sterilized and added to medium after autoclaving and cooling.

Follow the protocols listed in Procedure 33.2 to complete this experiment.

Procedure 33.2
Regeneration of transgenic chrysanthemum plants.

Step	Instructions and comments
1	Complete steps 1 to 4 as outlined in Procedure 33.1 except inoculate leaf sections with the disarmed A281 stain harboring the NPTII gene (EHA 105).
2	Complete step 5 as outlined in Procedure 33.1.
3	Transfer all leaf sections to MS medium without PGRs, but containing 100 µg/ml cefotaxime and 50 µg/ml kanamycin. Incubate sections as before for an additional 21 days. Count the number of white and green shoots for each pair.

Anticipated results

Shoots should have formed after 35 days from both the bacteria-treated leaf sections and those treated with AB medium only. Shoots originating from sections treated with AB medium only should appear bleached or white. Experiments in our laboratory have shown that shoots formed from untransformed cells of 'Iridon' are sensitive to as little as

5 µg/ml kanamycin. If any green shoots are present in this treatment, they are probably escapes and should be retested on medium without PGRs, but containing kanamycin and cefotaxime. If sensitive, most of the newly produced leaves will be bleached white and the shoots will not produce roots.

Most shoots formed from explants treated with bacteria should be bleached, but putatively transformed shoots will remain dark green (Figure 33.3). Some of these will be escapes and all green shoots should be excised and recultured to medium containing cefotaxime and kanamycin for an additional 3 weeks. If these shoots are escapes, then the new growth will be white and roots will not be formed. If the NPTII gene has been successfully incorporated, then the shoots will remain dark green and the shoots will produce roots. From our experience, about 2 to 5% of the shoots formed will be transgenic. Remember to autoclave all plants and materials from this experiment when finished.

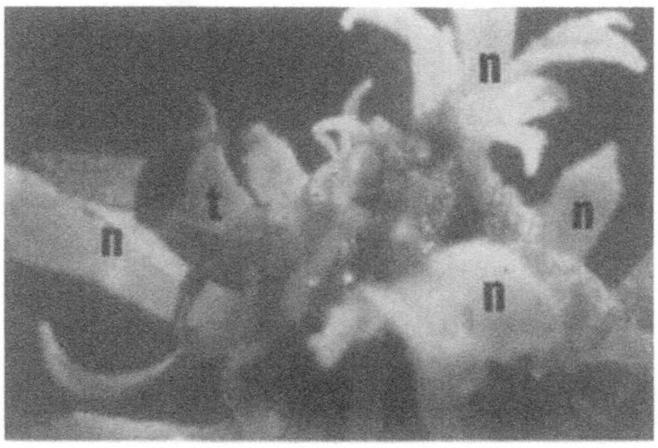

Figure 33.3 Transformation of leaf explants of 'Iridon' using a disarmed A281 and harboring the NPTII gene. Green shoots (t) may have been transformed (putative) to contain the NPTII gene. White shoots (n) are not transformed. From Trigiano, R. N. and Gray, D. J., Eds., *Plant Tissue Culture Concepts and Laboratory Exercises, First Edition*, CRC Press LLC, Boca Raton, FL, 1996.)

Questions

- Why are there "escape" shoots in the kanamycin-treated explants?
- Why is it necessary to provide physiological/molecular evidence in addition to data from kanamycin selection to conclusively demonstrate transformation?

Alternative Experiment 2. Regeneration of transgenic chrysanthemum plants

A more complete version of the above experiment is to use the following four treatments of leaf sections that are incorporated into a randomized complete block design: (1) co-cultivation with bacteria, cultured on MS plus antibiotics; (2) without bacteria (immersed in AB medium), incubated on medium lacking antibiotics; (3) without bacteria, cultured on MS amended with cefotaxime; and (4) without bacteria, incubated on MS supplemented with kanamycin. Place one leaf section per 60 × 15-mm petri plate.

White and green shoots are obtained after 35 days with treatment 1. Green shoots should be excised and cultured on medium containing both antibiotics. Only green shoots are produced in treatments 2 and 3 and generally more shoots are produced in treatment 3 (cefotaxime) than treatment 2 (without the antibiotic). Green and white shoots are produced in treatment 4, but when the green shoots are transferred to fresh medium containing kanamycin, all new growth will be white and rooting will not occur.

Literature cited

de Jong, J., W. Rademaker, and M. F. Van Wordragen. 1993. Restoring adventitious shoot formation on chrysanthemum leaf explants following co-cultivation with *Agrobacterium tumefaciens*. *Plant Cell Tissue Organ Cult*. 32:263–270.

Hood, E.E., G. L. Helmer, R.T. Fraley, and M. D. Chilton. 1986. The hypervirulence of *Agrobacterium tumefaciens* A281 is encoded in a region of pTiBo542 outside of t-DNA. *J. Bacteriol*. 168:1291–1301.

Lowe, J.M., M. R. Davey, J. B. Power, and K.S. Blundy. 1993. A study of some factors affecting *Agrobacterium* transformation and plant regeneration of *Dendrathema grandiflora* Tzvelev (syn. *Chrysanthemum morifolium* Ramat.). *Plant Cell Tissue Organ Cult*. 33:171–180.

Trigiano, R.N. and R.A. May. 1994. Laboratory exercises illustrating organogenesis and transformation using chrysanthemum cultivars. *HortTechnology* 4:325–327.

Urban, L.A., J.M. Sherman, J.W. Moyer, and M.E. Daub. 1994. High frequency shoot regeneration and *Agrobacterium*-mediated transformation of chrysanthemum (*Dendranthema grandiflora*). *Plant Sci*. 98:69–79.

White, F.F. and E.W. Nester. 1980. Hairy root: plasmid encodes virulence traits in *Agrobacterium rhizogenes*. *J. Bacteriol*. 141:1134–1141.

chapter thirty-four

Construction and use of a simple gene gun for particle bombardment*

Dennis J. Gray, Michael E. Compton, Ernest Hiebert, Chia-Min Lin, and Victor P. Gaba

As described in previous chapters (thirty-one, thirty-two, and thirty-three), there are several techniques that have been utilized successfully to introduce DNA into plant cells. Of these methods, particle bombardment, wherein microscopic metal particles coated with genetically engineered DNA are explosively accelerated into plant cells, has become the second most widely used vehicle for plant genetic transformation, after *Agrobacterium*-mediated transformation (Gray and Finer, 1993). Several distinct "particle guns" have been described, including the Biolistic PDS 1000/He (Kikkert, 1993), which is the only commercially available device. The most attractive of the noncommercial devices is the particle inflow gun (PIG) (Finer et al., 1992), which is based on a flowing helium device described by Takeuchi et al. (1992), since it can be fabricated from a steel plate with readily available parts and offers performance on par with the Biolistic PDS 1000/He (Brown et al., 1994).

This exercise is a modified version of an article published by Gray et al. in *Plant Cell Tissue and Organ Culture* (1994) and is arranged into two parts. In the first part, complete directions for constructing a simplified version of the particle inflow gun are provided. The device is termed a "Plastic PIG," because although it is based on the PIG, the steel specimen chamber of the PIG is replaced with a plastic vacuum jar and other parts, which dramatically simplifies construction (Gray et al., 1994). In the second part, directions for operating the Plastic PIG are provided, including preparation of the DNA-particle mixture and methods to obtain transient expression of the β-glucuronidase (GUS) gene, a convenient reporter gene, in plant tissues. Fabrication of this particle gun is an ideal laboratory exercise or demonstration for the entire class because it can be assembled in less that 40 min using only simple hand tools from parts that are readily available from technical equipment supply companies and hardware stores. The only drawback to its construction is that the components will cost \$400 to \$500. However, since it offers performance on par with other devices and is durable, once constructed, it can be used in subsequent class exercises or research programs.

* Florida Agricultural Experiment Station Journal Series No. R-07030.

0-8493-2029-1/00/\$0.00+\$.50

General considerations

Construction of the Plastic PIG

For convenience, specific brands and suppliers for the following required components are specified (Table 34.1); however, this does not constitute an endorsement of brands or suppliers, since others may be as suitable. Although 115 VAC versions of electrical parts are specified here, 220 VAC valves and timers are available through the same suppliers. Similarly, the closest metric-equivalent fittings may be substituted for the English measurements given below. Electrical connection to supply current should be through a ground fault-protected circuit. The setup must be examined and approved by a qualified electrician before it is plugged into supply current source. Tools required to accomplish fabrication are listed in Table 34.2. Figures 34.1 and 34.2 detail construction and use of the Plastic PIG and the assembly steps are provided in Procedure 34.1.

Table 34.1 Parts List for the Plastic PIG and Specimen Holders

Part	Catalog no.	Supplier
Vacuum chamber	5305-0910	Nalge Co., Rochester, NY
Solenoid valve	S24C-4V, NC 115/60V, 1/4 in.	Atkomatic Valve Co., Inc., Indianapolis, IN
Interval timer	G-08683-90	Cole-Parmer Instrument Co., Niles, IL
Apparatus positioner	G-08056-10	
Vacuum pump	G-07055-04	
Gas pressure regulator	G-98200-70	
Barbed fitting	G-06362-40	
Tee connector	G-06455-15	
Tubing clamp	G-06833-00	
13-mm plastic filter holder	4312	Gelman Sciences, Inc., Ann Arbor, MI
0.2-μm air filter	4464	
Plastic tubing	01T252PE-RDT	Ark-Plas Products, Inc., Flippin, AR
Helium tank, lab grade		Local supplier
Miscellaneous hardware		Local hardware supply
Metal hose clamps		
Two-prong electrical appliance cord		
Brass close nipple with flange, 1/4in. — 18 SAE		
Brass cap, 1/4 in. — 18 SAE		
Brass bushing, 1/4 in. × 1/8 in. NPT		
Steel jam nut, 9/16 in. SAE		
Teflon and electrical tape, rosin-core solder, silicone sealing glue		
Specimen holders		
PC petri dishes	5502-0010	Nalge Co., Rochester, NY
Spectra/mesh PP screen	08-670-185	Fisher Scientific, Pittsburgh, PA
Magnets 3 mm diam. (6)		Radio Shack, Inc.
Permatex high temp RTV silicone gasket maker		Local auto parts store

From Trigiano, R. N. and Gray, D. J., Eds., *Plant Tissue Culture Concepts and Laboratory Exercises, First Edition*, CRC Press LLC, Boca Raton, FL, 1996.

Table 34.2 Tools Required to Assemble Plastic PIG and Specimen Holders

Electric hand-held drill
7/64 in. and 1/4 in. drill bits for metal
7/16 in. drill bit for masonry
5/16 in. — 24 NF and 1/4 in. — 18 NPT thread taps
5/16 in. — 24 NF thread die
Adjustable hole saw
Hobby vise
Adjustable wrench
Pliers
Soldering iron

From Trigiano, R. N. and Gray, D. J., Eds., *Plant Tissue Culture Concepts and Laboratory Exercises, First Edition*, CRC Press LLC, Boca Raton, FL, 1996.

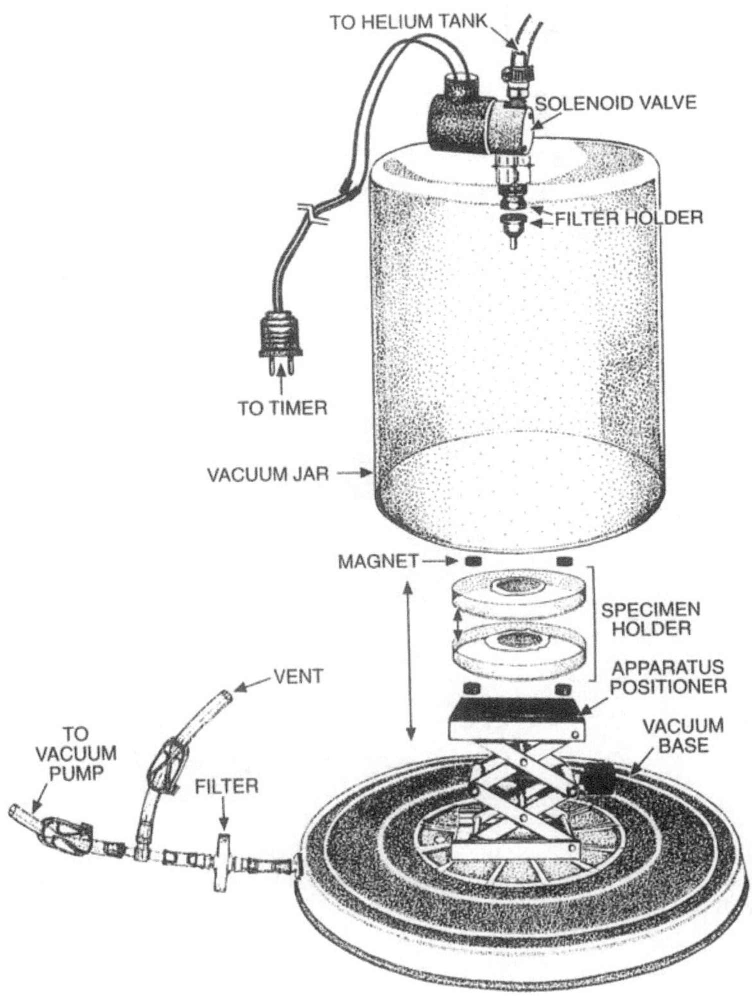

Figure 34.1(a)Construction and setup of the Plastic PIG. (a) Exploded view of Plastic PIG and specimen holders. (From Gray, D.J. et al. 1994. *Plant Cell Tissue Organ Cult.* 37:179–184. With permission.)

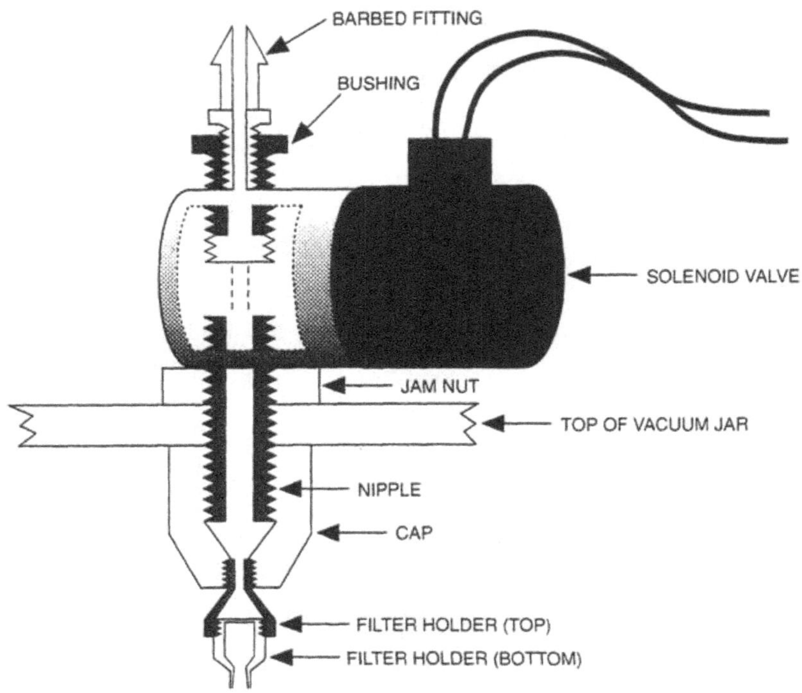

*Figure 34.1(b)*Construction and setup of the Plastic PIG.Detail of solenoid valve assembly. (From Gray, D.J. et al. 1994. *Plant Cell Tissue Organ Cult.* 37:179–184. With permission.)

Figure 34.2 Setup of Plastic PIG and timer. (From Gray, D.J. et al. 1994. *Plant Cell Tissue Organ Cult.* 37:179–184. With permission.)

Construction of specimen holders

Specimen holders consist of the top and bottom of autoclavable plastic petri dishes, in which holes have been drilled to accept screens (see Figure 34.1a and Table 34.1 for parts list). Follow the instructions provided in Procedure 34.2 to construct the specimen holders.

Procedure 34.1
Assembly of the Plastic PIG.

Step	Instructions and comments
1	Drill pilot hole in center top of vacuum jar. Enlarge hole with 7/16-in. masonry bit and thread the hole with 1/4 in. — 18 NPT tap.
2	Drill pilot hole in center of brass cap, then enlarge to 1/4 in. Use 5/16 in. — 24 NF tap to cut threads into hole.
3	Screw jam nut tightly onto close nipple, add teflon tape to threads on short end of nipple, and screw tightly into the #2 port of the solenoid valve. Screw the brass bushing, then the barbed fitting into the #1 port of the valve, using teflon tape to seal.
4	Add a small amount of silicone glue to the exposed threads of the close nipple and screw the valve assembly snugly into the top of the vacuum jar. Add more silicone glue to the exposed threads protruding into the jar before screwing the brass cap tightly onto the nipple (see Figure 34.1b).
5	Use the 5/16 in. — 24 NF die to carefully cut threads over the luerlock end of the plastic filter holder. Add teflon tape to threaded end and screw the filter holder into the brass cap. Allow silicone glue on the assembled valve-vacuum jar unit to cure overnight before subjecting to vacuum.
6	Solder the two-prong electrical appliance cord onto exposed wires of valve and wrap securely with electrical tape. Plug cord into interval meter.
7	Clamp one end of plastic tubing to barbed fitting and the other end to the helium regulator.
8	Attach one end of vacuum-vent tubing to vacuum plate as shown in Figure 34.1a and 34.2 and the other end to the vacuum pump.
9	Place the apparatus positioner in the center of vacuum plate.

Procedure 34.2
Construction of specimen holders.

Step	Instructions and comments
1	Use hole saw on drill to cut 4-cm-diameter hole in center of petri dish tops and bottoms.
2	Autoclave screen material to preshrink. Cut into 5-cm-diameter disks. Use high-temperature silicone to attach screen to the inside surface of the bottom and the outside surface of the top. The petri dish surface under the screen should be roughened with sand paper prior to attachment.
3	Cover glued surface with plastic wrap and add a weight to hold screen tightly in place. Allow to cure overnight.

For bombardment, specimens are held between screens by nesting the top piece of the dish over the inverted bottom, both of which are then clamped together with two to three pairs of magnets (Figure 34.1a).

Preparation of plasmid-particle mixture

The plasmid used is pBI221 (Clontech Laboratories, Inc., CA), which consists of a 3kb HindII-EcoRI fragment of pBI121 containing the CaMV 35S promoter, GUS gene, and NOS-terminator (Jefferson et al., 1987) cloned into pUC19. The plasmid can be maintained in DH5alpha bacterial host and purified according to the QIAGEN plasmid purification kit. Note that any plasmid that expresses the GUS gene in dicotyledonous plants may be substituted. Since instruction on plasmid purification and scale-up is beyond the scope of this book, we recommend that instructors who are unfamiliar with these procedures obtain purified plasmids, which are becoming increasingly available, from colleagues. Please see Chapters thirty-one and thirty-two for further discussion.

Follow the protocols in Procedures 34.3 and 34.4 to complete this part of the exercise.

Procedure 34.3
Preparation of plasmid/particle mixture (as modified from Finer et al., 1992).

Step	Instructions and comments
1	Autoclave M17 tungsten particles (Bio-Rad Laboratories, Inc., Hercules, CA) and place 50 mg in 0.5 ml 95% ethanol, incubate 20 min, then vortex well.
2	Centrifuge at 10,000 rpm for 5 min and wash with 0.5 ml sterile water; repeat five times.
3	Vortex, then pipette 25 μl of tungsten stock into eppendorf tube.
4	Add 5 μl DNA (1 μg/μl) and mix well. Add 25 μl of 2.5 M $CaCl_2$, then quickly add 10 μl of 100 μm spermidine (free base) and vortex. Incubate on wet ice (important to maintain cold temperature) for 5 min.
5	Without centrifugation, remove about 50 μl of solution, leaving the particles. Use immediately.

Procedure 34.4
Operation cycle of the Plastic PIG.

Step	Instructions and comments
1	Clamp tissue in a specimen holder using magnets; center on the positioner adjusted to the desired height. It is imperative that tissue not be "wet," i.e., not covered with a film of liquid, in order for the procedure to be successful.
2	Place a 2-μl drop of DNA/particle mixture on the middle of a plastic filter holder screen and screw tightly into the vacuum jar.
3	Place vacuum jar on base and evacuate the chamber to 90-kPa (27 in. Hg) vacuum.
4	Set timer to 0.1 sec and fire valve, then vent the chamber and replace specimens on culture medium.

Exercise

Experiment 1. Comparison of helium pressures on transient expression of the GUS gene

Particle bombardment is affected by many parameters, which include biological (i.e., plasmid construct, type, and physiological state of target tissue) and physical (i.e., particle size, vacuum level, and helium pressure) factors. In this experiment, the effects of two tissue types and three helium pressures will be compared by assessing differences in transient GUS reaction. Since GUS reaction can be assessed as soon as 48 h after bombardment, this experiment can be completed within 1 to 2 weeks.

Materials

The following materials are required for each student or team:

- Explant tissue, consisting of (1) approximately 100 basal cotyledonary explants of melon, cultured for 4 days on embryo induction medium as described in Chapter twenty-one and (2) the innermost leaves from fresh cabbage, purchased from a supermarket (approximately one half cabbage for each student or team)

- DNA/particle mixture, prepared continuously, immediately before every four bombardments
- Eight sterile specimen holders
- Three sterile moist chambers; i.e., empty 100 × 15-mm petri dishes containing moist filter paper (to incubate cabbage tissue after bombardment)
- 10 ml of X-gluc reagent (see Chapter thirty-two concerning *Agrobacterium*-mediated transformation)

The experimental design compares three helium pressures (410 kPa [60 psi], 620 kPa [90 psi], or 830 kPa [120 psi]) (Figure 34.3) and two explant tissues (melon cotyledons and cabbage leaves). Each treatment (pressure-explant combination) is replicated four times. Follow the protocol outlined in Procedure 34.5 to complete this experiment.

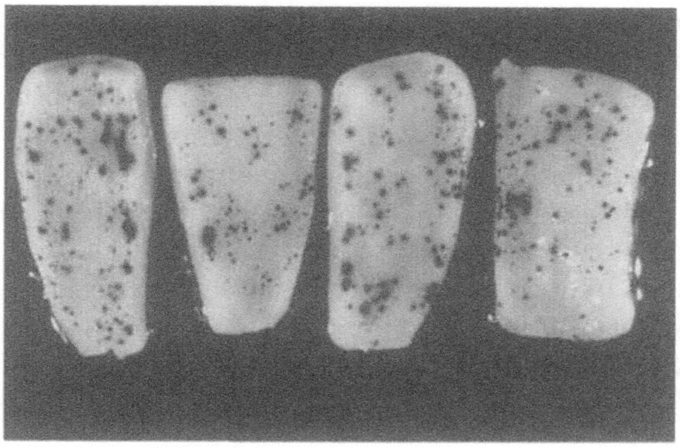

Figure 34.3 Transient expression of GUS in four basal cotyledon quarters bombarded at 830 kPa with a 15-cm particle-to-specimen gap. Discrete GUS foci, shown in this photograph as black spots of various sizes, were actually dark blue in coloration. (From Gray, D.J. et al. 1994. *Plant Cell Tissue Organ Cult*. 37:179–184. With permission.)

Procedure 34.5
Comparison of helium pressures on transient expression of the GUS gene.

Step	Instructions and comments
1	Set the particle holder-to-specimen gap at 15 cm by adjusting the apparatus positioner.
2	Arrange eight cotyledon explants as close together as possible, but not overlapping, in the center of a specimen holder and carefully center the specimen on the apparatus positioner relative to the bombardment pattern (to properly "aim" the gun, the actual particle pattern can be determined by first bombarding filter paper and viewing the resulting pattern).
3	As rapidly as possible, perform a bombardment as described above and place the explants back onto the culture medium. Repeat four times.
4	For cabbage, dissect young leaves into pieces approximately 2 cm in diameter and bombard. Place the four replicate leaves from one treatment into sterile moist chambers.
5	Incubate explants for 48 h and then place into X-gluc reagent in dark at 35°C for 12 to 24 h as described in Chapter thirty-two.
6	Count the number of distinct blue spots per replicate (specimens can be preserved indefinitely in cold 70% ethanol).

Procedure 34.5 continued Comparison of helium pressures on transient expression of the GUS gene.	
Step	Instructions and comments
7	Statistically analyze the data as a 2 × 3 factorial and perform a mean separation test as described in Chapter seven.

Anticipated results

The GUS reaction seen in this experiment is due to the direct gene activity in bombarded cells. For such a response to occur, at least one intact DNA molecule must enter a cell and the cell must survive the impact. However, this response does not denote stable transformation (i.e., integration of the GUS gene into the cell's chromosomes or genome). Dramatic differences in the number of blue spots will be seen due to different pressure treatments, with the highest pressure typically producing the greatest response in melon cotyledons; however, different tissues would be expected to respond differently. Differences in GUS reaction due to tissue type are more difficult to predict due to wide variation in available cabbage tissue. However, when successful, bombarded cabbage leaf tissue demonstrates the "blast" pattern very well. Experiments comparing the other parameters mentioned above can be conducted in a similar manner.

While transient expression of GUS is relatively simple to achieve, obtaining stable expression is a more difficult task, which involves careful selection over time. Selection usually is accomplished by incorporating a selectable marker gene, such as NPT II (neomycin phosphotransferase II), which confers resistance to the antibiotics kanamycin and gentamycin (see Chapter thirty-three).

Questions

- What accounts for the differences in GUS expression due to helium pressure? Given these results, what might be expected by varying the particle holder-to-specimen gap?
- Similarly, why did the two explants exhibit different GUS reactions?

Literature cited

Brown, D.C.W., L. Tian, D.J. Buckley, M. Lefebvre, A. McGrath, and J. Webb. 1994. Development of a simple particle bombardment device for gene transfer into plant cells. *Plant Cell Tissue Organ Cult.* 37:47–53.

Finer, J.J., P. Vain, M.W. Jones, and M. McMullen. 1992. Development of the particle inflow gun for DNA delivery to plant cells. *Plant Cell Rep.* 11:323–328.

Gray, D.J. and J.J. Finer. 1993. Editorial introduction. Special section: development and operation of five particle guns for introduction of DNA into plant cells. *Plant Cell Tissue Organ Cult.* 33:219.

Gray, D.J., E. Hiebert, C.M. Lin, M.E. Compton, and D.W. McColley. 1994. Simplified construction and performance of a device for particle bombardment. *Plant Cell Tissue Organ Cult.* 37:179–184.

Jefferson, R.A., T.A. Kavanagh, and M.W. Bevan. 1987. GUS fusions: β-glucuronidase as a sensitive and versatile gene fusion marker in higher plants. *EMBO J.* 6:3901–3907.

Kikkert, J.R. 1993. The biolistic PDS-1000/He device. *Plant Cell Tissue Organ Cult.* 33:221-226.

Takeuchi, Y., M. Dotson, and N.T. Keen. 1992. Plant transformation: a simple particle bombardment device based on flowing helium. *Plant Mol. Biol.* 18:835–839.

chapter thirty-five

Germplasm preservation

Leigh E. Towill

The concept of germplasm is inherently simple, yet extremely complex in scope. Germplasm, as defined for a crop, represents all the different genotypes that could be used for improvement. Improvement is accomplished traditionally by making sexual crosses between individuals and then selecting plants with the desired characteristics such as disease or stress tolerance or resistance. With the advent of more exotic techniques, for example, transformation and protoplast fusion, virtually any species may be modified by incorporating a gene or genes from unrelated species (see Chapter thirty-one). Thus, germplasm for a crop becomes more expansive. Nevertheless, we usually consider germplasm in its less cosmic size and to be the genepool from those individuals that are related and may intercross. Germplasm is valuable because it contains the diversity of genotypes that is needed to develop new and improved lines.

Germplasm diversity for any crop must be protected from loss to ensure its availability for future plant improvement (Stuessy and Sohmer, 1996). This diversity was collected from centers of crop origin and from landraces developed over many years. In the case of crops, preservation of this diversity is not in the wild, but in what are termed repositories or genebanks (Table 35.1). Chapters in a book describing the U.S. National Plant Germplasm System (NPGS) (Janick, 1989) are an excellent source for those interested in the details of a comprehensive system.

A key theme for safe preservation of diversity is use of two sites. One site, the active collection or genebank, has responsibilities for obtaining diversity (either by collection or by exchange with other genebanks worldwide), propagation, distribution to users, and characterization. The second site, the base collection or genebank, is for long-term preservation and utilizes the best storage systems available. This site provides a crucial function of being a safety net should materials be lost at the active collection. It is important to have the distinction between these two sites firmly in mind in discussing the issue of preservation and in deciding what methods are useful for each collection.

This chapter describes germplasm preservation for crop species and emphasizes where and how in vitro culture and cryopreservation are used for efficient and effective management. Germplasm preservation for most crop plants does not involve in vitro culture but utilizes seeds, and a brief discussion is included to provide a complete picture of preservation. The laboratory exercises following this chapter (Chapter thirty-six) emphasize cryopreservation of shoot tips.

0-8493-2029-1/00/$0.00+$.50

Table 35.1 Major Crops Maintained Clonally at Different Repositories Within the U.S. National Plant Germplasm System

Location	Species maintained clonally
Corvallis, OR	Pear, filbert, mint, strawberry, raspberry, blackberry, currants, hops
Davis, CA	*Prunus* species, *Vitis vinifera* and related species
Geneva, NY	Apple, *Vitis* species
Brownwood, TX	Pecans, walnuts
Hilo, HI	Guava, passion fruit, papaya, pineapple, macadamia, litchi, carambola
Griffin, GA	Sweet potato
Miami, FL/Mayaguez, PR	Avocado, banana, plantain, coffee, cocao, mango, sugarcane
Riverside, CA	*Citrus* species and related species, date palm
Orlando, FL	*Citrus* species and related species

From Trigiano, R. N. and Gray, D. J., Eds., *Plant Tissue Culture Concepts and Laboratory Exercises, First Edition*, CRC Press LLC, Boca Raton, FL, 1996.

Preservation of seed-propagated species

Seeds are classified as either being desiccation tolerant (capable of retaining viability after being dried to virtually any moisture content — also termed "orthodox" seed) or desiccation sensitive (seeds losing viability after being dried below a critical limit, usually about 12 to 30% moisture — also termed "recalcitrant" seed). This classification is a great simplification and a number of intermediate types exist. Most temperate crop species have desiccation-tolerant seeds and many studies have shown the importance of seed moisture and storage temperature for retaining viability (see review by Roos, 1989). Recent studies suggest that the optimum seed moisture content for obtaining the greatest longevity depends on storage temperature (Walters, 1998). In a practical sense, seeds can be equilibrated with a 25% relative humidity at 5°C and then stored between –5 and –20°C.

Most desiccation-tolerant seeds survive cryogenic storage, although some physical damage (seed coat or cotyledon cracking) occurs in some species (Stanwood, 1985). Temperatures less than about –130°C are desired for cryopreservation because of very low molecular kinetic energies, the absence of liquid water, and extremely slow diffusion. Thus, many reactions leading to deterioration are minimal and longevities are postulated to be extremely long (centuries) and limited only by the buildup of genetic lesions resulting from background irradiation. In practice, cryogenic storage is in liquid nitrogen (LN, –196°C) or in the vapor above LN (*circa* –150 to –180°C) because it is a relatively cheap and available cryogen.

Desiccation-sensitive seeds are produced by aquatic species, large-seeded species, some species native to tropical areas, and some temperate zone species of trees. Coconut, cocoa, mango, nutmeg, and rubber plants are examples of economically important species with recalcitrant seed. To retain viability, recalcitrant seeds are stored at as low a temperature as possible under conditions that retain relatively high seed moisture levels and assure a supply of oxygen for respiration. These seeds have short life spans, ranging from a few weeks to several months. These are too short for adequate germplasm preservation and these species usually are stored as vegetative plants.

In vitro technologies are becoming important in maintaining crops with desiccation-sensitive seed. In vitro plants may be maintained for active collections. Embryo culture may be used in preservation of desiccation-sensitive seeds. Isolated embryonic axes (cotyledons removed) from mature seeds can survive partial desiccation and cooling to low temperatures. The level of desiccation tolerated and whether this is sufficiently low to avoid damage from ice formation during cooling must be defined for each species. Upon retrieval from LN, axes are cultured and can develop into plants. Treatments involving application of cryoprotectants to excised embryos may also prove to be useful, but as yet no procedure has been developed for practical, reproducible cryopreservation of recalcitrant seeds using this strategy. Axillary shoot tip or bud cryopreservation from in vivo or in vitro plants also is useful for base-collection storage of these species.

Preservation of pollen

Pollen may also be used to store genes from crops (Hanna and Towill, 1995). The major use of germplasm is for improvement of plants by sexual processes and, thus, requires making controlled crosses between desired individuals. Supplying pollen to the user facilitates this process; the user does not have to grow the plant and wait for flowering of the male parent. This is of obvious benefit for those species which often take many months to flower and, therefore, is of potential value for some species that are vegetatively propagated. Pollen preservation, thus, could be integrated into the active genebank, where it would be collected, stored, and distributed upon request. Pollen from many species is desiccation tolerant and exhibits storage characteristics similar to desiccation-tolerant seed. Pollen from some species, however, is desiccation sensitive and short-lived. Cryogenic storage of this type of pollen is feasible but only within a narrow range of moisture contents. Pollen preservation is a supplement to seed or clone storage, not a substitute for them.

Preservation of vegetatively propagated species

Many crops either have an extremely long juvenile phase (no flowers), produce seed with extremely short longevities, do not easily produce seed, or do not produce viable seed (Towill, 1988). The vegetative plant for these species is stored either in the field or in greenhouses and is regularly propagated asexually to rejuvenate the line and prevent loss. These species are extremely diverse, comprising herbaceous and woody growth habits from tropical to temperate zones. These also differ greatly in stress tolerances to temperature, salt, moisture, and others as well as to disease.

Active collections

It is impossible to grow all species at one location and thus several repositories, termed clonal repositories, exist throughout the world. The vast number of different clones held precludes holding very many replicates, especially under field conditions. Maintenance is expensive and loss could easily occur due to disease or catastrophe.

The use of tissue culture in germplasm preservation usually is for species that are vegetatively propagated. In vitro culture of plants can be used to gain more efficient preservation (Withers, 1991). A number of advantages for using in vitro plants can be listed including small space needed for storage, ease of shipping, year-round availability, potential for eliminating disease and for maintaining disease-free plants as well as rapid propagation, and labor and cost savings (Towill, 1988). Many species can now be propagated in culture, but methods are often genotype specific and may require considerable modification. All in vitro lines held in one location are also susceptible to loss from catastrophe.

The advantages of maintaining in vitro vs. in vivo plants often relate to what is desired by the user community. For example, if crosses are to be made between apple lines, apple budwood from ex vitro sources is preferred, since a grafted bud can lead to a flowering shoot within about 18 months. An in vitro apple plant, by comparison, might take more than 3 years to flower after transplanting. However, for potato, flowering times from planting a tuber or transplanting an in vitro plant are similar. The advantage here for the in vitro system is ease in maintaining the disease-free status.

In vitro and in vivo systems are not mutually exclusive. To guard against loss, in vitro plants may serve at the same location as a backup to field or greenhouse lines. Again, recall that replication is often minimal; often only two individuals are held for a clone.

Maintenance of in vitro cultures occurs in the active collection, often under normal growth conditions. If material is requested by the user community, it can be quickly propagated and distributed. Backup materials at the active collection site may be maintained under conditions that minimize growth (Withers, 1991). This usually means holding the in vitro plant at lower temperatures — for example, potato at 10°C or mint at 4°C. Growth is retarded and cultures may be held for a few years, provided precautions are taken to avoid extensive water loss. Elevated osmotic pressures and other growth inhibitors may also be used to reduce the frequency of transfer needed to maintain a viable culture. Empirical studies are required to apply this method to different species.

Base collections: cryopreservation

It is prohibitively expensive to maintain duplicate collections for either ex vitro or in vitro plants at a second location. Hence, cryopreservation is desired for the base collection and is applied to axillary or apical shoot tips from either ex vitro or in vitro plants. Storage at low temperatures, such as –20°C or even –80°C, results in very short longevities with these propagules, for reasons which will not be discussed here. Cryogenic storage provides long-term storage at a reasonable cost with a low labor input. As with seed, storage is either in LN or in the vapor phase over LN.

It should be emphasized that, whereas cryopreservation for germplasm purposes uses shoot tips or buds, cryopreservation of protoplasts, cells, tissues, and somatic embryos is also needed for many "tissue culture" operations to guard against loss, to assure line purity and performance, and to avoid costly, frequent transfers. Production of synthetic seed from somatic embryos (see Chapters nineteen and twenty) and their cryopreservation may some day be used for germplasm preservation once induction from all genotypes is feasible, off-type production is minimized, and maturation efficiency is improved.

Shoot tips have higher water contents than seeds and cannot withstand extreme desiccation. In the hydrated state, exposure to low temperatures leads to damage due either to intracellular ice or to consequences of desiccation during extracellular ice formation. For these reasons, a more detailed protocol (Figure 35.1) is needed. Virtually all cryopreservation protocols follow these steps, but details vary with species. It is usually necessary to use cryoprotectants and to control cooling and warming rates. A theoretical treatment of rate phenomena is beyond the scope of this article; excellent reviews address this and related topics (Mazur, 1984; Fahy et al., 1987). A recent book provides technical information on application to cells, protoplasts, and shoot tips from many different plant species (Bajaj, 1995).

Pretreatment of the plant and isolated propagule

The physiological condition of the plant used for cryopreservation is important. Shoot tips isolated from plants that have been cold acclimated survive the cryogenic protocol in

in vitro or ex vitro plant	
↓	temperature/light treatment
conditional plant	
↓	excision
isolated shoot tips/buds	
↓	pretreatment {conc., time, temp.}
↓	application of cryoprotectants {conc., time, temp.}
↓	cooling {nucleation, rates}
↓	storage {time, temp.}
↓	warming {rates}
↓	removal of cryoprotectant {conc., time, temp.}
isolated shoot tips/buds	
↓	post-thaw treatment (culture)
↓	viability assessment
growth	

Figure 35.1 Generalized flow chart for cryopreservation of shoot tips. (From Trigiano, R. N. and Gray, D. J., Eds., *Plant Tissue Culture Concepts and Laboratory Exercises, First Edition*, CRC Press LLC, Boca Raton, FL, 1996.)

higher percentages than those not acclimated. This is useful only for those species that have the genetic capability to cold acclimate. Survival of excised shoot tips also can be enhanced by exposure to elevated sucrose levels. The exact effect of treatments on the plant or of the excised shoot tip is uncertain, but they probably alter intracellular contents of critical metabolites (particularly sugars), cell water content, and physiological characteristics of the cell (e.g., membrane permeabilities).

Cryoprotectant(s) application and cooling rate

After the pretreatment phase, shoot tips are exposed to cryoprotectants. The concentration of cryoprotectant(s) used and the time and temperature of exposure vary with the cooling strategy employed. Two cooling regimes are mainly used with shoot tips. The first is termed either two-step, slow, or optimum rate cooling and employs cooling of the cryoprotected shoot tip at about 0.25 to 1.0°C/min to about –35°C prior to immersion in LN. For this method shoot tips are incubated in either 1 to 2 M concentrations of a suitable permeating cryoprotectant, such as dimethyl sulfoxide (DMSO) or ethylene glycol. The second regime uses rapid cooling and is often termed a vitrification method. Here, shoot tips are exposed to much more concentrated solutions of cryoprotectants and then are immersed directly from about 22 or 0°C into –196°C.

Research with plant systems initially emphasized two-step cooling techniques to attain cryopreservation. Useful levels of survival were obtained for some species, but not for others. Reasons for this have not been elucidated. Summaries of this method can be found in recent reviews (Sakai, 1986; Towill, 1990). Attention now has focused on vitrification for cryopreservation.

The term vitrification refers to transformation of a liquid to a glass during cooling. Tissue culture literature has used the term vitrification in an entirely different sense to refer to plants that have an abnormal, water-soaked, or glassy appearance. However, "hyperhydricity" is now the preferred term for this abnormal appearance of plants in tissue culture.

There are increasing numbers of theoretical and practical studies examining vitrification in animal and plant systems (Fahy et al., 1987; Steponkus et al., 1992; Mehl 1996). Vitrification requires application of suitable concentrations of compounds, such that, with usually rapid cooling, the system forms a glass (i.e., the solution vitrifies). Information about physical aspects of the process is described in MacFarlane et al. (1992). Although it is difficult to prove that the cell itself vitrifies, circumstantial evidence suggests that this happens under defined conditions. Indeed, vitrification probably occurs naturally in extremely cold-hardy plants exposed to very low temperatures.

The advantages of vitrification are that it is a simple technique that does not require an expensive apparatus, it may avoid the damaging consequences of ice formation, and it may be applicable to larger pieces of tissue than are usually used for two-step cooling methods. Survival after cryogenic exposure has been reported with several different vitrification procedures for protoplasts, cells, somatic embryos, and shoot tips.

The general strategy for vitrification is to expose the shoot tips (or protoplasts, cells, somatic embryos, etc.) to low concentrations of cryoprotectants such that permeable components have sufficient time to penetrate (the loading phase). Next, the shoot tips are exposed to a vitrification solution (the dehydration phase). Since water permeability is much greater than solute permeability, the shoot tip dehydrates in the very concentrated solution, and endogenous solutes and any permeating cryoprotectants are concentrated. Shoot tips are then rapidly cooled to facilitate glass formation.

It has become apparent that loading is not necessary for shoot tips from some species and, indeed, may be detrimental (Steponkus et al., 1992). The cell contents alone, when sufficiently dehydrated, apparently vitrify when cooled at a rapid rate. Here preculture for 1 to 3 days in a moderate osmotic solution (0.3 to 1.0 M) prior to dehydration is necessary to achieve survival after vitrification in such shoot tips.

A major variation of vitrification uses an encapsulation/dehydration sequence (Fabre and Dereuddre, 1990). Shoot tips are encapsulated in an alginate gel, cultured in sucrose solutions, dehydrated in an air flow, and then usually rapidly cooled. Encapsulated shoot tips vitrify under the conditions employed. Survival has been found in shoot tips from several unrelated species such as potato, grape, carnation, and pear.

Warming rate and recovery

Rapid warming from –196°C to about 22°C is generally beneficial for obtaining high levels of survival for samples cooled either by two-step cooling or vitrification. Rapid warming minimizes any destructive ice crystal growth or devitrification as the cellular mileau gains more thermal motion. The application of cryoprotectants and the excursion of shoot tips to and from very low temperatures lead to sublethal damage which, in concept, can be repaired by cellular processes and can be influenced by environmental conditions. The subsequent culture of the retrieved shoot tip may require conditions that control shoot tips do not require. Treated shoot tips are usually cultured in the dark or dim light for a few days to minimize any oxidative stress that might occur due to damaged mitochondria or chloroplasts. Brief culture on elevated osmotica may be beneficial and may use low ammonium levels. Sometimes different levels of growth regulators are required. Eventually the shoot tip can be placed back under "normal" growth conditions.

Conclusion

Seeds and plants are used to preserve germplasm of most crops to assure that materials are available for future breeding to produce crops with specified characteristics. Whereas desiccation-tolerant seeds are simply stored, desiccation-sensitive seeds present challenges. Preservation of clones also is challenging due to high costs of plant maintenance and difficulties in using cryopreservation. Increasingly, plant culture allows for efficient and effective preservation for these latter two groups. Many species now may be propagated and maintained in culture. Embryonic axes and shoot tips may be regrown in culture after cryogenic treatment. Some species are more problematic, but with advances in developing cryopreservation methods, such as vitrification, practical use is now feasible.

Literature cited

Bajaj, Y.P.S., Ed. 1995. *Cryopreservation of Plant Germplasm* I. Biotechnology in Agriculture and Forestry Series No.32. Springer, Berlin.

Fabre, J. and J. Dereuddre. 1990. Encapsulation-dehydration: a new approach to cryopreservation of *Solanum* shoot tips. *Cryo-Letters* 11:413–426.

Fahy, G.M., D.I. Levy, and S.E. Ali. 1987. Some emerging principles underlying the physical properties, biological actions, and utility of vitrification solutions. *Cryobiology* 24:196–213.

Hanna, W.W. and L.E. Towill. 1995. Long-term pollen storage. *Plant Breeding Rev.* 13:179–207.

Janick, J., Ed. 1989. The National Plant Germplasm System of the United States. *Plant Breeding Rev.* Vol. 7, 230 pp. Timber Press, Portland.

MacFarlane, D.R., M. Forsyth, and C.A. Barton. 1992. Vitrification and devitrification in cryopreservation. In: *Advances in Low-Temperature Biology,* Vol. 1. P.L. Steponkus, Ed. pp. 221–278. JAI Press Ltd., Greenwich, CT.

Mazur, P. 1984. Freezing of living cells: mechanisms and implications. *Amer. J. Physiol.* 247:c125–c142.

Mehl, P. 1996. Crystallization and vitrification in aqueous glass-forming solutions. In: *Advances in Low Temperature Biology,* Vol. 3. P.L. Steponkus, Ed. pp. 185–255. JAI Press Ltd., Greenwich, CT.

Roos, E.E. 1989. Long-term seed storage. *Plant Breed. Rev.* 7:129–158.

Sakai, A. 1986. Cryopreservation of germplasm of woody plants. In: *Biotechnology in Agriculture and Forestry,* Vol. 1. Trees, I. and Y.P.S. Bajaj, Eds. pp. 113–129. Springer-Verlag, Berlin.

Stanwood, P.C. 1985. Cryopreservation of seed germplasm for genetic conservation. In: *Cryopreservation of Plant Cells and Organs.* K.K. Kartha, Ed. pp. 199–226. CRC Press, Boca Raton, FL.

Steponkus, P.L., R. Langis, and S. Fujikawa. 1992. Cryopreservation of plant tissues by vitrification. In: *Advances in Low-Temperature Biology,* Vol. 1. Steponkus, P.L., Ed. pp. 1–62. JAI Press, Greenwich, CT.

Stuessy, T.F. and S.H. Sohmer, Eds. 1996. Sampling the green world. In: *Innovative Concepts of Collection, Preservation and Storage of Plant Diversity*. 289 pp., Columbia University Press, New York.

Towill, L.E. 1988. Genetic considerations for germplasm preservation of clonal materials. *HortScience* 23:91–97.

Towill, L.E. 1990. Cryopreservation. In: *In Vitro Methods for Conservation of Plant Genetic Resources.* J. H. Dodds, Ed. pp. 41–70. Chapman and Hall, London.

Walters, C. 1998. Understanding the mechanisms and kinetics of seed aging. *Seed Sci. Res.* 8:223–244.

Withers, L.A. 1991. In-vitro conservation. *Biol. J. Linnean Soc.* 43:31–42.

chapter thirty-six

Vitrification as a method to cryopreserve shoot tips

Leigh E. Towill

The following experiments demonstrate the use of vitrification as a method for cryopreservation. Vitrification does not require a controlled cooling rate apparatus and easily can be performed at any location. Both a solution-based method and an encapsulation–desiccation method are described. Shoot tips are emphasized because they are the propagules used for preservation of clones. Preservation of cell suspensions, callus, protoplasts, or somatic embryos would follow a similar protocol, but durations and concentrations will need to be altered. As with any method, details often differ with species, cell type, and tissue complexity. Solution-based vitrification methods have been used successfully by many researchers and have been applied in my laboratory to several species of mint and various cultivars of potato, as well as apple, *Prunus* sp., endod, garlic, papaya, strawberry, carnation, and sweet potato. The encapsulation–dehydration method has been extensively studied by Dereuddre and associates (Fabre and Dereuddre, 1990) and has been applied to several species (e.g., Malaurie et al., 1998). Various modifications of both methods exist and features of both are sometimes combined in a single protocol (Hirai et al., 1998).

General considerations

Growth of plants and preparation of media

Stock plants of either carnation, chrysanthemum (see Chapter fifteen), mint, or potato (see Chapter eleven) are useful for these experiments since they grow rapidly in vitro, but almost any species may be utilized if a suitable shoot tip growth medium has been defined. In vitro-multiplied plants are preferred as sources of shoot tips because the incidence of contamination will be reduced; however, greenhouse or growth-chamber plants can be used. If the latter are used, be sure that the surface disinfestation procedure is effective. A cold period (e.g., 2 weeks at about 2°C under short days) for the in vitro stock plant is often beneficial for those species that have the ability to cold acclimate (carnation and chrysanthemum). Details for micropropagation of plants by nodal sections and meristem tip or shoot tip culture are found in Chapter eight.

0-8493-2029-1/00/$0.00+$.50

Either apical or axillary shoot tips may be used for the experiments. The former often show higher levels of survival after cryogenic treatments. The size of the shoot tip is also important. Shoot tips about 1 mm in length are best.

Both liquid and solidified media are required. The growth medium used for excised shoot tips of potato and mint contains Murashige and Skoog (MS) minerals (see Chapter three for composition) and vitamins, 100 mg/L (0.55 mM) myo-inositol, 2% (59 mM) sucrose, and, if solidified, 0.7% agar (pH 5.7). The plant growth regulators (PGRs) for mint are 0.5 mg/L (2.2 μM) benzyladenine (BA) and 0.1 mg/L (0.4 μM) indolebutyric acid (IBA), and for carnation and potato are 0.01 mg/L (0.04 μM) BA, 0.001 mg/L (0.005 μM) naphthaleneacetic acid (NAA), and 5 mg/L (14.4 μM) gibberellic acid (GA_3). These media facilitate shoot development and minimize callus formation, but may require modification for specific culture lines. If other species are used, prepare media as described in the their respective literature for shoot tip culture. Media are autoclaved and stored in small aliquots in a refrigerator.

Exercises

The following two experiments demonstrate the use of vitrification for shoot tip cryopreservation.

Experiment 1. Cryopreservation of shoot tips using a solution-based vitrification method

This experiment, illustrated in Figure 36.1, uses a vitrification solution devised by Sakai and researchers for plant cells and shoot tips (Sakai et al., 1991). Excising and manipulating shoot tips is time consuming. It is advisable to excise a few shoot tips and follow the experiment to determine how long it takes to do each task. All solutions and glassware are sterile and all manipulations should be done in a laminar flow hood. Cooling and warming the straws or cryotube need not be done in the flow hood.

Materials

The items below are needed for each student or team of students for the following two experiments:

- Solutions as described below
- Large sturdy forceps with insulated tips
- Wide-mouth liquid nitrogen dewar or large Styrofoam cups
- 0.25 or 0.5-ml semen straws or plastic cryotubes
- Straw closures or heat sealing devise for semen straws
- 12 × 75-mm glass test tubes
- Safety glasses or mask
- Timer

Vitrification solutions

All the solutions below can be stored at 4°C for at least 8 weeks with no apparent loss in effectiveness. PVS2 solution contains (wt/v) 30 glycerol, 15% dimethyl sulfoxide (DMSO), and 15% ethylene glycol (EG) (Sakai et al., 1991). The aqueous volume added contains MS salts (see Chapter three for composition), vitamins, and 0.4 M sucrose without PGRs. This solution is 100% PVS2. Sterilize by filtration. Also prepare a 2 M glycerol, 0.4 M sucrose aqueous solution (Matsumoto et al., 1994) in MS medium and autoclave.

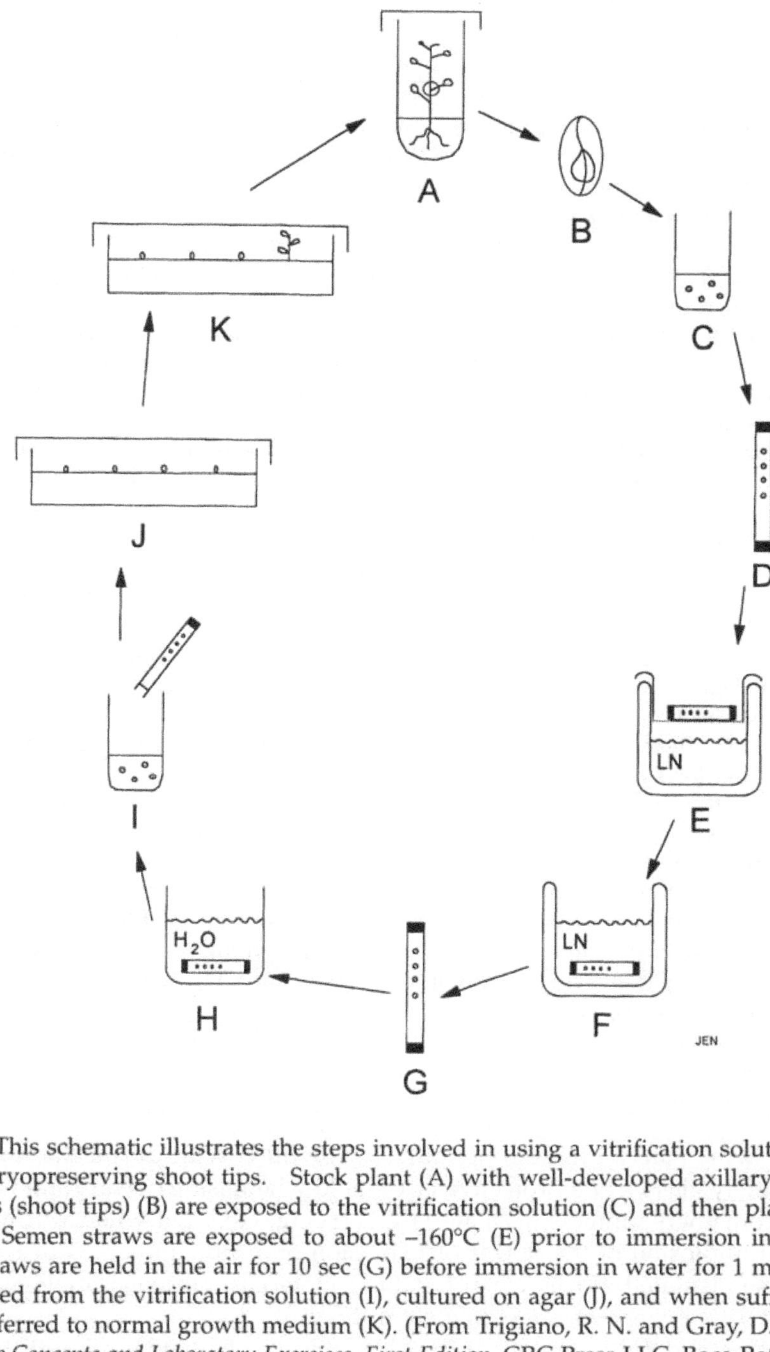

Figure 36.1 This schematic illustrates the steps involved in using a vitrification solution and rapid cooling for cryopreserving shoot tips. Stock plant (A) with well-developed axillary buds (circle). Excised buds (shoot tips) (B) are exposed to the vitrification solution (C) and then placed in semen straws (D). Semen straws are exposed to about −160°C (E) prior to immersion in LN (F). For warming, straws are held in the air for 10 sec (G) before immersion in water for 1 min (H). Shoot tips are diluted from the vitrification solution (I), cultured on agar (J), and when sufficient growth occurs, transferred to normal growth medium (K). (From Trigiano, R. N. and Gray, D. J., Eds., *Plant Tissue Culture Concepts and Laboratory Exercises, First Edition*, CRC Press LLC, Boca Raton, FL, 1996.)

Dilution medium

The dilution medium is used for removing the vitrification solution from shoot tips and is composed of MS salts and vitamins, 1.2 M sucrose, and 100 mg/L myo-inositol (pH 5.7).

The steps for the experiment are summarized in Procedure 36.1. Some additional details or suggestions are described below.

Procedure 36.1
Cryopreservation of shoot tips using a solution-based vitrification method.

Step	Instructions and comments
1	Excise shoot tips with a razor fragment, scalpel, or edge of a hypodermic needle. The isolated shoot tip should be about 1 mm in length and contain about four leaf primordia. Work quickly to minimize desiccation of the shoot tips. Culture a few shoot tips as untreated controls (see step 9).
2	Incubate shoot tips in a small petri dish or flask with liquid growth medium containing 0.3 M sucrose for 1 to 2 days at room temperature.
3	Remove incubation medium with a Pasteur pipette and add the 2 M glycerol + 0.4 M sucrose solution to each tube (*circa* 0.25 to 0.5 ml). Incubate at room temperature for 1 h. After the 1 h, remove a few shoot tips, dilute (step 8), and culture (step 9) to determine if the 2 M glycerol + 0.4 M sucrose is injurious.
4	Remove the glycerol solution completely from the remaining shoot tips and add about 0.5 ml of ice-cold PVS2 vitrification solution. Incubate on ice for 30 min.
5	During the last few minutes of the incubation above, transfer five shoot tips to each straw or cryotube. Fill the straw completely with PVS2; for cryotubes add about 0.25 ml of the PVS2 solution.
6	Wear safety glasses or face shield! At a total incubation time of 30 min in the PVS2, use four replicates for exposure to liquid nitrogen and use four replicates as solution-exposed, uncooled samples (step 9). For cooling, place the straw directly into the vapor phase above liquid nitrogen. Allow to sit for 15 to 30 min and then transfer into the liquid nitrogen. For cryotubes, directly plunge into the liquid nitrogen.
7	Wear safety glasses or face shield. Warm straws by holding in room temperature air for 10 sec, then place in +22°C water for 1 min. Warm vials by placing in +40°C water for 1 min, then place into +22°C water for 1 min.
8	Blot off water from exterior of straw or vial and wipe with 70% ethanol. Immediately open vial or cut off end of semen straw and remove shoot tips from the vitrification solution using either a Pasteur pipette or forceps. Shoot tips are placed into the dilution medium and incubated for 30 min.
9	Remove shoot tips from dilution medium and place on solidified growth medium in small petri dishes. Remove all excess liquid from around the shoot tip. Be careful not to embed the shoot tip within the agar. Wrap petri dishes with parafilm to retard moisture loss.
10	Incubate petri dishes in dim light for about a week prior to placing in normal light conditions.
11	Transfer shoot tips at regular intervals (1 to 3 weeks) to fresh growth medium.

The following safety precautions must be complied with during the exercise. You must wear safety glasses or a protective face shield when using dewars and working with LN (filling containers, immersing materials, carrying, etc.) and when warming samples in case of an implosion of the dewar or explosion of a tube immersed in LN and being warmed. Use insulated gloves when handling materials and wear a lab coat to protect arms. These safety concerns are extremely important!

For excision and culture of shoot tips, select stock plants that are healthy and sufficiently old such that axillary shoot tips are well developed but not elongated. Transfer isolated shoot tips into a 12 × 75-mm test tube containing a small amount of growth medium without PGRs. Place shoot tips into a series of test tubes in random order to minimize the effect of original shoot tip position on the shoot axis. Four shoot tips are placed in each test tube with five replicate tubes per treatment.

For application of the vitrification solution to shoot tips, completely remove preculture liquid with a Pasteur pipette from the test tubes containing shoot tips and add approximately 0.25 ml of the 2 M glycerol + 0.4 M sucrose solution at room temperature. Swirl the tubes periodically to ensure good mixing of the shoot tips with the medium. After 60 min, completely drain this solution using a Pasteur pipette, add about 0.25 ml of ice cold 100% PVS2 solution, and swirl. Place tubes in an ice bath. After about 10 min in this solution, transfer, using a long-tipped Pasteur pipette, shoot tips and a portion of the PVS2 solution to a semen straw that was previously heat sealed at one end. Add the solution such that there is no air pocket at the sealed end. Heat seal the other end of the straw (usually a small air pocket remains). The total time in the 100% PVS2 solution is 20 min prior to exposure to liquid nitrogen (LN) (or to dilution in the case of controls exposed to just the vitrification solution). Practice transferring shoot tips with liquid to straws to assure that you can load all the straws within the 10-min time frame. Be careful not to draw shoot tips too far into the Pasteur pipette during transfer or they may adhere to the glass and be difficult to expel. Straws can also be sealed with cotton or small polypropylene closures supplied by some manufacturers.

Plastic cryotubes can be used in place of semen straws, although their use slows the cooling rate during immersion into the vapor phase of LN. Be sure that the tubes are specified for use with LN. Include about 1 ml of vitrification solution with the shoot tips in the cryotube. Since ability to form a glass within cells is cooling rate dependent, use of cryotubes may result in less survival in some systems. Tests with mint lines have shown similar levels of survival using straws and cryotube, but in potato survival is lower when using cryotubes.

Cooling and warming

A temperature of –160°C is approximated by filling a wide mouth dewar flask about 1/3 full with LN and suspending a screen slightly above the liquid level. A loose-fitting insulated lid is placed on the dewar. Transfer of straws to –160°C reduces the incidence of glass fracturing within the straw during subsequent transfer to LN. Keep the dewar loosely covered when not manipulating the straws or vials. Alternatively, if a LN refrigerator is available, store the straws/vials directly in the vapor phase above the LN (*circa* –180°C). Instead of a dewar flask, Styrofoam cups may be used to hold LN. Briefly surface disinfest the cup with a 70% ethanol solution, allow to dry, and then add LN. The LN evaporates faster in a cup than in a dewar so be sure to keep the straws or cryotubes within the LN. If necessary, samples can be stored until the next laboratory session, but be sure that they will not experience any warming above about –130°C.

Transfer of the straw or cryotube from LN directly into water often causes excessive glass cracking, which can reduce viability in some systems. If cryotubes are used, LN may seep into the vial during immersion in LN. Although infrequent, during warming the rapid expansion of the liquid to a gas may cause the tube to shatter or explode. Wear your safety glasses. Only thaw a few samples at one time; once they are warm the shoot tips are exposed again to vitrification solution and the potentially toxic conditions it imposes. Hence, you must move quickly to the removal phase.

Removal of vitrification solution and culture of shoot tips

Dispense growth medium containing PGRs into 100 × 15-mm petri dishes and allow to solidify. Transfer shoot tips from dilution medium to the agar surface using a Pasteur pipette.

Anticipated results

Viability. Observe whether or not the shoot tips are green immediately after plating. Make growth observations using a dissecting microscope periodically over 4 to 6 weeks.

After LN treatment, the shoot tips should be green upon plating, but color may be lost within a day or so. Shoot tip expansion in controls often begins within 3 days, but vitrified shoot tips develop slower, often after 7 days or more. If they are brown immediately, they probably are not viable. Tips that slowly lose their color may still be alive and often redevelop color or show some signs of growth within 2 to 4 weeks. Transferring the shoot tips to fresh medium can enhance the growth rate.

Experimental parameters to examine. A number of variables can be identified in the experiment described above and could be examined to optimize or improve survival. When making comparisons, each treatment should comprise at least five replicate tubes of four buds/tube. A number of parameters are important. In order to test preculture of the isolated shoot tips, compare shoot tips that have been cultured for 1 or 2 days in growth medium with 0.3 M sucrose with those cultured in growth medium alone. Examine a range of sucrose concentrations between 0.3 and 0.7 M. To evaluate the effect of time in vitrification solution, which is toxic, examine times ranging from 15 to 90 min. Toxicity is a kinetic phenomenon, influenced by concentration, exposure time, and temperature. Hence, the temperature of application of the full-strength vitrification solution may be examined, for example, using 0 and 22°C applications. To test whether loading with the 2 M glycerol + 0.4 M sucrose solution is necessary, compare samples incubated with it to ones not exposed. Most other factors are somewhat minor, but may be important for some species or cultivars. These factors include time in the dilution medium and concentration of sucrose in the dilution medium. Be sure to include controls to determine whether or not the dilution medium itself is injurious.

Cryo-treated shoot tips sometimes need to be cultured for a brief time (days) on a modified growth medium before transfer to a normal medium. Compare media with 0.3 M sucrose to 2% sucrose; also compare a medium without ammonium ion to the usual MS medium.

Questions

- What advantages and disadvantages are there to storing germplasm in LN vs. tissue cultured plants?
- What is the purpose of the cryoprotectant? Could distilled water be substituted?

Experiment 2. Cryopreservation using the encapsulation–dehydration method of vitrification

Materials

Encapsulation media are required. A liquid medium is prepared containing 3% sodium alginate (medium viscosity) in a MS minerals and vitamins solution (see Chapter three for composition) modified by omitting calcium. A second medium contains MS minerals and vitamins solution modified by increasing the concentration of $CaCl_2$ to 100 mM. Both media contain 102.6 g/L (0.3 M) sucrose.

This experiment is depicted in Figure 36.2 and the steps are described in Procedure 36.2. Because drying rates vary with bead size and humidity, it is desirable in a preliminary experiment to measure changes in bead moisture content with time of drying in the laminar flow hood. You do not need shoot tips within these beads. You should also include some shoot-tip-free beads during each experiment such that when you remove a sample of beads with shoot tips for cooling, you can use the shoot-tip-free beads for percent moisture. Beads are weighed, then heated overnight at 85 to 90°C, and reweighed

Procedure 36.2
Cryopreservation of shoot tips using an encapsulation–dehydration method for vitrification.

Step	Instructions and comments
1	Excise shoot tips with a razor fragment, scalpel, or edge of a hypodermic needle. Store briefly in growth medium until all are harvested. If the shoot tips cannot be used the same day, they may be held overnight in growth medium with 0.3 M sucrose.
2	Drain medium from shoot tips and resuspend in modified MS medium containing alginate and without calcium.
3	Using a Pasteur pipette draw one shoot tip into the tip of the pipette along with some medium. Expel the tip and medium as a drop into the MS medium containing 100 mM $CaCl_2$. The size of the drop can be adjusted by using pipettes of different internal diameter (bores). Three to 5-mm-diameter beads are desired. Allow the gelled drops (beads) to sit about 30 min.
4	Drain the liquid from the beads and add 0.5 M sucrose in MS; gently shake overnight at room temperature.
5	Drain the liquid from the beads and add 0.75 M sucrose in MS; gently shake overnight at room temperature.
6	Remove the beads from the liquid and blot with sterile tissue paper to remove excess moisture.
7	Place the beads either in a laminar flow hood on a sterile glass surface (for example, the outside of a petri dish so air flows freely around the beads) or in a small vessel with a desiccant.
8	When the moisture content of the beads has decreased to 25% moisture (on a fresh weight basis), place five beads per cryotube. Four vials are transferred to liquid nitrogen (step 9) and four vials are directly plated (step 11) to determine the effect of desiccation alone. Repeat this sequence for moisture contents of about 22, 18, and 15%; eight vials per moisture level: four for cooling, four for desiccation.
9	Wear safety glasses. Immerse cryotube directly into liquid nitrogen. Vials should be stored at least 1 h before warming.
10	Wear safety glasses. Warm vials by removing them from the liquid nitrogen using forceps and placing them directly into 40°C water for 2 min. Wipe the excess water from the vial and briefly wipe with 70% ethanol.
11	Remove beads from the vials and directly place on the surface of solidified growth medium.
12	Culture in dim light for about 1 week and then transfer dish to normal light and growth conditions.

the next day. For cooling and warming, follow the same safety precautions for working with LN listed previously. Use beads that have been dried to a moisture content of about 15 to 25% (wet weight basis). The number of moisture levels to test depends on the number of shoot tips harvested.

Experimental parameters to examine

Several variables are important for optimizing survival. The major ones to examine are the preculture phase, including time in the sucrose solution and concentration of sucrose, and the time for dehydration of the bead in the air flow. Time for drying is empirically determined; plunge beads of about 25, 22, 18, and 15% moisture into LN to determine the best moisture level. Be sure to always include desiccated, but uncooled, samples for each moisture level to determine the effect of desiccation alone on survival. In most cases rapid cooling is desired, but for some species slower cooling (0.25 to 1°C/min) is beneficial.

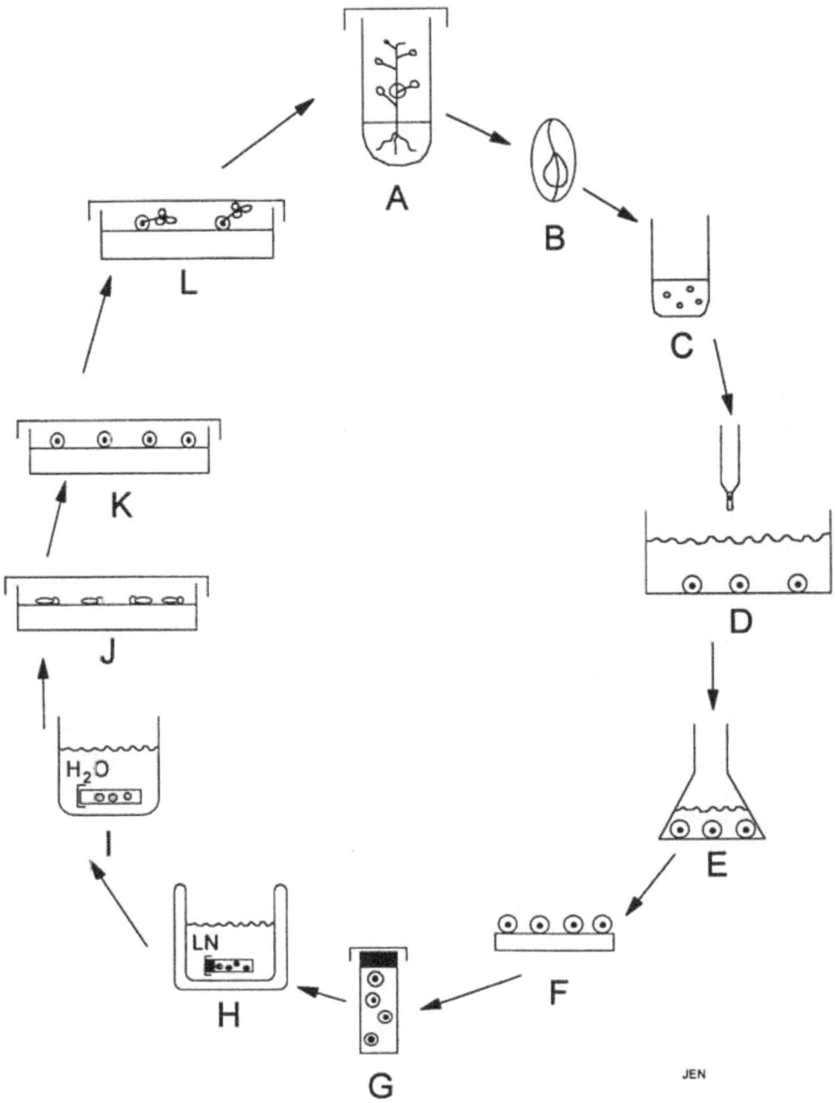

Figure 36.2 This schematic illustrates the steps performed in using encapsulation–dehydration and cooling as a method for cryopreserving shoot tips. A stock plant (A) is selected and shoot tips (circled area in A) are excised (B) and placed in a calcium-free alginate solution (C). Shoot tips are dropped into a calcium solution (D) to form the bead. Beads are incubated in sugar solutions (E) and then partially desiccated (F). Beads are placed in cryotubes (G) and the cryotubes directly immersed in LN (H). Warming is by immersion in water (I). Beads are directly cultured on agar (J) during which the bead rehydrates (K). When sufficient growth has occurred, shoots can be transferred to normal growth medium (L). (From Trigiano, R. N. and Gray, D. J., Eds., *Plant Tissue Culture Concepts and Laboratory Exercises, First Edition*, CRC Press LLC, Boca Raton, FL, 1996.)

Questions

- Why is it necessary to rapidly cool the beads?
- How does moisture level affect survivability of shoot tips stored in LN?

Literature cited

Fabre, J. and J. Dereuddre. 1990. Encapsulation-dehydration: a new approach to cryopreservation of *Solanum* shoot tips. *Cryo-Letters* 11:413–426.

Hirai, D., K. Shirai, S. Shirai, and A. Sakai. 1998. Cryopreservation of in-vitro grown meristems of strawberry (*Fragaria* × *ananassa* Duch.) by encapsulation-vitrification. *Euphytica* 101:109–115.

Malaurie, B., M.-F. Trouslot, F. Engelmann, and N. Chabrillange. 1998. Effect of pretreatment conditions on the cryopreservation of in vitro-cultured yam (*Dioscorea alata* 'Brazo Fuerte and *D. bulbifera* Noumea Imboro') shoot apices by encapsulation-dehydration. *Cryo-Letters* 19:15–26

Matsumoto, T., A. Sakai, and Y. Nako. 1994. Cryopreservation of in-vitro grown apical meristems of wasabi (*Wasabia japonica*) by vitrification and subsequent high plant regeneration. *Plant Cell Rep.* 13:442–446.

Sakai, A., S. Kobayashi, and I. Oiyama. 1991. Survival by vitrification of nucellar cells of Navel Orange (*Citrus sinensis* var. *brasiliensis* Tanaka) cooled to –196°C. *J. Plant Physiol.* 137:465–470.

Steponkus, P.L., R. Langis, and S. Fujikawa. 1992. Cryopreservation of plant tissues by vitrification. In: *Advances in Low-Temperature Biology,* Vol. 1. Steponkus, P.L., Ed. pp. 1–62. JAI Press, Greenwich, CT.

chapter thirty-seven

Secondary product expression in vitro

Mary Ann L. Smith

The accumulation of secondary products in plant cell cultures in response to deliberate treatments to the chemical or physical microenvironment provides students with striking and explicit experimental evidence to confirm basic physiological principles. Secondary products such as vivid pigments, aromatic compounds, flavors, and bioactive phytochemicals can now be successfully accumulated in many plant in vitro cultures. In light of the current consumer demands for natural food products and plant-derived medications, in vitro production of valuable plant secondary products (phytochemicals) has become an industrially promising alternative to synthetic compounds. Metabolite production in vitro also provides an excellent forum for in depth investigation of plant metabolic pathways under highly controlled conditions. In the past, many tissue culture courses have concentrated on in vitro propagation strategies and breeding/selection, including biotechnology. However, this overlooks the entire realm of in vitro work concerned more with valuable plant products rather than with plant production.

Production of secondary products from plant tissue culture is based on the premise that the same valuable product produced in nature in an organ, fruit, or other tissue from a plant can be stimulated to accumulate in undifferentiated cells. Researchers have succeeded in producing several valuable secondary phytochemicals in unorganized callus or suspension cultures. In some cases where the metabolite is only produced in specialized plant tissues or glands, production has not been successfully stimulated in cell culture. A prime example is ginseng (*Panax ginseng* C.A. Meyer). Since saponin and other valuable metabolites are specifically produced in ginseng roots, root culture is required in vitro. As another example, biosynthesis of lysine to anabasine occurs in tobacco (*Nicotiana tabacum* L.) roots, followed by the conversion of anabasine to nicotine in leaves. Callus and shoot cultures of tobacco can produce only trace amounts of nicotine because they lack the organ-specific compound anabasine. In other cases, at least some degree of differentiation in a cell culture must occur before a product can be synthesized (e.g., vincristine or vinblastine from *Catharanthus roseus* [L.] G. Don; madagascar periwinkle). Reliance of a plant on a specialized structure for production of a secondary metabolite, in some cases, is a mechanism to keep a potentially toxic compound sequestered. One goal of genetic engineering research is to provide the gene encoding a limiting enzyme to all tissues in vitro in order to uncouple synthesis of secondary metabolites from differentiation.

0-8493-2029-1/00/$0.00+$.50

Current trends for in vitro secondary metabolite production

There are several distinct advantages to producing a valuable secondary product in plant cell culture, rather than in vivo in the whole crop plant. Plant cell culture eliminates potential political boundaries or geographic barriers to production of a crop, such as the restriction of natural rubber production to the tropics, or anthocyanin pigment production to climates with high light intensity. When a valuable product is found in a wild or scarce plant species, intensive cell culture is a practical alternative to wild collection of fruits or other plant materials. Secondary products in plant cell culture can be generated on a continuous year-round basis (no seasonal constraints), production is reliable and predictable (independent of ambient weather), and in at least some cases, the yields per gram fresh weight may exceed that found in nature. Plant cell lines can be recurrently selected to amplify the productivity of the cell culture. Shikonin production in vitro at levels over 800 × that available from plant root is a case in point. Disagreeable odors or flavors associated with the crop plant can be modified or eliminated in vitro. Extraction from the in vitro tissues is much simpler than extraction from organized, complex tissues of a plant. Plant tissue culture techniques offer the rare opportunity to tailor the chemical profile of a phytochemical product, by manipulation of the chemical or physical microenvironment, to produce a compound of potentially more value for human use.

The three long-standing, classic examples of commercially viable production of a secondary metabolite in vitro — ginseng saponines, shikonin, and berberine — each feature products that have diversified uses including medicinal applications. Ginseng is produced in large-scale root cultures, whereas the other two products are produced in highly colored cell cultures. A tremendous research and development effort has advanced a number of other in vitro-derived secondary products to semicommercial status, including vanillin and taxol production in cell cultures. In a myriad of other cases, the in vitro processes for secondary metabolite production have fallen far short of expectations and have never approached commercializable status. Still, the arena of secondary product formation in cell cultures remains as an industrial pursuit. Engineers and biologists are currently joining forces on a global scale to develop new strategies for streamlining the critical bioprocesses (Smith et al., 1995). Research efforts on a broad range of plant cell culture-derived extracts can be cited in each of these major product categories: flavors (onion and garlic, peppermint and spearmint, fruit flavors, chocolate aroma, seaweed flavors, vanilla, celery, coffee, spice, sweeteners, etc.); edible colors for foods and medicines (mainly betalains and anthocyanins); nonfood pigments for cosmetics and textiles (shikonin, berberine, and various other products); several examples of fragrances, essential oils; and bioactive natural insecticides and phytoalexins useful in current integrated pest management programs. Of course, intensive activity has centered on production of natural drugs or chemoprotective compounds from plant cell culture. Some of the most prominent pharmaceutical products in this latter category include ajmalicine (a drug for circulatory problems) from *C. roseus*, and taxol (a phytochemical effective in treatment of ovarian cancer) from *Taxus* spp.

Because different levels of secondary metabolite production can be found within a cell line (e.g., sectoring of colored and noncolored callus in a culture), significant headway in the productivity of plant cell cultures has followed on the heels of intensive selection for high-producing cell lines. Single cell cloning and recurrent selection of the most pigmented cells and sectors has significantly enriched shikonin- and anthocyanin-producing cultures, for example.

Another target for research has been the development of improved instrumentation for cultivation of plant cells en masse. Because intensive cultivation of plant cells differs from that of animal or microbial cells on a number of key fronts, the innovations have revolved around specific adaptations for the unique needs of plant cells in vitro. For example, improved bioreactors have been specifically designed to permit light to reach plant cells during division and product accumulation. In many cases, the infiltration of light at optimum intensities and spectral quality is prerequisite to successful synthesis and maximal expression of secondary metabolites. Modifications in aeration, agitation, and nutrient supply have similarly led to enhanced plant product yields in recent years. Elicitors (compounds that "stress" cells, leading to formation of a secondary product) from biotic and abiotic sources have been added to culture media to stimulate production, but this strategy depends on a strong understanding of the biochemical pathway leading to secondary product synthesis. Another approach is biotransformation — the deliberate feeding of metabolic precursors to a cell culture that already contains the necessary enzymes to convert them to product.

Pigments from cell culture

Natural pigments including anthocyanins, betalains, shikonin, and other pigmented phytochemicals are conspicuously accumulated in some cell cultures. Some of these products are of interest commercially as replacements for synthetic food dyes that have fallen "under the gun" recently as adverse health and safety concerns have emerged. Negative reports and new medical studies in the popular press have made synthetic additives a target for contention, and have created a strong market demand for safe, natural ingredients in foods. As an added bonus, significant chemoprotective benefits (including cardioprotective and anticarcinogenic properties) have been attributed to natural anthocyanin pigments and related flavonoid phytochemicals (Rimm et al., 1996; Wang et al., 1997). These health-enhancing properties are unrelated to the nutritive value of the plants. Even natural nonfood colorants are sought as replacements for environmentally damaging synthetic dyes fabricated with toxic solvents and heavy metals. The myriad of microenvironmental control tools available to in vitro production makes this alternative more promising than harvest from the intact plant in nature (Havkin-Frenkel et al., 1997).

There may be little or no correlation between the level of valuable natural pigment typically produced in a whole plant (in fruit or foliage), and the level that can be induced to accumulate in tissue culture. As an example, a cranberry (*Vaccinium macrocarpon* Ait) cultivar selected in the field for intense color may not produce much pigment in callus, whereas a cranberry that is characterized by duller color in vivo may accumulate a greater concentration of complex anthocyanins in callus and suspension cultures. This type of in vitro production system offers a superb model to investigate control over anthocyanin production by introducing gene constructs using anthocyanin regulatory genes to the cell culture tissues. When excessively high anthocyanin levels interfere with cell culture growth, gene constructs may use an inducible promotor system. In this case, the bright red pigment is an obvious marker for research on manipulating flavonoid production and modifications in the simple cell culture system. These lines of research would be complex and unwieldy in whole plant models due to the restrictions of organ-specific product accumulation and interactions with other plant tissues.

In many ways, natural plant pigments are ideal target compounds for laboratory classroom or demonstration experiments on metabolite production, because their accrual in vitro is quick, visible, easy to detect, and fairly simple to quantify. In practice, and for validation of research results, pigment accumulation is typically measured by extraction, then quantification/separation with TLC (thin layer chromatography), HPLC (high performance liquid chromatography), NMR (nuclear magnetic resonance), and/or absorbance (spectroscopy). However, in a classroom, when traditional methods of pigment quantification are prohibited by the lack of sophisticated laboratory analytical equipment and/or insufficient student expertise, the production of a pigment in vitro can still be very easily detected and quantified by visual means alone, or by coupling visual observation with very simple absorbance measurements using a spectrophotometer.

Many of the best-known in vitro pigment production systems are actually difficult to demonstrate in a classroom setting because they either require cultures derived from specific, hard-to-initiate explants, or the high yielding models described in research reports have only been achieved after years of laborious repetitive subculture and selection of only highly pigmented sectors. The experimental model introduced in this chapter is not limited by these constraints and, therefore, is readily adaptable to classroom research.

Ajuga reptans N. Tayl. (Bugleweed) and the related species *A. pyramidalis* L. 'Metallica Crispa' are both widely distributed groundcover plants that grow rapidly and spread via stolons. *Ajuga* tolerates adverse conditions so well that it is considered an intractable weed when it oversteps its boundaries in the landscape, and stock plants can be maintained long term in a greenhouse with minimal care. *Ajuga* cultivars are typically propagated by division of mother plants. A few select cultivars of *A. reptans*, which feature purple or bronze foliage, include 'Purpurea', 'Giant Bronze', and 'Burgundy Glow'. The latter cultivar has a characteristic chimeral green, creamy white, and dark pink variegated foliage, but produces unicolor bronze, albino, or green variants from adventitious buds in vitro. *A. pyramidalis* 'Metallica Crispa', which is frequently confused with purple forms of *A. reptans* in the trade, is characterized by savoy-type, glossy, bronze-purple foliage.

For a number of reasons, selections in the genus *Ajuga* are particularly well suited for use in plant tissue culture exercises. Like many members of the mint family (Labiatie), microplants flourish in culture and may be held in cold storage successfully for up to a year, which helps cut back on the need for repetitive subculture maintenance between semesters. *A. reptans* proliferates readily in vitro from axillary or adventitious buds, exhibits classic, predictable responses to gradient increases of cytokinin concentration, and is a good, rapidly growing model plant on which to illustrate stages in an in vitro production cycle through acclimatization within a single semester (Preece and Huetteman, 1994).

The purple or bronze leaves that distinguish some varieties of *Ajuga* are readily expressed also in microcultured plants and even undifferentiated callus can exhibit pigment expression. Callebaut et al. (1993) reported that primary callus from flowers, petioles, leaves, and stems of *A. reptans* developed into a fairly homogeneous, blue-colored cell culture after over a year of selection and subculture. These suspensions produced anthocyanin pigments in the dark and pigment expression responded to changes in the light regime and to chemical medium treatments. In our alternative approach, vegetative disks are directly explanted from *Ajuga* foliage and induced to produce callus, which rapidly acquires a vivid purple hue when exposed to illumination (Madhavi et al., 1996). Intense, uniform pigment expression is routinely achieved without selection and friable callus colonies can be explanted into liquid suspension culture. Cell division and pigment accumulation are not mutually exclusive processes; both cell biomass and pigment accumulation increase steadily throughout each subculture cycle. The growth and pigment expression are so predictable and consistent that this system was used as a model to test the application of video image analysis to regulate bioprocesses by visual characterization of cell size, shape, aggregation, and color intensity (Smith et al., 1995).

Current publicity surrounding "Functional Foods" and natural/medicinally active phytochemicals is gaining momentum in the popular press, and a renewed interest in natural product synthesis is emerging. Since production of these valuable plant metabolites in vitro is a viable alternative production route with notable commercial advantages, it is likely that industry and research opportunities will continue to expand. Students can be introduced to the natural products research arena through the laboratory exercises in the following chapter (featuring *Ajuga* pigments in cell culture). These exercises have been repeatedly practiced by novice plant tissue culture students with satisfying and predictable outcomes. This experimental model is also amenable to individualized, student-directed experimentation with a vast range of physical or chemical microenvironmental parameters (treatments) that can be deliberately varied to determine the effects on metabolite accumulation. The experimental results also help students to appreciate the value of chemical analytical methods for measuring metabolite production as an adjunct to strict visual gauges for pigment production in a callus or suspension culture.

Literature cited

Callebaut, A., M. Decleire, and K. Vandermeiren. 1993. *Ajuga reptans* (Bugle): in vitro production of anthocyanins. In: *Medicinal and Aromatic Plants V.* Y.P.S. Bajaj, Ed. pp. 1–22. Springer-Verlag, Berlin.

Havkin-Frenkel, D., R. Dorn, and T. Leustek. 1997. Plant tissue culture for production of secondary metabolites. *Food Technol.* 51:56–61.

Madhavi, D.L., S. Juthangkoon, K. Lewen, M.D. Berber-Jiménez, and M.A.L. Smith. 1996. Characterization of anthocyanins from *Ajuga pyramidalis* 'Metallica Crispa' cell cultures. *J. Agric. Food Chem.* 44:1170–1176.

Preece, J.E. and C.A. Huetteman. 1994. An axillary shoot proliferation laboratory exercise — micropropagating *Ajuga reptans* L. *HortTechnology* 4:312–314.

Rimm, E., M. Katan, A. Ascherio, M. Stampfer, and W. Willett. 1996. Relation between intake of flavonoids and risk for coronary heart disease in male health professionals. *Ann. Intern. Med.* 125:384–389.

Smith, M.A.L., J.F. Reid, A. Hansen, Z.-W. Li, and D.L. Madhavi. 1995. Non-destructive machine vision analysis of pigment-producing cell cultures. *J. Biotechnol.* 40:1–11.

Wang, H., G. Cao, and R. L. Prior. 1997. Oxygen radical absorbing capacity of anthocyanins. *J. Agric. Food Chem.* 45:304–309.

chapter thirty-eight

Pigment production in Ajuga *cell cultures*

Mary Ann L. Smith and Randy B. Rogers

The following laboratory exercises provide clear-cut demonstrations of secondary metabolite accumulation in unorganized (callus and suspension) cultures from different genotypes of *Ajuga.* Cell cultures from both *Ajuga reptans* and from purple foliage cultivars of the related species *A. pyramidalis* have been recognized as donors for flavonoids (including acylated anthocyanin pigments with higher stability than obtained from in vivo plants) and other valuable phytochemicals (Callebaut et al., 1997; Madhavi et al., 1996; 1997)

Experiment 1 (the main exercise) explicitly illustrates the influence of chemical microenvironmental factors, specifically, carbon source, on the levels of pigmentation produced by *Ajuga* callus (originating from leaf explants). The effect of the chemical microenvironment on secondary metabolite accumulation in a cell culture is well recognized, and medium composition is usually deliberately manipulated to enhance anthocyanin accumulation in vitro. While a number of parallels exist in different pigment production systems, determining the optimum chemical regime still requires empirical testing to some extent. Several research teams have realized analogous results when testing the effects of the culture medium composition on pigment expression, and these trends can be demonstrated readily in classroom experiments. For example, the culture medium that best supports cell growth may not be the same as the medium composition that induces the most intense pigment production. Several in vitro pigment production systems for callus or suspension cultures are characterized by dual phases in the culture protocol, usually a "growth phase" followed by a "product accumulation phase" on a production medium that favors accumulation of pigment. For some of these systems, in the latter stage, the cell cultures have reached the status of little or no new growth during product accumulation. However, there is no question that the microenvironmental conditions imposed on cells during the growth phase exert substantial control over secondary metabolite production in the latter stage. Deliberate chemical modifications made to the culture medium have resulted in significant improvements in product recovery or stability; for example, cinnamic acids fed to wild carrot cell suspensions resulted in production of more stable, monoacylated anthocyanin pigments produced in vitro (Dougall et al., 1997).

Anthocyanin production in vivo and in vitro can be provoked by a great number of environmental factors. Elevated levels of sucrose or other sugar (as a carbon source and/or osmotic agent), a change in the nitrate-to-ammonium ratio, and changes in the growth

0-8493-2029-1/00/$0.00+$.50

regulator composition are frequently needed in the production phase of a two-phase culture system to stimulate pigments in vitro.

Experiment 2 demonstrates the influence of explant size and source on productivity and pigment expression in amorphous masses of callus. Experiment 3 features a variegated chimeral *Ajuga* genotype, and illustrates the variation in productivity and performance of callus isolated from different sectors of this chimeral leaf. A final exercise, Experiment 4, examines the influence of physical microenvironmental treatments (light wavelengths) on anthocyanin production from cell cultures. Pigment accumulation in nature and in vitro is often a response to environmental stress and consequently pigment expression in cell cultures can conspicuously mark the outcome of experiments that impose a chemical or physical pressure on cultured cells.

General considerations

Plant materials

Ajuga pyramidalis 'Metallica Crispa', and the variegated chimeral selection *A. reptans* 'Burgundy Glow' are both readily available at most garden centers as flats of stoloniferous, spreading plants ready to plant in the landscape. To avoid confusion, identified stock plants should be obtained from a reputable source, because many retail shops sell flats labeled as "Purple Ajuga," without clear distinction between *A. pyramidalis* 'Metallica Crispa' and the phenotypically similar bronze/purple selections of *A. reptans*. *A. reptans* and *A. pyramidalis* freely hybridize, resulting in much confusion in the trade. Again, although any of these selections will yield a good experimental response in the exercises below, it is important to make sure that all stock plants are of a single genotype to avoid a source of uncontrolled variation unrelated to treatment effects.

Stock plants are planted on 5-in. centers in a 1:1:1 mix (peat/perlite/soil) in a shallow flat and fertilized once per week with a 20:20:20 fertilizer. Only natural light (no supplemental lighting) is used in the greenhouse, which is maintained at 25 to 28°C. Both greenhouse plants and microcultured stock plants can be used as explant sources in these exercises. Approximately two to three healthy plants per student should be maintained to ensure sufficient new foliage for explanting trials. Greenhouse plants require low maintenance and may be divided and replanted approximately once per year to maintain stock. However, maximal success in explanting from the greenhouse is achieved when young, vigorously growing new leaves from new transplants are used as donor material.

Microcultured stock plants are initiated by explanting meristem tips (devoid of all leaves) or nodes from runners, each approximately 1 to 2 cm long. Any explant material taken from the greenhouse stock plants should not be in contact with the soil; runners or plant overhanging the sides of the flats are ideal for this purpose. Alternatively, *Ajuga* plants can be potted and placed in a growth chamber to avoid any overhead watering in the greenhouse until new growth is initiated; then explants can be taken from the new growth. Surface disinfest the explants by treatment in 15% commercial bleach with 0.1% Tween 20 (a few drops per liter). Agitate explants in the sterilant for 15 min followed by three rinses in sterile double-distilled water. After removing injured tissue, the explants can be placed in 25 × 150-mm culture tubes containing 10 mL of basal media composed of Woody Plant Medium (WPM) basal salts (Lloyd and McCown, 1980; see Chapter three for composition) amended with 88 mM (30 g) sucrose, 0.55 mM (0.1 g) myo-inositol, 0.28 mM (50 mg) L-ascorbic acid, 1.48 μM (0.5 mg) thiamine-HCl, 2.44 μM (0.5 mg) pyridoxine-HCl, 4.06 M (0.5 mg) nicotinic acid, 26.7 μM (2 mg) glycine, 1 μM (0.203 mg) 2iP, 150 mg L^{-1} PVP-T (polyvinylpyrrolidone-10), and 7 g/L of agar; pH 5.7 prior to autoclaving. After cultures begin to produce new growth, four to five explants are subcultured into each GA-

7 vessel (Magenta Corp, Chicago) with 50 ml of basal medium. The microcultured plants are maintained indefinitely at 25°C under 80 $\mu mol \cdot m^{-2} \cdot s^{-1}$ irradiance as shoot cultures, by subculture of meristems every 4 to 6 weeks to fresh medium. As an added advantage for use in yearly plant tissue culture classes, microcultured stock plants can be maintained in long-term cold storage to minimize the need for subculture upkeep. After subculture, microplants are placed in the normal culture room (25°C) described above for approximately 3 weeks, until new growth and reestablishment is visible. Subsequently, these cultures are parafilmed around the cap and can be stored at 4°C in the dark for 4 to 6 months. When removed from cold storage and reintroduced in the culture room, the *Ajuga* plants resume growth readily.

Exercises

Experiment 1. Effect of the carbon source on pigment expression in Ajuga *cell cultures*

This experiment was designed to illustrate the effects of carbon source and level on anthocyanin production in callus of *Ajuga pyramidalis* 'Metallica Crispa'.

The experiment begins at the explanting stage (initiation of callus from vegetative leaf disks) and continues through a final phase of pigment accumulation in callus colonies. Greenhouse leaves may be surface disinfested and used as a source of explants, or presterile microcultured leaves may be explanted (Figure 38.1). Alternatively, students can begin with precultured callus already growing in the dark (label C on Figure 38.1), to shorten the exercise. When starting with dark-grown callus, students can proceed immediately to the subculture of callus to different treatments of the production medium. The same experiment can be performed by subculturing the white friable callus to liquid production media without agar, and producing pigment in suspension culture.

Materials

The following items are needed for each team:

- One or two shoot cultures of *Ajuga pyramidalis* 'Metallica Crispa' in GA-7 vessels
- 1 cm diameter cork borer
- Fifteen GA-7 vessels containing a callus induction medium, which substitutes 2.3 µM 2,4-D and 3.5 µM kinetin as the growth regulators
- Eight GA-7 vessels, each containing experimental pigment production media, all with 2.26 µM IAA and 3.49 µM zeatin, but modified so that two vessels each contain either 10, 30, or 70 g/L sucrose or 30 g/L galactose

Follow the protocols listed in Procedure 38.1 to complete this experiment.

Anticipated results

After as little as 4 days under lights, callus in each of the four carbon source treatments should exhibit marked differences in anthocyanin production. The information given in Figure 38.2 is typical of the effect of sucrose concentration in the production medium on yield of anthocyanins after 2 weeks under lights. The galactose treatment will support cell growth as well as sucrose at the same concentration, as illustrated by similar increases in FW and DW, but will stimulate significantly greater anthocyanin expression. Typically, 15 to 20% greater anthocyanin concentrations will be produced in the galactose treatment as compared to sucrose at 30 g/L. It is important that anthocyanin concentrations be

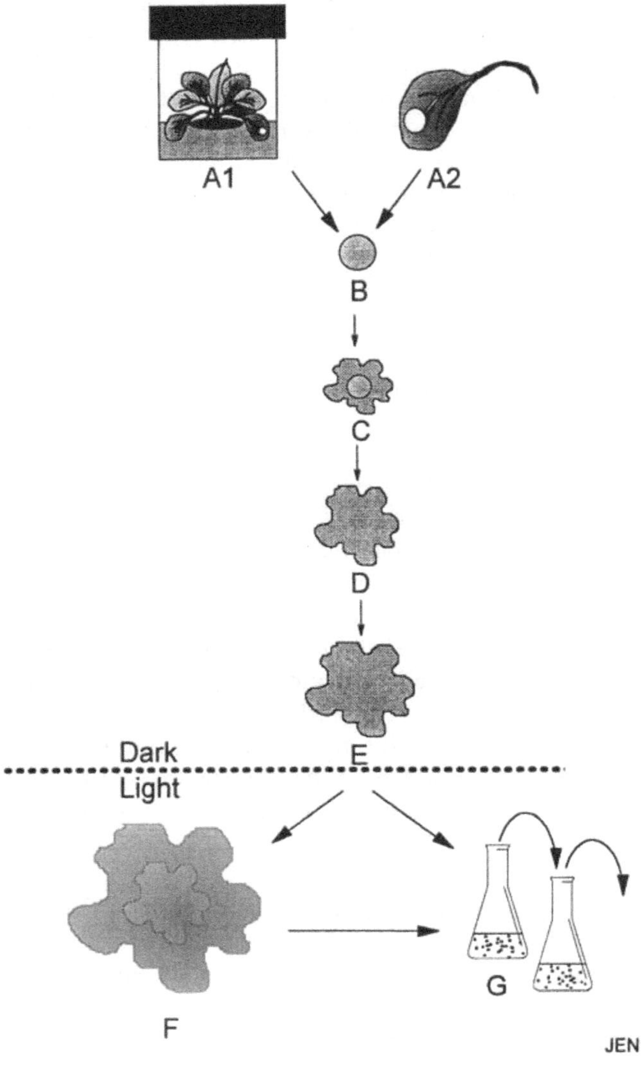

Figure 38.1 Schematic representation of the pigment production protocol for *Ajuga.* (A) Explants obtained from either presterile microcultured plants, or from greenhouse or growth chamber specimens. (B) Callus is excised from mother tissue and cultured independently for two additional rounds of culture growth in the dark. (C) Friable, ivory-colored callus is gently lifted from the callus induction medium, subcultured to pigment production medium, and placed on illuminated culture room shelves. (D) Friable, colored callus may be maintained on solid pigment production medium for subsequent subcultures, or introduced into liquid suspension cultures (G). (From Trigiano, R. N. and Gray, D. J., Eds., *Plant Tissue Culture Concepts and Laboratory Exercises, First Edition,* CRC Press LLC, Boca Raton, FL, 1996.)

compared on the basis of yield per treatment, rather than on a milligram of anthocyanin/gram of FW or DW basis, because different carbon sources and levels will support different callus growth rates as well as different levels of anthocyanin accumulation. The treatment containing low sucrose concentrations (10 g/L) will exhibit significantly less growth during the 2 weeks on pigment production medium. Callus in this treatment will be visibly duller, with less pigment concentration.

Procedure 38.1
Pigment expression in *Ajuga* callus and chemical microenvironmental effects.

Step	Instructions and comments
1	Collect five identical leaf disks using a 1-cm-diameter corkborer from presterile *Ajuga pyramidalis* 'Metallica Crispa' shoot cultures. (Alternatively, surface-disinfested leaves from greenhouse specimens can be used.) Place each disk on the surface of callus induction medium (WPM amended with 88 mM [30 g/L] sucrose, 0.55 mM [0.1 g/L] myo-inositol, 0.28 mM [50 mg/L] L-ascorbic acid, 1.48 µM [0.5 mg/L] thiamine-HCl, 2.44 µM [0.5 mg/Ll] pyridoxine-HCl, 4.06 µM [0.5 mg/L] nicotinic acid, 26.7 µM [2 mg/L] glycine, 2.3 µM [0.5 mg/L] 2,4-D, 3.5 µM [0.75 mg/L] kinetin, 150 mg/L PVP-T [polyvinylpyrrolidone-10], and 7 g/L of agar; pH 5.7 prior to autoclaving).
2	Incubate cultures in darkness at 25°C for 3 weeks.
3	Separate ivory-white friable callus from mother tissue and transfer to fresh induction media. Continue incubation in darkness, subculturing the callus to fresh media at 2-week intervals for 4 weeks.
4	After a total of 7 weeks from initiation, transfer callus masses to pigment production media treatments (callus induction media substituting 2.26 µM [0.4 mg/L] IAA and 3.49 µM [0.76 mg/L] zeatin as PGRs). The four treatments are sucrose levels of 10, 30, or 70 g/L or a galactose level of 30 g/L. For each treatment, five 2.5-cm-diameter callus masses are transferred to each of two GA-7 vessels containing treatment media (two replicate cultures/treatment, with five subreplicates (callus masses)/vessel).
5	Incubate the cultures under cool white fluorescent lamps providing a PPF of 150 $\mu mol \cdot m^{-2} \cdot sec^{-1}$ at 25°C for 2 weeks.
6	Evaluate growth and pigment production with an initial visual assessment of pigment content in the intact culture. Follow with fresh and dry weight analysis of two subreplicates from each culture. Grind the remaining three subreplicates with a mortar and pestle, store overnight in 10 ml ice-cold acidified methanol (1% HCl), filter, then measure the absorbance of the solution at 535 nm with a spectrophotometer to assess differences in the relative anthocyanin content between treatments.

Questions

- What function does the carbon source (sugar) play in the pigment production stage of culture for *Ajuga* callus? Is the sugar acting mostly as a metabolic agent or as a stress-evoking (osmotic) agent at the 30-g/L concentration? What about at the 70-g/L concentration?
- Why was callus initiated in darkness? How might initiation of callus from explants in the light change the outcome of these experiments?

Experiment 2. Explant size and source effects on in vitro pigment production

The feasibility of stimulating in vitro secondary metabolites from unorganized cells can hinge on the ability to produce uniform, prolific, rapidly growing callus cells, of good quality, from tissues of the target species. Explant size and donor plant physiology can influence the initiation and growth rate of parenchymous callus cells, the friability of the callus masses, and the propensity for the tissue to revert to regeneration (organogenesis) instead of callus production. Characteristics of the callus depend on a complex relationship between the explant tissue used to induce callus, the composition of the medium, and the microenvironmental conditions. In this experiment, a range of explant sizes from

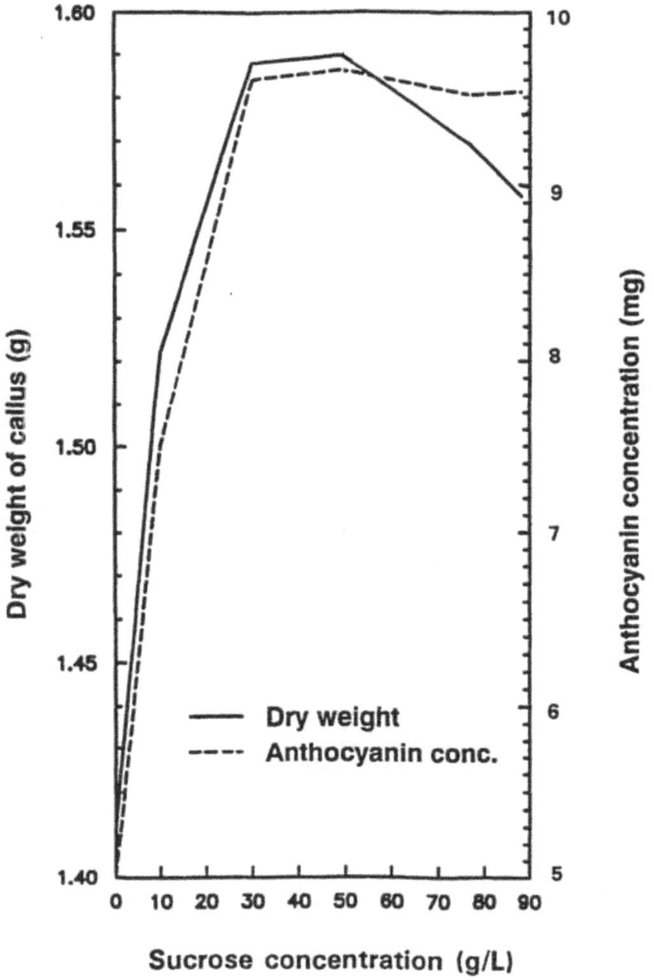

Figure 38.2 Dry weight and anthocyanin concentrations of *Ajuga* callus at various sucrose concentrations in the pigment production medium. (From Trigiano, R. N. and Gray, D. J., Eds., *Plant Tissue Culture Concepts and Laboratory Exercises, First Edition*, CRC Press LLC, Boca Raton, FL, 1996.)

two sources (greenhouse plants and microcultured plants) are compared in terms of callus induction rate, overall callus production, and the ability of the callus to express anthocyanin pigments. The experiment results in similar trends when using *A. pyramidalis* 'Metallica Crispa', or when using purple or bronze-leafed cultivars of *A. reptans*.

Materials

The following items are needed for each team:

- One or two shoot cultures of *Ajuga* in a GA-7 vessel
- One or two greenhouse plants of *Ajuga*
- 10% commercial bleach solution and volumes of sterile double-distilled water
- 250-ml screw top glass jars (Bellco™)
- 2-, 5-, 7-, and 10-mm-diameter cork borers

- Sixteen GA-7 vessels containing callus initiation medium (2.3 μM 2,4-D and 3.5 μM kinetin)
- Eight GA-7 vessels containing pigment production medium (2.26 μM IAA and 3.49 μM zeatin with 30 g/L sucrose)

Follow the procedures outlined in Procedure 38.2 to complete this experiment.

Procedure 38.2
Explant size and source effects on in vitro pigment production.

Step	Instructions and comments
1	Surface disinfest leaves of greenhouse plant material with a 15-min immersion in 10% commercial bleach with a few drops per liter of Tween 20 followed by three 5-min rinses with sterile distilled water.
2	Using cork borers, collect 2-, 5-, 7-, and 10-mm-diameter leaf disks (five replicates of each size) surface disinfested greenhouse plant material. Repeat leaf disk collection from presterile shoot cultures. Place disks on the surface of callus induction medium (see Procedure 38.1, step 1).
3	Incubate cultures in darkness at 25°C for 4 weeks. Inspect frequently for the appearance of callus along the cut edges of leaf sections to determine the effects of explant size and type on the rate of callus initiation. Record onset of callus initiation.
4	For one half of the subreplicates, separate callus from leaf tissue and collect fresh and dry weights. Transfer remaining subreplicates to pigment production medium (callus induction medium substituting 2.26 μM [0.4 mg/L] IAA and 3.49 μM [0.75 mg/L] zeatin as PGRs).
5	Incubate cultures under cool white fluorescent lamps at a PPF of 150 $\mu mol \cdot m^{-2} \cdot sec^{-1}$ for 2 weeks at 25°C. Inspect daily for pigment production.
6	Evaluate growth and pigment production. Evaluations may include: visual, fresh weight, dry weight, and/or relative anthocyanin content via extraction and measurement of absorbance by a spectrophotometer as described previously (Procedure 38.1, step 6).

Anticipated results

Callus initiation rate, overall productivity, and tissue quality are all affected by the size and source of leaf disk explant. Callus growth should be directly proportional to explant size for both greenhouse and micropropagated explants. Small explants produce substantially less biomass within a typical culture period. Although more callus is eventually produced from greenhouse explants, the rate of initiation is slightly faster from microcultured explants. Uniform callus tissue should be apparent after only 3 days on 7- to 10-mm microcultured leaf disks, but callus appears only sporadically around the perimeter of the same-sized greenhouse disks and not later. This result may be because any tissues damaged during surface disinfestation may grow less actively.

Pigment production is usually markedly better from callus generated on medium-sized greenhouse explants (5- or 7-mm disks) as compared to the same diameter disks from microcultured explants. Residual influence of the elevated levels of stored carbohydrates in greenhouse explants may be responsible for this response. Pigment intensity is not significantly different for any of the microcultured explant sizes, although significantly more color-producing biomass is generated from larger explant sizes. Callus initiates rapidly from the larger 10-mm explants from both the microcultured and greenhouse explants; however, pigment expression is apparent earlier from the smaller explant sizes.

Questions

- Why does explant size and source (in vivo stock plant vs. in vitro microplant) affect the productivity in a pigment production system?
- Pigment produced in a callus often has a simpler profile (a less complex anthocyanin structure) than pigment in tissues of the plant in vivo. How do you think pigments from *Ajuga* callus would compare to pigments found in leaves? Flowers?

Experiment 3. Productivity of variant Ajuga *callus lines*

To amplify levels of production for pigments and other secondary metabolites, highly productive callus sectors may be preferentially selected at the time of subculture, while remaining callus is excluded. The selection process and the repeated subcultures needed to stabilize a callus line tend to require long periods of time and do not fit within the time constraints of a typical semester. To illustrate the potential variability of callus lines generated from a single plant, the variegated chimeral *A. reptans* 'Burgundy Glow' may be used to quickly generate two or three callus lines with variable growth rates and pigment production levels. This cultivar of *A. reptans* and its variants perform quite well in the shoot culture and cold storage regimes outlined for *A. pyramidalis* 'Metallica Crispa', with the exception that growth regulators are omitted from the shoot proliferation media to minimize production of adventitious shoots.

Ajuga reptans 'Burgundy Glow' cultures can also provide excellent illustrations of other tissue culture topics. Adventitious shoot production from leaves, roots, and shoots occurs rapidly within 3 weeks when 5 mg/L BAP is included in the media. The separation of the chimera from leaf segments results in as many as three distinct variants (bronze, pink, and occasionally a pale-green, nearly albino plant). The nonphotoautotrophic variants (pink and pale green) have low levels of chlorophyll and do not survive outside of culture (Lineberger and Wanstreet, 1983), so they provide excellent visual examples during discussions of carbon sources and photosynthesis in culture. Additionally, shoots arising from axillary buds typically exhibit the chimeral variegation and adventitious buds produce variants, providing visual markers for an easily evaluated demonstration of the potential somaclonal variability that may occur in chimeral plants in culture. While time constraints would likely prohibit the production callus from these variants for use in subsequent experiments within a single semester, incorporating this adventitious/axillary bud exercise would reinforce precisely how the variants were produced.

In this experiment, leaf disks from individual colored sectors in the chimera are used as explants for either callus production directly, or for generation of adventitious shoots which in turn can be used to generate variant callus lines (Figure 38.3). The visibly different callus lines are compared in terms of growth rate and pigment production on solid media. It should be emphasized that this is an illustrative example of variation that might be achieved through selection.

Materials

The following items are needed for each team:

- Three or four shoot cultures of *A. reptans* 'Burgundy Glow' in GA-7 vessels
- Twelve GA-7 vessels of 50 ml of callus induction media per variant
- Four GA-7 vessels of 50 ml of pigment production media per variant

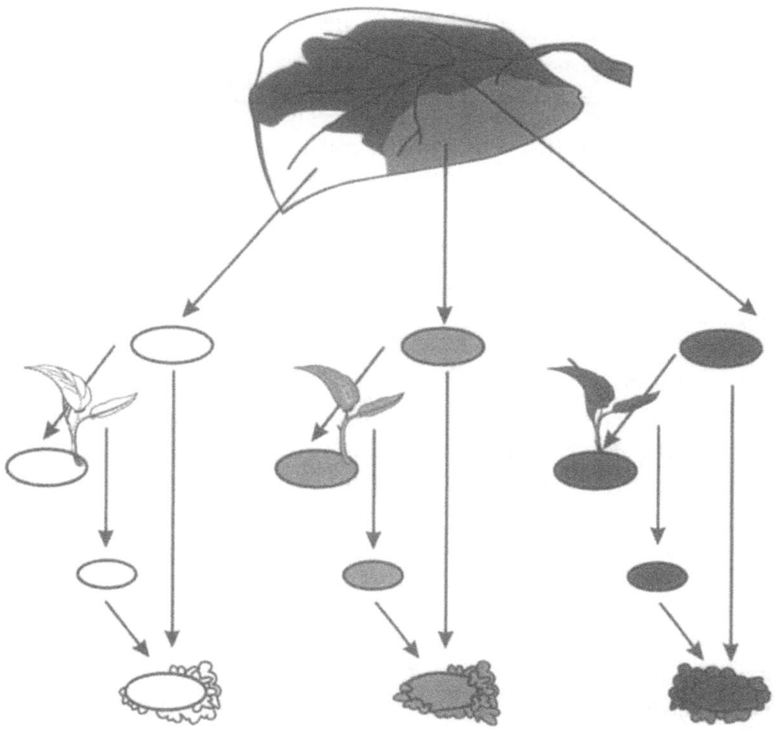

Figure 38.3 Callus production from variant cell lines obtained from the variegated chimeral *Ajuga reptans* 'Burgundy Glow'. Sectors from microplant leaves may either be used directly for callus generation, or to produce adventitious plantlets which in turn can be used as explants for callus generation.

Follow the instruction in Procedure 38.3 to complete this experiment.

Procedure 38.3
Productivity of variant *Ajuga* callus lines.

Step	Instructions and comments
1	Explants which are exclusively pink, pale green, or bronze can be preferentially excised from individual leaves of *A. reptans* 'Burgundy Glow' to initiate callus, or, alternatively, shoots of uniform pink, pale-green, or bronze phenotype may be regenerated from selected disks, and subsequently used as the explants for callus initiation. Initiate four cultures of each variant callus line by collecting five leaf disks 1 to 1.5 cm in diameter from these presterile ajuga shoot cultures. Place each disk on the surface of callus induction medium (see Procedure 38.1, step 1).
2	Incubate cultures in darkness at 25°C for 3 weeks.
3	Separate ivory white friable callus from mother tissue and transfer to fresh induction media. Continue incubation in darkness, subculturing the callus to fresh media at 2-week intervals for 4 weeks.
4	After a total of 7 weeks from initiation, transfer five 2.5-cm-diameter callus masses to pigment production media (callus induction medium substituting 2.26 µM [0.4 mg/L] IAA and 3.49 µM [0.76 mg/L] zeatin as PGRs).
5	Incubate the cultures under cool white fluorescent lamps providing a PPF of 150 $\mu mol \cdot m^{-2} \cdot sec^{-1}$ at 25°C.

Procedure 38.3 continued Productivity of variant *Ajuga* callus lines.	
Step	Instructions and comments
6	Harvest one culture for each variant on a weekly basis for 4 weeks. Evaluate growth and pigment production with an initial visual assessment of pigment content of the intact culture followed by fresh and/or dry weight analysis of two subreplicates in each culture to compare growth. Grind the remaining three subreplicates with a mortar and pestle, store overnight in 10 ml ice cold acidified methanol (1% HCl), filter, then measure the absorbance of the solution at 535 nm with a spectrophotometer to assess differences in the relative anthocyanin content between treatments.

Anticipated results

After 3 weeks of growth, callus of the bronze variant will have accumulated approximately twice the fresh weight mass of the slower growing pink and pale-green variants. By plotting the growth curves over a slightly longer (4-week) period, not only the differences in growth rate, but differences in shapes of the growth curves become evident. After 3 weeks, the growth rate of the bronze variant callus begins to decelerate, whereas production in the callus of the pink and pale-green variants continues to climb.

Each variant also exhibits a different level of pigment production. The bronze variant exhibits the highest anthocyanin concentration, while the pink and pale-green variants have concentrations approximately 80 and 40% that of the bronze variant, respectively.

Since both the growth rate and anthocyanin concentration of these callus lines vary, an even more striking example can be presented by estimating overall yield (mass × anthocyanin concentration). Such a comparison should indicate the potential yield from the bronze variant callus to be 2.5 and 5 times that of the pink and pale-green lines, respectively.

Questions

- When these variant callus lines are placed in solution culture and subcultured at the same frequency, the pink and pale-green lines tend to die off after the first cycle. What would cause this? How could the regime be altered to maintain these variants in solution culture?
- Could the productivity of the variant callus lines be improved?

Experiment 4. Influence of light wavelengths on growth and pigment production of Ajuga *cell cultures*

Secondary products, including anthocyanins and other pigments, are often preferentially produced at specific ranges of light wavelengths. Optimum lighting conditions and requirements are often underestimated factors of in vitro pigment production regimes and present particular problems when scaling an existing protocol up to larger volumes. In this experiment, narrow wavelength lamps are used as light sources. Growth and pigment accumulation are evaluated for *A. pyramidalis* 'Metallica Crispa' callus cultures on solid media.

Materials

The following are needed for each team or student:

- Standard broad-band cool white fluorescent lamps

- Narrow wavelength lamps in red (660 nm), blue (480 nm), and ultraviolet (313 nm) corresponding to Sylvania fluorescent lamps 2364, 2440, and 2096
- Shielding for each light source to prevent contamination from other light sources
- One callus culture of *Ajuga* in a GA-7 vessel per light source to be tested
- One GA-7 vessel containing pigment production media per light source to be tested
- UV-B transparent plastic film (Reynolds 910, Richmond, VA, U.S.) to wrap tops of GA-7 vessels (replacing standard lids)

Follow the protocols outlined in Procedure 38.4 to complete this experiment.

Procedure 38.4
Influence of light wavelengths on growth and pigment production of *Ajuga* cell cultures.

Step	Instructions and comments
1	Initiate one culture per light source to be tested by transferring five 2.5-cm-diameter clumps of callus from existing cultures to fresh pigment production media (WPM amended with 88 mM [30 g/L] sucrose, 0.55 mM [0.1 g/L] myo-inositol, 0.28 mM [50 mg/L] L-ascorbic acid, 1.48 μM [0.5 mg/L] thiamine-HCl, 2.44 μM [0.5 mg/L] pyridoxine-HCl, 4.06 μM [0.5 mg/L] nicotinic acid, 26.7 μM [2 mg/L] glycine, 2.26 μM [0.4 mg/L] IAA and 3.49 μM [0.76 mg/L] zeatin, 150 mg/L PVP-T [polyvinylpyrrolidone-10], and 7 g/L of agar; pH 5.7 prior to autoclaving).
2	Incubate cultures 10 to 12 cm below the selected light sources at 25°C for 2 weeks providing a PPF of 50 to 100 $\mu mol \cdot m^{-2} \cdot sec^{-1}$.
3	Inspect cultures frequently (daily, if possible) for pigment production and browning.
4	Harvest cultures from each light source for data collection. First, visual data regarding the pigment content and fresh weight data are collected to compare growth. Then grind the callus with a mortar and pestle, store overnight in 10 ml ice-cold acidified methanol (1% HCl), filter, then measure the absorbance of the solution at 535 nm with a spectrophotometer to assess differences in the relative anthocyanin content between treatments.

Anticipated results

In as little as 3 days, callus under ultraviolet light will be the first treatment to exhibit pigment production followed by the blue, white, and red light treatments, respectively. Indicative of a slower growth rate, the fresh weight of callus from the ultraviolet light treatments will be approximately 75% that of all other treatments after 14 days. Other treatments will not differ significantly in fresh weight. Pigment production will be significantly different for each light treatment, with cultures in the blue light producing the most pigment and the white, red, and ultraviolet producing approximately 65, 55, and 25%, respectively, of the amount of pigment measured in the blue light treatment. Brown necrotic areas should become visible in the ultraviolet light treatment by the tenth day.

Questions

- Which light source is the best promoter of pigment production? Why would this type of light have a negative impact on growth? Which would be the best choice for a production regime?
- Adequate light is relatively easy to provide in this exercise with solid media or in liquid culture involving small volumes. What problems would you envision

in the scale-up of larger volumes of liquid media such as in a bioreactor for commercial production? How easy would it be to enhance certain wavelengths of light?

Acknowledgments

Many thanks to Drs. Marie-France Pépin and L.A. Spomer for constructing the schematic diagrams illustrated for Experiments 1 and 3.

Literature cited

Callebaut, A., N. Terahara, M. de Haan, and M. Decleire. 1997. Stability of anthocyanin composition in *Ajuga reptans* callus and cell suspension cultures. *Plant Cell Tiss. Organ Cult.* 50:195–201.

Dougall, D.K., D.C. Baker, E. Gakh, and M. Redus. 1997. Biosynthesis and stability of monoacylated anthocyanins. *Food Technol.* 51: 69–71.

Lineberger, R.D. and A. Wanstreet. 1983. Micropropagation of *Ajuga reptans* 'Burgundy Glow'. Research Circular # *27419-22*. Ohio Agricultural Research and Development Center, Wooster, OH.

Lloyd, G. and B. McCown. 1980. Commercially feasible micropropagation of mountain laurel, *Kalmia latifolia*, by use of shoot tip culture. *Intl. Plant Prop. Soc. Proc.* 30:421–427.

Madhavi, D.L., S. Juthangkoon, K. Lewen, M.D. Berber-Jiménez, and M.A.L. Smith. 1996. Characterization of anthocyanins from *Ajuga pyramidalis* 'Metallica Crispa' cell cultures. *J. Agric. Food Chem.* 44:1170–1176.

Madhavi, D.L., M.A.L. Smith, A.C. Linas, and G. Mitiku. 1997. Accumulation of ferulic acid in cell cultures of *Ajuga pyramidalis* 'Metallica Crispa'. *J. Agric. Food Chem.* 45:1506–1508.

chapter thirty-nine

In vitro plant pathology

C. Jacyn Baker, Norton M. Mock, and Elizabeth W. Orlandi

Tissue culture has been an essential tool in the field of plant pathology for many decades. One of the first uses of tissue culture was in 1941 when White and Braun cultured tumors induced by the pathogen *Agrobacterium tumefaciens* on sunflower that were devoid of viable bacteria (White and Braun, 1941). In 1973, Peter Carlson was the first to demonstrate that tissue culture could be used to select for pathogen resistance in plants (Carlson, 1973). Using a structural analog of the fireblight toxin he selected resistant cells, developed callus, and, eventually, regenerated whole plants that were resistant to fireblight. In 1986, Roger Beachy's laboratory successfully incorporated a coat protein of Tobacco Mosaic Virus (TMV) into tobacco cells, endowing the plants with resistance to subsequent viral infections and significantly delaying disease development (Abel et al., 1986). More recently, protoplast fusion techniques have facilitated the transfer of resistance between plant species (Garg et al., 1988). Many of these techniques have been discussed in detail in previous chapters.

In addition to providing novel methods to develop disease-resistant plants, tissue culture has become an important tool in the study of the molecular aspects of plant/pathogen interactions. Plant cells react to environmental and pathogen-related stresses with a variety of responses (i.e., changes in ion fluxes across the plasma membrane, increased synthesis or activation of specific enzyme systems, and increased production of defense-related products). The use of plant cell suspension cultures or individual protoplasts has enabled easier and more accurate monitoring of these cellular processes at the biochemical level. Many of these important biochemical processes cannot be monitored in whole plants. Unlike intact plants, plant suspension cultures provide homogeneous populations of cells that are immediately accessible to treatment with pathogens or pathogen-related products.

This chapter will focus on the use of suspension cells to monitor plant cell responses to pathogenic bacteria. The ability of plants to recognize and respond to external stimuli, such as invading pathogens, is essential for their existence in nature. Successful recognition of a potential pathogen triggers various inherent defense mechanisms and in many cases results in a hypersensitive response (HR). The HR results in rapid plant cell death (9 to 20 h after inoculation) that essentially localizes the pathogen and prevents its spread throughout the plant. Pathogens that trigger the HR in a particular plant species or cultivar are considered to be "incompatible" on that plant. "Compatible" pathogens avoid recognition by the plant and, therefore, do not trigger the HR. This results in more extensive invasion of plant tissues and, ultimately, disease symptoms.

0-8493-2029-1/00/$0.00+$.50

Studies on the bacteria-induced HR have identified early plant responses, which indicate recognition of the pathogen (Atkinson et al., 1985; Baker et al., 1991; 1993). One of these responses is the K^+/H^+ exchange response involving a net uptake of H^+ from the extracellular space and a corresponding efflux of K^+ from the plant cell. This is a reversal of the normal cell ion transport in which H^+ is pumped out by ATPase activity and K^+ is taken up. The K^+/H^- response can be measured in disks cut from whole leaves, but it is more easily and rapidly monitored in plant cell suspensions. The K^+/H^+ exchange response precedes hypersensitive cell death by several hours in incompatible interactions.

The following laboratory exercises will focus on the measurement of the K^+/H^+ response and the subsequent hypersensitive cell death of suspension cells inoculated with bacteria (Figure 39.1). The experiments have been designed to demonstrate the use of cell suspensions to monitor plant recognition responses.

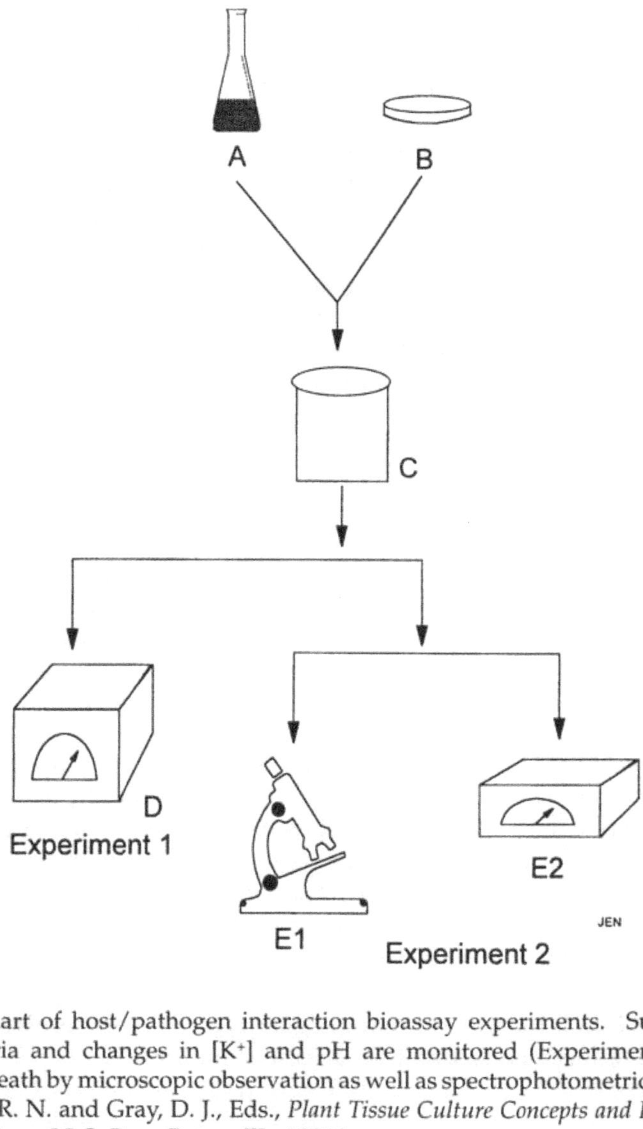

Figure 39.1 Flowchart of host/pathogen interaction bioassay experiments. Suspension cells are treated with bacteria and changes in [K^+] and pH are monitored (Experiment 1) followed by estimations of cell death by microscopic observation as well as spectrophotometric assay (Experiment 2). (From Trigiano, R. N. and Gray, D. J., Eds., *Plant Tissue Culture Concepts and Laboratory Exercises, First Edition*, CRC Press LLC, Boca Raton, FL, 1996.)

General considerations

Maintenance of plants

Tobacco and soybean plants are used for initiating suspension cell cultures and for testing the pathogenicity of bacteria used in subsequent suspension cell assays. Both tobacco and soybean plants should be grown under greenhouse conditions of approximately 28 ± 5°C and 12 h of light. Disperse seed from *Nicotiana tabacum* L. 'Hicks' in moistened vermiculite, cover with a thin layer of dry vermiculite, and moisten, being careful not to disturb the seeds. Seedlings will emerge within 1 week and should be thinned to approximately 1 in. apart. After 3 to 4 weeks the seedlings will be 1 to 2 in. tall and should be transplanted to 10-cm-diameter pots using Promix supplemented with osmacote 14-14-14 (both available through E.C. Geiger Co., Harleysville, PA). Once established (about 2 to 3 weeks later), plants are transplanted to 20-cm-diameter pots in Promix supplemented with osmacote. Tobacco plants should be watered only when the soil is dry since they are prone to root rot. It takes about 9 weeks from seeding to the time plants are ready to be used for most of these experiments.

Soybean (*Glycine max* L. Merr. 'Mandarin' or 'Harosoy 63') seeds are planted 5 cm deep in Promix. Cover these pots with dark plastic to facilitate uniformity in germination. Remove the cover after approximately 2 to 3 days or once seedlings have emerged. Soybean plants should be ready to use approximately 12 days after seeding.

Preparation and maintenance of plant suspension cells

Tobacco suspension cells are initiated from callus derived from the pith of the plant. Remove a 10-cm section from the growing tip of a tobacco plant and cut off all of the leaves. Immerse the section into 95% ethanol for 15 sec. Using a sterile scalpel cut off and discard the tip. Next, peel away and discard the outer vascular tissue to expose approximately 6 cm of the pith. Place the exposed pith into a sterile petri dish, section the pith into 5-mm portions, and place sections on Murashige and Skoog solidified medium (MS) (made with purified agar such as Difco "Agar Noble") (Murashige and Skoog, 1962). Incubate the pith at 27°C in the dark. Callus will begin to grow within 1 week. Once callus is established, after approximately 1 month, move it to Schenk and Hildebrandt solidified medium (SH) (Schenk and Hildebrandt, 1972) and routinely transfer it each month. The callus will proliferate and can be transferred in small chunks (8 to 10 mm diameter) into 40 ml of liquid SH media in a 125-ml culture flask fitted with a sterile sponge closure (Bellco Biotechnology, Vineland, NJ). The flasks should be placed on a rotary shaker in the dark at 27°C and 150 rpm. Within a couple of weeks these cultures will have thickened and can then be transferred routinely. A typical transfer routine consists of aseptic transfer every 4 days of 15 ml of culture into 75 ml of fresh liquid SH media contained in a 250-ml flask. It will take approximately 2 months to establish suspension cells beginning with the initiation of callus from pith tissue.

Soybean suspension cells are started from callus, which is initiated from the hypocotyl. Surface disinfect soybean seeds with 95% ethanol for 1 min and then rinse with sterile distilled water. Soak the seeds in 25% commercial bleach (diluted with sterile distilled water) for 5 min. Rinse seeds well three times with sterile distilled water. Place seeds in a sterile petri dish, which contains filter paper moistened with sterile distilled water. Do not seal the petri dishes as this will reduce gas exchange and negatively affect germination. Place the petri dish into a glass dish with moistened paper towels and cover with plastic wrap to hold in the moisture. Incubate in the dark at 27°C for 4 to 5 days. Once the seeds have germinated, aseptically excise the green hypocotyl, which is 1 to 2 cm below the cotyledon, and cut into 2- to 4-mm sections. Place these sections on Gamborg's B-5 solid media (Gamborg et al., 1968) supplemented with 1 mg/L (4.5 μM) 2,4-dicholorphenoxyacetic acid (1B-5) and place in dark incubator at 27°C. Callus will begin to form within a few days. In about 2 to 3 weeks transfer callus to fresh 1B-5 solidified medium, then transfer routinely every month. Once callus is established, place small chunks (8 to 10 mm diameter) into 40 ml liquid 1B-5 media in a 125-ml flask with a sponge closure and incubate on shaker at 27°C and 150 rpm in normal laboratory light. Once cultures have thickened, suspension cells can be routinely transferred aseptically every 4 days by diluting 1:1 with fresh media for a total of 80 ml per 250-ml flask. Soybean suspension cells require approximately 2 months for establishment from excised pith sections.

Maintenance of bacteria

The bacteria that are recommended for use in these experiments are listed in Table 39.1. All bacteria isolates are aseptically loop-inoculated onto Kings B agar (KB) medium in petri plates, sealed with parafilm, and allowed to grow 1 to 2 days in an incubator set at 30°C. Plates are then kept at 4°C and fresh cultures are prepared at monthly intervals. Bacteria listed in Table 39.1 are all grown on Kings B agar (KB) consisting of 20 g proteose peptone #3 (Difco), 1.5 g K_2HPO_4, 6.14 g $MgSO_4 \cdot 7H_2O$, 15 ml glycerol, and 20 g Bacto agar (Difco) brought to 1 L with distilled water.

Table 39.1 Recommended Pathavars of *Pseudomonas syringae* for HR Tobacco and Soybean Experiments

	Tobacco	Soybean
Compatible bacteria (disease causing)	*Pseudomonas syringae* pv. *tabaci*	*P. syringae* pv. *glycinea* race 4
Incompatible bacteria (HR causing)	*P. syringae* pv. *tomato* *P. syringae* pv. *syringae* 61 *P. syringae* pv. *glycinea* race 4	*P. syringae* pv. *tomato* *P. syringae* pv. *syringae* 61 *P. syringae* pv. *tabaci*

From Trigiano, R. N. and Gray, D. J., Eds., *Plant Tissue Culture Concepts and Laboratory Exercises, First Edition*, CRC Press LLC, Boca Raton, FL, 1996.

Initial testing of bacteria isolates by inoculation into whole plants

It is essential to test the bacterial isolates on plants prior to bioassays to ensure that their pathogenicity is properly characterized. This is best done by infiltrating healthy plants with a 2×10^8 colony-forming units (cfu)/ml bacterial solution (bacteria are prepared as described below in "Preparation of bacteria for bioassay"). Prick the leaf with a tuberculin syringe needle. Place the barrel of a needleless syringe containing bacteria against the hole in the leaf and with one finger on the other side of the leaf, gently infiltrate. Lightly outline and label the infiltrated water-soaked area on the underside of the leaf using a permanent marker. The leaf areas that were inoculated with incompatible bacteria should turn necrotic overnight (Figure 39.2a and c), whereas the areas inoculated with compatible bacteria should be symptomless or display yellowing over the next few days (Figure 39.2b and c). For tobacco, best results are achieved when the third fully expanded leaf from the top is inoculated; for soybean, inoculate the fully expanded primary leaves. It should be noted here that soybean leaves are more difficult to infiltrate than tobacco leaves and the infiltrated leaf area of soybeans will be considerably smaller than that of tobacco leaves.

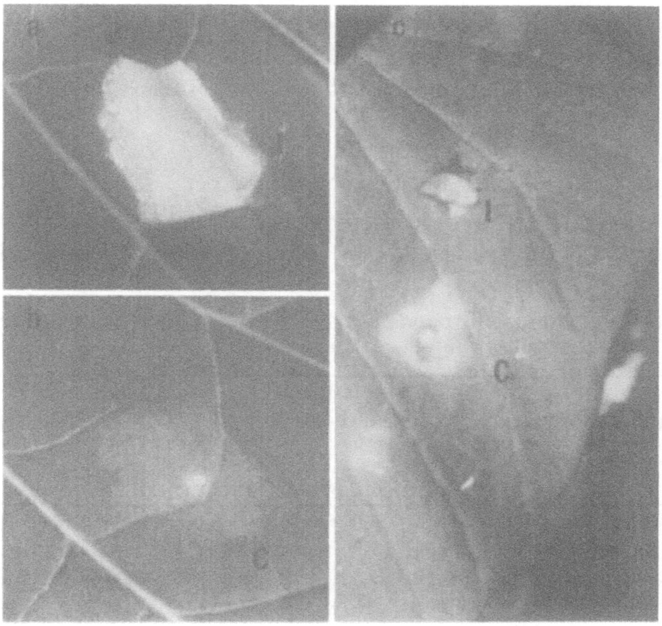

Figure 39.2 Typical symptoms produced on tobacco (a and b) and soybean (c) by incompatible (I) and compatible (C) bacterial pathogens infiltrated into leaf tissue. (From Trigiano, R. N. and Gray, D. J., Eds., *Plant Tissue Culture Concepts and Laboratory Exercises, First Edition*, CRC Press LLC, Boca Raton, FL, 1996.)

Exercises

Experiment 1. Correlation of the K^+/H^+ response with the incompatible plant/bacteria interaction

As discussed previously, the K^+/H^+ exchange is a recognition response indicating that the plant has detected the presence of an incompatible pathogen. In most cases this will result in the triggering of plant defense mechanisms resulting in a resistant reaction. In this experiment the increase in external pH and $[K^+]$ will be monitored using pH and conductivity electrodes, respectively. The response should begin approximately 1.5 to 3 h after inoculation with incompatible bacteria. Previous studies have demonstrated that the increase in conductivity is due to K^+ efflux and not due to a general leakage of ions through the membrane (Atkinson et al., 1985). Membrane breakdown occurs at a later stage of the HR.

Preparation of suspension cells for bioassay

Materials. The following materials and equipment will be needed for each student or team of students:

- 500 ml assay buffer (0.5 mM MES; 0.5 mM $CaCl_2$; 0.5 mM K_2SO_4; 175 mM mannitol), pH 6.0
- Six 50-ml disposable polystyrene beakers
- One 25-ml plastic cylinder (or 25 ml large tip opening serological glass pipette)

- One 100-ml cylinder
- One glass funnel (75 mm diameter)
- One 7-in. square piece of grade 10 cheesecloth
- One 5-in. square piece of miracloth (obtained from Calbiochem, Corp.; LaJolla, CA 92037)
- One rubber band
- One stainless steel microspatula
- Waterbath shaker fitted to hold the disposable beakers and calibrated to 27°C and 160 rpm, or flat bed rotary shaker in temperature-controlled environment

Follow the protocols outlined in Procedures 39.1 through 39.3 to complete this experiment.

Procedure 39.1
Preparation of tobacco and soybean suspension cells for bioassay.

Step	Instructions and comments
1	Pour log phase tobacco or soybean suspension cells (2 to 3 days after transfer) onto a single layer of cheesecloth, secured with a rubber band to mouth of a glass beaker. Gently pour 100 to 200 ml of assay buffer over the cells that remain on top of the cheesecloth, collecting fine aggregates that filter into the beaker. Pour collected cells onto miracloth-lined funnel placed on top of the 1000-ml flask. Wash cells with 100 to 200 ml assay buffer. Drain well, but do not allow the cells to dry.
2	Weigh out 5 g of cells into a 250-ml flask and slowly add 100 ml assay medium for a final concentration of 0.05 g/ml. If medium is added too rapidly, cells may form clumps and will need to be gently separated with a spatula.
3	Resuspend cells by gently swirling flask and dispense 15 ml of the cell suspension into each of the 50-ml beakers, using a plastic 25-ml cylinder (the cells do not stick to the plastic).
4	Place the beakers in the water bath shaker adjusted to 27°C and 160 rpm.

Preparation of bacteria for bioassay

Materials. The following materials and equipment will be needed for each student or team of students:

- Kings B (KB) medium
- Sterile glass 5-ml pipettes (one for each bacterium)
- 1-ml Pipetman and tips
- 200-µl Pipetman and tips
- 50-ml centrifuge tubes (one for each bacterium)
- Disposable 3-ml cuvettes (one for each bacterium, plus one for blank)
- Vortex-mixer
- Low-speed centrifuge
- Spectrophotometer

Choose bacterial isolates that will give compatible and incompatible reactions on the suspension cells being used (Table 39.1) and follow the methods listed in Procedure 39.2 to complete this portion of the experiment.

Procedure 39.2
Preparation of bacteria for bioassay.

Step	Instructions and comments
1	Choose bacterial isolates that will give compatible and incompatible reactions on the suspension cells being used (Table 39.1).Transfer bacteria from the stock plates to fresh media by spreading a loop of the bacteria over the surface of the agar media. Incubate the dishes at 30°C for 18 to 24 h.
2	Gently scrape the bacterial lawns off the agar using 5 ml distilled water and a sterile glass 5 ml pipette. Transfer bacteria to a 50-ml centrifuge tube. Cover the tube with parafilm and vortex gently to suspend bacteria. Bring volume in tube to 2/3 full using distilled water. Centrifuge for 3 min at 8000 rpm at room temperature. Decant and discard supernatant being careful not to disturb the bacterial pellet. Suspend bacterial pellet in 8 ml distilled water.
3	Prepare a 1:30 bacterial dilution in a 3-ml cuvette to determine the optical density units (ODU) at 500 nm.
4	The final bacterial concentration for inoculating plant leaves should be approximately 2×10^8 colony-forming units/ml (cfu). This normally corresponds to about 0.2 $ODU_{500\ nm}$ for most bacteria. This can easily be obtained by the following steps: (1) determine the ODU of the concentrated bacterial solution, which is 30 × the ODU of the 1:30 dilution obtained above; (2) divide this number by 0.2 ODU (the desired final concentration of bacteria) to obtain the final volume to which 1 ml of the concentrated bacteria should be diluted with distilled water. For example, if the ODU of the 1:30 dilution is 0.600, bring 1 ml of the concentrated bacterial solution to 90 ml, which is [(0.600 × 30)/0.2].
5	For the suspension cell assays described in the following procedure, the final bacterial concentration should also be 0.2 ODU/ml in each beaker. For a 15-ml beaker, bring 1 ml of the concentrated bacteria to 1/15 of the total volume determined above for the plant inoculations. Using the example above, if the ODU of the 1:30 dilution is 0.600 and the final volume of the solution prepared for inoculation into the plant is 90 ml, 1 ml of the concentrated bacteria should be diluted with distilled water to 6 ml (90/15). One milliliter of this diluted bacterial solution will be added to beakers as described in Procedure 39.3.

Bioassay for the K^+/H^+ response

Materials. The following materials and equipment will be needed for each student or team of students:

- 200-µl Pipetman and tips
- pH meter (must be digital and capable of reporting pH to two decimal places)
- Conductivity meter

Procedure 39.3
Bioassay for the K^+/H^+ response.

Step	Instructions and comments
1	Add 1 ml of the appropriate bacterial suspension prepared above to pre-incubated tobacco or soybean suspension cells. For controls, add 1 ml of assay buffer.
2	Using calibrated meters measure the pH and conductivity of the suspension cells every half hour up to at least 5 h. Store electrodes in assay buffer between readings. To assure consistency of the pH readings, beakers must be gently swirled by hand during readings. Alternatively, the pH and conductivity electrodes can be directly immersed into the beakers while they are rotating on the shaker.

Anticipated results

Figures 39.3A and B depict typical changes in pH of soybean and tobacco suspension cells treated with compatible and incompatible bacteria. The pH of the tobacco controls usually decreases (perhaps as low as 5.1) over the 5-h period, while the soybean controls maintain a more consistent pH level during the experiment. Both tobacco and soybean suspension cells treated with compatible bacteria will undergo pH changes similar to their respective controls; suspension cells treated with incompatible bacteria will show an increase in pH which will begin within 3 h. Changes in conductivity should be similar to changes in pH. Changes in pH can be more accurately accessed if the results of the control are subtracted from the bacteria treatment (Figures 39.3C and D).

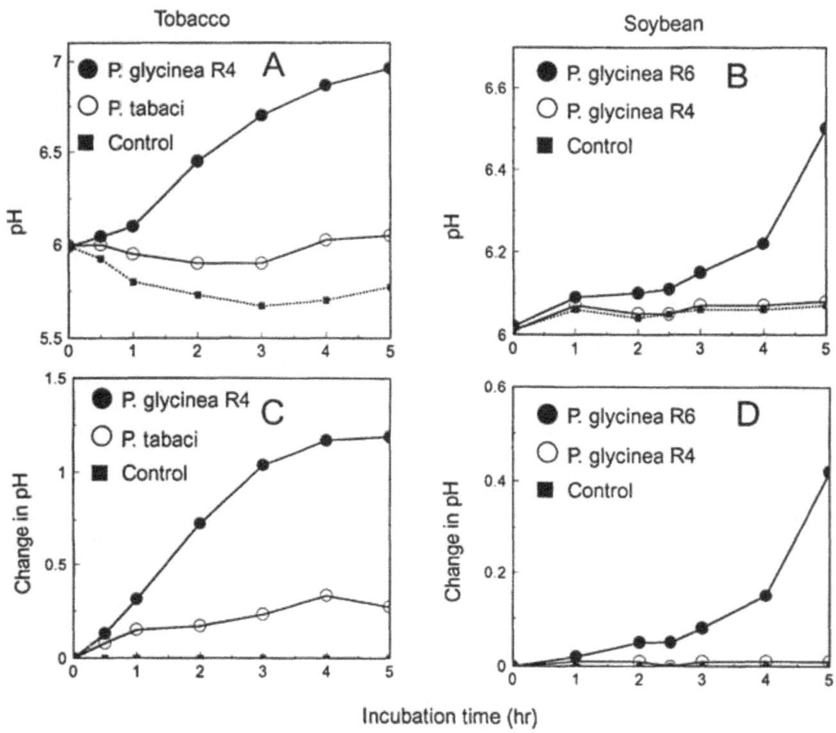

Figure 39.3 Anticipated change in pH for tobacco and soybean suspension cells inoculated with compatible and incompatible pathogens. A and B show typical actual pH readings from a typical experiment; C and D show the change in pH for the various treatments with changes in the control subtracted from the data for the treatments. (From Trigiano, R. N. and Gray, D. J., Eds., *Plant Tissue Culture Concepts and Laboratory Exercises, First Edition*, CRC Press LLC, Boca Raton, FL, 1996.)

Questions

- Define compatible vs. incompatible plant/pathogen interactions.
- *Pseudomonas solanacearum* is a pathogen of potato and causes a HR on tobacco. What kind of changes in pH and K^+ would you expect if (1) tobacco suspension cells or (2) potato suspension cells were treated with *P. solanacearum*?

Experiment 2. Measuring suspension cell death in response to compatible and incompatible bacteria

The following exercise will demonstrate hypersensitive cell death due to the inoculation of suspension cells with incompatible bacteria. Many viability assays that are currently being used in the plant sciences are either relatively laborious or time-consuming or measure the cell's temporary metabolic state rather than irreversible cell death (Keppler et al., 1988). Evans blue has been a reliable stain for the microscopic determination of cell death (Figure 39.4) (Turner and Novacky, 1974) and a spectrophotometric procedure using this stain has been developed that allows rapid, reproducible quantification of the stain retained by dead cells (Baker and Mock, 1994). Evans blue stain penetrates through ruptured membranes of dead cells but not live cells; therefore, the amount of stain retained by the cell fraction reflects the amount of cell death. This will be verified by microscopic observation.

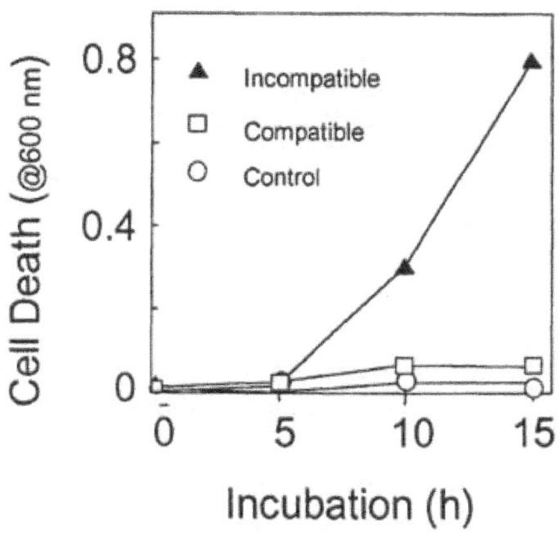

Figure 39.4 Anticipated results for cell death as monitored by extraction of retained Evans blue and spectrophotometric measurement at 600 nm. Compatible and buffer control will retain a background level of dye, which will not increase substantially compared to the incompatible treatments. (From Trigiano, R. N. and Gray, D. J., Eds., *Plant Tissue Culture Concepts and Laboratory Exercises, First Edition*, CRC Press LLC, Boca Raton, FL, 1996.)

Materials

The following materials and equipment will be needed for each student or team of students:

- 25 ml of 0.25% Evans blue in distilled water (Evans Blue, #E2129, Sigma Chemical Co., St. Louis, MO)
- 10 ml of 1% SDS in distilled water
- Carborundum (approximately 180 mesh)
- 1000-ml beaker to collect wash
- Twelve 35 × 10-mm petri dishes
- Two 1-ml cuvettes
- Twenty-four 1.5-ml capped eppendorf tubes
- One plastic eppendorf tissue grinder/pestle
- Six cell strainers, 70-µm nylon mesh (Falcon #2350)
- One stainless steel microspatula

- One 1000-µL pipetman and tips (tips should be cut off diagonally to allow passage of larger clumps of cells — use one per beaker)
- Pipette tip filters (optional)
- Fuchs-Rosenthal or other suitable counting chamber (Hausser Scientific Partnership, Horsham, PA)
- Tabletop centrifuge
- Spectrophotometer
- Compound microscope

Caution: Sigma Scientific classifies Evans blue as a possible carcinogen. Take appropriate precautions to limit exposure to this chemical — wear gloves, goggles, protective clothing, and a particle mask.

Follow the protocol listed in Procedure 39.4 to complete this exercise.

Procedure 39.4
Measuring cell viability (cell death) with Evans blue.

Step	Instructions and comments
1	Prepare suspension cells and bacteria as described earlier.
2	Inoculate the soybean or tobacco suspension cells with bacteria as in Experiment 1 (Procedure 39.3, step 1).
3	At 0 and 18 to 24 h after inoculation cell death estimates should be made. Remove 2 ml from each treatment and put into a labeled cell strainer. Wash the suspension cells lightly with distilled water to remove the assay medium. Place individual cell strainers into petri dishes with 4 ml each of 0.25% Evans blue. Add additional Evans blue into strainers to cover cells. Incubate for 5 min. Remove one cell strainer at a time and gently rinse dye from the cells with distilled water, collecting waste in waste beaker. Place the strainers into petri dishes containing 4 ml distilled water for 5 min. Again, remove one cell strainer at a time and wash any remaining dye from the cells, collecting waste.
4	Using a spatula, gently transfer the cells into labeled eppendorf tubes containing a small amount of carborundum (<0.1 mg) and 0.5 ml 1% SDS. Grind cells with the plastic pestle and then add 0.5 ml distilled water. Centrifuge for 3 min.
5	Remove 0.7 ml of the supernatant taking care not to disturb the pellet or remove any particulates that may be floating on the surface. (Use pipette tip filter if necessary.) Place into 1-ml cuvette and measure at 600 nm on spectrophotometer zeroed with water blank.
6	To observe cell death microscopically, repeat the above procedure to treat cells with Evans blue.
7	After cells have been incubated with Evans blue and rinsed with distilled water, remove the cells from the strainers and put them into small test tubes with 0.5 ml distilled water. Place a drop of cells onto microscope slide and apply cover slip.
8	Observe cells using 100 to 200 × magnification and determine the percentage of cells, which retain Evans blue.

Anticipated results

At 0 hours the amount of Evans blue extracted from the cells in all treatments should be relatively low. At 18 to 24 h the readings from suspension cells treated with either buffer or compatible bacteria should still be relatively low, indicating low levels of Evans blue retention. However, the suspension cells treated with incompatible bacteria should have higher levels of extractable dye from the cells. Beyond this time, the amount of Evans blue retained in dead cells may decrease with time due to the complete disintegration of the cellular structures and membranes.

Questions

- Why does the spectrophotometer reading increase in cells treated with incompatible bacteria?
- Why is there more cell death in an incompatible (resistant) reaction?

Literature cited

Abel, P. P., Nelson, R. S., De, B., Hoffman, N., Rogers, S. G., Fraley, R. T., and R. N. Beachy. 1986. Delay of disease development in transgenic plants that express the tobacco mosaic virus coat protein gene. *Science* 232:738–743.

Atkinson, M. M., Huang, J.S., and J.A. Knopp. 1985. The hypersensitive reaction of tobacco to *Pseudomonas syringae* pv *pisi*: activation of plasmalemma K^+/H^+ exchange mechanism. *Plant Physiol.* 79:843–847.

Baker, C.J., Mock, N.M., Glazener, J. A., and E. W. Orlandi. 1993. Recognition responses in pathogen/nonhost and race/cultivar interactions involving soybean (*Glycine max*) and *Pseudomonas syringae* pathovars. *Physiol. Mol. Plant Pathol.* 43:81–94.

Baker, C. J., O'Neill, N. R., Keppler, L. D., and E. W. Orlandi. 1991. Early responses during plant-bacteria interactions in tobacco cell suspensions. *Phytopathology* 81:1504–1507.

Baker, C. J. and N. M. Mock. 1994. An improved method for monitoring cell death in cell suspension and leaf disc assays using Evans blue. *Plant Cell Tissue Organ Cult.* 39:7–12.

Carlson, P. S. 1973. Methionine sulfoximine-resistant mutants of tobacco. *Science* 180:1366–1368.

Gamborg, O. L., Miller, R. A., and K. Ojima. 1968. Nutrient requirements of suspension cultures of soybean root cells. *J. Exp. Cell Res.* 50:151–158.

Garg, G. K., Singh, U. S., Khetrapal, R. K., and J. Kumar. 1988. Application of tissue culture in plant pathology. In: *Experimental and Conceptual Plant Pathology.* Vol. 1. Techniques. Singh, R. S., Singh, U. S., Hess, W. M., and D. J. Weber, Eds., pp. 83–119.

Keppler, L. D., Atkinson, M. M., and C. J. Baker. 1988. Plasma membrane alteration during bacteria-induced hypersensitive response in tobacco suspension cells as monitored by intracellular accumulation of fluorescein. *Physiol. Mol. Plant Pathol.* 32:209–219.

Murashige, T. and F. Skoog. 1962. A revised medium for rapid growth and bioassays with tobacco tissue cultures. *Physiol. Plant.* 15:473–497.

Schenk, R. U. and A. C. Hildebrandt. 1972. Medium and techniques for induction and growth of monocotyledonous and dicotyledonous plant cell cultures. *Can. J. Bot.* 50:199–204.

Turner, J. G. and A. Novacky. 1974. The quantitative relationship between plant and bacterial cells involved in the hypersensitive reaction. *Phytopathology* 64:885–890.

White, P. R. and A. C. Braun. 1941. Crown gall production by bacteria-free tumor tissues. *Science* 94:239-241.

Part VI

Special topics

chapter forty

Variation in tissue culture*

S. Jayasankar

As discussed in previous chapters concerning genetic engineering, plant cell culture is an important technique for plant genetic improvement. Historically, and as implied throughout this book, plant cell culture has been viewed by most to be a method for rapid cloning. In essence, it was seen as a method of sophisticated asexual propagation, rather than a technique to add new variability to the existing population. For example, it was believed that all plants arising from such tissue culture were exact clones of the parent, such that terms like 'calliclone,' 'mericlone,' and 'protoclone' were used to describe the regenerants from callus, meristems, and protoplasts, respectively. Although phenotypic variants were observed among these regenerants, often they were considered as artifacts of tissue culture. Such variation was thought to be due to 'epigenetic' factors such as exposure to plant growth regulators (PGRs) and prolonged culture time. As more and more species were subjected to tissue culture, reports of variation among regenerants also increased. In a historically significant review, Larkin and Scowcroft (1981) proposed the more general term 'somaclones' to the regenerants coming out of tissue culture, irrespective of the explant used. Variation displayed by such regenerants from tissue culture would then be 'somaclonal variation.' Tissue culture studies in the 1970s and early 1980s started to focus their attention toward this variation and it was soon recognized that somaclonal variation exists for almost all the phenotypic characters.

To a plant scientist, somaclonal variation is perhaps the best route to study somatic cell genetics. In contrast to the earlier opinion of 'true-to-type regeneration' among tissue culture-derived plants, the frequency of genetic variation may actually be quite high. In some species — e.g., oil palm and banana — variation among tissue culture-derived progenies is higher than one would expect to occur in vivo. In perennial crops that are asexually propagated, somaclonal variation offers excellent opportunity to add new genotypes to the gene pool. In such cases it is important to understand and identify the causal mechanism behind the variations, so that we can effectively control them to our advantage.

It is important to differentiate between 'variation' and 'mutation' at this point. In the literature, 'variant' is rather loosely used to depict any type of phenotypic change that is present either in cell culture or the eventual plants regenerated. Often, these descriptions are not accompanied by a clear evidence of the cause and most of them are non-Mendelian in their inheritance. 'Mutation,' on the other hand, should be exclusively used only for cases where clear evidence is presented for genetic alteration. Mutants need not necessarily present

* Florida Agricultureal Experimental Station Journal Series No. R-06862.

0-8493-2029-1/00/$0.00+$.50

an altered phenotype, as there are point mutations that are not expressed phenotypically. Jacobs et al. (1987) lists several criteria for a mutant, among which sexual transmission of the trait to the offspring and molecular evidence for the alteration are noteworthy.

Major causes of somaclonal variation

Over the years considerable research on trying to identify the causes of variation in culture has been conducted. Seemingly, every possible factor that could result in a genetic change has been accounted for as a cause for somaclonal variation. For convenience, we can pool these factors that contribute to somaclonal variation into three major categories, i.e., physiological, genetic, and biochemical.

Physiological causes of variation

Variation induced by physiological factors can be identified quite early on and often without the aid of any tools. Classic examples of such variation are those induced by habituation to PGRs in culture and culture conditions. Often such variations are 'epigenetic' and may not be inherited in a Mendelian fashion. Prolonged exposure of explanted tissue to powerful auxins such as phenoxyacetic acids (e.g., 2,4-D or 2,4,5-T) often results in variation among the regenerants. For example, oil palm (*Elaeis guineensis* Jacq.) plants generated from long-term callus cultures in the presence of 2,4-D showed significant amounts of variability in the field. In grapevine (*Vitis vinifera* L.), embryogenic cells that have been maintained in culture for several years gradually lose their ability to differentiate and regenerate into plants over time.

Genetic causes of variation

Genetic variation occurs among tissue culture regenerants as a result of alterations at the chromosomal level. Although the explanted tissue may be phenotypically similar, plants often have tissue comprised of diverse cell types of cells. In other words, there are cytological variations among the cell types within an explanted tissue. Such a pre-existing condition often results in plants regenerated from the tissue that are dissimilar. These species are referred as 'polysomatic' species. Species such as barley (*Hordeum vulgare* L.) and tobacco (*Nicotiana tabacum* L.) have been documented to possess such polysomatic tissues.

Chromosomal rearrangements such as deletion, duplication, and somatic recombination are the chief sources of genetic variations exhibited by somaclones. Extensive studies have been conducted using cultured cells of both plant and animal species to demonstrate chromosomal rearrangements. Lee and Phillips (1988) have described in detail the possible mechanisms that lead to these chromosomal changes. They point out that late replicating heterochromatin is the primary cause of somaclonal variation in maize (*Zea mays* L.) and broad beans (*Vicia faba* L.). Transposable elements are activated during culture in explanted tissue and this results in altered genotypes among the regenerated plants. The well-known transposable element complex of *Ac-Ds* in maize has been shown to be activated following in vitro culture.

Biochemical causes of variation

Biochemical deviations are the most predominant type of variation in tissue culture and many of them are barely noticeable, unless a specific test is performed. To date, several biochemical variations have been identified in various crop plants in tissue culture. While some of these variants show Mendelian inheritance, many may be 'epigenetic' and may be lost in the plants regenerated. Examples of such biochemical variation include alterations in carbon metabolism leading to lack of photosynthetic ability ('albinos' in cereals such as rice), starch biosynthesis, carotenoid pathway, nitrogen metabolism, and antibiotic resistance. In contemporary plant science, any variation in antibiotic resistance/susceptibility has significant implications, since antibiotic resistance is a vital marker in transformation studies.

Genomic DNA exhibits normal methylation patterns. Methylation is a process where a particular nucleotide — usually adenine (A) or cytosine (C) — has a methyl group (CH_3) attached to it. Prolonged exposure of plant tissues to in vitro culture has resulted in the alteration of normal methylation patterns. When such methylation occurs in a region of DNA that encodes an active gene, it prevents the gene from further processing and the gene is silenced. Results of such gene silencing due to methylation may not be noticed phenotypically. Although methylation due to tissue culture has been shown in several species such as maize, potato (*Solanum tuberosum* L.), and grapevine, at present we do not know why this process happens.

Induced (or directed) causes of variation

In conventional plant improvement, a plant breeder will have to screen an extensive number of plants in the greenhouse or field if selecting for a particular trait of interest. For instance, if the goal is to develop a salt-tolerant line of a particular species, a large number of individual plants must be grown and subjected to various doses of salt to eventually identify the plants that can withstand the screening process. In doing so, the amount of material that can be tested will be limited by availability of space and time. In addition, environmental factors will also interfere with the selection process. Cell culture systems provide the breeder with the ability to select from a very large amount of genetically uniform material and to conduct the screening quickly in a few petri dishes or flasks. This provides much greater control over the selection process. As cell/tissue culture systems were developed for various species, the potential of in vitro selection was quickly recognized.

In vitro mutation

Certain crop plants such as bananas and plantains (*Musa* spp.) do not have a large genetic base and have been obligately propagated by asexual means for thousands of years. Genetic improvement in these species is cumbersome as they seldom produce fertile seeds. In such cases, the only available options are to look either for natural somatic mutations, which do occur at an extremely low frequency, or induce mutations. In many vegetatively propagated crops, even inducing mutations is difficult as their propagules are quite large as in the case of banana suckers. Attempts to irradiate such large vegetative propagules result either in mosaics or fatalities. Development of a cell regeneration system, such as somatic embryogenesis, provides an opportunity to expose a large number of regenerative cells to either gamma irradiation or chemical mutagen such as ethylmethyl sulfonate etc., in a very controlled manner, and thus widen the existing germplasm base (Novak, 1992).

Use of selection agents

Since the first report of successful regeneration of methotrexate-resistant (a toxin produced by *Pseudomonas tabaci*, the causal bacterium of tobacco wild fire disease) tobacco plants from protoplast culture by Carlson (1973), the use of phytotoxins for developing disease resistance in plants from tissue culture has been pursued. Fungal and bacterial toxins have been used to select disease-resistant lines of rice (*Oryza sativa* L.), maize, tobacco, and alfalfa (*Medicago sativa* L.). Identification and isolation of phytotoxins from plant pathogenic bacteria and fungi during the 1970s greatly aided resistance breeding through tissue culture. Selection was carried out at the protoplast, cellular, and tissue level in various plant species with varying degree of success. About the same time, resistance for salts (sodium chloride) and heavy metals (aluminum) was also attempted through in vitro selection in herbaceous monocots and dicots.

The promise of 'genetic engineering' in the 1980s slowed in vitro selection research, as emphasis was shifted to the identification and cloning of useful genes that could confer resistance to biotic and abiotic stresses. As genetic engineering technology began to mature in the 1990s (see Chapter two), research into in vitro selection effectively ceased. Our understanding of plant resistance to stress factors has advanced tremendously due to development of molecular tools. These molecular tools can identify the changes that occur at DNA, RNA, or protein levels when plant cells are exposed to challenging environments. The use of molecular tools such as randomly amplified polymorphic DNA (RAPD), restriction fragment length polymorphism (RFLP), DNA amplification fingerprinting (DAF), amplified fragment length polymorphism (AFLP), etc. allows rapid and accurate study of the occurrence and nature of variation.

Applications of tissue culture-derived variation

Cell culture systems offer plant breeders a well-defined environment where selection pressures can be imposed on thousands of genetically uniform single cells, each capable of growing into a whole plant. The effect of environmental variation is minimized, so that escapes or adaptations that can revert back to original genetic background are also reduced. This controlled growth atmosphere in a minimal space provides the plant breeder new options for introducing variation. In addition, a scientist can study a tropical species in a temperate region or vice versa because specialized environmental conditions can be provided anywhere. Although it is not studied much today, culture-derived variation holds potential for crop improvement, especially in perennial species, which are hampered by a narrow germplasm base and long regeneration cycles. Some of the important applications for induced variation are discussed below with one or two classical examples from the literature.

Development of disease-resistant plants from tissue culture

Hammerschlag (1992) points out the following as the most important factors among the several criteria to be met before effective in vitro selection for disease resistance can proceed.

1. An effective selection agent, that can be produced and utilized in an in vitro system, must be identified. The identified selection agent should act at the cellular level and should be an important factor in the disease process.
2. There must be a reliable protocol for regenerating whole plants from single cells for the species in question. The protocol must allow the cells to withstand several cycles of selection in a stringent environment and still be able to regenerate whole plants.

In addition to these important factors, effective tools to determine if selected cells are truly resistant to the pathogen at the cellular level and whole plant level are necessary. For example, a bioassay that can be performed on selected cells or PCR analysis to identify genetic changes is a possible diagnostic tool. Greenhouse and/or field exposure to the pathogen are required tests.

During late 1970s and early 1980s, phytotoxins were employed as selection agents to impart disease resistance. A partial list of important work done in this area is summarized in Table 40.1. However, information regarding the heritability or stability (in perennial species) of such in vitro-derived resistance is lacking. Often, enhanced resistance is accompanied with other undesirable characters such low yield etc. For instance, continued selection with a phytotoxin resulted in potato regenerants that were resistant to, but the tubers were small and inferior in edible qualities. One possible reason is that these phytotoxins are produced by the pathogen in a very timely and specific manner and in very low quantities during the disease process. When plant cells are subjected to higher doses of these toxins, they not only affect the ability of the cells to resist the phytotoxin, but also cause some unwarranted genetic damage to the cells. Another problem is that sometimes regenerants resistant to the phytotoxin were not resistant to the pathogen. These results exposed problems of using phytotoxins to select for disease resistance.

Table 40.1 In Vitro Selection for Disease Resistance Against Phytotoxins

Crop	Pathogen	Phytotoxin	Results
Alfalfa *Medicago sativa*	*Colletotrichum gloeosporioides*	Culture filtrate	Enhanced disease resistance
Asparagus *Asparagus offcinalis* L.	*Fusarium oxysporium* f. sp. *asparagi*	Pathogen inoculation	Increased disease resistance
Banana *Musa* spp.	*Fusarium oxysporum* f.sp. *cubense race* 4	Fusaric acid	Resistance clones to the disease
Citrus/lemon *Citrus limon* L.	*Phoma tracheiphila*	'Mal secco' toxin	Resistant embryogenic cultures
Coffee *Coffea arabica* L.	*Colletotrichum kahawae*	Partially purified culture filtrate	Plants with increased resistance
Maize *Zea mays*	*Helminthosporium maydis*	T-toxin	Phytotoxin-resistant cells and plants
Mango *Mangifera indica*	*Colletotrichum gloeosporioides*	Colletotrichin and culture filtrate	Resistant embryogenic cultures
Oat *Avena sativa* L.	*Helminthosporium victoriae*	Victorin	Inheritable disease resistance
Peach *Prunus persica* L.	*Xanthomonas campestris* pv. *pruni*	Culture filtrate	Resistant clones to the disease
Potato *Solanum tuberosum*	*Phytophthora infestans*	Culture filtrate	Phytotoxin-resistant plants
Rice *Oryza sativa*	*Xanthomonas oryzae*	Culture filtrate	Filtrate-resistant plants
Strawberry *Fragaria* sp.	*Fusarium oxysporum* f.sp. *fragariae*	Fusaric acid	Resistant shoots
Sugarcane *Saccharum officinarum* L.	*Helminthosporium sacchari*	Culture filtrate	Disease-resistant clones
Tobacco *Nicotiana tabacum*	*Pseudomonas tabaci*	Methionine sulfoximine	Phytotoxin-resistant plants
Tomato *Lycopersicon esculentum* Mill.	*Fusarium oxysporum*	Fusaric acid	Plants with elevated resistance

Another strategy involved the use of crude fungal or bacterial culture filtrate as a selection agent to impart disease resistance. In most cases, the level of resistance expressed in the regenerants was higher than the original parent and these changes were also genetic, as the disease resistance was transmitted to the sexual offsprings in a Mendelian fashion. However, some of the early studies such as the use of *Phytophthora citrophthora* culture filtrate to select citrus nucellar calli of orange gave results that did not concur with what is known about the host's natural ability in the field. These studies showed that field resistance and in vitro resistance were in reverse order (Vardi et al., 1986)! As there were more failures than successes in using crude culture filtrate for developing disease resistance, this approach was not preferred. Several reasons can be attributed to the ineffectiveness of crude culture filtrate as selection agent. The most important reasons are (1) involvement of secondary metabolites such as PGRs produced by the pathogen, which enhance growth of plant cells during selection, (2) differential response of the host plant to the culture filtrate at the cellular (in vitro) and whole plant level (in vivo), and (3) ineffective selection methods that allow more 'escapes' (Daub, 1986). However, another problem for the failures of in vitro selection may have been due to use of an inadequate culture environment. For example, most of the in vitro selection studies were carried out on a semisolid medium, supplemented with a known dose of the crude culture filtrate or phytotoxin. This likely allowed the formation of a gradient in actual selection pressure within the explanted tissue such that those cells farthest from the medium surface were able to escape selection and/or become adapted to the culture environment. Often, these adapted cells were regenerated into plants in such medium containing the culture filtrate and then reported as resistant genotypes. Resistance in such regenerants usually break down at the whole plant level, as it is not genetic resistance. A solution to this problem has been to use cell suspension cultures bathed in selection medium so that all cells are subjected to complete and rigorous selection. Escapes or epigenetic variations following in vitro selection using suspension cultures are very minimal compared to semisolid medium. Cells and plants regenerated from such cultures following selection in suspension culture show elevated levels of resistance against which they were selected.

Our understanding of the plant-pathogen interaction during the disease process has increased tremendously in recent years. One of the main findings is that there are compounds other than phytotoxins produced by the pathogen that are involved in the disease process. For instance, 'harpin' proteins produced by the pathogen have been shown to elevate plant resistance against a diverse group of pathogens. Using these compounds as a whole unit (as in a crude culture filtrate) in suspension cultures could be a better approach to bring out the true genetic resistance of the plant.

Induction of salt and heavy metal tolerance through tissue culture

A considerable area of land is unfit for agriculture because the soil is high in salts such as sodium chloride or heavy metals like aluminum. Development of crop plants suitable for these regions is still a high priority as the availability of arable land is continuously shrinking. Tolerance to sodium chloride using in vitro selection has been achieved in several crop species, such as rice, potato, sugarcane, and tomato (Table 40.2). However, in most cases the resistance was epigenetic. In some instances, salt tolerance was also accompanied by other undesirable characters such as low yield or lack of fertility. Only a few studies have been carried out to assess the use of somaclonal variation for heavy metal tolerance. Since there were many more failures than success, current research is more directed toward the understanding of molecular mechanisms involved acquiring such tolerance. The identification and cloning of genes that could elevate resistance to salt and heavy metals is a major breakthrough for plant geneticists.

Table 40.2 In Vitro Selection for Abiotic Stress

Crop	Selection factor	Result
Almond *Prunus amygdalus* Batsch.	Iron deficiency	Resistant shoots
Citrus/sweet orange *Citrus sinensis* (L.) Osbeck	Sodium chloride	Resistant plants
Flax *Linum usitatissimum* L.	Sodium sulfate	Resistant plants; inheritable
Grape *Vitis rupestris* Scheele	Sodium chloride	Resistant embryos
Potato *Solanum tuberosum*	Sodium chloride; seawater	Resistant callus
Rice *Oryza sativa*	Seawater; sodium chloride	Resistant plants; not inheritable
Sugarcane *Saccharum officinarum*	Sodium chloride	Resistant callus
Tobacco *Nicotiana tabacum*	Sodium chloride	Tolerant plants; inheritable

Identification of variation in tissue culture

It is important to identify desirable variation in culture as early as possible so that the long waiting period before the regenerated plants can show the phenotypic variability can be avoided and the possible accumulation of additional undesirable traits due to stringent selection can be reduced. If the variation is directed against a particular trait of interest, a specific test or bioassay will be helpful. A classic example is the selection of putative transformants following a gene transfer experiment. Gene transfer experiments routinely utilize a selectable marker, such as antibiotic resistance, to identify the transformed cells. Explants are placed in a medium containing the particular antibiotic to which resistance was introduced. Only those cells that have acquired the new gene will show resistance to the antibiotic and grow, whereas all other nontransformed cells will succumb (see Chapters thirty-one and thirty-three).

In vitro bioassays are very easy to perform and the results are seen rather quickly. For instance, if a perennial crop is subjected to in vitro selection for disease resistance, it may take several weeks or months before the regenerated plants are ready to be tested for the particular disease. However, a dual culture assay, where the pathogen is grown in the same culture plate along with the selected cells, might determine if the selected cells there express resistance. When successful, such an assay shows the pathogen growth to be significantly reduced by the selected cells when compared to the unselected cells (Jayasankar and Litz, 1998). This assay has worked quite effectively against fungal pathogens in several fruit crops such as *Citrus* spp., mango (*Mangifera indica* L.), and grapevine. It will also be easy to test if the induced resistance is specific against the host alone or if it is broad-spectrum resistance. However, there are certain limitations to this approach. For example, the pathogen may not grow well in the plant growth medium. So far this has been demonstrated effectively only with fungal pathogens.

In the past decade, several molecular techniques that can distinguish minor changes at the nucleic acid level have been developed. Amplification-based scanning techniques such as RAPD, AFLP, DAF, and arbitrary signatures from amplification profiles (ASAP) are used to identify if there is a permanent genetic change in the regenerants. In contrast to RFLP and other hybridization-based techniques, amplification-based scanning or profiling techniques can distinguish individuals and closely related organisms. Hence, these techniques can identify variation induced in tissue culture more precisely and quickly. The major advantage of these techniques is that they need only a small amount of tissue, from which 'information-rich' nucleic acids can be used to generate critical data for analyses. Further, these techniques require much less effort than performing a phenotypic or cytological analysis (Caetano-Anolles and Trigiano, 1997).

For instance, RAPD has been used effectively to identify the genetic changes that were induced following in vitro selection of mango embryogenic cultures against *Colletotrichum gloeosporioides* phytotoxins (Jayasankar et al., 1998). Resistant cultures exhibited significant differences in the genetic markers generated after RAPD analyses in comparison with the unselected controls as well as the original parent trees. The phytotoxins may also act as mild mutagen to warrant permanent genetic changes in the host cells and these changes can be specific such as the activation of the promotor in resistance genes. RAPDs have also been used to identify somaclonal variation in asparagus, beet, Norway spruce, peach, sugarcane, and wheat. AFLPs are more recent inventions and can identify variations more rapidly and in greater number. Furthermore, AFLP markers can be easily located in the genome, since only a known fragment of the genomic DNA is amplified for analysis. AFLPs are relatively easy to perform and are more reproducible than RAPDs. Currently, AFLPs have been successfully used to identify somaclonal variation in pecan (*Carya illinoinensis* Wangenh.), lettuce (*Lactuca sativa* L.), and *Chrysanthemum* species. DAF is perhaps the most useful tool in identifying variants in perennial species that are regenerated through tissue culture. Minor variants that are poorly characterized phenotypically can be precisely identified using DAF. This helps breeders to eliminate the unwanted rogues very early on. Currently, DAF is being successfully used in dogwood (*Cornus florida* L.) breeding.

Future prospects

It is evident that, despite its potential, variation from tissue culture has not been fully exploited. Induced variation still is the best route in perennial crop improvement, although one can argue in favor of the currently untapped potential of genetic transformation. However, it must be noted that tissue culture-induced variation does not have the socio-ethical hurdles like GM crops (see Chapter two). In addition, there are not any significant technology ownership issues as has become problematic with genetic engineering. Further, gene transfer techniques, though successful in herbaceous species, still have not been commercialized in perennial and woody species. Molecular techniques have greatly aided in understanding the plant cell response to biotic and abiotic stresses at the subcellular level. The future lies in utilizing these techniques to induce the species' own resources, such as disease resistance, to our benefit.

Acknowledgments

I thank Dr. A.K.Yadav, Fort Valley State University, for critically reading the manuscript and offering suggestions.

Literature cited

Caetano-Anolles, G. and R.N. Trigiano. 1997. Nucleic acid markers in agricultural biotechnology. *Agric. Biotech. News Inf.* 9:235–242.

Carlson, P.S. 1973. Methionine-sulfoximine resistant mutants of tobacco. *Science* 180:1366–1368.

Daub, M.E. 1986. Tissue culture and selection of resistance to pathogens. *Annu. Rev. Phytopathol.* 24:159–186.

Hammerschlag, F.A. 1992. Somaclonal variation. In: *Biotechnology of Perennial Fruit Crops*. Hammerschlag, F.A. and Litz, R.E., Eds. pp. 35–56. CAB International. Wallingford, Oxon.

Jacobs, M., I. Negrutiu, R. Dirks, and D. Commaerts. 1987. Selection programmes for isolation and analysis of mutants in plant cell cultures. In: *Plant Tissue and Cell Culture*. Green, C.E., Somers, D.A., Hackett, W.P., and Biesboer, D.D., Eds. p.243–264. Alan R. Liss, New York.

Jayasankar, S. and R.E. Litz. 1998. Characterization of embryogenic mango cultures selected for resistance to *Colletotrichum gloeosporioides* culture filtrate and phytotoxin. *Theor. Appl. Genet.* 96:823–831.

Jayasankar, S., R.E. Litz, R.J. Schnell, and A.C. Hernandez. 1998. Embryogenic mango cultures selected for resistance to *Colletotrichum gloeosporioides* culture filtrate show variation in random amplified polymorphic DNA (RAPD) markers. *In Vitro Cell. Dev. Biol. Plant* 34:112–116.

Larkin, P.J. and W.R. Scowcroft. 1981. Somaclonal variation — a novel source of variability from cell cultures for plant improvement. *Theor. Appl. Genet*. 60:197–214.

Lee, M. and R.L. Phillips. 1988. The chromosomal basis of somaclonal variation. *Annu. Rev. Plant Physiol. Plant Mol. Biol.* 39:413–437.

Novak, F.J. 1992. *Musa* (bananas and plantains). In: *Biotechnology of Perennial Fruit Crops.* Hammerschlag, F.A. and Litz, R.E., Eds. p.449–488. CAB International. Wallingford, Oxon.

Vardi, A., E. Epstein, and A. Breiman. 1986. Is the *Phytophthora citrophthora* culture filtrate a reliable tool for the *in vitro* selection of resistant *Citrus* variants. *Theor. Appl. Genet.* 72:569–574.

chapter forty-one

Short-circuiting the fern life cycle: apospory in Ceratopteris richardii

Leslie G. Hickok and Thomas R. Warne

Sexual plant life cycles are typically characterized by a distinct alternation of generations. This involves the presence of two generations or phases within the life cycle, sporophytes and gametophytes. The structure and level of complexity associated with each of these phases vary considerably from group to group. For instance, bryophytes have a predominant gametophytic phase with a small and typically ephemeral sporophytic phase. In contrast, flowering plants and gymnosperms have a dominant sporophytic generation with much reduced gametophytes. Homosporous ferns present an interesting intermediate situation in which both gametophytes and sporophytes are well defined, independent, and free living.

Alternation between the gametophyte and sporophyte involves a corresponding change in chromosome number, from n to 2n. This change is accomplished through the processes of meiosis in sporophytes, in which the chromosome number is reduced to n in spores, and fertilization in gametophytes, which reconstitutes the 2n condition in zygotes. What role does chromosome number play in the differentiation process? Does the haploid condition itself determine the differentiation of gametophytes and, correspondingly, the diploid condition determine differentiation of sporophytes, or is control of differentiation independent of chromosome number? This laboratory exercise provides an opportunity to investigate this question by manipulating the sexual cycle of the fern *Ceratopteris richardii* Brongn. This fern is becoming widely used in various educational settings under the cultivar name, *C-Fern*™. Spores of wild type and a variety of mutant *C-Fern* stocks, along with culture supplies and a detailed manual on its biology, teaching applications, and culture methods, are available through Carolina Biological Supply Co., Burlington, NC (U.S.).

Overview of the life cycle of C. richardii (C-Fern)

Ceratopteris is a genus of homosporous ferns found in most tropical and subtropical areas of the world (Lloyd, 1974). Species grow as either aquatics or subaquatics in ponds, rivers, or other intermittent wet areas. In some environments *Ceratopteris* sporophytes can become an aggressive weed by clogging up freshwater streams and drainage systems. Buds found in the axes of subdivisions of the frond (leaf) can develop rapidly into plantlets and are likely the reason for its weedy nature in some habitats.

0-8493-2029-1/00/$0.00+$.50

Like all homosporous ferns, *C. richardii* has two independent, autotrophic phases: a developmentally simple haploid gametophyte and a vascular diploid sporophyte. The gametophytic phase develops mitotically after germination of the single-celled spore and can be cultured axenically on a simple inorganic nutrient medium. Development of this haploid phase is very rapid. Germination occurs 3 to 4 days following inoculation and full sexual maturity of gametophytes is attained 12 to 14 days from inoculation (Hickok and Warne, 1998a).

As in all homosporous ferns, *Ceratopteris* spores are all alike (Figure 41.1a). Although isolated spores individually develop into hermaphrodites, two morphologically distinct gametophyte types, males and hermaphrodites, develop in populations (Figure 41.1b and c). Differentiation of sexual types in populations is associated with a pheromone-like system. Mature hermaphroditic gametophytes are 2 to 3 mm in diameter, morphologically simple, essentially two-dimensional plants with rhizoids, a meristem, vegetative cells, and both types of sexual organs (archegonia and antheridia). Males are smaller than hermaphrodites (1 mm), lack a meristem, are thumb-shaped, and are covered with numerous antheridia. In the presence of water the male sex organs (antheridia) burst open and discharge swimming sperm (spermatozoids). The sperm swim rapidly in the water and are attracted chemotactically to archegonia that contain mature eggs. After fertilization of the egg, the resulting diploid zygote develops rapidly by mitotic cell division and forms an embryo. Embryos are clearly visible after a few days (Figure 41.1d), and in 1 to 2 weeks small diploid sporophytes can be identified by the presence of a root and leaves (Figure 41.1e). The gametophyte soon dies and the sporophyte grows to maturity to undergo meiosis and produce spores to begin the life cycle again.

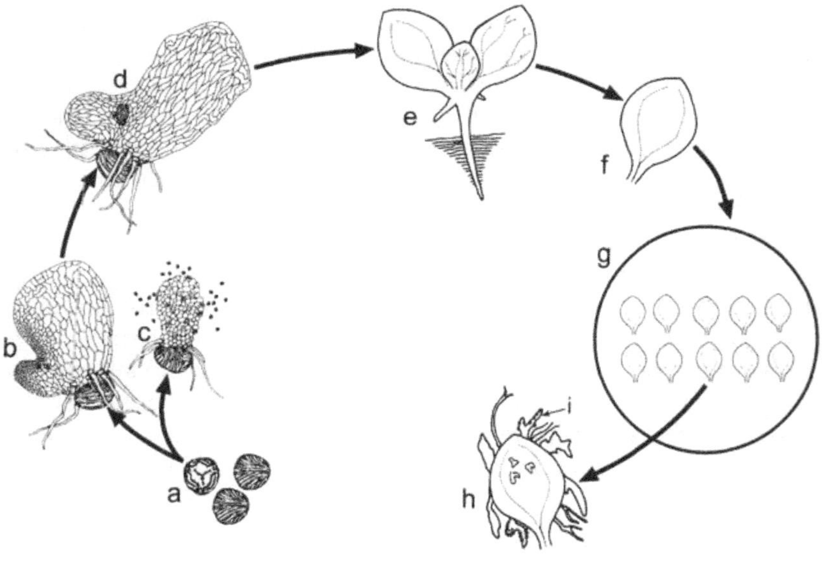

Figure 41.1 (a) Spores of *Ceratopteris richardii*, distal and proximal views. (b and c) 12-day- old hermaphrodite and male gametophytes, respectively. (d) Young diploid sporophyte embryo behind notch area of hermaphrodite. (e) Young sporophyte at three-leaf stage. The senescing gametophyte to which the sporophyte is normally attached is not shown. (f) Detached first leaf. (g) Culture of detached leaves on basic *C-Fern* medium supplemented with 10 M GA. (h) First leaf 2 weeks after transfer. Outgrowths of various tissue types are evident (see text). (i) Aposporous gametophytic outgrowth showing antheridia and sperm production. Scale for (a) through (d): spore diameter = 120 μm. Scale for (e) through (i): width of first leaf with bifurcate vein = about 2 mm. (Modified from Hickok, L.G. and T.R. Warne. 1998b. *C-Fern Manual for Teaching and Research*. Carolina Biological Supply Co., Burlington, NC.)

Overview of exercise

This exercise, which explores the phenomenon of apospory in the fern *C. richardii*, takes place over several weeks and includes a minimal amount of manipulation. Initially, gametophyte cultures are established from spores and, when mature, these cultures are watered to produce sporophytes. After 2 weeks' growth, sporophytes are harvested for leaves to initiate the portion of the exercise that directly deals with apospory. All necessary methods are presented below. Additional details on all aspects of gametophyte and sporophyte development and culture can be found in Hickok and Warne (1998a and b). An overview to the exercises is presented in Table 41.1.

Table 41.1 Establishment of *Ceratopteris* Gametophytes and Sporophytes and Initiation of Aposporous Outgrowths from Leaves

Week	Overview of manipulations and observations
1	Prepare Basic *C-Fern* media and, if not using presterilized spores, surface sterilize spores. Inoculate cultures with spores and maintain cultures at 28°C.
2 (optional)	Observe gametophyte growth and initial sexual differentiation using a stereomicroscope and transmitted illumination.
3	Add sterile dH_2O to gametophyte cultures and observe fertilization events.
4 (optional)	Observe young sporophytes that are still attached to gametophytes.
5	Harvest leaves from sporophytes and transfer to Basic *C-Fern* medium supplemented with 10 μM GA.
7–8	Observe aposporous outgrowths on detached leaves. Identify gametophytic and sporophytic cell types. Observe swimming sperm produced from aposporous gametophytic tissue. Subculture gametophytic tissue to sexually produce tetraploid sporophytes.

Basic culture methods are adapted from Hickok and Warne (1998) and routine sterile technique should be followed for all procedures. Although use of a laminar flow chamber can facilitate procedures and observations, its use is optional as long as proper precautions are taken to minimize airborne contaminants.

Ceratopteris richardii gametophytes and sporophytes can be grown in a simple mineral nutrient medium. The medium is available in powdered form from Carolina Biological Supply Co. (#15-6782). Alternatively, it may be prepared from stock solutions as detailed below. For this exercise, gametophyte cultures should be initiated on Basic *C-Fern* medium. After sporophyte production, leaves harvested from sporophytes should be cultured on basic *C-Fern* medium supplemented with 10 μM gibberellic acid (GA). GA should be added by sterile filtration after autoclaving. Although the use of GA-supplemented medium is not mandatory, results are improved with its use.

Materials for preparation of nutrient stock solutions

- Macro- and micronutrient and Fe salts (refer to Table 41.1 for listing)
- 1-L volumetric flask(s)
- Microbalance
- Magnetic stir plate and stir bars
- For Fe stock solution: hot plate, 2-L erlenmeyer flasks, watch glass
- Storage bottle(s)

Prepare macro- and micronutrient stock solutions separately by dissolving all listed quantities of components (Table 41.2), individually, in sequence, into about 800 ml high-quality, distilled water and bring to 1 L. Both macro- and micronutrient stock solutions can be autoclaved. Autoclaved stock solutions will keep for over 6 months and should be stored in glass at 4°C.

Table 41.2 Composition of Basic *C-Fern* Medium Stock Solutions and Final Medium[a]

Nutrient components	Stock solutions g/L	Final medium (ml stock/L)	Final medium (mg/L)
10 × Macronutrients		100	
NH_4NO_3	1.25		125
KH_2PO_4	5.00		500
$MgSO_4 \cdot 7H_2O$	1.20		120
$CaCl_2 \cdot 2H_2O$	0.26		26
200 Micronutrients		5	
$MnSO_4 \cdot H_2O$	0.0500		0.250
$CuSO_4 \cdot 5H_2O$	0.0740		0.370
$ZnSO_4 \cdot 7H_2O$	0.1040		0.520
H_3BO_3	0.3720		1.860
$(NH_4)_6Mo_7O_{24} \cdot 4H_2O$	0.0074		0.037
100 Chelated iron solution		10	
$FeSO_4 \cdot 7H_2O$	2.78		27.8
$Na_2EDTA \cdot 2H_2O$	3.73		37.3

[a] Higher concentrations of macronutrients in the stock solution are unstable and may form precipitates as will most combinations of macronutrient, micronutrient, and chelated iron stock solutions.

Modified from Klekowski, E.J. 1969. *Bot. J. Linn. Soc.* 62:361–377.

Prepare chelated Fe-EDTA stock solution by dissolving each component separately into about 450 ml of distilled water. On a hot plate, heat EDTA solution to boiling and add hot EDTA solution to the $FeSO_4$ solution. Boil combined solutions for 1 h, cool completely, then bring to 1 L. Store chelated Fe-EDTA solution in glass at 4°C.

Materials for preparation of final medium

- Macro- and micronutrient and Fe stock solutions
- Volumetric glassware: 100-ml graduated cylinder, 10-ml pipette, 1-L volumetric flask
- 2-L Erlenmeyer flask for each liter of medium
- Difco Bacto Agar
- Sterile 60 × 15-mm petri dishes

To prepare Basic *C-Fern* nutrient medium, add appropriate volume of each stock solution (Table 41.2) to about 800 ml distilled water in a volumetric flask and bring to a 1-L final volume. For agar-solidified medium, transfer nutrient solution to two l-L Erlenmeyer flasks and add 10 g (1% w/v) Difco-Bacto Agar. Note: Some plant tissue culture grade agars and agar substitutes can result in inhibited or abnormal growth of gametophytes or sporophytes and should therefore be avoided. Adjust nutrient medium to pH 6.0 using 0.1 N NaOH. Dispense about 15 ml of sterilized medium in 60 × 15-mm dishes and about 40 ml in 100-mm dishes. This ensures an adequate nutrient and water supply through to the young sporophyte stage. One liter of nutrient medium should pour about 55 60 × 15-mm dishes and 20 100-mm dishes. Allow dishes to cool completely undisturbed. Water condensation on petri dish lids is minimized if dishes are poured and cooled in stacks.

Follow Procedure 41.1 to complete this exercise with *C. richardii*.

Procedure 41.1
Initiating and maintaining cultures of *Ceratopteris richardii.*

Step	Instructions and comments
1	Spores: Wild type *C-Fern* spores are available from Carolina Biological Supply Co. in presterilized 10-mg units (#15-6728) that are sufficient to sow approximately 35 60×15-mm petri dishes. Alternatively, spores can be purchased in 40-mg bulk units (#15-6729) and surface disinfested according to the supplied directions. If using unsterilized spores, surface sterilize and follow the above proportions (10 mg spores/4 ml/35 dishes) to obtain the standard sowing density.
2	Culture inoculation with presterilized spores: Before opening the spore vial, be sure that all spores are at the bottom of the vial by tapping the bottom of the vial on a hard surface. Remove the cap and transfer 4 ml sterile distilled water to the spore vial using a sterile transfer pipette. The standard presterilized spore unit will sow about 35 dishes at a standard density of 300+ spores per dish when 4 ml sterile distilled water is added. Allow spores to wet completely by firmly attaching the cap and inverting the vial two to three times. With the cap on, check the bottom of the vial to be sure that all the spores have been suspended.
3	Sowing spores: To achieve consistent sowings, spores must be thoroughly suspended between each sowing. Suspend spores gently by drawing the liquid along with the spores in and out of the pipette. To sow, withdraw a small amount of the spore suspension into the pipette and immediately dispense the appropriate number of drops — not squirts! — onto the agar surface. A standard sowing density of 300+ spores per petri dish requires three drops of spore suspension. When sowing, tilt the lid of the petri dish upward only just enough to permit access of the pipette tip. To insure a consistent drop size, hold the pipette at a constant angle (45°) throughout the sowing. Do not touch the agar surface with the tip of the pipette. Resuspend spores between each sowing by gently squeezing and releasing the pipette bulb.
4	Spreading spores. Use a plastic or glass spreader (bacterial type). Spreaders may be sterilized by wiping or dipping in alcohol, e.g., 70% isopropanol or ethanol, and allowing to completely air dry. Metal wire spore spreaders may be sterilized by heating in a flame and allowing to cool. When distributing spores, allow the spreader to just rest on the surface without pressure. Move the spore spreader rapidly but gently back and forth across the surface of the agar while rotating the petri dish slowly with the other hand. The goal is to uniformly distribute spores over the surface of the medium.
5	Culture maintenance. Once cultures are inoculated with spores, place them into Culture Domes consisting of clean plastic greenhouse trays covered with transparent humidity domes. For best results, Culture Domes should be thoroughly clean. Culture domes are available through Carolina Biological Supply Company (#15-6792) or other sources. Culture Domes serve to reduce the possibility of contamination, variations in temperature, and humidity and permit easy handling of a larger number of dishes. Do **not** tightly seal the petri dishes, for example with Parafilm®, as this results in poor growth presumably due to ethylene buildup.

Helpful hints

The temperature optimum for spore germination and gametophytic development is about 28°C (82°F) as measured **inside** the Culture Dome. This is somewhat higher than for many other plants. Similar growth and development of gametophytes can be obtained at 26 to 30°C (79 to 86°F). However, lower temperatures will substantially alter developmental timing; for example, at 20°C (68°F) development times will be increased by about twofold. It is a good idea to monitor and record the temperature inside Culture Domes daily. If controlled temperature growth chambers are not available, some control of the temperature inside the Culture Domes can be achieved by adjusting the distance between the light source and Culture Domes. Once a suitable temperature is achieved, the height of the lights should remain constant during all phases of culture.

Water condensation on petri dish lids may occur during culture if sufficient temperature variation occurs between the lid and agar surface. Moderate amounts of condensation, i.e., fogging on the lid, are not a problem. However, excessive condensation can result in free water falling onto the culture surface, i.e., "raining," which may result in uncontrolled release of spermatozoids from antheridia. A constant temperature within the Culture Domes reduces the chances of condensation on petri dish lids. If condensation is a problem, cultures may be grown upside down once the sowing water has been absorbed into the agar medium.

Continuous illumination by two 40-W cool white fluorescent tubes at a distance of 45 cm or less from the cultures (about 80 µmol photosynthetically active radiation $\cdot$ m^{-2} $\cdot$ sec^{-1}) will accommodate two standard Culture Domes (54 × 27 cm). This will provide enough space for up to 64 individual 60 × 15-mm petri dishes. Smaller or larger setups can be used to serve individual needs. Temperature inside the Culture Dome is more important than light level, so the distance between the Culture Dome and the light source can be varied considerably to obtain a temperature near the optimum.

Anticipated results and some additional protocols

Culture week 1 — culture inoculation

Spores of *Ceratopteris* are approximately 120 µm in diameter (Figure 41.1a). They contain a single haploid nucleus that can sometimes be seen quite clearly under a compound microscope. If extra spores are remaining after inoculation of the plates, a simple wet mount can be made and spores observed under magnification of 200× or higher. Spores have fingerprint-like markings on their surface except for a small, smooth area containing a trilete ridge in the center. Often, the haploid nucleus can be seen as a distinct light-colored sphere near the center of this trilete ridge. By 3 to 4 days following inoculation, spore germination should be evident. Cracking of the spore wall is followed by extension of a clear rhizoid and development of photosynthetic tissue.

Culture week 2 — observation of gametophyte cultures (optional)

Under appropriate culture conditions, *Ceratopteris* gametophytes show initial differentiation of sexual types after 1 week. Use a stereoscope with bottom illumination to observe the cultures. Although they become more distinct with age, both males and hermaphrodites can be distinguished at this time. Males are tongue shaped and may be observed to have many small round structures (bumps) on their surface, the developing antheridia. At maturity, each antheridium will contain 16 sperm. Other gametophytes developing as immature hermaphrodites are somewhat larger than males and may show a slight lateral indentation which is the developing meristem. At this time, there may be some archegonial initials located immediately behind the lateral meristem. Antheridia do not develop on hermaphrodites until approximately 12 days from inoculation.

Culture week 3 — fertilization of gametophytes

At this time gametophytes should be fully mature sexually. Males and hermaphrodites (Figure 41.1b and c) should be apparent. Using a sterile transfer pipette, add approximately 3 mL of sterile water to a culture dish. Observe the culture using bottom illumination of a stereo microscope. Within a few minutes many swimming sperm should be visible, even under low magnification (e.g., 20×). Sperm rapidly locate receptive archegonia and will swim toward and then down the small necks of archegonia to fertilize the eggs. Note: Water should be added to the cultures immediately after they are removed from the Culture Dome. A sudden change in temperature can initiate premature release of sperm, even in the absence of water, and consequently reduce the number of successful fertilizations.

Culture week 4 — observation of young sporophytes (optional)

Young sporophytes can be clearly observed developing in the region behind the notch meristem of hermaphrodites (Figure 41.1d). The first sporophyte leaf may be visible. As the sporophytes grow, the attached gametophyte will turn brown and senesce. At this time it may be possible to distinguish gametophyte and sporophyte cell types. Gametophyte cells are rectangular, whereas epidermal cells of sporophyte leaves are puzzle shaped. This distinction is useful when analyzing aposporous outgrowths on older cultures.

Culture week 5 — harvesting and transferring leaves from young sporophytes

After 2 weeks culture, two to three young leaves should be visible on the young sporophytes (Figure 41.1e). A small root may also be visible growing into the agar. The expanded first and second leaves from sporophytes can be harvested using a sterile needle and fine forceps. Gently hold the petiole of a leaf with the forceps and, using the needle, gently detach it from the rest of the sporophyte. Transfer the detached leaf to a petri dish containing the GA-supplemented medium (Figure 41.1f and g). Try to work quickly so that the sporophyte leaves do not become desiccated. Place each transferred leaf flat on the agar surface, ventral (abaxial) side down. Each initial gametophyte culture should contain enough individual sporophytes to inoculate several GA-supplemented petri plates with 10 to 15 leaves each. After transfer, place each of the new dishes in the Culture Dome along with the original gametophyte cultures.

Culture week 6 — observations (optional)

Initial swellings or outgrowths may be visible on the margins of leaves. However, significant growth is typically not observable until 2 weeks after transfer.

Culture weeks 7 and 8 — observation

Two to three weeks after transfer significant marginal growth should be observed on many of the leaves (Figure 41.1h). In addition, some cell proliferation may be observed on the laminar portion of the leaves. At this time, it should be possible to distinguish several cell types. Clear hair-like rhizoids may be visible and will appear very similar to typical gametophytic rhizoids. Tissue showing two-dimensional growth with rectangular cell types very similar to those on gametophytes should also be present. As cultures continue to mature, some of these tissues (aposporous gametophytes) can develop a meristem-like region and continue to grow into a functional gametophyte. In these cases, there will be archegonia produced immediately behind the meristem, just as in typical gametophytes. Eggs contained within the archegonia are diploid (2×) in contrast to the normal situation in gametophytes. In addition, regions of aposporous gametophyte tissue often show a bumpy appearance associated with the formation of numerous antheridia (Figure 41.1i). As the antheridia mature (e.g., 3 weeks from initial leaf transfer), it is possible to remove tissue from one of the leaves or remove the entire leaf along with its outgrowths and place it in a small drop of water on a microscope slide. Within 5 min, swimming sperm may be observed in the area directly adjacent to the proliferating tissue. This is best observed with transmitted (bottom) illumination and at a magnification of 30 to 50×. Because the sperm have been produced from aposporous gametophytic tissue, they are diploid (2×) with a chromosome number equivalent to the sporophytic number of 2n = 78. Note: Since n and 2n designations are typically reserved to designate the normal alternation between haploid gametophyte and diploid sporophyte phases, these diploid gametophytes or gametes are appropriately designated as n = 2× = 78. If it is difficult to observe sperm after a few moments, the leaf tissue may be removed from the drop of water and the drop closely observed to determine if any swimming sperm can be seen. If sperm are not observed, try another leaf.

As an option for further observations and manipulations, the aposporous 2× gametophyte tissue can be isolated and subcloned on petri dishes containing either minimal or GA-supplemented medium. As these mature and form both antheridia and archegonia, add a drop of sterile water to the cultures to allow self-fertilizations to occur. It may be necessary to repeatedly add a small amount of water to the cultures over a period of 1 or 2 weeks. Any resulting self-fertilizations will result in the production of autotetraploid (2n = 4× = 156) sporophytes since the gametophyte tissue and, consequently, gametes that they were derived from were 2×. The sporophytes may have some difficulty in establishing normal growth, but with time some should develop sufficiently so that they can be cultured in the greenhouse. Because of multivalent formation during meiosis, autotetraploid sporophytes may show reduced spore viability. They may produce some viable spores with variable numbers of chromosomes.

In addition to the aposporous gametophytic tissue described above, other types of tissues will also be present. Terete three-dimensional outgrowths may show characteristics unlike typical gametophytes or sporophytes. However, some of this tissue may contain sporophyte-like epidermal cell types as well as vascular tissue. As cultures mature, small sporophytic outgrowths or buds may form and give rise to young sporophytes. In these situations, the sporophytes are not the result of a fertilization event. Rather, they have just redifferentiated from the leaf tissue and would be expected to have a normal sporophytic (2n = 2× = 78) number of chromosomes. In some instances, these buds may differentiate from tissue that initially shows gametophytic characteristics.

Questions for thought and discussion

Apospory (without spores) is defined as the direct production of gametophytic material from sporophytic tissues, which short circuits the typical pathway of meiosis and spore production in a normal sexual cycle. Resulting gametophytes are unreduced chromosomally and provide a possible means of producing polyploid sporophyte progeny if fertilizations take place subsequently. Although a large number of ferns have been shown to produce aposporous gametophytes naturally or in culture, the significance and mechanisms controlling this process in ferns are not well understood (Raghavan, 1989).

Apogamy (without gametes) is another type of short circuit of the sexual cycle. Apogamous growth of sporophytic tissue from gametophytic, without gamete production and fertilization, is a widespread phenomenon among plants, including angiosperms. As with apospory, apogamy has been documented in many different ferns (Raghavan, 1989), including *Ceratopteris* (Hickok, 1977; 1979). In a broader context, these and other types of asexual modifications of the life cycle are referred to as forms of apomixis (Reiger et al., 1976). The possible use of some types of apomixis in crop production has received recent attention, since the ability to asexually produce genetically uniform seed has important agronomic possibilities (Vielle Cazada et al., 1996).

Based on the observations that have been made on the aposporous outgrowths from *Ceratopteris* sporophyte leaves, what is an appropriate answer to the question that was posed initially? Is chromosome number alone responsible for determining the gametophytic or sporophytic state of development? What other factors are likely to be involved? How likely is it that apomictic processes can be controlled or manipulated through genetic changes? To date, relatively little information on the genetic basis of apomixis is available, but renewed interest in possible agronomic applications may encourage additional research. In addition to applied studies with crop plants, other organisms that readily show various apomictic behaviors, such as ferns, can be used as models on which to base angiosperm studies (Vielle Cazada et al., 1996).

Acknowledgments

Development of *C-Fern* has been supported, in part, by the National Science Foundation and the University of Tennessee. We thank Stephenie Baxter for technical support and artwork.

Literature cited

Hickok, L.G. 1977. An apomictic mutant for sticky chromosomes in the fern *Ceratopteris*. *Can. J. Bot.* 55:2186–2195.

Hickok, L.G. 1979. Apogamy and somatic restitution in the fern *Ceratopteris*. *Amer. J. Bot.* 66:1074–1078.

Hickok, L.G. and T.R. Warne. 1998a. Sex and the *C-Fern*: not just another life-cycle. *BioScience* 48:1031–1037.

Hickok, L.G. and T.R. Warne. 1998b. *C-Fern Manual for Teaching and Research*. Carolina Biological Supply Co., Burlington, NC.

Klekowski, E.J. 1969. Reproductive biology of the Pteridophyta. III. A study of the Blechnaceae. *Bot. J. Linn. Soc.* 62:361–377.

Lloyd, R.M. 1974. Systematics of the genus *Ceratopteris* Brongn. (Parkeriaceae). II. Taxonomy. *Brittonia* 26:139–160.

Raghavan, V. 1989. *Developmental Biology of Fern Gametophytes*. Cambridge University Press, New York.

Reiger, R., A. Michaelis, and M.M. Green. 1976. *Glossary of Genetics and Cytogenetics*. Springer-Verlag, Berlin.

Vielle Cazada, J.-P., C.F. Crane, and D.M. Stelly. 1996. Apomixis: the asexual revolution. *Science* 274:1322–1323.

chapter forty-two

Commercial laboratory production

Gayle R. L. Suttle

Over the past 15 years, the horticultural nursery trade has found ever-increasing ways to utilize micropropagation as a practical and cost-effective production tool. This chapter will describe some of the basic steps involved in plant micropropagation, discuss some of the practical aspects of the commercial micropropagation business, and give some examples of how and why micropropagation is being used by growers today.

Many "scientific" papers have been published describing methods, formuli, and techniques for the successful micropropagation of a multitude of plant species. In theory, once one has the right "recipe," all that is needed is a well-equipped kitchen, some trucks to haul the goodies to market, and a good bank to hold all the money when it comes rolling back in. In reality, successful micropropagation on a commercial scale requires an understanding of two equally complex and dynamic factors — the plants and the marketplace. It is not enough to produce high quality, healthy plant material (a major challenge in itself); one must also produce it in a timely fashion giving customers what they want, when they want it, for a price they can afford.

The facility

Well-designed laboratories generally have a good work flow allowing for the actual movement of supplies, people, cultures, and the finished product in an easy and logical pattern. Many small laboratories, unable to build a facility from scratch, use existing residential housing or mobile homes and remodel them to accommodate their needs. In order to keep airborne contaminants to a minimum, the ideal facility has very few entry sites (doors) and often has large areas designated as "clean rooms" with purified (HEPA filtered) air flowing under positive pressure. The basic areas in any laboratory are listed below (Figure 42.1).

Media preparation/dishes. This area is set up similar to a kitchen with lots of counter space and shelves for chemicals and stock solutions, a pH meter, scales, autoclaves (usually more than one), media-dispensing equipment, and a dishwashing area. Also important here is adequate storage space for empty culture vessels. Autoclaves create a tremendous heat load and usually are placed on an exterior wall with good clean ventilation.

Media storage. Once media is "cooked" (sterilized) and dispensed into culture vessels, it must be cooled down (again with good clean ventilation) and stored until needed.

0-8493-2029-1/00/$0.00+$.50

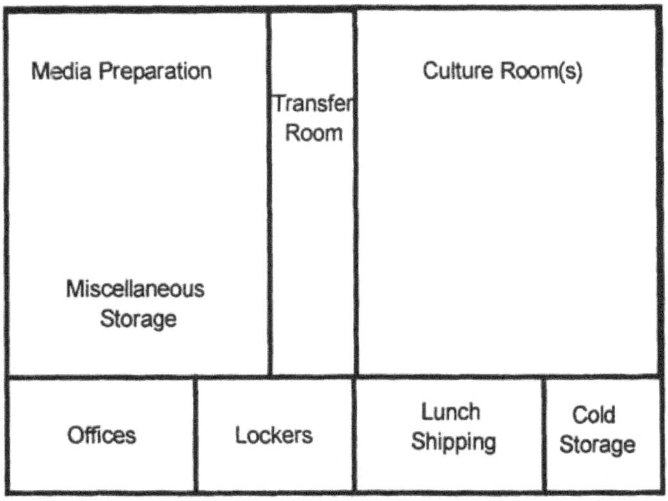

Figure 42.1 A design of a commercial tissue culture laboratory facility. (From Trigiano, R. N. and Gray, D. J., Eds., *Plant Tissue Culture Concepts and Laboratory Exercises, First Edition,* CRC Press LLC, Boca Raton, FL, 1996.)

Often this area is combined either with the media preparation room or in a section of the transfer room.

Transfer room. This is the heart of any laboratory, filled with clean benches (laminar flow hoods) where all of the culture initiation and subsequent subculturing takes place. "Ripe" cultures (cultures with usable shoots) are harvested and cultures replanted (transferred or reinitiated) to clean fresh media. Up to 80% of the labor force may be involved in this one room, so it must be a comfortable and pleasant working environment. Tools are sterilized at the hoods using glass bead sterilizers, bacticinerators, or alcohol and flame. Often transfer rooms tend to be cramped and the heat load generated by people, hoods, and sterilizers are compensated for with a good air conditioning system. People and plants suffer when it is hot and stuffy, and productivity declines.

Culture room(s). Usually the largest area of any laboratory, cultures are incubated on shelves under specific light and temperature regimes. Light quantity and quality can be key factors in culture success. Generally, regular or high output cool white fluorescent lighting is used, with higher intensities achieved by using more lights per square foot or by bringing cultures closer to the lights. Once again, heat buildup can be a major problem. Many laboratories remove the ballasts from the fluorescent light fixtures and mount them in an external area for easier cooling. Good air movement is critical in equalizing temperatures throughout the culture room. Temperature requirements are often crop specific, but generally range from 18 to 26°C. Different light or temperature regimes may require separate culture rooms for different crops. Usually air to the culture room is purified by HEPA filters.

Cold storage. Cold storage is an absolutely essential component of any commercial micropropagation facility growing woody plants. Cold storage (2 to 4°C) is used for the following purposes: (1) to maintain stock blocks in culture for plants that are produced on a seasonal basis, (2) to allow production to continue during the cold winter months by "banking" finished plant material ahead of time, and (3) to provide a chilling requirement for crops such as apple (*Malus*) and pear (*Pyrus*) species that respond in the greenhouse much more rapidly if given a "rest" or dormant period before out-planting.

Shipping/lunch area. Ripe product (either microcuttings or rooted plantlets) must be harvested, boxed, packed, and otherwise prepared for shipping. This activity requires sinks to rinse the plants and also should be near the cold storage unit for holding until shipping. This area may serve double duty as a lunch or break area.

Locker room. When first entering any commercial facility, employees usually must change shoes and may be required to wear additional clean clothing such as laboratory coats, hairnets, etc. into production areas. This locker area may also, as above, be incorporated with a lunch room and/or rest rooms.

Offices. Perhaps the least important place in a facility from an operations standpoint, but essential for running the minor details such as sales, production planning, accounting, and, of course, payroll, are the offices.

Miscellaneous storage. No laboratory ever has enough general storage space. Supplies, chemicals, shipping containers, extra equipment, spare parts, and items such as vessels used on a seasonal basis all require space. Logical locations for storage are near an external door (for receiving) or near the media preparation and shipping prep areas where supplies are most often used.

The process

What to grow

There are two different approaches in determining product focus. The first is to grow what you think customers might want (perhaps a novel plant) based on what you hope is good market research and analysis, and then working hard to convince potential customers that they really do want it. While this approach is the very basis of our country's free enterprise system, it can be very risky if you happen to guess wrong. The most common reason for the failure of commercial laboratories over the past decade has been growing plants that the market either doesn't want or need. The second approach is to work with the customer on more of a contractual basis, asking what their problems are, what they need, and when they need it, then going to work to try to solve the problem and fill the order. Arranging customers in advance is a much safer route to success considering the costs involved. Regardless of approach, every plant must be evaluated for suitability for micropropagation by asking the following questions:

1. What are the minimum number of plants needed? — Most laboratories have a minimum requirement (5,000 to 10,000 or more), although smaller facilities may be willing to have lower minimums.
2. How many customers are interested in a specific product? — The more the better, but sometimes exclusivity is important.
3. How long is the product needed? — Is it just for mother block establishment, seasonal, or will it be needed on a continuous basis?
4. How difficult is the plant to culture? — Is there existing literature or knowledge or will the technology need to be developed?

5. Who "owns" the plant? — Are there patent or customer restrictions on marketing the plants?
6. What is the estimated value of the plant once produced? — Can you charge enough to recover costs and make a profit?

All of these factors must be considered before starting any project. Often the price of a product cannot be defined until considerable research has been done on protocols for the given plant. Guarantees of success are impossible to make. The up-front cost of protocol development may or may not be shared with the customer. If exclusivity is needed, then the customer usually funds the work. Once a plant is deemed worthy for an attempt at micropropagation, the process of producing it may begin.

Stock plant preparation

Healthy well-labeled stock plants are essential. The most important job the customer must perform is to make absolutely sure that the stock wood sent to the laboratory is exactly the named material that will be wanted in large numbers when finished. This step sounds simple, but in reality, mistakes on labeling can and do happen far too frequently. This point cannot be overemphasized.

Ideally, 10 to 15 young, juvenile stock plants are placed under an intense fertilization and pesticide regime to provide rapid, healthy shoot growth. The plants are watered from the base to avoid getting moisture on the foliage, a breeding ground for contaminants. Sometimes there may be a single very old tree located thousands of miles away as the only source or a customer may have only a small amount of stock and does not want to risk sending it away. In these cases, wood must be collected periodically throughout the growing season and sent by overnight express.

Stage I: Culture initiation

The process begins by carefully harvesting soft, active shoot growth from the stock plant. Usually 5- to 15-cm cuttings are snipped early in the morning while the plants are fully turgid. The cuttings are immediately placed in a cooler until disinfestation takes place. The leaves, and most if not all of the petiole, are removed and each stem is cut into one to two nodal pieces.

The disinfestation or cleanup process usually involves several cold soapy water washes, then soaking the shoots in a 10 to 20% commercial bleach (sodium hypochlorite) solution along with a surfactant such as Tween 20 for 10 to 20 min followed by a series of sterile water rinses. Damaged tissue is trimmed away using aseptic techniques. The treated cuttings are then placed individually in test tubes filled with a nutrient gel (medium) composed of a "best guess" formula or range of formulas that may allow the buds to grow. The tubes are usually incubated in a culture room under 16 h light at 25°C.

Visual screening takes place every 3 to 5 days and the dead or dirty (visibly contaminated) cultures are discarded. Visibly clean, viable buds are transferred individually to fresh medium as new growth appears, usually 2 to 6 weeks after disinfestation. A portion of the stem piece is placed in a rich liquid or agar-based medium to detect bacterial contaminants that may be "invisible" or "unnoticeable" to the naked eye. This contaminant screening process (also called indexing — see Chapters nine, forty-three, and forty-four) allows us to discard any additional and unseen dirty material from the pool of viable shoots. This procedure may be repeated periodically, however; unfortunately, even this procedure does not catch everything and growing "clean" stock is a continuous struggle.

Culture initiation can be more of an art than a science. Often the concentration of bleach and/or time of disinfestation must be adjusted based on the woodiness, hairiness, and source of the stock tissue. Several variations may be tried if there is an abundance of source material, fine tuning the process based on previous efforts and the time of year. If materials are limited, the best guess is tried and hopefully will succeed.

Timing

The best success for getting plants into culture is typically realized by taking the first flushes of vegetative growth in the early spring. A very successful second option is to force dormant buds from sticks in the winter after the full chilling requirement has been satisfied. These sticks are placed in a "forcing solution" of 30 g/L Floralife Floral Preservative (Floralife, Inc., 120 Tower Drive, Burr Ridge, IL 60521), incubated at approximately 20°C with a 10- to 12-h photoperiod (typical office conditions). Floral preservatives generally have sugar, an antimicrobial agent, and an acidifier to prolong bloom life. The dormant buds of many plant species will begin to sprout and grow under such conditions. Flower buds are removed and discarded when they appear to allow for better vegetative growth. When these soft shoots reach the length of 2 to 4 cm, they are easily removed from the stem and processed. They tend to be extremely clean so that surface disinfestation and establishment in culture are usually easier.

One theory explaining why such early growth seems to go into culture so well is that endogenous plant growth regulators (PGRs) and growth factors may have primed the plants to "spring" into rapid growth phase (inhibitory factors are at a low ebb). Fortunately, many times this new growth is very "clean" — it hasn't had time to acquire a load of contaminating organisms. Whatever the reason, early spring growth is without question usually the best starting material. By starting early in the growing season, it also gives more time to generate numbers and bring more cuttings into culture if needed.

Late summer or outdoor field-grown stock posed particular problems from increased contamination and a lack of turgidity due to heat stress that causes plant tissues to be less tolerant of bleach disinfestation. Success rates for establishing material in Stage I can range from 0 to 95% due to all these factors and many more we do not yet comprehend. The ultimate goal of Stage I is to get the plant material clean and actively growing in culture. Once this is achieved, Stage II becomes the next challenge.

Stage II: Shoot proliferation

The primary method of increase in woody plant micropropagation is by axillary shoot proliferation, although adventitious shoot proliferation may occur with some kinds of plants, such as species of the family Ericaceae. Axillary shoot proliferation is preferred in order to avoid the potential of off-types (somaclonal variation — see Chapter forty) from occurring. Genetic mutation can occur by any vegetative propagation method and if the characteristic is based on a relatively unstable chimera, such as some variegations are, micropropagation can enhance reversion. Micropropagators must continuously monitor the quality of plants they produce to ensure trueness. In our 15 years of experience of micropropagating trees, genetic mutation has not been even a minor problem. Useful strategies we employ to help reduce the potential of off-types are limiting the number of subcultures or length of time in culture, keeping PGR levels in the culture medium as low as possible and always going back to the original stock source to start new lines.

Once a plant is clean and actively growing in culture, the goal of Stage II is to induce all (or at least some) of the axillary buds along the new, young, aseptically growing shoot to sprout and grow. In other words, we want the shoot to branch. Every 2 to 8 weeks (culture cycles vary depending upon the type of plant), we can subculture or harvest the new branches, "replant" or transfer them to fresh medium, and wait another 2 to 8 weeks until each of these new branches sprouts its own set of branches. There are rarely any roots involved during Stage II.

With woody plant micropropagation, there are two main methods of achieving multiplication of new axillary buds depending on the type of plant material involved. The first is by promoting many branches on a single stem piece resulting in a "clump" of new shoots. Species of apple and cherry (*Prunus*) would be two examples of plants best micropropagated by the clump method. During subculturing, the nice tall shoots are removed and either replanted in fresh multiplication medium to produce their own clumps or planted into a rooting medium or shipped out as microcuttings to be rooted in the greenhouse. The basal clump mass is left intact and replanted to yield another crop of branches. The clumps get thicker and thicker with new branches and can be used as a mini stock block. Eventually the clumps decline in quality and quanity, usually after two to five culture cycles, and must be discarded and replaced.

The second form of axillary shoot proliferation can be described as the single-node method. Plants exhibiting strong apical dominance such as lilac (*Syringa*) and maple (*Acer*) species are best micropropagated by the single-node method. A single shoot grows up straight and tall, producing two to ten sets of leaves. One- or two-node sections are separated and replanted whereby each of these runs straight up again and is now ready to be chopped into nodal sections again. During any subculture, the nice terminals or tips can be harvested and planted into rooting medium or shipped out as microcuttings to be rooted in the greenhouse.

Media formuli

The actual media formuli used during Stages I and II are often the same. There are three or four basic formuli that are adapted and modified as necessary, sometimes on a cultivar-by-cultivar basis. The proper ratios of the inorganic components in a formula actually are the main key to developing a successful protocol for a given plant. PGRs are also a key factor, but secondary to proper nutrition. It is not uncommon for a commercial laboratory to have over 50 modifications of basic media on file for multiplication alone and another 40 or so Stage III modifications. When a new plant is brought in, one takes an educated guess as to what might work based on past experience and literature searches, but usually modifications must be made by a systematic trial-and-error method of marching through the various components. In general, Stages I and II media usually have higher levels of salts, sugars, and cytokinins. Stage III media generally have lower levels of salts, sometimes lower sugars, and usually lacks cytokinins, but includes high levels of auxins.

Stage III: Rooting

When sufficient numbers of healthy shoots have been generated by one of the proliferation processes described above, healthy active terminal buds are selected either for in vitro rooting or for shipping directly to customers as microcuttings for rooting and acclimatization in the greenhouse. The decision whether or not to root in vitro is quite often made by the customer based on several factors. In vitro rooting adds about 30% to the final cost of the product and some crops, such as birch (*Betula*) and lilac species, root so well in the greenhouse as microcuttings that it simply does not make sense to root them in vitro. Other crops such as *Malus, Pyrus, Prunus,* and *Acer* are much easier to acclimatize as in vitro rooted plantlets.

Perhaps the most important factor of all for the grower, other than survival, is that in vitro rooted plantlets require 2 to 5 weeks less time in the greenhouse to reach a size suitable for planting in the field. Many growers have limited greenhouse space and do not want to have to "babysit" tender microcuttings for such a long time. A grower may be able to process two, three, or four crops through his greenhouse in the same time it takes to raise one crop from microcuttings. Often, it pencils out as more economical to buy rooted plantlets.

Stage IV: Greenhouse acclimatization

Fresh Stage II microcuttings (without roots) or Stage III in vitro rooted plantlets are both planted in any of several well-drained soilless potting mixes in the greenhouse under high humidity. Various types of fogs, tents, domes, and mist systems are used to wean the tender shoots into the real world by gradually reducing the humidity. The weaning process takes from 3 days to 6 weeks depending on what the plants are, the weather, and if they were rooted in vitro. Fertilization is not recommended until the plantlets become established and begin to grow. Some crops such as *Betula* can become stunted if overfertilized.

Production scheduling

When preparing an actual production schedule for any given plant, many factors play a role. How many are ordered and in what form (microcuttings or rooted plantlets)? When are they wanted? Can the order be prepared ahead of time (does it require or tolerate cold storage)? How does it multiply (nodal or clump)? What are the multiplication rates during buildup? What is the number of rootable or harvestable microcuttings (yield) and does it change over time? What rooting percentage can be expected? Can smaller batches be produced and stored or must it be produced in lump sum? How much time is required between cycles (during buildup and during harvest cycles)? What are the estimated labor requirements for each cycle? Once these factors are determined, one works backward from delivery date, calculating the number of cultures required at each of the various cycles along the way. Information on when each cycle must take place, what activities must be done at each step, and estimated labor required are all plotted on a production planning calendar. This planning process must be done for each cultivar on an individual basis since multiplication rates, yields, number ordered, and dates ordered will vary from plant to plant. Critical labor peaks must also be smoothed out as well as possible ahead of time. Simply put, not everything can come out at once.

One of the most frustrating factors in planning a production schedule is the very dynamic and very unpredictable nature of the plant material itself. Plants are living systems and we simply don't control or even understand all the factors involved in in vitro propagation. Multiplication rates and harvestable yields, for example, can only be best guesstimated. Any plan has to be continuously updated and revised.

Seasonality

While a laboratory can pump out material on a year-round basis, anyone working on temperate deciduous crops knows that there is a season for everything. Plants can be fooled into thinking it is summertime in culture by continuously being subjected to long (16-h) days in the culture room, but as fall approaches, the rate of growth in the greenhouse slows considerably. To take maximum advantage of the growing season, customers want to get their product in the early spring through early summer. Use of cold storage to store crops ahead of time allows laboratory production of some plant material to take place in the winter, but much must come out fresh, making labor needs seasonal as well.

Reasons why micropropagation is used

While micropropagated material represents only a tiny percentage of the total number of plants produced in America today, there is no doubt that many of the largest and smallest nurseries in the U.S. view micropropagation as an essential tool that helps them maintain their competitive edge by growing better plants more efficiently. The remainder of this chapter will be devoted to a discussion of some of the advantages that in vitro micropropagation offers over conventional production systems.

New introductions

Perhaps the most obvious use for micropropagation is to get a "jump start" on growing the newest and hottest items quickly. Micropropagation cuts 3 to 10 years off the time it takes to bulk up new selections and get them to market. For example, micropropagation was used by one grower to establish layer beds of a new apple understock. He sold over 1 million rootstocks in the same amount of time his competition had bulked up to only a few thousand using conventional propagation methods. The market is always looking for something new and exciting. Nurseries on the forefront of introducing new plant material build their reputation as leaders in the industry causing customers to come back year after year to find out what else is new.

In some cases, growers use micropropagated plant material to establish mother blocks and to fill in production shortages while mother blocks are too young to be productive. Conventional propagation methods such as hardwood or softwood cuttings, layer beds, or budding and grafting may take over long-term production needs once enough wood becomes available. In many other cases, micropropagation remains the method of choice for a variety of reasons.

Rapid response to market demand

Large mother blocks or scion orchards are time consuming and expensive to establish and maintain. It is often difficult for growers to adjust quickly to the rise and fall in popularity of a given plant. With micropropagation, the stock block is maintained in a 10 × 10-ft cold storage unit. If a customer gets a call for an additional 10,000 liners of a particular blueberry, for example, they simply call up and ask when is the earliest they can take delivery on the additional microcuttings. They then add the time they need for the greenhouse growing and call their customer back with a delivery date.

Clean plants

Micropropagation is inherently a cleaner system for producing plants compared to traditional production methods. Since the plants are grown in culture, diseases are not transmitted from the field into the greenhouse and on to subsequent generations. A single disease-free mother plant can theoretically produce unlimited disease-free daughter plants without the possibility of reinfection. Conversely, individual mother plants in a traditional virus-free cutting block must be tested again every year in order to maintain and ensure virus-free status (see Chapter forty-three). Testing fees add significantly to the expense of maintaining large mother blocks in the field.

Ease of propagation

Bud incompatibility on budded or grafted stock, and poor rooting percentages with softwood or hardwood cuttings make micropropagation the method of choice or the only option on many difficult-to-propagate plants such as *Syringa* and redbud (*Cercis*). Some red maple (*A. rubrum* L.) cultivars, such as 'Karpick' and 'Bowhall', are absolutely impossible to root from cuttings. Unreliable seedling availability and poor or unpredictable bud stands on *Betula*, *Tilia*, and *Morus* are problems growers are able to avoid by planting micropropagated material. Growing plants on their own roots offers major advantages for plants, such as contorted Filbert (*Corylus avellana* 'contorta'), where suckering of understocks can be a major problem.

Sometimes micropropagated material provides the grower with a nucleus of "juvenile" material from which additional cuttings can be more easily rooted. Each year or two, the customer starts over with a fresh batch of starter material from the laboratory.

Speed

Research done at various universities indicates that it is possible to grow plants much more quickly to size than is traditionally seen in nurseries today, regardless of how the plants are propagated. Such rapid growth requires optimization of all growing conditions including fertilizer, light, and temperature. However, it is amazing the results that can be achieved with even modest adjustments of growing practices. Several field growers are now producing well-formed, small branched trees of *Prunus* 'Kwanzan', *M. alba* 'Chaparral' (mulberry), and others in 1 year instead of 2. Blueberry plant production can be dramatically speeded up. Use of micropropagated cherry understock yields increased vigor and earlier fruitset (and earlier payback) for the orchardist.

Better branching

Because the internode length is greatly reduced on micropropagated plants, there is generally more opportunity to develop a fuller head on the growing plant. Indeed, this is also one of the reasons why survival is often greater. If something happens to destroy the terminal bud (for example, damage caused by freezing, hungry rabbits, or poor pruning), there are other buds below available to choose from. One grower accidentally sprayed a young block of *A. rubrum* liners twice with Surflan, an herbicide. The stems of the young micropropagated liners were girdled right at soil level. All was not lost, for the grower dug the soil out from around the base of the plants, removed the damaged tops, and the buds below the girdling all pushed out again. The grower lost some height on his crop (about 1 to 2 ft), but shorter plants are better than no plants at all. Having more buds to choose from also allows a grower to cut back closer to the ground producing straighter trees.

The better branching is a real advantage when it comes to growing well-formed shrubs. *Hydrangea quercifolia* 'Snow Queen' (PP4458) produced from rooted cuttings tends not to throw many branches at an early age, whereas micropropagated plants are easily developed into a bushy habit with routine pinching. The increased branching of blueberry plants is seen as a great advantage by some growers; others prefer the more traditional vase-shape formed from rooted cuttings.

Greater survival and uniformity

While cultural practices play an important role in the ultimate performance of any block of plant material, the two key attributes most often given in describing micropropagated material are superior survival rates and much greater uniformity. Better survival is explained, in part, by heavier root systems and more buds to choose from on the top. The fact that the root-to-shoot ratio is more balanced from the very beginning and that the small plantlets are highly uniform right from the start helps to explain why subsequent growth is more reliable and consistent.

Conclusion

As the use of micropropagation becomes increasingly varied and widespread, growers will find new and ingenious ways to take advantage of its power. Even so, growers must weigh the pros and cons of this relatively new technique against the old methods and determine for themselves, on a case by case basis, whether it is worth the trouble to change. It must help them solve problems, find new markets, save time or money. The successful micropropagator will be one who fills these needs and many more.

chapter forty-three

Indexing for plant pathogens

Michael J. Klopmeyer

Flowering plants are capable of reproducing either by seed or by vegetative propagules such as budwood, leaves, stems, or cuttings. When plants are reproduced via vegetative propagules, there is a higher risk of transmitting plant pathogens from the original mother plant to its offspring compared to seed propagation. This problem can be circumvented by establishing methods to eliminate these pathogens and maintaining pathogen-free stock through a production cycle.

This chapter will outline the various procedures used to eliminate fungal, bacterial, and viral pathogens from vegetatively propagated plants. These basic methods can be applied toward plant species propagated in tissue culture (micropropagation), in greenhouses, and in the field.

What is indexing?

Indexing, simply stated, is testing various plant parts (i.e., cuttings) for the presence of any plant pathogen. If plant pathogenic fungi, bacteria, viruses, or viroids are present in the plant part tested, that plant is either destroyed or subjected to various procedures designed to eliminate the pathogen. Only plant samples that test negative for any given plant pathogen are considered to be disease free and these are increased in large numbers under strict sanitary conditions.

Specific crops targeted

Indexing programs are being used in many university, government, and commercial laboratories around the world. The major vegetatively propagated crops that are indexed include the following: fruit trees, such as stone fruits (i.e., apricot, cherry, peach, and plum); pome fruits (apples and pears) and citrus (orange, lemon, and lime); vegetables, such as the Irish potato and sweet potato; fruits such as banana and papaya; and floricultural crops, such as foliage plants, bulbs (Easter lilies, Asiatic lilies, iris, gladiolus), chrysanthemum, carnation, New Guinea impatiens, and geranium.

Each crop is unique in regards to the pathogens that infect it, what techniques are efficient in eliminating these pathogens, and what are the most effective methods for verifying the pathogen-free status after pathogen elimination. Since many of these elimination techniques require passage of the plant through tissue culture, careful evaluation of an artificial growth medium is also required prior to initiation of the elimination procedures.

0-8493-2029-1/00/$0.00+$.50

Indexing techniques

There are various indexing techniques utilized to eliminate plant pathogens from vegetatively propagated plants and to verify that the pathogens have been eliminated. Indexing strategies revolve around the host crop studied as well as the specific pathogens affecting that crop. Indexing methods need to be developed that (1) exploit the biology of the pathogen within the host, (2) utilize a detection and/or eradication technique that is sensitive and economically feasible, (3) are repeated over months to reduce the probability of pathogen infection, and (4) are performed under strict sanitation in the field, greenhouse, and/or laboratory.

The remainder of this chapter will focus on strategies employed in the floriculture industry to eliminate the major plant pathogens from the Florists' geranium (*Pelargonium* × *hortorum*). The principles discussed are applicable to other crop/pathogen systems including plants maintained in tissue culture.

Bacterial and fungal elimination

One method for the elimination of systemic fungal and bacterial plant pathogens from geraniums is accomplished using a procedure called culture indexing. This procedure was initially developed in the 1940s to free chrysanthemums of the fungal wilt pathogen, *Verticillium* (Raju and Olson, 1985). Due to the tremendous success with chrysanthemum, culture indexing was utilized to eliminate the major systemic plant pathogens of geranium including *Xanthomonas campestris* pv. *pelargonii* (Xcp), a bacterium, and *Verticillium* spp. (Oglevee-O'Donovan, 1993).

Bacterial blight (caused by Xcp) is the limiting factor in the successful production and finishing of zonal, ivy, and regal geraniums. Under optimum environmental conditions, this bacterial pathogen causes plant death in a relatively short time by physically blocking the xylem vessels (water-conducting tissue) in the stem. Understanding the biology of the pathogen allows for culture indexing to select plants free of the pathogen. Culture indexing is simplistic in its method, yet very effective in the selection of plants free of systemic fungal and bacterial pathogens. Briefly, culture indexing entails the following six steps:

1. Establishment of mother plants in an incubation greenhouse at temperatures that are optimum for disease development. In this instance, bacterial blight caused by Xcp develops quickly at 25 to 30°C by day, 20 to 25°C at night.
2. Top cuttings (5 to 8 cm long) are removed from the mother plant where they are individually numbered and brought to the laboratory (without storage) for testing.
3. The bottom 2-cm portion of the cutting is excised with a sterile knife and surface disinfected with a 10% bleach solution for approximately 10 min.
4. The surface-disinfested stem section is aseptically sliced into thin (2 mm) sections, placed into a test tube containing sterile nutrient broth, and incubated at room temperature (23 to 25°C) for 10 to 14 days.
5. After incubation, if the nutrient solution appears cloudy, due to bacterial or fungal contamination emanating from the plant's vascular system, the original numbered cutting is discarded. Only those cuttings that yielded no growth of any microorganism in the nutrient broth tube are retained (Figure 43.1). The original mother plant is also destroyed.
6. All cuttings that culture index free of pathogens are planted in the incubation block (at temperatures optimum for disease development) and retested after 3 to 4 months. If cuttings from these mother plants test clean, they are retained, whereas the original mother plant is again destroyed. This process is repeated at least one

Figure 43.1 Culture indexing test tubes containing nutrient broth plus surface-disinfested geranium stem sections. Tube containing cloudy medium (arrow) indicates growth of microorganisms and a positive finding. (From Trigiano, R. N. and Gray, D. J., Eds., *Plant Tissue Culture Concepts and Laboratory Exercises, First Edition*, CRC Press LLC, Boca Raton, FL, 1996.)

more time for a total of three times. At least one complete year of testing (three to four culture indexings) should be done to assure freedom from all systemic bacterial and fungal pathogens.

This method is easily adaptable to the monitoring of plants maintained in tissue culture on a year-round basis. It also provides a means to detect both pathogenic and nonpathogenic bacterial and fungal pathogens and/or contaminants. Theoretically, a single bacterial cell or fungal spore could be detected. A combination of sequential mother plant incubation and culture indexing repeated over at least 1 year reduces the probability of fungal and bacterial contamination to near zero. Culture indexing is also a simple, inexpensive test that does not require sophisticated laboratory equipment and supplies.

Virus indexing

Plant pathogenic viruses are submicroscopic particles consisting of nucleic acid (RNA or DNA) surrounded by a protein coat. Their effects on plants are varied, ranging from no symptoms, to necrotic or chlorotic ringspots and mosaic patterns on the foliage, to severe plant stunting and death (Agrios, 1988). These symptoms do not generally correlate with virus concentration or titer. For example, some viruses may be present at very high titer and not cause severe symptoms on susceptible hosts.

Virus indexing can accurately verify the cleanliness of a given variety and also determine if one or more viruses are present. Unfortunately, in many cases, one single testing method may not assure virus-free plants. Plant sampling methods, seasonality, testing method sensitivity, and specificity all affect virus indexing results.

The two major methods that are commonly used to detect plant pathogenic viruses in clean stock programs are serology and bioassays (biological indicator plant inoculation). Alternate indexing techniques include the use of nucleic acid probes, specific for the virus tested (Miller and Martin, 1988), or visualizing the replicative form of the viral RNA isolated from infected plants (Valverde et al., 1990).

Serology

The most common serological method used is enzyme-linked immunosorbent assay (ELISA) (Miller and Martin, 1988). ELISA has become the predominant technique used since it is reliable, sensitive, and relatively inexpensive on a per-sample basis. It is also more sensitive than bioassays, but does require some specialized laboratory equipment.

An ELISA detection technique is developed by administering purified virus, i.e., Tobacco Mosaic Virus (TMV), to rabbits, sheep, or mice to induce the formation of antibodies against TMV. After 6 to 12 weeks, the antibodies are isolated from the animal's blood serum and purified and prepared for the ELISA test. ELISA is performed in commercially available, 96-well, polystyrene plastic plates (Figure 43.2). First, the viral antibodies are coated onto the surface of each well. Next a macerated plant extract of the suspected virus-infected plant is incubated in the plate well. After incubation, the plant extract is removed and the well thoroughly rinsed of all remaining plant debris.

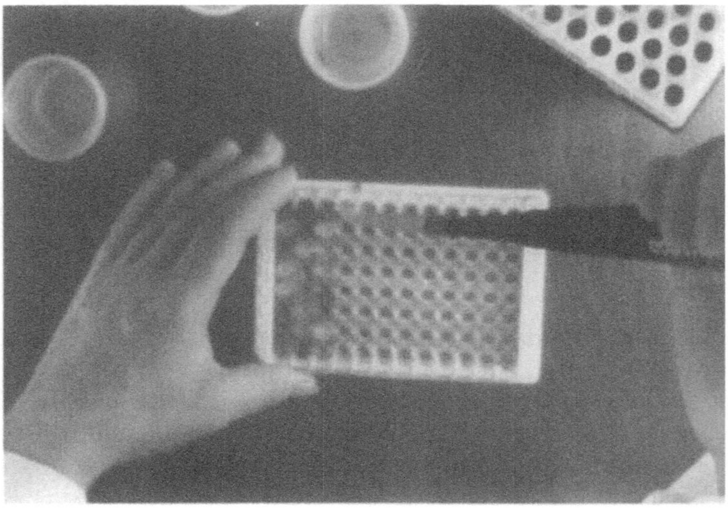

Figure 43.2 Enzyme-linked immunosorbent assay (ELISA) used for the detection of plant pathogenic viruses. (From Trigiano, R. N. and Gray, D. J., Eds., *Plant Tissue Culture Concepts and Laboratory Exercises, First Edition*, CRC Press LLC, Boca Raton, FL, 1996.)

During the initial incubation period, any virus particles present in the extract bind exclusively to the antibodies coating the plate well. Next, additional antibodies (antibody conjugate), which are attached to a marker enzyme such as alkaline phosphatase or horseradish peroxidase, are added to the well. During another incubation period, this antibody conjugate will bind to any virus particles present in the well that were previously bound by the coating antibody. This double binding, termed double antibody sandwich, is verified by adding an enzyme substrate that will indicate the presence of the antibody conjugate (attached to the enzyme). Many tests utilize colorless substrates that convert to a distinctive color after enzyme reaction. The presence and the amount of color development in each well signify the presence and relative concentration of the target virus.

Bioassays

Bioassays are accomplished by mechanically inoculating an indicator plant with sap extracted from the test plant. The species of indicator plant used is dependent on the types of viruses that are capable of infecting the test plant. The most common indicator plants used are members of the Solanaceae or tobacco family, including *Nicotiana tabacum* L., *N. benthamiana* Domin, and *N. clevelandii* Gray. Other popular and useful indicator plants are lambsquarters, *Chenopodium quinoa* Willd. and/or *C. amaranticolor* Coste and Reyn.

Infected plant sap is prepared by macerating 5 to 10 g of plant tissue in a specially prepared buffer solution at approximately a 1:10 dilution. This buffer usually contains anti-reducing agents to prevent the suspected virus(es) from being denatured during maceration of the plant tissue. The sap/buffer extract is gently rubbed onto the leaf surface of the indicator plant trying to inflict as little damage as possible. The intent of the inoculation is to wound the leaf cell wall slightly, which allows for virus entry into the cell with subsequent cell wall healing. Since viruses are obligate parasites, the indicator plant cells cannot be destroyed during inoculation. Often an abrasive such as carborundum or celite (diatomaceous earth) is added to the sap/buffer extract at inoculation to facilitate indicator plant cell wall wounding. Virus symptoms may develop on the indicator plant leaves anywhere from 3 to 21 days following inoculation. Typically, two major symptoms are observed. The first, a local lesion response, is typified by small necrotic or chlorotic lesions (1 to 2 mm in diameter) on inoculated leaves (Figure 43.3). This local lesion response may occur soon after inoculation. The second, a systemic response, is typified by mosaic or vein-clearing symptoms on newly emerging leaves. This systemic response usually appears 7 to 21 days after inoculation. As with all virus indexing methods, repeated testing of individual plants at different times of the year is necessary for accurate results.

Figure 43.3 Severe local lesion (white areas) response caused by an unknown plant pathogenic virus 16 days after inoculation into the bioindicator plant, *Chenopodium quinoa*. (From Trigiano, R. N. and Gray, D. J., Eds., *Plant Tissue Culture Concepts and Laboratory Exercises, First Edition*, CRC Press LLC, Boca Raton, FL, 1996.)

Virus elimination

The virus indexing procedures described above may reveal complete virus infection. If a positive reaction occurred in an ELISA test, the specific virus name, or class, is known and elimination procedures can be specifically designed. However, if virus symptoms appear on indicator plants, the identification of the specific virus may still be unknown.

If visual selection of virus-free clones is not possible, there are several techniques available to eliminate viruses, including the following: (1) exploiting erratic virus distribution within plants, (2) chemotherapy, (3) heat or thermotherapy, and (4) meristem tip culture (Matthews, 1991). Sometimes a combination of the above techniques may achieve greater success.

Erratic virus distribution

Some viruses are found erratically located within plants, and shoot tips from rapidly growing stems on these plants may be propagated and found to be virus free. Virus titer may change depending on the time of year. For example, during the summer months, high temperatures may slow virus replication and movement into the youngest growing points. Shoot tips from these rapidly growing shoots may be virus free.

Chemotherapy

Some limited success has been reported utilizing chemicals to cure plants of pathogenic viruses. Ribavirin, a potent viricide, has been reported to eliminate major viral pathogens from potato and other crops (Griffiths and Slack, 1988; Hollings, 1965). In most instances, chemotherapy is more effective if incorporated into a tissue culture medium that allows for the plant to uptake the chemical uniformly through the root system. Improved virus elimination can occur if chemotherapy is also combined with thermotherapy with subsequent meristem tip culture. The use of chemotherapy is difficult since virus metabolism is so closely integrated with plant metabolism.

Thermotherapy

The most commonly employed method for elimination of viruses from infected plants is a combination of thermotherapy and meristem tip culture (Horst and Klopmeyer, 1993). The success of thermotherapy depends on the target virus(es) inability to multiply and move readily within plants being exposed to air temperatures of 35 to 40°C (Grondeau and Samson, 1994). Successful treatment times may vary from weeks to months depending on the plant cultivar and the specific virus.

Thermotherapy is very effective in reducing virus titer in geraniums. Geraniums, in general, are not very heat tolerant, since plants will die after 5 to 6 weeks at 37°C. One technique that can overcome this problem is to place plants in a temperature-controlled growth chamber at lower temperatures (24°C by day, 21°C at night) initially. Each day, the temperatures are raised approximately 3°C until a final setting of 38°C by day and 33°C at night. This gradual acclimatization period allows for the plant to adjust to the higher temperatures. Plants can then be maintained under these conditions for as long as possible (3 to 4 weeks), after which time meristem tips are removed for tissue culture.

Meristem tip culture

The removal of meristem tips from plants grown under high temperatures is very effective in virus elimination. The excision of the meristem tip essentially excludes the virus from the "new" plant since this tip developed on the mother plant under the influence of virus-inhibiting high temperatures. The size of the meristem tip is critical for successful virus elimination. Obviously, the smaller the tip that is taken, the better chance for virus exclusion. However, as smaller tips are taken, meristem tip survival on tissue culture media is also reduced.

Meristem tip removal is accomplished by removing plant tissue surrounding the meristematic region and carefully excising the apical dome and one to two leaf primordia (Figure 43.4). Survival and growth are greatly improved if the tissue taken includes slightly expanded leaf primordia. After 1 to 2 weeks, the excised meristem tip expands, and the leaf primordia develop and differentiate into mature leaves. At this point, the young plant can be transferred into a tissue culture medium containing IBA or IPA to initiate root development. After 8 to 12 weeks, the young plants are ready for transfer back into the greenhouse. This is accomplished by carefully washing the agar medium from the roots and placing the plant into a soilless greenhouse medium under intermittent mist for 5 to 7 days. Misting is important to help harden off the plant by allowing the formation of cuticle on the leaf surfaces.

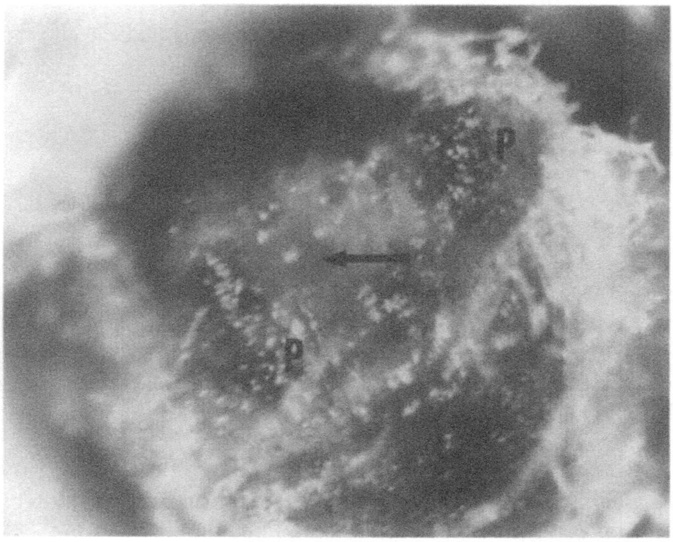

Figure 43.4 Chrysanthemum meristem tip (arrow) and two young leaf primordia (P). (From Trigiano, R. N. and Gray, D. J., Eds., *Plant Tissue Culture Concepts and Laboratory Exercises, First Edition*, CRC Press LLC, Boca Raton, FL, 1996.)

After establishment in the greenhouse, the plants are ready for virus indexing by ELISA and bioassay to verify that all viruses have been eliminated. If virus elimination was not successful, thermotherapy and meristem tip culture will need to be repeated.

Clean plant production

The development, maintenance, and production of certified stock require the strict adherence to the following four fundamental principles of clean plant production (Figure 43.5).

Annual renewal. At the completion of each production year all old stock should be destroyed, the greenhouses disinfected, and the stock base renewed with clean plants.

Unidirectional flow. During annual renewal of the stock base, the new stock should always come from a cleaner source. Cuttings from old stock should never be used to reestablish a new stock base.

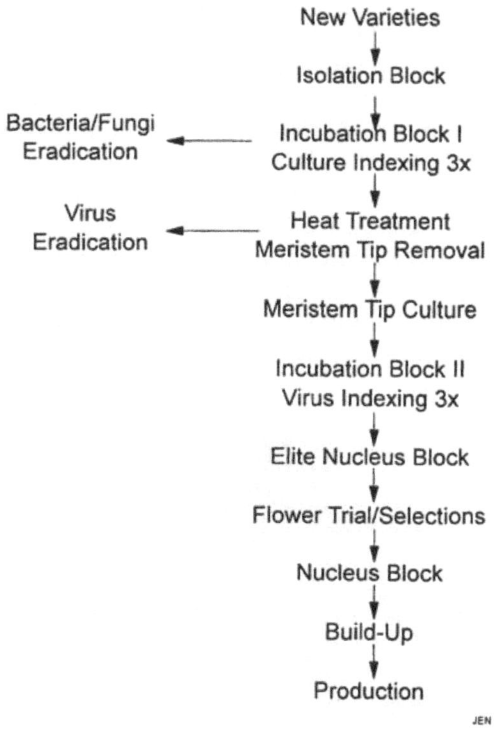

Figure 43.5 Flow chart for the production of bacterial, fungal and virus-free geraniums. (From Trigiano, R. N. and Gray, D. J., Eds., *Plant Tissue Culture Concepts and Laboratory Exercises, First Edition*, CRC Press LLC, Boca Raton, FL, 1996.)

Repeated testing. Indexing of certified stock needs to be repeated over a period of at least 1 year. This will allow for maximum opportunity to detect plant pathogens under different environmental conditions.

Clonal selection. After intensive indexing procedures including tissue culture maintenance, the certified stock base should be flower trialed under controlled conditions. This will allow for selection of superior clones that flower earlier and more often on a plant with an excellent habit. Since most vegetatively propagated ornamental varieties are protected by a plant patent, clonal selection criteria should also closely approximate the variety description outlined in the patent application.

Summary

Vegetatively propagated crops are constantly threatened by plant disease. Since their method of reproduction is amenable to spread of bacterial, fungal, and viral diseases from generation to generation, the development of reliable clean plant production procedures is required. These procedures include effective elimination techniques and precise detection techniques to verify that all plants are clean. Finally, since all plants are just as susceptible to these pathogens after elimination, careful and controlled conditions must be maintained to successfully produce these commercially important crops.

Literature cited

Agrios, G.N. 1988. Plant diseases caused by viruses. In: *Plant Pathology.* Academic Press, New York, pp. 622–702.

Griffiths, H.M. and S.A. Slack. 1988. Potato virus elimination by heat and ribavirin. *Phytopathology* 52:1230–1241.

Grondeau, C. and R. Samson. 1994. A review of thermotherapy to free plant materials from pathogens, especially seeds from bacteria. *CRC Crit. Rev. Plant Sci.* 13:57–75.

Hollings, M. 1965. Disease control through virus-free stock. *Annu. Rev. Phytopathol.* 3:367–396.

Horst, R.K. and M. J. Klopmeyer. 1993. Controlling viral diseases. In: *Geraniums IV.* White, J., Ed. Ball Publishing, West Chicago, IL, pp. 289–292.

Matthews, R. E. F. 1991. Control measures. In: *Plant Virology.* Academic Press, New York, pp. 596–634.

Miller, S.A. and R. R. Martin. 1988. Molecular diagnosis of plant disease. *Annu. Rev. Phytopathol.* 26:409–432.

Oglevee-O'Donovan, W. 1993. Culture indexing for vascular wilts and viruses. In: *Geraniums IV.* White, J., Ed. Ball Publishing, West Chicago, IL, pp. 277–286.

Raju, B.C. and C. J. Olson. 1985. Indexing systems for producing clean stock for disease control in commercial floriculture. *Plant Dis.* 69:189–192.

Valverde, R. A., S. T. Nameth, and R. L. Jordan. 1990. Analysis of double stranded RNA for plant virus diagnosis. *Plant Dis.* 74:255–258.

chapter forty-four

Culture indexing for bacterial and fungal contaminants

Michael E. Kane

Rapid production of specific pathogen-eradicated plants is a fundamental goal of the micropropagation process (see Chapter forty-three). Whether for commercial in vitro propagation or more fundamental metabolic, genetic, or morphogenetic research, it is desirable to establish and maintain plant cultures that are also free of nonpathogenic microbial contaminants (Cassells, 1997; Knauss, 1979). The surfaces of plants are naturally populated with a diverse microflora consisting of bacteria, fungi, yeast, and other organisms. A primary objective of Stage I is the elimination of this microflora and the subsequent establishment of aseptic cultures (see Chapter eight). This is usually accomplished through surface disinfecting explants (e.g., meristem tips, shoot tips, stem, or leaf tissue) with alcohol and/or sodium hypochlorite prior to culture inoculation (see Chapter three). Once explants are established in vitro, it is essential that cultures be indexed (screened) for the presence of microbial contaminants. Contaminated cultures may exhibit no symptoms, variable growth, regeneration, reduced shoot proliferation, rooting, or poor survival (Leifert et al., 1989). Many culture contaminants are not pathogenic to plants under field conditions but become pathogenic in vitro often due to the release of toxic secondary metabolites into the medium (Leifert et al., 1994).

Research and commercial tissue culture laboratories often only use visual methods to index for culture contamination (the so-called *EBD* or eyeball determination). There are several important reasons why screening only for visible microbial contamination is not adequate. Visible growth of microbial contaminants may be suppressed in plant culture media (Leifert and Waites, 1992). Although many fungal contaminants quickly become visible, bacterial infections may be latent and difficult to detect (Tanprasert and Reed, 1997). Microbial contaminants such as *Methylobacteria* have evolved close biochemical relationships with the plant epidermis in vivo and will not grow independently on standard plant culture media (Holland and Polacco, 1994).

Indexing for microbial contaminants is usually accomplished by inoculating tissue sections or intact shoots into enriched selection medium that will promote the visible growth of bacteria, filamentous fungi, yeast, or other contaminants (see Chapter forty-three). Since secondary culture contamination can occur as a result of poor aseptic technique or contaminant vectors such as mites, cultures should be routinely reindexed.

Many specific procedures are available to screen for culture contaminants. Knauss (1976) screened for the presence of systemic bacteria and fungi by culturing 0.5- to 1.0-mm-thick

0-8493-2029-1/00/$0.00+$.50

internodal and nodal stem cross sections from Stage I plantlets in four enriched microbiological media that promoted microbial growth. This procedure was effective in establishing cultures of foliage plants such as *Dieffenbachia* that were free of specific pathogens such as *Xanthomonas dieffenbachiae* and *Erwinia chrysanthemi*. Other procedures use a single indexing medium to screen for culture contamination in explants following surface disinfestation (Leifert et al., 1989; Viss et al., 1991).

The primary objective of this exercise is to illustrate indexing procedures for cultivable contaminates using a single indexing medium. This exercise will familiarize students with the following concepts: (1) importance and limitations of culture indexing; (2) selection of indexing media; and (3) tissue sampling and inoculation techniques.

General considerations

Indexing media selection

Without specific knowledge of the actual microbial contaminants present, limitations in detecting contamination can occur if inappropriate indexing media are used. One approach is to use several highly enriched indexing media such as: (1) Sabouraud dextrose medium; (2) yeast extract dextrose broth; and (3) AC broth (Knauss, 1976). Sabouraud dextrose medium is formulated for the growth of fungi including yeast, filamentous fungi, and aciduric microorganisms. Yeast extract dextrose broth medium is used to stimulate bacterial growth, whereas AC broth is a sterility test medium for a variety of microorganisms. Contaminated cultures will index positive in all three media about 98% of the time (M.E. Kane, unpublished). These and many other sterility test media are manufactured by Difco Laboratories (Detroit, MI) and distributed through Fisher Scientific (Norcross, GA). For this laboratory exercise it is probably best to select a single broad-spectrum sterility test medium.

Several media have been formulated to also sustain the plant tissue being indexed. This increases efficiency by allowing tissues, such as shoot tips, first to be placed in the selection medium and then transferred to fresh tissue culture medium only if they index as negative. Murashige and Skoog (1962) medium enriched with yeast extract, peptone, or glucose may serve as an adequate indexing medium (Tanprasert and Reed, 1997). A highly effective and recommended indexing medium is Leifert and Waites Sterility Test Medium (Leifert et al., 1989). This medium both sustains plant tissue and promotes growth of a broad range of latent bacterial contaminants with the exception of some *Xanthosomonas*, *Methylobacterium*, and *Hyphomycrobium* species and all mycoplasma-like organisms and is available from Sigma Chemical Company, St. Louis, MO (L-1906).

Indexing medium preparation and plant cultures

For this exercise, both semisolid and liquid Leifert and Waites Sterility Test Medium will be used. The semisolid medium is prepared by the addition of 10 g/L agar. Medium is prepared without pH adjustment and dispensed as 10 ml into 20-ml glass scintillation vials covered with autoclavable screw caps (Figure 44.1c). Scintillation vials are both shallow, which facilitates inoculation, and reusable. The cardboard tray in which vials are shipped makes a convenient rack to hold 100 vials during incubation. After autoclaving, the agar-supplemented medium is cooled and solidified as 45° slants (Figure 44.1c). Although indexing of Stage I shoot cultures is described in this exercise, almost any type of plant culture can provide tissue for indexing. This indexing exercise can be readily incorporated into any of the laboratory exercises outlined in this text that use a surface-disinfecting procedure.

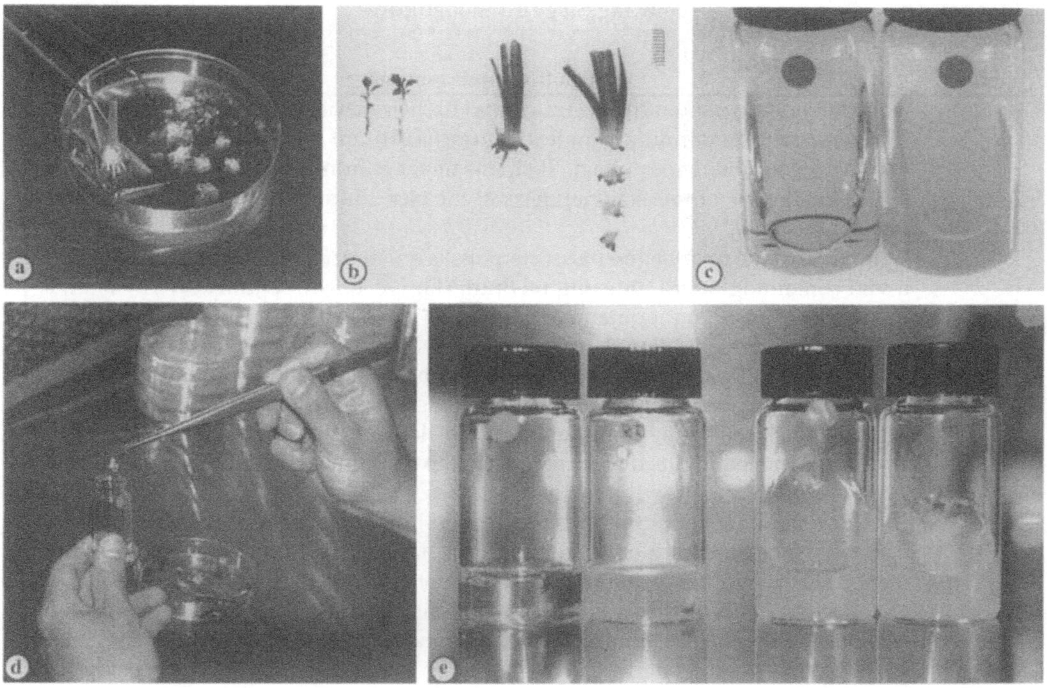

Figure 44.1 Culture indexing for cultivable contamination. (a) Aseptic technique is use to obtain tissue sections from the Stage I shoot culture. (b) Shoots with upright stems (left) are sectioned into nodal segments. The shoot tip is either cultured in the indexing medium or transferred to fresh Stage I medium. Cross-sectional stem disks are made from shoots with basal meristems (right). Scale bar = 1 cm. (c) Indexing medium is prepared as a liquid and agar-solidified slants. (d) Several tissue samples are directly placed in the liquid indexing medium. (e) Clouding of inoculated liquid indexing medium is a positive indication of the presence of cultivable contaminants in the tissue sample. Development of colonies on the surface of the medium and/or halo in the medium where the tissue sample was stabbed indicates culture contamination.

Exercise: Indexing Stage I shoot cultures

Materials

The following are needed for each student or student team:

- Five established Stage I shoot cultures
- Seven scintillation vials each of liquid and agar-solidified Leifert and Waites Sterility Test Medium
- Five culture vessels containing fresh Stage I medium

Follow the protocol in Procedure 44.1 to complete this experiment.

Anticipated results

Rapid clouding of inoculated liquid indexing media is a positive indication of the presence of cultivable contaminants in a tissue sample (Figure 44.1e). The presence of contaminants on agar-solidified indexing media is usually confirmed by development of colonies on the surface of the medium and/or development of a halo in the medium where the tissue

Procedure 44.1
Indexing Stage I shoot cultures.

Step	Instructions and comments
1	Aseptically remove shoots from culture and place in sterile petri dish (Figure 44.1a). If a shoot consists of multiple nodes (Figure 44.1b), cut into single node segments leaving the terminal apex intact. If shoots develop from a basal meristem (Figure 44.1b), make thin cross-sectional disks at the base. Be sure not to cut through the basal meristem.
2	Slightly crush the node segments or stem disks and inoculate several segments into a vial containing liquid indexing medium (Figure 44.1d). Lightly swirl vial and replace cap but do not tighten completely.
3	Inoculate the agar-solidified slant by first gently stabbing the tissue into the medium at the base of the slant (try not to leave a bubble in the stabbed medium). Slowly lift tissue out of the medium and then lightly streak (drag) the tissue up along the medium surface leaving it at the top of the slant (Figure 44.1e). This stabbing procedure inoculates the deeper regions in the medium where lower oxygen levels might promote the growth of some microbes.
4	Transfer remaining shoot tip onto fresh Stage I medium (or inoculate it into the same vial inoculated in step 3). For reference, label both scintillation vials with the same culture number as the transferred shoot culture.
5	Repeat steps 1 to 5 for the four remaining Stage I cultures.
6	Do not inoculate the remaining vial of liquid and semisolid medium and label each as "Control."
7	Inoculated indexing medium and control vials are maintained in the dark at 22 to 30°C for 3 weeks and then screened for visible contamination. Autoclave vials with contamination before disposing.

sample was stabbed (Figure 44.1e). Detection of slow-growing contaminants is usually facilitated on agar-solidified media. Students should make close observations for the presence of slowly developing colonies on the surface of the indexed tissues. Control vials should not display microbial growth. It should also be noted that systemic contamination does not occur uniformly throughout the plant and may be localized in specific regions or tissues that are not indexed. This results in false-negative indexing. It is important that cultures be reindexed after a period of growth. The occurrence of contaminated cultures provides an opportunity to discuss what to do with them. Further options to produce negatively indexed cultures include redisinfestation and isolation of meristem tips or larger shoot tips and the use of antibiotics (Falkiner, 1997).

Many indexing media will promote the growth of diverse microorganisms. However, growth of many microbial species will not be supported on these media. Because of this uncertainty, it is recommended that indexed negative plant cultures should not be described as being aseptic, sterile, or free of contaminants (Leifert et al., 1994). These cultures can only be described as being indexed negative or free of detectable (cultivable) contaminants if such statements are qualified by a description of the microbiological indexing methods used (Leifert et al., 1994).

Questions

- Why is it important to stab and then streak the Stage I tissue sample across the surface of the agar-solidified indexing medium?
- How is it possible to have a contaminated Stage I shoot culture that indexes negative for contamination?
- Why is it recommended to index a culture more than once?

Literature cited

Cassells, A.C. 1997. Pathogen and microbial contamination management in micropropagation — an overview. In: *Pathogen and Microbial Contamination Management in Micropropagation*. A.C. Cassells, Ed., pp. 1–13. Kluwer Academic Publishers, Dordrecht.

Falkiner, F. 1997. Antibiotics in plant tissue culture and micropropagation — what are we aiming at? In: *Pathogen and Microbial Contamination Management in Micropropagation*. A.C. Cassells, Ed., pp. 155–160. Kluwer Academic Publishers, Dordrecht.

Holland, M.A. and J.C. Polacco. 1994. PPFMs and other covert contaminants: is there more to plant physiology than just plant? *Annu. Rev. Plant Physiol. Mol. Biol.* 45:197–209.

Knauss, J.F. 1976. A tissue culture method for producing Dieffenbachia picta cv. `Perfection' free of fungi and bacteria. *Proc. Fla. State Hort. Soc.* 89:293–296.

Knauss, J.F. and M.E. Knauss. 1979. Contamination of plant tissue cultures. *Proc. Fla. State Hort. Soc.* 92:341–343.

Leifert, C., W.M. Waites, and J.R. Nicholas. 1989. Bacterial contaminants of micropropagated plant cultures. *J. Appl. Bacteriol.* 67:353–361.

Leifert, C. and W.M. Waites. 1992. Bacterial growth in plant tissue culture media. *J. Appl. Bacteriol.* 72:460–466.

Leifert, C., C. E. Morris, and W. M. Waites. 1994. Ecology of microbial saprophytes and pathogens in tissue culture and field grown plants: reasons for contamination problems in vitro. *CRC Crit. Rev. Plant Sci.* 13:139–183.

Murashige, T. and F. Skoog. 1962. A revised medium for rapid growth and bioassays with tobacco cultures. *Physiol. Plant.* 15:473–497.

Tanprasert, P. and B. Reed. 1997. Detection and identification of bacterial contaminants of strawberry runner explants. In: *Pathogen and Microbial Contamination Management in Micropropagation*. A.C. Cassells, Ed., pp. 139–143. Kluwer Academic Publishers, Dordrecht.

Viss, P.R., E.M. Brooks, and J.A. Driver. 1991. A simplified method for the control of bacterial contamination in woody plant tissue culture. *In Vitro Cell Dev. Biol.* 27P:42.

Appendix I

List of suppliers

Appendix I

List of some suppliers, addresses, and toll-free telephone numbers*

Amresco Inc.
30175 Solon Industrial Parkway
Solon, OH 44139-4300
(800) 829-2805

Carolina Biological Supply Co.
P. O. Box 6010
Burlington, NC 27216-6010
(800) 334-5551

Electron Microscopy Sciences
321 Morris Road, P. O. Box 251
Fort Washington, PA 19034
(800) 523-5874

Fisher Scientific
3970 John's Creek Court
Suwanee, GA 30024
(800) 766-7000

Gibco BRL
Gaithersburg, MD 20897-8406
(800) 828-6686

Lab Safety Supply
401 S. Wright Road
Janesville, WI 53546
(800) 356-0783

Select Tech
Gaithersburg, MD 20897-8406
(800) 828-6686

Sigma Chemical Company
P. O. Box 14508
St. Louis, MO 63178
(800) 325-3010

Thomas Scientific
P. O. Box 99
Swedesboro, NJ 08085-6099
(800) 345-2100

VWR Scientific Products
1750 Stoneridge Drive
Stone Mountain, GA 30083
(800) 234-5227

Ward's
P.O. Box 92912
Rochester, NY 14692-9012
(800) 892-3583

* Reference to the suppliers listed above does not imply endorsement of them nor does it imply criticism of those not listed.

0-8493-2029-1/00/$0.00+$.50

Index

A

AB medium, 323–326
ABA, *see* Abscisic acid
Abies spp., *see* Firs
Abiotic stress, 393
Abscisic acid (ABA), 13, 28, 186, 240
AC Broth indexing medium, 90, 93, 94, 428
Acclimatization
    chrysanthemum, 142
    commercial laboratory production, 412–413
    lilac, 116
    miniature roses, 122–123
    orchardgrass, 199
    peanuts, 219–220
    potato, 107
    stage IV micropropagation, 84
    Syngonium podophyllum, 92–94
    Torenia fournieri, 162
    watermelon, 152, 154, 156
Acetocarmine stain, 289, 293–294
Aceto-orcein stain, 244
Act in trans, 305
Activated charcoal, 27, 228
Active collections, 339–340
Adorn, 139, 142–145, *see also* Chrysanthemum
Adventitious apical meristems, 125, *see also* Apical meristem
Adventitious bud
    organogenesis, 130–133
    micropropagation, 101
Adventitious roots, 133
Adventitious shoots
    micropropagation, 139
    organogenesis and darkness effect, 155–157
    regeneration potential, 150–153
    stage II micropropagation, 82
AFLP, *see* Amplified fragment length polymorphism
Agar, 30, 50–51, 287
AgarGel, 112
Agarose, 30, 251, 273
Age, plants and tissue culture, 24
Agglutination, 255–256
Agnet, 7
Agrobacterium tumefaciens
    genetic transformation
        chrysanthemum leaf explants, 323–327
        mediation and protoplast use for plant improvement, 259
        plants, 298–299, 300, 305, 307–309
    transformation of carrot hypocotyls and somatic embryogenesis, 315–319
    tumor induction, 373
Ajmalicine, 356
Ajuga reptans, *see* Bugleweed
Alcian blue stain, 49–50
Alcohol, 22, *see also* Ethanol
Aldehydes, 46
Alfalfa, 128, 186, 390
Alginic acid, 202–203
Aluminum foil, 35
Amino acids, 27, 200
4-Amino-3,5,6-trichloro-2-pyridinecarboxylic acid, *see* Picloram
Ammonium nitrate, 252, 254
Amplified fragment length polymorphism (AFLP), 390, 393–394
Anabasine, 355
Analysis of variance (ANOVA)
    completely randomized design, 63, 64
    darkness effect on adventitious shoot organogenesis, 156
    factorial experiments, 66
    multiple comparison tests, 68
    statistical analysis, 62
    sucrose effect on embryo growth and regression analysis, 69
Androgenesis, 285, 287–288
Aneuploid regenerants, 263
Angiosperms, 12, 39, 176, 240
Annual renewal, 423
ANOVA, *see* Analysis of variance
Anther culture
    factors affecting, 286–287
    haploid culture, 285, 292–296
Antheridia, 398, 402, 404
Anthocyanin pigment
    production in cell culture, 356–357
    Bugleweed, 358, 361–371
    selection of hybrid cell lines and protoplast use, 258
Antibiotic resistance, 258, 297
Antibiotics, 259, *see also* Individual entries
Anticlinical division, 133
Antioxidants, 80
APHIS guidelines, 310
Apical dominance, 76–77, 114
Apical meristem, 75–79, 90–91
Apomictic embryos, 175
Apospory, 399–405
Arabidopsis spp., 135
Araceae, *see Dieffenbachia*
Arachis hypognaea, *see* Peanut
Arbitrary signatures from amplification profiles (ASAP), 393
Arch McClean, 116, *see also* Lilac

Archegonia, 398, 403
Aronia arbutifolia, 78, 82
Arrowhead vine, *see Syngonium podophyllum*
ASAP, *see* Arbitrary signatures from amplification profiles
Aseptic cultures, 80–81, 160–161, 250
Asexual plant propagation, 97, 125
Assay buffer, 377, 378
Asteraceae, *see* Somatic embryogenesis, cineraria flower receptacles
Asymmetric hybrids, 263
Asynchronous growth, 185
AT127, 213, *see also* Peanuts
ATPase activity, 374
Autoclave, 34–35
Autofluorescence, 251, 282
Autotetraploidy, 404
Auxin
  embryogenic culture initiation, 180
  history of plant tissue culture, 12
  melon culture media, 206
  micropropagation, 81–82, 84
  organogenesis, 129, 169–170
  protoplast growth and development, 254
  somatic embryogenesis, 218, 240, 245
  photoperiod effect, 216–218
  tissue culture media, 27–29
Axillary buds, 111–117
Axillary shoots
  commercial laboratory production, 411–412
  micropropagation of miniature roses, 121–122
  propagation, 78–79, 81–83

B

B vitamins, 12
B5 medium, 311, 317–318
2B5 medium, 317
BA, *see* Benzyladenine
Backcrossing, 263
Backward step procedure, 68
Bacteria, 379, 418–419
Bacteriophages, 297
Banana, 387
Barley, 388
Barriers, 300
Basal cotyledonary explants, 207–208
Basal medium (BM)
  chrysanthemum growth, 140
  cineraria, 225–226, 227
  miniature roses, 120
  Torenia fournieri,160–161
  yellow polar, 236
Base collection, 337, 340
Beads, vitrification process, 351
Benzyladenine (BA)
  direct somatic embryogenesis 225
  embryogenic culture initiation, 180
  factorial experiments, 66–67
  genotype effect and shoot formation, 154
  media for tissue culture, 28–29
  micropropagation, 106, 114
  stages, 81–83
  organogenesis, 131, 152, 169–171
  direct, 141–143, 146
  protoplast growth and development, 254
  vitrification, 346
Benzylaminopurine, 161
Berberine, 356
Betalains, 356
Binary vector system, 299–300, 309–313
Binucleate microspore, 289, *see also* Microspores
Bioassay
  bacterial isolates for pathogenicity, 376, 379
  K^+/H^+ exchange response, 379–380
  variation identification in tissue culture, 393
  virus indexing, 419–421
Biolistic PDS 1000/He, 329
Bioreactors, 240, 357
Biotechnology, 14–15
Biotin, 26
Biotransformation, 357, *see also* Secondary products
Bismarck brown stain, 49–50
Black and white film, 58
Blight, 255
BM, *see* Basal medium
Branching, 415, *see also* Micropropagation
Brassica campestris see Rape
Brightfield microscopy, 45
Broad beans, 388
Browning, 208
Bryophytes, 397
BT proteins, 302
Buds, 128, 294, 414, *see also* Individual entries
Budding, 119, 282
Bugleweed, 358, *see also* Anthocyanin pigment; Pigment production
Buoyant density, 249
Burgandy glow, 368–369, *see also* Bugleweed

C

Cabbage, 334, *see also* Gene gun, particle bombardment
Calcium, 252, 254, 351–352
Calcium washing medium, 269, 272–274
Calcium phosphate, 300–301
Calcofluor White stain
  protoplast isolation, 251
    from leaves, 278, 281–282
    tobacco culture, 270–271
  regeneration of plants from protoplasts, 251
Calliclone, 387
Callus
  cultures, nutrition
    general considerations, 40
    growth in serially composed media, 42–44
    growth potential, 40–42
  diploid and androgenesis, 288
  genetic transformation, 325

media for tissue culture, 24
morphology and protoplast fusion, 275
organogenesis, 126–128
Tortenia fournieri, 162, 164
pigment production, 368–370
plant tissue culture, 4
history, 12–13
regeneration, 170–173
suspension cells, 375–376
Callus induction medium
direct/indirect organogenesis in *Torenia fournieri*, 160, 164
pigment production of Bugleweed, 365, 367–368
Calyx, 293–294, 296
Camera, 35 mm single reflex, 56–57
Camile, 100, *see also Dieffenbachia*
Campthotheca acuminata, 312
Canalization, 128
Cancer, 14
Cantaloupe, 183–184, *see also* Dicots; Somatic embryogenesis, melon
Carbenicillin, 311, 316–317, *see also* Antibiotics
Carbon, sources, 254, 361, 363–365
Carrot, 40–41, 175, 263, *see also* Genetic transformation
Carryover effects, 83
Carya illinoinensis, *see* Pecan
Casein hydrolysate
embryogenic cultures of orchardgrass, 192, 200
media for tissue culture, 27
somatic embryogenesis, 236–237, 241
Casting specimens, 47–48
Categorical data, 69–70
Categorization, 210–211
Cefotaxime
genetic transformation
chrysanthemum, 322–323, 325
tobacco, 309–310, 312
regeneration of transgenic plants, 326–327
use for contaminated potato plants, 265, 267
Celastris, 115
Cell death, 381–383
Cell division, 28, 180, 252
Cell lines, high-producing, 356
Cell regeneration system, 205
Cell sorter, 259
Cell wall, 249, 251–252, 254
Cellulase, 249, 277–279
Central American pine, 130–133
Centrifugation techniques, 249, 279, 283
Ceramic tiles, 36
Ceratopterus richardii, 397–405
Cereal grains, 181, 183, 300, *see also* Embryogenic culture, orchardgrass
C-Fern, 397–398, *see also Ceratopterus richardii*
C-Fern medium, 399–400
Cfu, *see* Colony-forming units
Charles Joly, 111, *see also* Lilac
Chemoprotective compounds, 356
Chemotherapy, 422
Chimeras, 111
Chi-square, 70
Chlorophyll, 251, 282
Chloroplasts, 259
Chromosomes, 263, 388, *see also* Individual entries
Chrysanthemum morifolium, *see* Chrysanthemum
Chrysanthemum, 394, 418, *see also* Direct organogenesis, chrysanthemum; Genetic transformation; Protoplasts, isolation from chrysanthemum and orchardgrass leaves
Cindy cultivar, 223
Cineraria, *see* Somatic embryogenesis
Citrullus lanatus, *see* Watermelon
Citrus spp., 393
Clean plants, 414, 423–424
Cleavage proembryogenesis, 245
Clonal orchards, 239
Clonal propagation, 76, 239
Clonal repositories, 339
Clonal selection, 424
Clones, 133–134, 192
Coconut milk, 27, 43, 254
Cocultivation, 259
Cointegration, 299–300
Colchicine, 289
Cold pretreatment, 294, 296
Cold storage, 408–409
Coleoptile, 182–183, 197, 198
Colony-forming units (Cfu), 324, 379
Color slide film, 57–58
Colorado potato beetle, 257
Coloration, 208
Commercial laboratory production
facility, 407–409
process, 409–413
reasons for micropropagation, 414–415
Compacta, 100, *see also Dieffenbachia*
Compatible pathogens, *see also* Incompatible pathogens
cell death response, 381–383
hypersensitive response, 373, 376–377
Competency, 122, 127–129
Completely randomized design (CRD)
factorial experiments, 66
statistical analysis, 62–63
explant tissue effect on somatic embryogenesis, 215
isolation of protoplasts, 279
regeneration potential, 152
Complex organic compounds, 27, *see also* Organic compounds
Composition, photography of plant/tissue, 58–60
Compound microscopes, 56–57
Concanavalin A, 255
Concentration units, 31–34
Condensation, 402
Conifers, 181–183, 188
Conjugation-proficient plasmid, 308
Constitutive promoter, 302, 309–313, *see also* Promoters
Contamination
direct/indirect organogenesis, 160, 164
media for tissue culture, 30

micropropagation
donor plant selection, 79
Dieffenbachia, 100
lilac, 113, 114
miniature roses, 121
shoot culture of *Syngonium podophyllum*, 91–92
somatic embryogenic cultures, 194, 207, 209
stored culture media, 37
protoplasts, 265
Controlled relative humidity dehydration (CRHD), 70, 71
Conventional breeding, 239
Conversions, molar solutions, 33
Convolvulus spp., 127–128, 135
Cooling, 349
rate, 341–342
Coplin jar, 22
Corn sugar, 26
Corn, 15, 178
Cornus florida, *see* Dogwood
Corolla, 293–294, 296
Corrective lens, 57
Cotyledons, 182–185, 213
CPW 13M medium, 269, 272–273
Cranberry, 357
CRD, *see* Completely randomized design
CRHD, *see* Controlled relative humidity dehydration
Crimson Sweet, 153–154, *see also* Watermelon
Critical point drying, 51, 53
Crops, 15, 263, 417
Crown gall, 12, 298
Cryogenic storage, 338
Cryopreservation, 340
Cryoprotectants, 340–342
Cryotubes, vitrification of shoot tips
encapsulation–dehydration, 351–352
solution-based method, 346, 348–349
Cucumis melo, 58–59
Cucurbitaceae, *see* Somatic embryogenesis, melon
Cultigen, 167
Cultivable contamination, 429
Culture filtrate, 392
Culture indexing, 418, 428–430, *see also* Indexing
Culture initiation, commercial, 410–411
Culture media
AB, 323–326
AC Broth indexing, 90, 93–94, 428
assay buffer, 377–378
B5, 311, 317–318
2B5, 317
basal
chrysanthemum growth, 140
cineraria, 225–227
miniature roses, 120
Torenia fournieri, 160–161
yellow poplar, 236
calcium washing, 269, 272–274
callus induction, 160, 164, 365, 367–368
C-Fern, 399–400
CPW 13M, 269, 272–273
Dieffenbachia, 99
dilution, 347–349
embryo development, 206–211
embryo initiation, 206–207, 209–210
enzyme, 265–267, 270–271
filter-sterilized, 267
flotation
protoplast fusion, 268, 272–274
protoplast isolation and culture, 270–271
regeneration of plants from protoplasts, 251
forcing solution, 411
Fusogen PEG 22.5, 269, 272–274
Gamborg B-5, 24–25, 376
Gel-gro, 214
germination, 151, 154, 156, 228
GUS-assay buffer, 310
half-strength Litvay (1/2 LV), 241
indexing, 90, 428–429
induction
hormone-containing, 130–131
somatic embryogenesis, 232, 234, 236–237
isolation, 279
Kings B agar (KB), 376, 378
LB, 307–308
Leifart and Waites Sterility test, 428–429
Linsmaier and Skoog, 24
M9, 308
maturation, 228
melon, 206
microtuber induction, 105–107
Murashige and Skoog (MS), *see* Murashige and Skoog (MS) medium
N6, 287
Nitsch and Nitsch, 24–25, 287, 292
node culture, 105–107
pigment induction, 367
pigment production, 362, 368, 371
plasmolysis, 273
potassium phosphate buffer, 52
potato, 265, 267–269
predigestion, 279
preplasmolysis, 267, 270, 274
propagation, 266–267
protoplast
fusion, 268, 272–275
isolation, 265–266, 270–271, 277
growth and development, 252, 254
regeneration, 151–152, 154, 156
rinse, 268, 270, 272, 274
rooting, 151, 154, 156
rose establishment/multiplication, 120, 122
Sabouraud dextrose, 428
Schenk and Hildebrandt (SH)
direct somatic embryogenesis, 225, 226–228
embryogenic cultures, 192, 201–202
in vitro plant pathology of suspension cells, 375
protoplast isolation, 277
tissue culture, 24–25
secondary, 218–219
SEM culture, 90–94

serially composed, 42–44
SH30, 194–197, 199–200
SH-30C, 192, 200
shoot elongation, 151, 154, 156
SMM culture, 90–94
steps in preparation, 33–34
tobacco, 265–266
TR, 310, 312
protoplast fusion in potato, 268, 272, 275
White's, 24–25
woody plant medium (WPM)
media for tissue culture, 24–25
micropropagation, 112–113
pigment production, 362, 371
Yeast dextrose, 428
Culture rooms, 23, 408
Culture-induced variants, 250
Cuttage, 134
Cuttings, 125
Cybrids, 263–264
Cyclosis, 282
Cytohistology, 133
Cytokinin, *see also* Individual entries
anther culture media, 287
apical dominance role, 77
embryogenic culture initiation, 180
formation of shoots and roots, 143–144
history of plant tissue culture, 13
media for tissue culture, 27–29
melon culture media, 206
micropropagation
Dieffenbachia, 100
potato, 104
stage I, 81
stage II, 82
organogenesis
auxin ratio, 129
shoot from leaf explants, 169–170
protoplast growth and development, 254
somatic embryogenesis, 240, 245
Cytoplasmic streaming, 245

D

2,4-D, *see* Dichlorophenoxyacetic acid
Dactylis glomerata, see Orchardgrass
DAF, *see* DNA amplification fingerprinting
Dark pretreatment, 155–157, *see also* Light conditions
Datura spp., 14, 285, 291
Daucus carota, see Carrot
Dedifferentiation, 126–128, 134
Degrees of freedom (DF), 62–63, 65–66
Dehydration, 47, 70–71, 186
Deionization, 22
Dendranthema grandiflora, see Chrysanthemum
Depth of field, 56, 58–59, 60
Desiccation, 351–352
Determinant growth, 76
Determination, 127, 128, *see also* Organogenesis
Developmental biology, 13
Developmental process, 126–127, *see also* Organogenesis
Devitrification, 342, *see also* Vitrification
DF, *see* Degrees of freedom
DH5 alpha bacterial host, 333
Dicamba, 180, 192
2,4-Dichlorophenoxyacetic acid (2,4-D)
embryogenic culture initiation, 180
explant size/source and pigment production, 367
genetic transformation, 316
in vitro plant pathology of suspension cells, 376
organogenesis, 128
direct/indirect, 161
regeneration and somaclonal variation, 171
somatic embryogenesis
cineraria, 225
conifers, 241
peanut, 215–219
Dichroic mirror, 282
Dicots, 75, 181–183, *see also* Monocots
Dieffenbachia, 97–101
Dielectrophoresis, 256
Differentiation
organogenesis, 127, 129–133
sexual types and apospory, 402
tissues and nonzygotic embryo development, 183
Digestive enzymes, 250
Dihaploid plants, 285, 289, 291
Dilution medium, 347–349
Dimethyl sulfoxide (DMSO), 341, 347
Direct organogenesis, chrysanthemum, 127, 131, *see also* Direct/indirect organogenesis; Organogenesis
general considerations, 139–140
pulse treatment with plant growth regulators and initiation/growth of shoots, 146–147
shoot/root
differences in genotype, 140–143
plant growth regulator effects, 143–146
Direct secondary embryogenesis, 175, 181
Direct transformation systems, 300–301
Direct/indirect organogenesis, *Torenia fournieri* shoots, *see also* Direct organogenesis; Organogenesis
callus formation, 162–164
explant preparation and basal medium, 160–161
growth of plants, 160
leaves, 161–162
Disarmed T-DNA, 300
Disease-free status, 340
Disease resistance, 255, 390–392
Disinfectants, 37–38
Disinfesation, 250, 288, 410
Disposable blades, 48, 51
Dissecting microscopy, 56
Dissection, specimens, 51–52
DMSO, *see* Dimethyl sulfoxide
DNA, 14, 260–261, 288, *see also* Individual entries
DNA amplification fingerprinting (DAF), 390, 393–394

DNA-coated particles, 301
DNA ligase, 297
DNMRT, *see* Duncan's New Multiple Range Test
Dogwood, 394
Dome structures, 132–133
Donor plant location, 80
Donor plants, 79–80, 286
Dormancy, 108, 186
Double antibody sandwich, 420
Doubled haploid, 285
Drugs, natural, 356
Dumbcane, *see Dieffenbachia*
Duncan's New Multiple Range Test (DNMRT), 68

E

EBD determination, 427
ED, *see* Embryo Development media
Eden Gem, 206
Edible colors, 356
EG, *see* Ethylene glycol
EI, *see* Embryo initiation medium
Elaeis guineensis, *see* Oil palm
Electrofusion, 256–257, *see also* Protoplasts
Electroporation, 260, 301, 307–308
Elicitors, 357
ELISA, *see* Enzyme-linker immunosorbent assay
Elite seed, 239
Embedded molds, 47–48
Embryo axis, 183, 198
Embryo Development (ED) media, 206–211
Embryo induction, 12
Embryo initiation (EI) medium, 206–207, 209–210
Embryogenesis, 14, *see also* Individual entries
Embryogenic callus, 191–198, 200, *see also* Callus
Embryogenic cells, 176, 178–180, *see also* Nonzygotic embryogenesis
Embryogenic culture
  common attributes, 177–178
  maintenance, 209–210
  orchardgrass
    encapsulation of somatic embryos to produce synthetic seed, 201–203
    establishment of culture, 192–194
    gradient embryogenic response, 195–196
    initiation and manipulation of suspension cultures, 199–201
    maintenance of callus, 196–197
    monocotyledonous somatic embryos, 197–198
    plant growth and culture media, 192
    somatic embryo germination and plant recovery, 199
  uses, 187–188
Embryogenic suspension cultures, *see* Embryogenic culture
Embryogen-P, 192, 193
Embryoids, 175, *see* Nonzygotic embryos
Encapsulation, 201–203, *see also* Vitrification
Endopolyploidization, 255
Endosperm, 177
Environment
  callus cultures, 39
  nonzygotic embryo development, 181, 185
  regeneration of plants from protoplasts, 250
  transfer to natural and stage IV micropropagation, 84–85
Enyzmic digestion, 279
Enzyme medium, protoplast fusion
  potato, 267
  tobacco, 265–266, 270–271
Enzyme-linker immunosorbent assay (ELISA), 420
Enzymes, 249, *see also* Individual entries
Epicormic shoots, 231
Epifluorescence microscopy, 282
Epigenetic factors, 387
Epinastry, 13
Erratic virus distribution, 422
Error rates, 66
Escapes, 327, 392
Escherichia coli, 297, 298, 308
Ethanol, 36–37
Ethylene, 13, 28
Ethylene glycol (EG), 341, 347
Ethylmethyl sulfonate, 389
Etiolation, 157
Evans blue stain, 251, 282, 381–382
Exine, *see* Pollen cell wall
Experimental unit, 62–63, 65
Explantation time, 80
Explants
  cineraria, 225
  maturity and petunia organogenesis, 168
  size
    micropropagation, 80–81
    source effect on pigment production, 365–368
  somatic embryogenesis, 214–216
    type and melon, 206–208
  tobacco and potato, 265, 292
Eyeball determination, 427

F

Factorial experiments, 66–67
False-negative indexing, 430
FDA, *see* Fluorescein diacetate
Fern, life cycle, 397–398
  apospory, 399–405
  stage II micropropagation, 82
Fertility, 255, 264
Fertilization, 168, 403
Fiber optic illuminator bundles, 56, 58
Filter paper bridges, 30
Filter-sterilized media, 267
Fireblight, 373
Firs, 241
First pollen mitosis, 289
FITC, *see* Fluorescein isothiocyanate
Fixation, 47
Fixatives, 46

Flame sterilization, 37
Flavonoids, 361
Flavors, 356
Float culture, 287
Floral buds, 288
Flotation medium
potato protoplast fusion, 268, 272–274
protoplast isolation and culture of tobacco, 270–271
regeneration of plants from protoplasts, 251
Flowering, 119, 121
Flowers, *see* Somatic embryogenesis
Fluorescein diacetate (FDA)
protoplasts isolation
leaves, 278, 282–283
tobacco culture, 270–271
regeneration of plants from protoplasts, 251
Fluorescein isothiocyanate (FITC), 258–259, 272, 274
Fluorescent lights, 23
Fluorescent microscopy, 45
Focal plane, 57, 60
Focusing, 57
Forcing solution, 411
Forest productivity, improvement, 239–240
Forward step procedure, 68
Fragrances, 356
Fructose, 26
Fruit ripening, 13
F-statistic, 63
Fuchs–Rosenthal counting chamber, 382
Fungi, 418–419
Fusarium wilt, 255
Fusogen PEG 22.5 medium, 269, 272–274
Fusogenic agents, 255, *see also* Electrofusion; Polyethylene glycol
F-value, 62–66

G

GA_3, *see* Gibberellic acid
Galactose, 363, 365
Gamborg B-5 media, 24–25, 376
Gametoclonal variation, 288
Gametophytes
Ceratopterus richardii, 398–404
haploid culture, 285
Gamma irradiation, 389
Gel-gro, 214
Gellan gums, 30
Gelling, 30, 34
Gelling agents, 287
Gene gun
direct genetic transformation of plants, 301
particle bombardment
helium pressures on transient expressions of GUS gene, 334–336
plasmid-particle mixture, 333–334
plastic particle inflow gun (PIG) construction, 330–334
specimen holder construction, 332–333
Gene transfer, 393
microprojectile-mediated, 231
Genebanks, 337
Genes, 179–180, 302, 306
Generation alteration, 397
Generative nuclei, 288–289
Genetic engineering, *see also* Genetic transformation
history of plant tissue culture, 14
origin of nonzygotic embryos, 178
pathogen resistance, 205
peanut, 213
plant improvement, 177
variation in tissue culture, 390
Genetic instability, 162
Genetic transformation, *see also* Genetic engineering
Agrobacterium system, 298–299
preparation, 307–309
somatic embryogenesis, 315–319
chrysanthemum
regeneration of transgenic plants, 326–327
wild-type *Agrobacterium tumefaciens*, 323–326
cointegration and binary vector system, 299–300
direct transformation systems, 300–301
germination of carrot seeds, 307
growth of tobacco plants, 306–307
GUS gene expression under regulation of promoters
regeneration of tobacco plants, 309–313
wounding effect, 313–315
plant viruses as vectors, 300
protoplast use for plant improvement, 259–261
reasons for plant transformation, 301–302
Genetic variation, 79
Genetically modified (GM) plants, 15
Genotypes
anther culture, 286
differences and direct organogenesis, 140–143
protoplast manipulation, 264
selection and embryogenic culture protocols, 178
shoot formation of watermelon cotyledons, 153–155
single and pigment production, 362
somatic embryogenesis, 209
Geranium, 418, 422
Germinants, 237
Germination
genetic transformation and carrot, 307
nonzygotic embryo, 186–187
petunia seeds, 168
somatic embryos, 199, 202, 219–220
suspension culture, 200
Germination medium, 151, 154, 156, 228
Germplasm, preservation
cryoprotectants and cooling rate, 341–342
pollen, 339
pretreatment of plant and isolated propagule, 340–341
seed-propagated species, 338–339
vegetatively propagated species, 339–340
warming rate and recovery, 342
Gibberellic acid (GA_3), 105, 346

Gibberellins, 13, 28
Ginseng, 355–356
GK7, 213, *see also* Peanut
Glass distillation, 23
Glass formation, 342, *see also* Vitrification
Glassware, 21
β-Glucouronidase (GUS) reporter gene
gene gun for particle bombardment, 329, 333–336
genetic transformation
tobacco and carrot, 306
regeneration of tobacco under regulation of promoters, 309–313
wounding effects, 313–315
Glutamine, 186
Glutaraldehyde, 52
Glycerol solution, 346, 348–349
Glycine max, *see* Soybean
Glycine, 27
GM, *see* Genetically modified plants
Goldmine, 139, 142, 323, 325
Grafting, 119
Grandiflora types, 168
Grape
nonzygotic embryo development, 181–185
sucrose effect on embryonic growth, 69–70
Grapevine, 388, 393
Grass, 181–183
Greenhouse acclimatization, 413, *see also* Acclimatization
Growth
abnormalities, 211
chrysanthemum cultivars, 139
nonzygotic embryo development, 183, 185
plant tissues and agar, 30
potential, 41–42
protoplast development, 252, 254
sucrose effect and regression analysis, 69–70
tobacco seeds, 291–292
Torenia fournieri, 160
Guard cells, 259
GUS, *see* β-Glucouronidase gene
GUS-assay buffer, 310
Gymnosperm, 182

H

Haberlandt, Gottlieb, 11–12
Half-strength Litvay (1/2 LV) medium, 241
Hansa, 223, *see also* Cineraria
Haploid breeding programs, 14
Haploid culture
androgenesis, 287–288
anther culture
factors affecting, 286–287
general procedures, 288–290
tobacco
cold pretreatment and plant yield, 294, 296
explant preparation, culture handling, and culture media, 292
microspore development and plant yield, 292–295
plant growth, 291–292
Haploid embryos, 175
Haploid phase, 398
Haploid plantlets, 289–290
Heavy metals, 46
tolerance, 392–393
Hekla, 139, 142, *see also* Chrysanthemum
Helianthus annuus, *see* Sunflower
Helium, 334–336
Helper plasmid, 308
Hemacytometer, 251, 270–271, 278, 281
Hematoxylin, 49
Hemi-cellulase, 249
HEPA filters, 13, 36
Herbicide resistance, 15, 258
Hermaphrodite, 398, 402–403
Hexademethrine bromide, 260
High performance liquid chromatography (HPLC), 357
Histochoice fixative, 46–47
Histological staining, 312
Histological techniques
casting specimens into blocks, 47–48
dehydration and infiltration, 47
equipment and supplies, 46
fixation, 47
microtomy and mounting sections on slides, 48–49
protocol for paraffin studies, 46–47
scanning electron microscopy, 51–53
specimen immobilization, 50–51
staining sections, 49–50
Histology, 130–131, *see also* Organogenesis
History, plant tissue culture, 11–15
HMGR, *see* 3-Hydroxy-3-methylglutaryl coenzyme A reductase
Homosporous ferns, 397, *see also Ceratopterus richardii*
Homozygous plants, 285
Hordeum vulgare, *see* Barley
Hormones, 27–28
HPLC, *see* High performance liquid chromatography
HR, *see* Hypersensitive response
Humidity, 141, *see also* Relative humidity
Hybrid cell lines, 257–259
Hydrahydricity, 82
Hydrolytic enzymes, 249, 279
3-Hydroxy-3-methylglutaryl coenzyme A reductase (HMGR), 312–313, 315, 318
Hyperhydric growth, 112, 116
Hyperhydricity, 120, 144, 147, 341
Hyperplasia, 321
Hypersensitive response (HR), 373
Hypocotyl, 183

I

IAA, *see* Indoleacetic acid
IBA, *see* Indolebutyric acid

Ice formation, 339
Image, cropping, 57
Immobilization, specimens, 48, 50–51
Immune antisera, 255
In vitro growth conditions, 177
In vivo growth conditions, 177
Incompatible pathogens, *see also* Compatible pathogens
  cell death response, 381–383
  hypersensitive response, 373, 376–377
  K^+/H^+ exchange response, 377–380
Indeterminant growth, 76
Indexing
  commercial laboratory production, 410
  micropropagation
    Dieffenbachia, 100
    shoot culture of *Syngonium podophyllum*, 90–94
  plant pathogens
    bacterial and fungal elimination, 418–419
    clean plant production, 423–424
    definition and use, 417
    techniques, 418
    virus elimination, 421–423
    viruses, 419–421
Indexing medium, 90, 428–429
Indicator plants, 420
Indirect organogenesis, 126–127, *see also* Organogenesis
Indoleacetic acid (IAA)
  Bugleweed cell cultures, pigment production
    carbon source, 363, 365
    explant size/source, 367
    light wavelength effect, 371
    variant, 369
  chrysanthemum organogenesis, 140–142, 146
  embryogenic culture initiation, 180
  factorial experiments, 66–67
  history of plant tissue culture, 12
  media for tissue culture, 28–29
  micropropagation, 81, 100
  regeneration and monitoring somaclonal variation, 171
  stock solution preparation, 31
Indolebutyric acid (IBA)
  cytokinin and auxin on leaf explants, 169–170
  direct/indirect organogenesis, 161
  embryogenic culture initiation, 180
  media for tissue culture, 28–29
  stage I micropropagation, 81
  stock solution preparation, 31
  vitrification of shoot tips, 346
  watermelon cotyledons, 152, 154
Induced variation, 389
Inducible promoter, 302, 333, *see also* Promoters
  GUS gene expression, 309–313, 315
Induction, organogenesis, 127–129
Induction frequency, 243
Induction medium
  hormone-containing and organogenesis effect, 130–131
  somatic embryogenesis, 232, 234, 236–237
Infiltration protocol, 47
Inorganic minerals, 24–26
Inositol, 26–27
Insects, predation and resistance, 15, 302
Integument removal, 206
Interface, media, 251
Internode, 28
Interspecific hybridization, 285
Inverted microscopes, 56–57
2-iP, *see* 2-Isopentenyl adenine
Iridon, 139, 142–143, 145, 323, 325, *see also* Chrysanthemum
Irish potato, 40–41
Isolation, protoplasts, 249
Isolation media, 279
2-Isopentenyl adenine (2-iP), 81, 83, 100–101
Isopropanol, 47

J

Jerusalem artichoke, 40–41
Jessica, 139, *see also* Chrysanthemum
Juvenile explants, 80, *see also* Explants
Juvenile stock plants, 410, *see also* Stock plants

K

K and N salts, 24
K^+ efflux, 377
K^+/H^+ exchange response, bacteria-induced
  bioassay, 379–380
  hypersensitive response in plants, 374, 377–380
Kanamycin, *see also* Antibiotics
  genetic transformation
    carrot and somatic embryogenesis, 316–317
    chrysanthemum with wild-type *A. tumefaciens*, 322
    tobacco and carrot, 305, 307–311
  regeneration of transgenic chrysanthemum, 326–327
KB, *see* Kings B agar medium
Kennebec, 105, *see also* Potato
Kennebec haploid, 264
Kinetin
  Bugleweed cultures, pigment production, 363, 367
  embryogenic culture initiation, 180
  history of plant tissue culture, 13
  micropropagation stages, 81, 83
  organogenesis, 128
Kings B agar (KB) medium, 376, 378
Klett–Summerson spectrocolorimeter, 311, 318
Kodak Ektachrome ASA 160 T film, 58
Kodak Ektachrome ASA 64 T film, 57
Kodak Technical Pan (TP 135), 58

L

Labeling, 35, 48, 114
Lack-of-fit (LOF), 68–69

Lactuca sativa, see Lettuce
Laminar flow hoods, 13, 36, 288
Laser beams, 260
Lateral buds
  apical meristem, 75
  culture, 78–79
  micropropagation, 98, 100, 121
  shoot tip culture, 88–89
LB medium, 307–308
Leaf abscission, 13
Leaf explants, 169–170, 193
Leaf mesophyll, 250
Leaf position, 195–196
Leaf primordia, 132–133
Leifart and Waites Sterility test medium, 428–429
Lens, close-up, 56
Lettuce, 394
Light conditions
  adventitious shoot organogenesis, 155–157
  anthocyanin production, 358
  apospory in *Ceratopterus richardii*, 402
  embryogenic culture protocols, 178
  genetic transformation, 307
  genotype differences and direct organogenesis, 142
  growth
    chrysanthemum, 140
    tobacco plants, 292
    Torenia fournieri, 160
  micropropagation, 113, 116, 120
  protoplast growth and development, 254
  regeneration of plants from protoplasts, 250
  secondary products expression, 357
  somatic embryogenesis, 207
Light source, photography, 56
Light wavelength, 370–371
Lignin, wood content, 239
Lilac, 111–116
Linsmaier and Skoog media, 24
Liposome-mediated genetic transformation, 260, *see also* Genetic transformation
Liquid nitrogen (LN)
  storage of seed-propagated species, 338
  vitrification of shoot tips, 347–349, 351
Liriodendron tulipifera, 232–238, *see also* Somatic embryogenesis, yellow poplar
Little Boy Blue, 116, *see also* Lilac
LN, *see* Liquid nitrogen
Loading phase, 342, 350
Local lesion response, 421
LOF, *see* Lack-of-fit
Ludwig Spaeth, 111, *see also* Lilac
Lupine, 12
Lycospersicon esculentum, see Tomato

M

M9 medium, 308
Macerozyme, 249
Macronutrients, 25
Macrophotography, 56
Macrosalts, 24–26
Madame Lemoine, 111, *see also* Lilac
Main plots, 65–66
Maize, 70–71, 388
Maltose, 26
Mangifera indica, see Mango
Mango, 393, 394
Mannitol, 251, 277–278
Marie Frances, 116, *see also* Lilac
Markers, 297, 300, 336
Market demand, 414
Maturation
  apospory in *Ceratopterus richardii*, 403
  nonzygotic embryo, 185–186
  somatic embryos, 226, 228–229
Maturation medium, 228
Maximum Likelihood, 70
Mean separation, 63
Mean square (MS), 63–64
Mean square error (MSE), 63–66
Media, serially composed, 42–44
Media formulations, 24–25, 412
Media preparation/dishes, 407
Media storage, 407–408
Medicago sativa, see Alfalfa
Megagametophyte, 242
Membrane filter, 35–36
Mericlone, 387
Meristem tip culture, 77–78, 103, 108, 422–423
Meristemoids, 126, 130, 132–133
Meristems, 12
Metabolic precursors, 357
Metabolite production, 355
Metallica Crispa, 362, 365–366, 370, *see also* Bugleweed
4-Methylumbilliferone glucuronide (MUG), 314
Methylation patterns, 389
Methylobacteria, 427
Microclear, 46
Microcuttings
  commercial laboratory production, 413
  lilac, 115–116
  micropropagation, 81, 107
  miniature roses, 122–123
  shoot culture, 79, 94
Micrograms per liter, calculation, 32
Microinjection, 260–261, 301
Microphotography, 56–57
Micropropagation
  definition, 75
  Dieffenbachia, 97–102
  Lilacs, 111–117
  miniature roses, 119–123
  potato, 103–109
  reasons for using, 414–415
  stages, 79–85
  Syngonium podophyllum, 88–94
Microsalts, 26
Microshoots, 126
Microspores, 288–289, 292–295
Microtomy, 48

Microtuber induction medium (MIM), 105–107
Microtuberization, 107–109
Microwaves, 34
Milligrams per liter, calculation, 32
MIM, *see* Microtuber induction medium
Mineral nutrition, 39
Mineral salts, 34–35
Miniature roses, 119–123
Miniaturized tubers, 103
Minilee, 151, 153–154, *see also* Watermelon
Mitosis, 252
Mitotic frequencies, 245
Mixed cultures, 92
Moisture levels, 338
Molar solutions, calculation/conversions, 31–33
Monocots, 181–183, 300–301
Morphogenetic plasticity, 128–129
Morphology, somatic embryos, 244–246
Mortality, 91
Mounting, specimens, 48
MS, *see* Mean square
MS, *see* Murashige and Skoog medium
MSE, *see* Mean Square error
MUG, *see* 4-Methylumbilliferone glucuronide
Multiple comparison test, 68, 156
Multiple-node culture, 103, 106
Multiple range tests, 68
Multiple shoot clusters, 81
Murashige and Skoog (MS) medium
  anther culture, 287
  enrichment and indexing, 428
  genetic transformation
    chrysanthemum, 323–325
    tobacco, 307, 310
  growth of tobacco callus, 42–44
  history of plant tissue culture, 14
  macro- and micronutrients, 32
  melon culture, 206
  micropropagation, 99, 105, 120
  organogenesis
  direct/indirect, 160
    chrysanthemum shoots, 140–143, 146
    shoot, 168
  plant pathology of suspension cells, 375
  protoplast isolation and fusion, 267
  regeneration, 152, 171, 326–327
  somatic embryogenesis
    direct of cineraria leaves/flowers, 225–229
    peanut, 214–215, 220
  tissue culture, 24–25
  vitrification of shoot tips, 346, 350–351
Musa spp., *see* Plantains
Mutations, 254, 387–389

N

2n plants, 285, 404
N6 media, 287
NAA, *see* Naphthaleneacetic acid
Nancy Hall, 120, *see also* Miniature roses
Naphthaleneacetic acid (NAA)
  embryogenic culture initiation, 180
  media for tissue culture, 28–29
  micropropagation stages, 81–82
  protoplast growth and development, 254
  regeneration of petunia and monitoring
    somaclonal variation, 171
  vitrification of shoot tips, 346
Naphthoxyacetic acid (NOA), 180
Nasturtium, 12
NCM, *see* Node culture medium
Necrosis, 376
Neomycin phosphotransferase (NPTII) gene, 308, 322,
  326, 336
Nephrolepsis, see Fern
Nicotiana tabacum, see Tobacco
Nicotine, 309, 355
Nicotinic acid, 26
Nightshade, 263
Nitrate-to-ammonium ratio, 361
Nitrogen, 25, 200
Nitsch and Nitsch media, 24–25, 287, 292
NMR, *see* Nuclear magnetic resonance
NOA, *see* Naphthoxyacetic acid
Nodal explants, 120–122
Node culture, 78–79, 103
Node culture medium (NCM), 105–107
Nonembryogenic cells, 179
Nonfood colorants, 357
Nonzygotic embryogenesis, 175–177
  embryo
    development, 180–185
    germination and plant development, 186–187
    maturation, 185–186
  history of plant tissue culture, 14
  in vivo versus in vitro growth conditions, 177–178
  inductive plant growth regulators, 180
  initiation of embryogenic cells, 178–180
  melon, 205, *see also* Somatic embryogenesis, melon
  origin, 178
  quiescence and dormancy, 186
  uses for, 187–188
Nonzygotic embryos, 76, 175
Nonzygotic nucellar seed tissue, 175
Nopaline, 298
Norway spruce, 182
NOS-terminator, 333
NPGS, *see* U.S. National Plant Germplasm System
NPTII, *see* Neomycin phosphotransferase gene
Nucellar embryos, 175
Nuclear magnetic resonance (NMR), 357
Null hypothesis, 61, 63
Nutrition, zygotic embryos, 177

O

Observational unit, 62–63, 66
Octopine, 298
ODU, *see* Optical density units
Off-types, 82

Oil palm, 387–388
Ontogenic pathways, 126
Opines, 298–299
Optical density units (ODU), 379
Optoporation, 260–261
Orchardgrass, 277–284
Orchestrating gene, 128
Orchids, 76
Organ, definition, 126
Organ primordium, 126
Organ type, 24
Organic compounds, 26–27, 254
Organogenesis, *see also* Direct organogenesis; Direct/indirect organogenesis
  dedifferentiation, 127–128
  developmental process, 126–127
  differentiation, 129–133
  induction, 128–129
  petunia shoots, 169–173
  root-specific, 133–135
  watermelon shoots, 64
    darkness and adventitious shoots, 155–157
    plant genotype, 153–155
    regeneration potential, 150–153
Organ-specific products, 357
Orientation
  explants, 82
  specimen and photography, 59–60
Orthodox seed, 338
Oryza sativa, *see* Rice
Osmium tetroxide, 52
Osmotic potential, 26
Osmotic pressure, protoplasts, 250–251, 279, 283
Osmotic stabilizers, 254
Osmoticum, 240, 252
Oxygenation, 23

P

Paired variate design
  direct somatic embryogenesis, 225–228
  genetic transformation, 323–327
Panax ginseng, *see* Ginseng
Paraffin, protocol, 46–50
Paraplast, protocol, 46–47
Parasexual hybrids, 250, 263
Parasitized plant tissue, 12
Parsnip, 40–41
Particle bombardment, 301, 329
Particle guns, 301
Particle inflow gun (PIG), 330–334
Parts per million (ppm), calculation, 32
Pathogen-eradicated plants, 76–77, 113, *see also* Micropropagation
Pathogen-free status, 417
pBI221 plasmid, 333–334
pBIN19 plasmid, 305, 308
PC, *see* Protoplast culture
p-Calli, *see* Protoplast-derived calli
PCR analysis, 390
Peanut, 216–220
Pearl chains, 256
Pecan, 394
Pectinase, 249
Pectolyase Y-23, 277–279
PEG, *see* Polyethylene glycol
Percentage based on volume percent, calculation, 31
Percentage based on weight percent, calculation, 31
Perfection-137B, 97, *see also Dieffenbachia*
Periclinical division, 133
Petri dish, 22
Petunia, 167–172
PGRs, *see* Plant growth regulators
pH, 34, 380
Phenocopy, 128
Phenocopying agents, 128
Phenocritical times, 128
Phenolic browning, 91
Phenolics, 27
Photoautotrophic state, 84
Photographic methods
  composition, 58–60
  film types, 57–58
  focusing and cropping the image, 57
  types used in plant/tissue culture research, 55–57
Photoperiod
  auxin concentration on somatic embryogenesis, 216–218
  direct/indirect organogenesis, 160
  micropropagation, 104, 108
Photosensitivity, 28
Photosynthetic competency, 84
Physiological stabilization, 81
Physiological status, donor plants, 80
Phytagel, 90, 99, 112, 120, 307
Phytotoxins, 390–391
PI, *see* Protoplast isolation medium
Picacea, *see* Somatic embryogenesis, conifers
Picea glauca, *see* White spruce
Picloram, 28–29
PIG, *see* Particle inflow gun
Pigment induction medium, 367
Pigment production, Bugleweed (*Ajuga*) cell cultures
  carbon source effect on expression, 363–366
  explant size and source effects on in vitro production, 365–368
  light wavelength effect on growth and production, 370–371
  plant materials, 362–363
  productivity of variant callus lines, 368–370
Pigment production medium, 368, 371
Pigments, 356–359
Pines, 135, 241, 243
Pinus oocarpa, *see* Central American pine
Pinus spp., *see* Pines
Pith, 375
Plant cell culture, 356–357, *see also* Secondary products
Plant cell suspensions, 374
Plant cells, totipotent, 249
Plant chimeras, 92

Plant development, 186–187, *see also* Nonzygotic embryos
Plant growth regulators (PGRs), *see also* Individual entries
- anther culture media, 287
- callus cultures, 39
- chrysanthemum
- genotype differences, 141–145
  - formation of shoots/roots, 143–146
  - genetic transformation, 325
  - pulse treatment and initiation/growth of shoots, 146–147
- commercial laboratory production, 411
- embryogenic culture protocols, 177
- history of plant tissue culture, 11–13
- induction and embryogenic culture initiation, 180
- media for tissue culture, 27–29
- micropropagation, 100–101, 112, 114
- organogenesis, 128
- direct, 161
- shoot, 167, 169–170
- protoplast growth and development, 254
- somatic embryogenesis, 214, 226, 228–229, 241
- stock solution preparation, 31, 35
- vitrification of shoot tips, 346

Plant improvement, 177
Plant pathology, in vitro
- correlation of K^+/H^+ exchange response with plant/bacteria interaction, 377–380
- initial testing of bacteria isolates by inoculation into whole plants, 376–377
- maintenance of bacteria, 376
- maintenance of plants, 375
- measuring suspension cell death in response to compatible/incompatible bacteria, 381–383
- preparation of plant suspension cells, 375–376

Plant tissue culture
- agar and other support systems, 30
- characteristics of more common culture media, 24
- components of culture media, 24–27
- culture room, 23
- disinfecting plant tissues, 37–38
- equipment and supplies, 21–22
- history, 11–15
- plant growth regulators, 27–29
- preparing the sterile transfer hood, 36–37
- steps in preparation of culture media, 30–31
- sterilizing equipment and media, 35–36
- stock solutions of mineral salts, 34–35
- storage, 37
- supporting methodologies, 4
- units of concentration, 31–34
- water, 22–23

Plant viruses, as vectors, 300, *see also* Viruses
Plantains, 389
Planting, somatic embryos, 219–220
Plant-TC listserv, 7
Plasmid-particle mixture, 333–334
Plasmids, 297–298
Plasmolysis, 250, 283
Plasmolysis medium, 273
Plating efficiency, 252
Poaceae, *see* Embryogenic cultures, orchardgrass
Polarized structures, 245
Pollen, 285, 339
Pollen cell wall, 287
Pollen embryos, 175
Polyamines, 27
Polybrene, *see* Hexademethrine bromide
Polycationic chemical, 259–260
Polyethylene glycol (PEG)
- genetic transformation, 300–301
- maturation of nonzygotic embryo, 186
- protoplast
  - fusion, 255–256
- mediation in potato, 274–275
  - use for plant improvement, 259–260

Polyethylene rafts, 30
Polyphenol oxidase, 80
Polyphenolic compounds, 27
Polyploidy, 210, 255, 263
Polysomatic species, variation, 388
Potassium phosphate buffer, 52
Potato
- cybrids, 264
- growth of plants, 265
- node culture, 79
- protoplast fusion, 256–257, 272
- protoplast isolation and culture, 264
- somaclonal variation, 255
- somatic hybridization, 263

ppm, *see* Parts per million
Precocious germination, 186, *see also* Germination
Preconditioning, 250
Predigestion media, 279
Preincubation, 286–287
Preplasmolysis medium, 267, 270, 274
President Grevy, 111, *see also* Lilac
Presterilized containers, 21
Pretransplant, 83–84
Pretreatment, anther culture, 286–287
Primary explants, 126–127, *see also* Explants
Primordia, 126, 133, 135
Product focus, commercial production, 409
Proembryonal complex, 178, 180–181, 185, 245
Proembryonic masses (PEMs), 231–233, 235–238
Progenitor cells, 180
Promoters
- genetic transformation, 302
  - tobacco and carrot, 306
  - tobacco and GUS gene expression, 309–313
- regulation of wounding effects on GUS gene expression, 313–315

Propagation, preexisting meristems
- micropropagation stages, 79–85
- shoot apical meristems, 75–79

Propagation medium, 266–267
Propiono-carmine, 289
Protoclones, 255, 387
Protoplast culture (PC), 231, 266, 270
Protoplast isolation (PI) medium, 265–266, 270–271, 277

Protoplast-derived calli (p-Calli), 253, 271–272, 275
Protoplasts
direct genetic transformation, 301
isolation from chrysanthemum and orchardgrass leaves, 277–284
isolation, culture, and fusion
culture medium, 265–269
explant preparation, 265
general considerations, 264
growth of plants, 265
potato, 272–275
tobacco, 270–272
plant improvement
factors influencing growth and development, 252, 254
fusion and somatic hybridization, 255–259
future prospects, 261
genetic transformation, 259–261
isolation, 249
procedures to regenerate plants, 250–253
somaclonal variation, 254–255
uses, 250
totipotent and history of plant tissue culture, 14
Pseudomonas solanacearum, 380
Pseudomonas syringae pathavars, 376
Pulse treatment, 146–147, 186
Putrescine, 27
Pyridoxine, 26

Q

Quality, evaluation, 281–282
Quiescence, 186

R

r^2, *see* r-Square values
Racks, 22
Randomization, 62
Randomized complete block design (RCBD), 65, 327
Randomized incomplete block design, 143–144
Randomly amplified polymorphic DNA (RAPD), 390, 393–394
RAPD, *see* Randomly amplified polymorphic DNA
Rape, 287
Rave, 139, 142, *see also* Chrysanthemum
RCBD, *see* Randomized complete block design
Recalcitrant seed, 338
Recessive mutations, 285, *see also* Mutations
Recombinant DNA, 297, 300
technology, 149
Recovery, 199, 342
Recovery rate, 186
Recurrent flowering, 122
Red Minimo, 120, *see also* Miniature roses
Red Pontiac, 105, *see also* Potato
Redetermination, 179
Reforestation programs, 239
Regeneration
ability and organogenesis, 129
callus and monitoring somaclonal variation, 170–173
indirect organogenesis, 163
maturation of somatic embryos, 226, 228–229
plants from protoplasts, 250–253
protoplast fusion, 263
Regeneration medium, 151–152, 154, 156
Regeneration potential, 150–153
Regression analysis, 68–69
Relative force, calculation, 279
Relative humidity, 84–85, 338, *see also* Humidity
REM, *see* Rose establishment/multiplication medium
Repetitive embryogenesis, 218–219
Replicates, 62, 65–66, 350
Reporter genes, genetic transformation, see also β-Glucouronidase (GUS) reporter gene
tobacco and carrot, 306, 309–313
monitoring in plants, 302
Resolution, 45
Restriction endonuclease recognition site, 297
Restriction enzymes, 297
Restriction fragment length polymorphism (RFLP), 390, 393
Reverse osmosis, 22–23
RFLP, *see* Restriction fragment length polymorphism
Rhodamine isothiocyanate (RITC), 258–259, 272, 274
Ribavirin, 422
Rice, 390
Rinse medium, 268, 270, 272, 274
RITC, *see* Rhodamine isothiocyanate
RNA polymerase, 302
Rocky Ford Green Flesh, 206, 209
Root apical meristem, 183
Root meristem, 135
Root tips, 12
Rooting, 83–84, 122–123, 412, 414
Rooting medium, 151, 154, 156
Roots, 28, 101, 128
Root-specific organogenesis, 133–135
Root-stimulated bacterium (RSB), 134
Rosa spp., *see* Rose
Rosa Rouletti, 120
Rose, 119–123
Rose establishment/multiplication (REM) medium, 120, 122
Rosmarin, 120, *see also* Miniature roses
Royal Ruby, 120, *see also* Miniature roses
RSB, see Root-stimulated bacterium
r-Square (r^2) values, 68–69
RubPcase activity, 84
Russet Burbank, 105, 264, *see also* Potato

S

Sabouraud dextrose medium, 428
Safety precautions, 348
Safranin O stain, 49–50
Salt tolerance, 392–393
Samsun, 264, *see also* Tobacco
Sangria, 153–154, *see also* Watermelon

Saponines, 356
Scanning electron microscopy (SEM)
  protocol for paraffin studies, 51–53
  nonzygotic embryo development, 182
  organogenesis, 131–133
Scheduling, commercial laboratory production, 413
Schenk and Hildebrandt (SH) media
  direct somatic embryogenesis, 225–228
  embryogenic cultures, 192, 201–202
  in vitro plant pathology of suspension cells, 375
  protoplast isolation, 277
  tissue culture, 24–25
Screening, *see* AC Broth indexing medium; Indexing
Scutellum, 182–183, 197–198
SE, *see* Standard error of mean
Seasonality, 413
Seaweed, 30
Secondary medium, 218–219
Secondary products
  expression in vitro
    current trends, 356–357
    pigments from cell culture, 357–359
  protoplast use for plant improvement, 255
Seed coat, 184
Seedlings, 232, 415
Seed-propagated species, 338–339
Seeds, 168, 232–235, 338
Self-fertilization, 404
SEM, *see* Scanning electron microscopy
SEM culture medium, 90–94
Semen straws, 347–349
Senescence, 13, 122, 398, 403
Serial dilution technique, 35
Serology, 419–420
Sexual hybridization, 178, 250
Sexual incompatibility, 263
SH, *see* Schenk and Hildebrandt media
SH30 media, 194–197, 199–200, *see also* Schenk and Hildebrandt media
SH-30C media, 192, 200, *see also* Schenk and Hildebrandt media
Shikonin, 356
Shipping, 409
Shoot apical meristem, 75–79, 183
Shoot clusters, 122
Shoot cultures, 429–430
Shoot elongation medium, 151, 154, 156
Shoot multiplication, 92–94
Shoot proliferation, 411–412
Shoot tips
  culture, 76, 78–79
  meristems and stage I micropropagation, 90
  cryopreservation, 340, 346
Silver nitrate, 28
Simple linear regression, 68–69
Single factor completely randomized design (CRD) with subsampling, 63–64
Single-node culture, 103, 106–107
Single-node method, 412
Sister halves, 226
Slants, 90, 93
Slides, protocol, 48–49
SMM culture medium, 90–94
SNK, *see* Student-Newman-Kuels test
Sodium alginate, 350, 352
Solanaceae, 167
Solanum nigrum, *see* Nightshade
Solanum tuberosum, *see* Potato
Solenoid valve assembly, 332
Solution-based vitrification, *see* Vitrification
Somaclonal variation
  causes, 388–389
  organogenesis, 126
  pigment production of variant Bugleweed callus lines, 368
  protoplast use for plant improvement, 254–255
  regeneration of petunia from callus, 170–173
  Torenia fournieri, 159
  watermelon organogenesis, 149
Somaclones, 387
Somatic cells, 179
Somatic embryo, 186
Somatic embryogenesis
  carrot hypocotyls transformed by *Agrobacterium tumefaciens*, 315–319
  conifers, 241–246
  cineraria flower receptacles/leaves, 222–229
  media for tissue culture, 28
  melon, 206–211
  peanut, 213–220
  regeneration of Poaceae, 191, *see also* Embryogenic cultures, orchardgrass
  yellow poplar, 232–238
Somatic embryos, 175
  nonzygotic embryo development, 181, 183, 185
  orchardgrass, 196–199
Somatic hybridization, 14, 257–259, 263
Somatic hybrids, 263
Somatic recombination, 388
Sonication, 260
Sorbitol, 26
Soybeans, 15, *see also* Plant pathology, in vitro
Special error term, 64
Specific pathogen-free tested (SPT) stocks, 103
Specimen holders, 332–333
Spectrophotometer, 371
Spectrophotometric method, 381
Spermatozoids, 398
Split-plot design, 65–66
Spontaneous mutation, 254, *see also* Mutations
Spores, 398, 401
Sporophytes
  Ceratopterus richardii, 397–404
  development in angiosperms, 176
Spruce, somatic embryogenesis, *see* Somatic embryogenesis, conifers
SPT, *see* Specific pathogen-free tested stocks
Squashing, 289
SS, *see* Sums-of-squares
Stability, media, 34
Staining, 49–50, 251

Stains
    Aceto-carmine, 289, 293–294
    aceto-orcein, 244
    alcian blue, 49–50
    Bismarck brown, 49–50
    Calcofluor White, 251, 270–271, 278, 281–282
    Evans blue, 251, 282, 381–382
    histological, 312
    Safranin O, 49–50
    triple, 49–50
    vital, 258
Standard error of mean (SE), 68
Standard error of response, 207
Statistical analysis
    completely randomized design, 62–63
    factorial experiments, 66–67
    methodologies associated with scientific research, 61–62
    multiple comparison and multiple range tests, 68
    randomized complete block design (RCBD), 65
    simple linear regression, 68–70
    single factor CRD with subsampling, 63–64
    split-plot design, 65–66
    standard error of mean, 68
Stem cells, 134
Stem pith, 40–42
Stem tips, 12
Stereomicroscopy, 56, 58–59
Sterile conditions, 13
Sterilization, 35–36
Stock cultures, 112
Stock plants
    Dieffenbachia, 99
    growth of orchardgrass, 192
    micropropagation, 79–80
    pigment production, 362–363
    potato, 103, 105
    preparation and commercial laboratory production, 410
    Syngonium podophyllum, 88–89
    vitrification, 345–346
Stock solutions, 30–31, 34–35
Stomata, 131
Storage, 37, 47, 338
Storage reserves, 186
Structural anomalies, 185
Student-Newman-Kuels (SNK) test, 68
Stump sprouts, 231
Subsampling, 63–65
Sucrose
    anther culture media, 287, 292
    carbon source and pigment production, 363, 365
    direct/indirect organogenesis, 160
    embryo growth and regression analysis, 69–70
    genetic transformation, 307
    media for tissue culture, 26
    micropropagation, 104, 108
    nonzygotic embryo maturation, 186
    protoplast fusion, 274
    somatic embryogenesis, 225, 241
    vitrification of shoot tips, 348–351
Sugar Daddy, 168, *see also* Petunia
Sugar Plum, 168, *see also* Petunia
Sugar, 26, 254, *see also* Individual entries
Sums-of-squares (SS), 63–66
Sunflower, 12, 129
Super Market, 206, 209
Superior, 105, *see also* Potato
Supermagic Pink, 168, *see also* Petunia
Suppliers, list, 435
Surface disinfesting, 37–38
Surfaces, sterile, 22–23
Survival, 351, 415
Suspension cells, 375–376, 378
Suspensor, 177, 183, 245
Sweet potato, 40–41
Syngonium podophyllum, 87–94
Syngonium podophyllum, micropropagation, *see* Micropropagation, *Syngonium podophyllum*
Syngonium wendlandii, 89
Synthetic food additives, 357
Synthetic seed
    cryopreservation, 340
    encapsulated somatic embryos of orchardgrass, 201–203
    somatic embryos as, 192
    technology, nonzygotic embryo, 187–188
Syringa, *see* Lilac

T

T value, 68–69
Taxol, 356
T-DNA, *see* Transfer DNA
TDZ, *see* Thidiazuron
Temperature
    anther culture, 287
    apospory in *Ceratopterus richardii*, 401–402
    embryogenic culture protocols, 178
    micropropagation, 113
    preservation of seed-propagated species, 338
    protoplast growth and development, 254
    regeneration of plants from protoplasts, 250
    tissue culture, 23
Terminal buds, 114
Terminal shoot tips, 88–89, 98
Terminal shoots, 76, 114
Testing, 15, 424
Thermotherapy, 422
Thiamine, 12, 26
Thidiazuron (TDZ), 28, 143, 180
Thin layer chromatography (TLC), 357
Ti genes, 305
Tillers, 193–194
Timing, cultures, 411
Tissue culture
    causes of somaclonal variation, 388–389
    variation, 389–394
Titer, virus indexing, 419
TLC, *see* Thin layer chromatography
TMV, *see* Tobacco Mosaic Virus